SAT
MATH
BIBLE

PowerScore
Test Preparation

Published by
PowerScore Publishing, a division of PowerScore Incorporated
57 Hasell Street
Charleston, SC 29401

Author: Victoria Wood

Manufactured in Canada
February 2010

ISBN: 978-0-9826618-1-9

Contents

Chapter One: Introduction to the SAT

Chapter Two: The Basics of SAT Math

Chapter Three: Operation Mastery

Chapter Four: Solution Strategies

Chapter Five: Arithmetic Mastery

Chapter Six: Algebra Mastery

Chapter Seven: Algebra II Mastery

Chapter Eight: Geometry

Chapter Nine: Coordinate Geometry Mastery

Chapter Ten: Data Analysis, Statistics, and Probability Mastery

Chapter Eleven: Test Readiness

Appendix: Glossary of Terms

Index

About PowerScore

PowerScore is one of the world's fastest growing test preparation companies. Headquartered in Charleston, South Carolina, PowerScore offers SAT, LSAT, GMAT, and GRE preparation courses in over 125 locations in the U.S. and abroad. For more information, please visit our website at www.powerscore.com.

Chapter One:
Introduction to the SAT

Welcome to the PowerScore SAT Math Bible! The purpose of this book is to provide you with a thorough review of tested math concepts and to teach you new strategies for approaching the math on the SAT. The math questions on the test are more like puzzles and less like typical classroom questions, so it is important to learn new techniques and solution strategies prior to taking the SAT. We are certain that you can increase your math score with close study and application of the PowerScore techniques.

In order to apply our methods effectively and efficiently, we strongly recommend that you carefully read and re-read each of the chapters regarding basic mathematical concepts. We also suggest that you closely examine all questions to find patterns among both the questions and their answer choices. By doing so, you will greatly increase your chances of recognizing these patterns on the actual test.

This book also contains a variety of drills and exercises that supplement the discussion of techniques and question analysis. The drills help strengthen specific skills that are critical for SAT excellence, and for this reason they are as important as the questions. In the answer keys to these drills we will often introduce and discuss important SAT points, so we strongly advise you to read through all explanations.

Because access to accurate and up-to-date information is critical, we have devoted a section of our website to SAT Math Bible students. This free online resource area offers supplements to the book material, answers questions posed by students, suggests study plans, and provides updates as needed. There is also an official book evaluation form that we strongly encourage you to use.

The exclusive PowerScore SAT Math Bible online area can be accessed at:

www.powerscore.com/satmathbible

The concepts and techniques discussed in this book are drawn from our live SAT courses, which we feel are the most effective in the world. If you find that you need a more structured learning environment, or would just like to participate in a live classroom setting, please visit our website to learn about our course offerings.

If we can assist you in your SAT preparation in any way, or if you have any questions or comments, please do not hesitate to contact us via e-mail at sat@powerscore.com. Additional contact information is provided at the end of this book. We look forward to hearing from you!

The first two chapters of this book cover the format of the SAT and its math sections. While the reading may be heavy at times, we recommend you thoroughly cover these chapters first so that you understand the construction of the test. However, if you are anxious to begin studying for the SAT, you can skip to Chapter 4, provided you return to the first two chapters at a later date.

Please do not hesitate to write in this book! Take notes, underline important sentences, and complete the problems presented. Your SAT success depends on your understanding of the concepts presented here, and achievement will only come with study and application.

The Format of the SAT

The SAT was first administered in 1926 and originally stood for "Scholastic Aptitude Test." Now the acronym has no meaning.

The SAT is a standardized test used for admission to thousands of colleges in America. The test assesses a student's ability to reason critically in math, reading, and writing, and predicts his or her success at the university level. It is written by high school teachers, college professors, and testing experts for a non-profit company called the College Board, a division of Educational Testing Services (ETS).

The SAT has 10 sections, given over three hours and 45 minutes:

Section 1 on every SAT is the essay. Sections 2 through 7 are the remaining 25-minute sections, which appear in random order. Sections 8 and 9 are the two 20-minute sections, again in random order. Section 10 is always the 10-minute writing section.

Section Type	Question Types	Time Limit	Section Number
Math	8 multiple choice questions 10 grid-in questions	25 minutes	2–7
Math	20 multiple choice questions	25 minutes	2–7
Math	16 multiple choice questions	20 minutes	8 or 9
Writing	1 essay	25 minutes	1
Writing	11 Sentence Improvement questions 18 Identifying Errors questions 6 Paragraph Improvement questions	25 minutes	2–7
Writing	14 Sentence Improvement questions	10 minutes	10
Reading	8 Sentence Completion questions 16 Passage-Based Reading questions	25 minutes	2–7
Reading	6 Sentence Completion questions 18 Passage-Based Reading questions	25 minutes	2–7
Reading	5 Sentence Completion questions 14 Passage-Based Reading questions	20 minutes	8 or 9
Experimental	An unscored version of any of the 25-minute sections	25 minutes	2–7

The test contains:
- 54 math questions
  - –44 multiple choice
  - –10 grid-in
- 1 essay
- 49 writing questions
- 67 reading questions

In all, there are 54 math question, 44 of which are multiple choice and 10 of which are grid-in. There is one essay and 49 multiple choice writing questions. There are 67 multiple choice reading questions, which are typically distributed as shown in the table above. Aside from the essay, which always occurs in Section 1, the 25-minute sections are randomly distributed from Sections 2 through 7. The two 20-minute sections are randomly assigned to Sections 8 and 9, and the 10-minute writing section is always Section 10.

Scoring the SAT

Each of the three content areas receives a scaled score from 200 to 800 on the SAT. The three scores are added together to produce a total score. The lowest score a student can receive is 600 (200 Math, 200 Writing, and 200 Reading) while a perfect score is 2400 (800 Math, 800 Writing, and 800 Reading). Consider an example:

Math	520
Writing	600
Reading	490
Total	1610

The average SAT score varies from year to year, but is usually around 1500.

You may hear some students refer to their score out of 1600. This is because a perfect score used to be 1600, when only Math and Reading were tested. In 2005, however, the College Board introduced the Writing section which added 800 more possible points to the test. If a friend says, "I am so excited! I got a 1400!" it is likely that she is using the old scale, where a 1400 is in the top 10 percent, rather than the new scale, where a 1400 is slightly below average. As each year passes and more data is available for the Writing section, more colleges will actively use the section as an admission criterion, and fewer people will refer to the old scale.

The SAT has been scored out of 2400 points since 2005.

Each multiple choice question that you answer correctly earns you 1 point toward a raw score. Each question you answer incorrectly costs you a quarter of a point (thus if you miss 4 questions you lose 1 point). Grid-in math questions earn you one point for every correct answer, but you are not penalized any points for incorrect answers. Questions that are omitted and left blank neither receive points nor lose points. Once a raw score is calculated from the points earned and lost, test makers use a conversion table to produce a scaled score for each section. Each SAT test has its own conversion table, but they are usually quite similar. Here is a partial conversion from a recent administration:

Grid-ins are the only questions not penalized for a wrong answer.

	Raw Score		
Scaled Score	Math (out of 54)	Reading (out of 67)	Writing* (out of 49)
800	54	64	49
750	52	60	46
700	49	57	43
650	44	52	38
600	39	45	32
550	32	38	25
500	26	30	19
450	20	22	13
400	14	15	7

*Assumes an essay score of 9.

To better understand the scoring system, let us analyze one student's reading test results:

> Of the 67 reading questions, Tyler answered 43 correctly and 21 incorrectly, and left 7 blank.
>
> Therefore, he earns 43 points for the questions he answered correctly:
>
> Correct: 43
>
> And he loses 5 points for the 21 he missed:
>
> Incorrect: $21 \times -0.25 = -5.25$ Rounded to nearest whole number: -5
>
> He does not gain or lose any points for the omitted questions.
>
> Omitted: $7 \times 0 = 0$
>
> So Tyler's raw score is:
>
> $43 - 5 + 0 = 38$
>
> According to the conversion chart on the previous page, a raw score of 38 equates to a scaled score of 550.

Rounding the value of the points lost to the nearest whole number has it advantages and disadvantages; you can miss 5 questions and only lose 1 point, but you also lose a whole point when you only miss 2 questions.

Calculating a student's math score is a little bit more difficult, as the Grid-in questions do not have a penalty for wrong answers:

> Of the 44 multiple choice math questions, Tyler answered 37 correctly and 6 incorrectly, and left 1 blank.
>
> Of the 10 grid-in questions, Tyler answered 4 correctly and 5 incorrectly, and left 1 blank.
>
> Tyler earns 37 points for the multiple choice questions he answered correctly and another 4 points for the grid-in questions he answered correctly:
>
> Correct: $37 + 4 = 37$
>
> He loses 2 points for the 6 multiple choice questions that were incorrect:
>
> Incorrect: $6 \times -0.25 = -1.5$ Rounded to the nearest whole number: -2
>
> He does not gain or lose any points for either omitted question:
>
> Omitted: $2 \times 0 = 0$
>
> He does not lose any points for the incorrect grid-in questions. His raw score, then, is the points earned minus the points lost:
>
> $41 - 2 = 39$
>
> As you can see on the conversion chart, a raw score of 39 converts to a scaled score of 600.

The writing score is a combination of a multiple choice raw score and an essay score.

Interpreting the SAT Score Report

An SAT score is often called "The Great Equalizer" among college admissions officers. It is the only tool in your admissions file that truly compares you to other applicants. Grade point averages, class ranks, and extracurricular activities are important, but they are difficult for anyone to judge in comparison with another student from another school. For example, the amount of work it takes to achieve an 'A' in Mrs. Jones' English class in San Diego may only equate to a 'B–" in Mr. Smith's English class in Boston. And Lincoln High School in Minnesota may only allow one student to be ranked number one in the class, while Washington High School in Louisiana allows all students with a GPA higher than 4.0 to receive a class ranking of one. While admissions officers maintain school files with information about class rankings and course difficulty in order to try to more adequately compare students from different schools, this system still presents great difficulty and potential unfairness. Therefore, because the SAT is standardized and taken by all applicants, it is used to directly compare students in math, reading, and writing.

Each scaled SAT score places a student in a certain relative position compared to other test takers. These relative positions are represented through a percentile that correlates to each score; the percentile indicates where the test taker falls in the overall pool of test takers. For example, a score of 650 on the math section represents the 85th percentile, meaning a student with a score of 650 scored better than 85 percent of the college-bound seniors who have taken the test in the last year. Only 15 percent of test takers scored better. The percentile is critical since it is a true indicator of your positioning relative to other test takers, and thus college applicants.

The average SAT score is around 500 in each of the three sections, for a 1500 composite average.

The College Board has also prepared a composite score percentile conversion. For example, a student who has earned a 1780 on all three sections receives an 80th percentile ranking. This student has performed better than 80 percent of test takers.

Percentile tables for individual sections and their composites are available on the College Board website. The percentiles may vary slightly from year to year, but a recent representation is available for your study on the following pages. First, though, consider a summary of some benchmark scores:

1% of test takers score 2200 or higher.
7% of test takers score 2000 or higher.
19% of test takers score 1800 or higher.
39% of test takers score 1600 or higher.
51% of test takers score 1500 or higher.
63% of test takers score 1400 or higher.
84% of test takers score 1250 or higher.

SAT Individual Section Percentile Table

Scaled Score	Critical Reading Percentile	Math Percentile	Writing Percentile
800	99	99	99+
790	99	99	99+
780	99	99	99
770	99	99	99
760	99	98	99
750	98	98	99
740	98	97	98
730	97	97	98
720	96	96	97
710	96	95	97
700	95	93	96
690	94	92	95
680	93	91	94
670	92	89	93
660	91	88	92
650	90	86	90
640	88	83	89
630	85	81	87
620	84	79	85
610	82	76	83
600	79	74	81
590	77	71	79
580	75	68	76
570	72	66	73
560	69	63	71
550	66	60	68
540	62	56	64
530	59	53	62
520	56	50	58
510	52	47	54
500	49	43	51
490	45	40	47
480	42	36	44
470	38	33	40
460	35	30	37
450	32	27	33
440	28	25	30
430	26	22	27
420	22	19	23
410	20	16	20
400	17	14	18
390	15	12	15
380	12	11	13
370	11	9	11
360	9	7	9
350	8	6	7
340	6	5	6
330	5	4	5
320	4	4	4
310	4	3	3
300	3	3	3
290	2	2	2
280	2	2	2
270	2	1	1
260	1	1	1
250 or less	1 or less	1 or less	1 or less
200	–	–	–
Average	**502**	**515**	**494**

Say you scored a 540 on the math section. This equates to the 56th percentile for math. But if you scored a 540 in writing, you earned a spot in the 64th percentile.

SAT Composite Percentile Table

Scaled Score	Composite Percentile	Scaled Score	Composite Percentile	Scaled Score	Composite Percentile
2400	99+	1850	85	1300	25
2390	99+	1840	84	1290	24
2380	99+	1830	84	1280	23
2370	99+	1820	83	1270	22
2360	99+	1810	82	1260	21
2350	99+	1800	81	1250	20
2340	99+	1790	80	1240	19
2330	99+	1780	80	1230	18
2320	99+	1770	79	1220	17
2310	99+	1760	78	1210	17
2300	99+	1750	77	1200	16
2290	99+	1740	76	1190	15
2280	99	1730	75	1180	14
2270	99	1720	74	1170	13
2260	99	1710	73	1160	13
2250	99	1700	72	1150	12
2240	99	1690	71	1140	11
2230	99	1680	70	1130	11
2220	99	1670	69	1120	10
2210	99	1660	68	1110	9
2200	99	1650	67	1100	9
2190	98	1640	66	1090	8
2180	98	1630	65	1080	8
2170	98	1620	64	1070	7
2160	98	1610	62	1060	7
2150	98	1600	61	1050	6
2140	98	1590	60	1040	6
2130	97	1580	59	1030	5
2120	97	1570	58	1020	5
2110	97	1560	57	1010	5
2100	97	1550	55	1000	4
2090	96	1540	54	990	4
2080	96	1530	53	980	4
2070	96	1520	52	970	3
2060	95	1510	51	960	3
2050	95	1500	49	950	3
2040	95	1490	48	940	3
2030	94	1480	47	930	2
2020	94	1470	46	920	2
2010	94	1460	44	910	2
2000	93	1450	43	900	2
1990	93	1440	42	890	2
1980	92	1430	41	880	1
1970	92	1420	39	870	1
1960	91	1410	38	860	1
1950	91	1400	37	850	1
1940	90	1390	36	840	1
1930	90	1380	35	830	1
1920	89	1370	33	820	1
1910	89	1360	32	810	1
1900	88	1350	31	800	1
1890	88	1340	30	790	1
1880	87	1330	29	780–600	1 or less
1870	86	1320	28		
1860	86	1310	26	***Average***	**1511**

A student who scored a 570 Math, 620 Writing, and 550 Reading would have a composite score of 1740 and rank in the 76th percentile.

Composition of the SAT

Although the math section is the focus of this book, let's briefly review the other sections on the SAT.

THE CRITICAL READING SECTIONS

The Critical Reading score is determined by 67 multiple choice questions from three reading sections.

Sentence Completion Questions

There are two types of Critical Reading questions: Sentence Completion and Passage-Based Reading.

The first type of Critical Reading question is the Sentence Completion question. These questions ask you to fill in one or two blanks in a sentence with a word or phrase from the answer choices:

1. Though its robust body looks extremely -------, the American alligator is actually ------- enough to run 35 miles per hour.

   (A) porcine . . sturdy
   (B) wieldy . . awkward
   (C) supple . . heavy
   (D) troublesome . . lively
   (E) cumbersome . . agile

There are 19 Sentence Completion questions on every SAT, comprising 28% of your Critical Reading score. There are many patterns and clues within these questions, from the use of certain words in the sentence to the types of wrong answer choices presented.

Passage-Based Reading Questions

The other type of Critical Reading question is the Passage-Based Reading question. Often called Reading Comprehension, these questions ask you to find answers based on a long or short passage:

1. The statement in lines 59–61 ("Not long...heart attacks") implies that

   (A) athletes have extremely long life spans
   (B) little is known about the EPO hormone
   (C) European athletes are highly susceptible to heart failure
   (D) a person's heart rate affects his or her risk of heart attack
   (E) the cyclists were using the EPO hormone

Again, there are many patterns among Passage-Based Reading questions, which you can learn from analyzing dozens of tests. PowerScore also offers courses and books to help you with your complete SAT preparation, which you can find online at powerscore.com.

THE WRITING SECTIONS

The Writing score is determined by one essay and 49 multiple choice questions from three writing sections.

The Essay

The Essay is always assigned the first section of the SAT. You are asked to write a persuasive essay in response to a prompt and assignment:

> It has long been said that money cannot buy happiness, yet millions of people still yearn for millions of dollars. Plato said, "Wealth is the parent of luxury and indolence, and poverty of meanness and viciousness, and both of discontent."

Assignment: Does a person's wealth determine his or her level of happiness? Write an essay in which you defend your opinion on this topic. Support your viewpoint with examples and evidence from your education, reading, and personal experience.

Did you know that it is perfectly acceptable—and even expected—for students to use the word "I" in their essays?

Improving Sentences Questions

There are three types of questions on the two multiple choice Writing sections. The first, and most prevalent, are the Improving Sentences questions. These questions ask a test taker to determine whether an underlined portion of a sentence contains an error, and if so, to choose an answer choice that corrects the error:

1. Not one of the juniors on the varsity tennis team <u>have won a single match on the road this year</u>, but the second half of the season has very few away games.

   (A) have won a single match on the road this year
   (B) have managed to win a single match on the road this year
   (C) have won a single match this year on the road
   (D) has won a single match on the road this year
   (E) has won being on the road this year a single match

There are 25 Improving Sentences questions on each SAT.

There are three types of multiple choice Writing questions: Improving Sentences, Identifying Sentence Errors, and Improving Paragraphs.

Identifying Sentence Errors Questions

The second Writing question type is Identifying Sentence Errors. These sentences have five underlined parts. You simply have to determine if there is an error, and if so, which underlined portion is the error. You do not have to correct the error:

2. Only after the tide had rose three feet were the volunteers
   A B

   able to free the stranded whale and her calf from the
   C D

   beach. No error
   E

Identifying Sentence Errors questions appear 18 times on the longest writing section of the test.

In the Improving Sentences and Identifying Sentence Errors questions, there are about 25 grammatical errors that frequently appear on the SAT. These errors can be found by studying multiple tests, or by preparing with a PowerScore course or publication.

Improving Paragraph Questions

The final type of Writing question tests a student's ability to organize and edit an existing essay. In these Improving Paragraphs questions, students are given a short rough draft of an essay and then asked how to correct portions of the essay:

1. In context, which is the best way to deal with sentence 5?

   (A) Replace "Sadly" with "Fortunately".
   (B) Change "its" to "their".
   (C) Change "who" to "whom".
   (D) Insert "It is with much regret that" at the beginning of the sentence.
   (E) Delete "in their address book".

There are only 6 Improving Paragraph questions, which appear at the end of the longest writing section. Two or three of the questions usually deal with the same errors covered by the first two types of questions.

THE EXPERIMENTAL SECTION

Each SAT administered by the College Board has one unscored 25-minute section. Sometimes called the variable or equating section, this section is used to collect data, such as difficulty level and validity, about specific questions that will be used on future SATs.

Since the College Board tests so many questions, test takers in the same location will have different experimental sections. The experimental section may be an additional Math, Critical Reading, or multiple-choice Writing section. It is never an additional essay, nor is it an additional 10- or 20-minute section. Therefore, it will never occur in Sections 1, 8, 9, or 10. Some students feel that the experimental section is more challenging than the other sections, but it is nearly impossible to determine which section is experimental while the test is occurring (especially since you are not allowed to look forward in the test booklet). You *must* attack all 25-minute sections as if they are being used to calculate your score.

The experimental section will never occur in sections 1, 8, 9, or 10.

However, you should be able to narrow down the possibilities for the experimental section to two sections. First, begin by analyzing which subject area had four sections. For example, if you had four Math sections, one of them was experimental. You can narrow it down even further using the following information:

During the test, it is almost impossible to know which section is experimental. Therefore, you should attack every section as if it were contributing to your score.

- If you have two 25-minute multiple choice Writing sections, one of them is experimental.
- If you have two Math sections with grid-ins, one of them is experimental.
- If you have two 25-minute Math sections without grid-ins, one of them is experimental.
- If you have two Critical Reading sections with long dual passages, one of them is experimental. If one of them is only 20 minutes long, then the 25-minute section is the experimental.
- If you have two Critical Reading sections with a single long passage, one of them is experimental. If one of them is only 20 minutes long, than the 25-minute section is the experimental.
- If you have two Critical Reading sections, with two separate, long passages, one of them is experimental. If one of them is only 20 minutes long, then the 25-minute section is the experimental.

Remember, even if you can narrow the experimental section to one of two sections, it is still extremely important that you try your best on both sections, as it is impossible to precisely determine the experimental section during the test without looking forward in the test booklet.

Testing Schedule

The SAT is offered seven times a year, usually in October, November, December, January, March, May, and June.

You should take your first SAT during the spring of your junior year, making sure that you have completed Algebra I, Geometry, and Algebra II. If your high school offers a grammar course, also take this class prior to registering for the SAT. Because the SAT is considered a junior-level test, it is wise to take the exam when junior-level material is still current and fresh in your mind. The test is offered in May and June; choose the administration that works best with your end-of-school-year schedule.

March of 2009 marked a new score reporting policy for College Board. By choosing "Score Choice," students can pick which single test score they would like to submit to colleges. This allows you to take the test multiple times without releasing multiple results to admissions officers. You must select this option when registering for the SAT; students who do not select it will have all scores from all administrations sent to prospective colleges.

Although PowerScore recommends taking the test during your junior year, it is extremely important that you are fully prepared prior to taking the SAT. Be sure to download and take the free practice test from the College Board website. The accompanying score report will help you determine if you are ready for the SAT. If you find that you need additional help, consider a prep course or private tutoring. PowerScore instructors are SAT experts who can help you unlock the secrets of the test.

Most test prep companies promise that your score will improve after completing a course. PowerScore offers a 300 point increase guarantee in our Full Length Course. This is a great option for seniors who are retaking the test.

If your spring score does not meet your expectations, you have several more opportunities to try again during the fall of your senior year. In fact, most students take the test two or three times. The test is offered four more times during the fall and winter of your senior year, although some colleges may not accept scores from the January administration.

The PSAT

The Preliminary SAT (PSAT) is a standardized test program used to help students gauge their performance on the SAT. Although co-sponsored by the College Board and the National Merit Scholarship Corporation (NMSC), the PSAT is administered by local high schools and is generally given to sophomores and juniors.

Most students take the PSAT in order to practice for the real SAT. Others register in order to submit their names to colleges as prospective students. For these reasons alone, the test is well worth the small fee charged by your high school. Additionally, students who take the PSAT are also qualified for several scholarships or recognition programs, including the National Merit Scholarship Program.

The PSAT is a shorter and slightly easier version of the SAT. Most of the concepts tested on the SAT are also tested on the PSAT:

	THE SAT	THE PSAT
Math Sections	Two 25-minute sections One 20-minute section	Two 25-minute sections
Math Concepts	Arithmetic, Algebra I, Algebra II, Geometry, Data Interpretation	Arithmetic, Algebra I, Geometry, Data Interpretation
Math Questions	44 multiple choice 10 grid-ins	28 multiple choice 10 grid-ins
Reading Sections	Two 25-minute sections One 20-minute section	Two 25-minute sections
Reading Concepts	Sentence Completion, Passage-Based Reading	Sentence Completion, Passage-Based Reading
Reading Questions	67 multiple choice	48 multiple choice
Writing Sections	Two 25-minute sections One 10-minute section	One 25-minute section
Writing Concepts	Identifying Sentence Errors, Improving Sentences, Improving Paragraphs	Identifying Sentence Errors, Improving Sentences, Improving Paragraphs
Writing Questions	49 multiple choice 1 essay	39 multiple choice

The two SAT content areas omitted on the PSAT are Algebra II and essay writing.

The PSAT is only given once a year, usually in the second week of October. Because the test contains sophomore-level material, you should take it after the completion of your second year of high school, in the fall of your junior year. Many students also take the PSAT in their sophomore year; if you choose to take the test as a sophomore, be sure to also take it as a junior in order to qualify for any junior-level scholarships or recognition programs.

Studying for the PSAT with the PowerScore SAT Math Section Bible

Because the SAT and PSAT are so similar, students studying for the PSAT Math section will greatly benefit from studying this book. Nearly all of the material covered in the following chapters may appear on the PSAT. However, for the Algebra II chapter, only study the section on functions; the other third-year Algebra II concepts will not appear on the test.

Rethinking the SAT

The SAT is unlike any classroom exam you have ever encountered, and your potential for success depends on your ability to change your mind-set about the test. Most students enter the test center with several assumptions that affect the way they approach the test. If believed, these toxic thoughts can undermine your confidence and poison your final results. It is imperative that you begin thinking about the SAT in a new way, and avoid saying or believing any of the following toxic thoughts:

Test anxiety is often caused by toxic thoughts.

Toxic Thought #1:
"I am not good at taking tests."

Tests from high school are designed to measure your knowledge of the material that was previously taught in school. Can you remember and apply a formula to find a numerical sequence? Can you memorize and define the word "pragmatic?" If you have a hard time with tests in school, it is because you haven't adequately learned the material presented.

The SAT is not really designed to test your knowledge of high school curriculum. There is not much memorization required. Instead, the SAT is used to assess your ability to solve problems using critical reasoning. Critical reasoning is a skill used to analyze a problem in order to find the best solution. It is a skill that is acquired through practice, and unfortunately, it is not practiced enough in most high schools. The College Board has designed a test in which common high school math symbols and concepts are used to disguise puzzle-like questions. This book is written to help you learn to think logically and reason critically when approaching the math section of the test. So even if you do not test well on regular high school tests, you can still do exceptionally well on the SAT.

Memorization is not required to earn a great score on the SAT. However, PowerScore does recommend certain formulas and shortcuts be memorized, which we note throughout this book. Memorization can help shave minutes off of a section, thus allowing you to check your answers before time is called.

It is important to stop saying, "I am not good at taking tests." This thought has more power than you know, and once spoken, its negative potential can spread like a virus. You must begin thinking and saying, "I am great at taking the SAT." Just as a negative statement can negatively influence your performance, a positive statement can help boost your achievement. If you slip and think or say "I am not good at taking tests," counter the negative energy by adding, "but I'm great at taking the SAT."

Toxic Thought #2:
"I am not good at math."

If you believe the toxic thought, "I am not good at math," it's quite likely that someone—a teacher, a parent, or even a peer—once said or indirectly told you that you were not good at math. These words are especially influential during middle school, when students are the most susceptible to evaluation by others. Maybe it was a 'C' you received from Mr. Peters in Algebra, or maybe it was your father saying, "I was already in Calculus when I was your age," or maybe it was even your best friend telling you she received a 98% on the same test on which you scored a 78%. Whatever they said, you heard, "You are not good at math." And, as we noted above, words have power. You internalized this sentence, and began your life as a person who is not good at math. You likely never did well in math class again and shied away

Confidence Quotation
"A successful person is one who can lay a firm foundation with the bricks that others throw at him or her."
—David Brinkley, newscaster

from any elective courses that focused on math. You believed you weren't good at math, and this toxic thought spread to your teachers, parents, and peers. They did not think you were very good at math either, and thus didn't expect very much from you when it came to math. One bad math experience led to many more, which led to other people telling you that you were not good at math. Someone might even be telling you right now to concentrate on the reading and writing sections of the SAT, because you have a better chance of improving in the verbal portion than in the math portion.

What a sad situation to have resulted from one simple statement so long ago! Can you imagine where you might be if someone had told you that you were good at math? Think about your classmates who *are* good at math—I bet you can remember specific instances in which people praised them for their abilities in mathematics. Thoughts and words are so powerful.

It is time to stop thinking that you are bad at math. For one thing, you may have misunderstood the person whose words or actions led you to believe they thought you were bad at math. What if Mr. Peters gave you a 'C' because you talked too much in class? What if your dad brought up Calculus simply to make a comment about how much school has changed since he was a student? What if your friend lied about her 98% because she was embarrassed about her real score?

One PowerScore student told a story of a misunderstanding that set her on a negative math journey. She had begun to believe she was bad at math in 6th grade, when her math teacher asked her to attend an after school study session to review the current lesson. Even though she was in the advanced math class, she had been so humiliated at being asked to attend a review that she skipped the review session and her math grade started to slip. She soon quit taking advanced math courses, and her parents and teachers quit expecting high math grades from her. Her own feelings about math changed given the negative experiences she had. When asked about the moment she was told she was bad at math, she admitted that years later she learned that the 6th grade math teacher invited both the highest and lowest scoring students in class; the teacher was hoping the advanced students could tutor and help bring up the grades of the struggling students. She had no idea if she had been asked to the review session as one of the highest or lowest scoring students, but at that point, it no longer mattered as the damage had been done and she firmly believed she was no longer good at math.

If you can pinpoint the moment that you were first "told" you were bad at math, can you think of any alternative meanings the speaker might have had? Can you look back and see how things might have been different if you had taken the comment a different way?

Another reason that you must stop putting down your math abilities is that SAT math is much different from high school math. Most of the mathematical concepts covered on the SAT come from elementary and middle school math courses. Seriously. Concepts such as fractions, percentages, decimals, lines and angles, remainders, and integers make up much of the math portion. Higher level math concepts are not covered. For example, your ability to find sine, cosine, and tangent will not be tested. These Algebra II concepts involve memorization and leave little room for critical reasoning, whereas a fraction involving variables tests a student's ability to think logically rather than his ability to crunch numbers in a formula. Remember, critical reasoning is a skill rather than a knowledge; even if you truly do struggle with math, you can still be good at SAT math.

You must banish all negative thoughts concerning your abilities in math. If you believe that you are not good at math, or at a particular branch of math, you will not be receptive of the new techniques covered in this book. For example, a previous PowerScore student simply did not understand basic functions. He had been told he was bad at math in 8th grade when his teacher irresponsibly assigned seats each week based on their Friday quiz scores. The students with the best scores sat in the front of class, and the students with the worst scores sat in the back. After sitting in the back a couple of weeks during the unit on functions, our former student started to believe he was bad at math, and particularly bad at functions. When he arrived in our office, he was convinced he could not learn how to solve basic functions. It was like a curtain dropped the minute a function was mentioned; his eyes sort of glazed over, and you could almost hear him thinking, "I can't learn this." And as long as he thought in this manner, he certainly could not learn functions.

Confidence Quotation
"It's not who you are that holds you back, it's who you think you're not." —Denis Waitley, productivity consultant

It wasn't until we got at the root of his dislike of functions that we were able to change his thinking. He realized the teacher's inappropriate actions still had total control over his math abilities, and he agreed to start thinking more positively about math. The curtain soon lifted, and he quickly learned to process functions. He commented later at how odd it was that he was unable to learn such a simple concept. His belief in the certainty of his failure had held his abilities hostage, and with a little positive thinking, he now ironically refers to functions as "simple."

You, too, must begin thinking positively about math. Replace any negative thoughts with "I am great at SAT math." Remember, SAT math is much more basic than high school math and it requires more skill than knowledge to solve. Anyone can be good at SAT math with a little practice. If you slip and think, "I am bad at math," add a positive clause at the end: "but I am great at SAT math."

Toxic Thought #3:
"I have to do well on the SAT."

The pressure surrounding your performance on the SAT can be colossal. It may seem like everything—from admissions to scholarships to pride—is riding on those few hours spent huddled over a bubble sheet on a desk. But the very thought that the test *can* cost you everything *will* cost you everything.

One of our instructors takes the SAT at least once a year in order to keep up with any new trends on the test. There is no pressure on her to do well, as she has already attended college and no one will ever see her results unless she chooses to share them. She usually scores above a 750 on each section of the test. One year, however, her friends started making wagers on how they thought she would do. One friend had so much confidence in her, he made a $100 wager that she would score a 2350 or better. He promised to give the instructor half his winnings if she could win the bet. She didn't think it would be too difficult, as she had already made that score several times. But on test day, the pressure started to get to her. She made it through the essay, and the first reading section, but when she reached the first math section, she panicked when she didn't know how to solve a question in the middle of the section. As she tried to find a solution, a little voice in the back of her head started saying, "There goes the money! He's going to kill me. I can't believe I can't solve this medium-level question!" The more she panicked, the more questions she started to misread or misunderstand. When time was called, she still had three questions unanswered and she had guessed on 4 others. She went into the hallway and regrouped; by leaving three questions blank and possibly missing more, it was impossible to now get a 2350. The money was gone and there wasn't anything she could do about it. She took a deep breath and went back in, knowing the pressure was off.

When she got her score report a few weeks later, she contacted us to share her story and her interesting results. She did indeed miss the 2350, as she suspected, but what was so intriguing was her location of errors. In the first reading section, which she thought she had aced, she missed three questions. In the first math section, she missed the three omitted, two of the four that she guessed on, and two more that she thought she had done correctly. After that section, the pressure had been relieved, and her remaining sections had a direct consequence: she did not miss another question on the test.

She wanted to share her story with us because it reminded her how students feel when taking the SAT, and revealed to her how the slightest bit of pressure can endanger an SAT score. Adults sometimes forget about the pressure surrounding this test, because they haven't taken it in so long and because they are no longer under the illusion that the test is the ultimate element for college admissions. If one of our instructors was so affected over $50, we can only imagine how our students feel when admissions and thousands of dollars of scholarship money are on the line.

You control the stress you feel surrounding the SAT. Learning to control the pressure takes effort, but will be rewarded on the test and throughout your life.

Many students want to blame their parents, their future college, their teachers, or even the test itself. But the truth is that you are the one creating the pressure. If you did not care about your future, nothing your parents or teachers said about the importance of the SAT would matter at all. You would simply shrug them off. In fact, you would not be reading this book right now. Your parents would have purchased it for you, and you would have thrown in on your desk or under your bed with no intention of reading it, because the SAT is not a priority for you. Just the fact that you are reading this sentence assures us that you are the creator of any pressure you are experiencing concerning college admissions. You want to go to college. You probably want to go to a specific college. You may even want to win scholarship money based on an SAT score. The very existence of these desires creates a fear of not fulfilling the desires, which in turn creates pressure. Marcus Aurelius the Wise, one of the last great Roman Emperors, wrote "If you are distressed by anything external, the pain is not due to the thing itself, but to your estimate of it; and this you have the power to revoke at any moment."

Marcus Aurelius knew the secret to SAT success over 1700 years before the SAT was even invented: you must revoke the power you have given the SAT in order to avoid the pressure associated with it. You cannot make it the single measure of your achievements and you cannot believe that it has the ability to ruin your future.

The SAT is only one component of your college admission folder. Admissions officers will also be looking at your application, transcript, other test scores, activity list, essays, recommendations, and possibly even an interview. A college is looking for the most dynamic individuals to fill their hallowed halls, and while test scores can reveal a person's potential academic success, they do little to show character and integrity. For this reason, some schools—especially those that specialize in liberal arts—no longer require the SAT. And in most schools where the SAT is required, students can offset a subpar test score or average grades by documenting initiative in an academic pursuit or passion for a single extracurricular activity.

For a list of colleges that do not require the SAT, visit www.fairtest.org.

You should also remember that this is not your only chance to take the SAT. If your score does not meet your expectations, you have options. Under the College Board program Score Choice™, which was designed to relieve test anxiety and maximize student achievement, you can choose which individual test administration you would like to send to colleges. This means that an uncharacteristic score never has to be revealed to any admissions program. Or, if you choose not to use Score Choice™, many colleges look at all of your tests, using only your highest scores on each section. For example, say you take the test in January and receive a 630 on Critical Reading, a 540 in Writing, and a 490 in math. You retake the test in June and score a 600 on Critical Reading, a 590 in Writing, and a 550 in math. Colleges that use this policy will only look at the 630 in Critical Reading, the 590 in Writing, and the 550 in math. The lower scores are ignored. To find out how your prospective colleges view multiple test scores, call their admissions departments. Knowing that you have other options and opportunities should relieve much of the stress surrounding the test.

Also, don't forget that the SAT is a standardized test, meaning that every test is similar. Similar tests have patterns, as there is a finite number of concepts tested on the SAT. If you learn all of these patterns and concepts, you will have no problem mastering the test. This knowledge should give you the confidence you need to do well and banish any fear or anxiety surrounding the test.

Finally, remember that the SAT is not an IQ score, nor is it a predictor of how well you will do in college. The makers of the test would like you to believe otherwise, but test prep instructors can give you many examples of students with average IQs and exceptional SAT scores. It is a beatable test. The concepts can be learned and mastered by every student. The SAT simply tests how well you will do on the SAT; it does not indicate how well you will do in the rest of your life. Do not overinflate its importance.

The moral of the story is that you do not *have* to do well on the SAT. It would be great if you did, but there is not an obligation to do so, and even if your score is below your expectations, you can take the test again. To obsess on a particular score will only keep you from reaching your ultimate potential on the test. You must change your mantra for the SAT. Instead of worrying about what you *have* to do, assert what you *will* do: "I will do well on the SAT." Recite this sentence several times a day throughout your preparation. Write it down and hang it from your bathroom mirror so you see it every morning and every night. If you begin to panic during the test, take several deep breaths while remembering you always have options. You will do well on the SAT.

Power Thoughts

As you work through this book, there are three thoughts on which you should stay focused. These thoughts empower you:

> *"I am great at taking the SAT."*
> *"I am great at SAT math."*
> *"I will do well on the SAT."*

If any toxic thoughts begin to seep into your conscious, immediately recite an empowering thought and then reread this section. If you begin to feel pressure, think about how you can gain control of the situation while reciting an empowering thought. We cannot adequately stress the power of positive thinking. You must believe in yourself and in your abilities if you want to do well on the SAT.

To aid you in your visualization of success, we have placed a series of quotations on positive thinking and confidence in the margins of this book. If certain quotations particularly inspire you, rewrite them on index cards and hang the cards on your mirror, locker, refrigerator, bulletin board, or somewhere else that you can read them every day before the SAT. Learning to visualize your full potential takes practice and repeated affirmations.

Confidence Quotation

"Confidence is preparation. Everything else is beyond your control." —Richard Kline, actor

About This Book

The PowerScore SAT Math Bible is organized in two main sections. The first four chapters offer an introduction to the SAT and to the SAT math sections, highlighting basic information and the fundamentals of success on the test. The remaining chapters discuss specific SAT content and strategy from math subject areas. It would be wise to return to the first four chapters after completing the book to review the basics once more.

Throughout the book, we use the margins to highlight important information. Some of these margin notes have specific topic names to help your organize your review of the book. The notes titled "Arithmetricks" call attention to shortcuts and tricks designed to save time on the test. The notes called "Caution: SAT Trap" warn students about common wrong answer choices and mistakes made by previous test takers. Finally, the margin notes designated "Memory Markers" show formulas that you will need to memorize before taking the SAT.

Because access to accurate and up-to-date information is critical, we have devoted a section of our website to SAT Math Bible students. This free online resource area offers supplements to the book material, questions posed by students and their answers, and updates as needed. In addition, every formula notated by a Memory Marker is provided in a free set of SAT Math Bible Flash Cards, only available on our website to book owners. There is also an official book evaluation form that we strongly encourage you to use.

The exclusive SAT Math Bible online area can be found at:

www.powerscore.com/satmathbible

We strongly recommend purchasing *The Official SAT Study Guide*, published by the College Board. This book has ten real practice tests and you should complete as many as possible prior to taking the SAT. To help you with your study of SAT math, we have mapped test questions in *The Official SAT Study Guide* by content area to offer you hours of practice with real math questions. You can find this map, called the Blue Book Database, in the SAT Math Bible online area.

If we can assist you in your SAT preparation in any way, or if you have any questions or comments, please do not hesitate to contact one of our helpful instructors via email at satmathbible@powerscore.com. Additional contact information is provided at the end of this book. We look forward to hearing from you!

ARITHMETRICK

Look to Arithmetricks for shortcuts and time-saving techniques.

CAUTION: SAT TRAP!

Don't fall into the traps set for you by the College Board. Learn to avoid these traps with these margin notes.

MEMORY MARKER:

These markers serve to remind you to remember! You may want to start a formula sheet that you add to whenever you see a Memory Marker.

Chapter Two:
The Basics of SAT Math

The SAT assesses a student's ability to reason by problem solving. These problems are drawn from basic math concepts, listed below along with the subject areas where you may have first encountered them:

SAT-TESTED CONCEPTS	SUBJECT AREAS
Operations	
Integers and Real Numbers	Arithmetic and Algebra
Fractions and Decimals	Arithmetic
Ratios, Proportions, and Percents	Arithmetic
Exponents and Roots	Arithmetic and Algebra
Counting Problems	Arithmetic and Statistics
Properties of Integers	Arithmetic and Algebra
Sets	Arithmetic and Algebra
Sequences	Algebra
Logical Reasoning	All math courses
Arithmetic Word Problems	Arithmetic
Algebra and Functions	
Linear and Quadratic Equations	Algebra
Inequalities	Algebra
Direct and Inverse Variation	Algebra
Functions	Algebra II
Functions as Models	Algebra II
Symbolic Functions	Algebra II
Algebraic Word Problems	Algebra
Geometry and Measurement	
Lines	Geometry
Circles, Squares, and Triangles	Geometry
Polygons and Quadrilaterals	Geometry
Solids	Geometry
Coordinate Geometry	Geometry
Area, Perimeter, and Volume	Geometry
Geometric Visualizations	Geometry
Transformations	Geometry and Algebra II
Data Analysis, Statistics, and Probability	
Data Interpretation	All math and science courses
Average, Median, and Mode	Arithmetic and Statistics
Probability	Arithmetic and Statistics

As you can see, many of these math concepts were covered in middle school. You are not required to have advanced math experience to do well on the SAT.

Math Questions

There are two types of math questions on the SAT: multiple-choice and grid-ins. The difference between these two question types lies in the way students answer them, which will be discussed in detail later in this chapter. However, the actual question does not vary from type to type. Consider the following example, which could be a multiple-choice or a grid-in:

16. If a is the least prime factor of 70 greater than 2, and b is the greatest prime factor of 90, what is the value of $b - a$?

Read Carefully!

Many students miss SAT math problems because they misread the actual question. Pay close attention to any rules established for a variable, such as $x > 0$, or x and y are odd consecutive integers. If any word or rule is omitted, you may incorrectly solve the question. It is also important to understand what the question is asking; notice in the example above that the question is requiring the value of $b - a$, rather than the value of a single variable. Reread the question after finding a solution to make sure you found the requested value or expression.

Paraphrase

If you have a hard time understanding the question as it is written, paraphrase it in your own words. For example, some students may paraphrase the first part of the question above as: "*a* is the smallest factor of 70 that is a prime number, but it must be larger than 2." Using your own words, rather than the confusing words of the test maker, may help you to understand a question more clearly.

☠ CAUTION: SAT TRAP!
The College Board is not completely diabolical; on occasion, they will use capitals, italics, or underlining to warn you that a question uses tricky language. Read these questions extra carefully to avoid a trap.

Beware of Emphasized Text

While the College Board does use some intentionally confusing phrases, they will sometimes alert you to their deceptive tricks. If a problem uses words in CAPITALS, *italics*, or underlining, pay close attention to the wording of the phrase. This is the test makers way of warning you that the text may be confusing. For example, a multiple-choice question about an equation may end with "All of the following could be the value of x EXCEPT". If you are not careful, you may pick the first answer that does work, rather than the one answer that does not solve the equation.

Understand the Nature of the Solution

If the question asks, "What is the value of x?" it is important to understand that there is only one value for x. On the other hand, some SAT math questions may ask, "What is one possible value of x?" which indicates that there is more than one solution. This may give you some insight into the question before you begin to solve.

Similarly, some questions use "must" and "could be" to cause confusion. If the question says, "then x must be...," x has a single answer, no matter what values are used to compute x. If the question muses "then x could be...," there may be multiple possibilities for x.

Section Format

As mentioned in the previous chapter, there are three scored math sections on the SAT. Each section has an assigned number of questions and a time limit for answering those questions:

Section A
- 18 questions (8 multiple-choice, 10 grid-in)
- 25-minute time limit
- May occur anywhere between Section 2 and Section 7 of the test
- Multiple-choice questions progress from Easy (E) to Medium (M) to Hard (H); grid-in questions also progress from Easy to Hard. Consider a general section map of difficulty:

Multiple-Choice								Student-Produced Response (Grid-Ins)									
1	2	3	4	5	6	7	8	9	10	11	12	13	14	15	16	17	18
E	E	E	M	M	M	M	H	E	E	E	M	M	M	M	M	H	H

In this section, question #8 is almost always considered "hard."

Section B
- 20 questions (all multiple-choice)
- 25-minute time limit
- May occur anywhere between Section 2 and Section 7 of the test
- Questions progress from Easy to Hard. Consider a general section map of difficulty:

Multiple-Choice																			
1	2	3	4	5	6	7	8	9	10	11	12	13	14	15	16	17	18	19	20
E	E	E	E	E	E	M	M	M	M	M	M	M	M	M	M	H	H	H	H

Section C
- 16 questions (all multiple-choice)
- 20-minute time limit
- May occur in Section 8 or Section 9 of the test
- Questions progress from Easy to Hard. Consider a general section map of difficulty:

Multiple-Choice															
1	2	3	4	5	6	7	8	9	10	11	12	13	14	15	16
E	E	E	E	E	M	M	M	M	M	M	M	M	M	H	H

<u>*Confidence Quotation*</u>
"Whatever the mind can conceive and believe, it can achieve."
—Napoleon Hill, author

These section maps offer a general indication of difficulty. A section may have more Easy questions and fewer Medium questions or vice versa. You may also find an Easy question among the Medium or a Medium question among the Hard, but for the most part, you can determine a question's difficulty by its placement in the section.

Difficulty Levels

The College Board has two systems to designate the difficulty level of questions. The first, which is seen most frequently in *The Official SAT Study Guide*, uses the labels "Easy," "Medium," and "Hard." The second, which is used on official score reports, assigns the numbers 1 (easiest), 2, 3, 4, and 5 (most difficult). We have chosen to use the first method, with Easy, Medium, and Hard, to most closely model *The Official SAT Study Guide*.

These labels are not assigned based on analyzed complexity, but rather on the number of students who correctly answered a particular question. Every SAT question has been used on a previous SAT in an experimental section; based on the number of students who successfully answered the question, it is given a difficulty level. If a large majority of students answered the question correctly, it is labeled Easy, and if a large majority of test takers answered the question incorrectly, it is labeled Hard. Therefore, Hard questions are not necessarily difficult to solve. They may just have a trick solution or misleading phrase that caused most of the students to miss it when it was previously tested. In this book, you will see many examples of Hard questions that take just seconds to solve once you learn to look for solution patterns.

Hard questions may actually be very easy to solve! Be sure to read all questions on the SAT.

So how will you know when a question is Easy, Medium, or Hard? On the SAT, math questions progress in order of difficulty. The first few questions are relatively easy to solve, the last few questions are always difficult, and the ones in between have a moderate level of difficulty. In the two multiple-choice sections (Sections B and C on the previous page), generally the first five or six questions are Easy and the last two to four questions are Hard. Note that even though the majority of questions are Medium, these Mediums still progress in difficulty; they start as easy Mediums, and progress to more difficult Mediums.

The section with the grid-ins (Section A on the previous page) is a bit different:

Multiple-Choice								Student-Produced Response (Grid-Ins)									
1	2	3	4	5	6	7	8	9	10	11	12	13	14	15	16	17	18
E	E	E	M	M	M	M	H	E	E	E	M	M	M	M	M	H	H

The first 8 questions are multiple-choice, and they progress from Easy to Hard. Then the difficulty progression starts over with question #9, the first grid-in. The first few grid-ins are Easy, and again progress to Hard. At the beginning of a math section, it is important to note whether it is the section with grid-ins, as this will affect your pacing and strategy.

ARITHMETRICK
In the grid-in section, always skip question #8 and come back to it after answering all of the easy and medium grid-in questions.

Even though there are three differentiated difficulty levels, each question is worth the same value: one point. An Easy question has the same value as a Medium or Hard question. There are no weighted questions. So it makes sense in a timed situation to answer all of the Easy and Medium questions, which you have a higher probability of answering correctly and thus earning points, before working on the Hard questions, where your odds are not as promising. Since question #8 in the grid-in section is almost always Hard, you should skip it in order to make sure that you answer all of the Easy and Medium grid-ins. If you fail to do so, you could get caught up trying to solve #8 and lose valuable time that would have gained you several Easy and Medium points.

Expect straightforward solutions and answers from Easy questions. There are no tricks here. If the question asks you to solve for x, use the formulas and solution strategies that you have been using in math class to find the answer. Students who miss these questions tend to say, "But that answer just looked too easy." It *is* that easy. Do not complicate the problem by trying to make it more difficult. These questions are designated Easy because they are easy.

Hard questions are a different story. Occasionally, a Hard question will be straightforward, but only because the solution involves intricate manipulation of variables or numbers, and thus most students make a mistake when solving. But most Hard questions will have a shortcut or a simple trick solution. Hard questions usually do not involve long calculations or common high school solution strategies, and most can be solved in seconds after determining the simplest solution. It's finding the solution that may take up your time on these questions.

If you think a high-numbered Medium or Hard question was too easy, you probably fell into a trap left by the test maker. After you solve a high-numbered question, you should think, "That was hard," or "That would have been hard if I missed the tricky solution."

Medium questions present a mix of straightforward and shortcut answers. The lower the question number, the more likely it is that the problem uses common solution strategies. As the Medium questions progress, they become more difficult, so the higher numbered Mediums may have some solutions that can be found through critical reasoning rather than long calculations.

Pacing and Time Management

It is important to plan an approach to each math section before arriving at the test center. Consider Section B again:

Multiple-Choice																			
1	2	3	4	5	6	7	8	9	10	11	12	13	14	15	16	17	18	19	20
E	E	E	E	E	E	M	M	M	M	M	M	M	M	M	M	H	H	H	H

If you were to spend an equal amount of time on each question in this section, you'd have a minute and 15 seconds per question. But consider the three levels of difficulty. It doesn't make sense to spend the same amount of time on an Easy question as you do on a Medium question, given that they are both worth the same amount of points. In order to maximize your score on the SAT, you should follow the PowerScore pacing plan:

1. **Take several passes through each section.**

   On your first journey through the section, answer questions that you can easily and confidently solve. This will probably include all of the Easy questions, such as #1 through #6 in the section above. It will also include a majority of the Medium problems. Skip any question that is immediately confusing, but circle it in your book. Be careful when transferring answers to the answer booklet that you skip the omitted question. Depending on your ability level, you may choose to read the Hard questions on this pass; if you immediately identify a shortcut solution, solve the problem. But if you feel that the best solution is a long calculation, skip the question. Most students, however, should probably save the Hard questions until after they answer as many Medium questions as possible. Remember to skip multiple-choice question #8 in the section with the grid-ins, no matter your ability level, as this question is always Hard and should be approached only after attempting all of the Easy and Medium grid-in questions. When you reach the end of the section, go back to the beginning and start a second pass. Work on the questions that you originally omitted, including any Hard questions, and now skip only the questions that absolutely stump you. If you complete your second pass before time is called, continue working on the questions that have caused you the most difficulty. You should plan at least three passes through the section.

   This method will ensure that you gobble up all of the "sure thing" points in the quickest amount of time. It prevents you from spending too much time on one particular question, only to have time called without getting to one or more questions you would have gotten right.

If you read a question and it immediately seems hard, skip it before attempting to solve it. You can come back later after answering all of the easier questions.

2. **Budget your time on each question.**

   The Easy questions should not take up much of your time because they are, by definition, easy. You should plan to spend 20 or 30 seconds on the easiest questions in the section. Then there are only a handful of Hard questions, most of which can be solved using quick logic rather than long calculations. Therefore, you should spend most of your time tackling the Medium questions because you have a good chance of answering each of them correctly. A student who answers every Easy and Medium question correctly will score well over a 600 on the test!

Allow yourself 60 to 70 seconds on each of the Medium and Hard questions. You will solve many of these much more quickly, which will give you a little more time on other questions. If you spend more than a minute and a half on your first attempt at any question, it is imperative that you make a guess and move on. You can mark the question in your test book and come back later if there is extra time at the end of the test, but you simply cannot waste any more time on a single point. Your subconscious will often continue working on the problem after you move on, and the problem may suddenly seem easier when you come back to it at the end of the testing period. It is also important that you make a guess, given that you have invested so much time in solving the question. After studying the problem so closely, it should be easy to eliminate at least one or two of the answer choices.

You should practice taking timed sections of real SATs (available in the *Official SAT Study Guide* or online at www.collegeboard.com) with a countdown timer so that you begin to get a sense of how long 30 seconds, a minute, and a minute and a half takes when testing. You can also wear a wrist watch with a second hand during the actual SAT, but it's better to have an innate sense so that you don't lose time watching the clock. If you take enough timed sections, you'll quickly realize when you are spending too much time on a question.

Silent countdown timers are available at www.powerscore.com. Although timers are not allowed at the official SAT administration, they are still valuable tools for practice.

3. **Never stop working before time is up.**

   Our instructors always come back from real SAT administrations with stories about students at the testing centers who finished a section and put their heads down on their desks. This is outright lunacy!

   You are never done on an SAT math section (or any section for that matter). If you finish your three passes before time is called and feel that you are finished answering questions, you should spend the remaining minutes looking for calculation mistakes. Double check every problem, but especially those that are designed to cause mistakes, such as problems with negative number multiplication and division, exponent distribution, absolute value calculation, inequality division, and negative distribution (see the next chapter for more on calculation errors). Then reread every problem to make sure that you found the requested value. It is a careless mistake to solve for x only to find the problem wanted y. Many of our students have reported finding errors when checking their work.

☠ CAUTION: SAT TRAP! Problems that are designed to cause calculation errors may test negative number multiplication and division, exponent and negative distribution, absolute value calculation, and inequality division.

It is important to practice time management before approaching the actual SAT. By taking several timed sections in the *Official SAT Study Guide*, you will become comfortable with making passes, budgeting time on each question, and double checking your calculations and the question's intended value.

Guessing Strategy

On the SAT, each incorrect multiple-choice answer is penalized a quarter of a point. This dissuades many students from making a guess, despite the time and effort they have put into a problem. However, there are two instances when guessing can increase your score.

You should always guess on the SAT when:

1. **You are working on a grid-in math question.**

   The grid-ins, formally referred to as Student Produced Response questions, do not have a penalty for a wrong answer. Therefore, you should bubble in an answer for all 10 of the grid-ins, regardless of the difficulty level. No grid-in questions should be left blank.

2. **You can eliminate one answer choice on any multiple-choice question.**

   For many math questions, it's easy to eliminate at least one answer choice. Maybe an answer is outlandishly large or mathematically impossible. If you can eliminate one answer choice, your chance of earning points by guessing outweighs the possibility of losing points by guessing. However, you must use this strategy EVERY time you eliminate one answer choice. It will not work if you guess on question #12 but decide to omit #16, even though you eliminated at least one answer choice in both questions. There is a long mathematical proof involving probability that explains this strategy, but your time will be better spent studying SAT material than why you must employ this strategy every time you can eliminate at least one wrong answer.

ARITHMETRICK

The integers 2 and 3 are common answers in the grid-in section, but only use them when they make sense. For example, if the question asks "What fraction of the marbles are blue?," answer with a fraction (such as 2/3) rather than an integer. ALWAYS guess on grid-ins!

Calculators and Scratch Paper

Good news first: your calculator is permitted for the SAT! And even better news next: unless you struggle with basic math facts, you may not even need it!

The College Board, the makers of the test, understands that not all students own a calculator. In order to be fair, they write the SAT so that every problem can be quickly solved without a calculator. Most calculations are very simple and involve small numbers that are easily divisible. For example, you will never be asked to solve 6^8, as it would take too long with pen and paper to find the answer. But you may be expected to find 6^2 or even 6^3, as you can quickly solve this, even without a calculator.

You are permitted to use a four-function, scientific, or graphing calculator. Because the SAT does involve coordinate geometry, we recommend that you take the test with a graphing calculator. However, it's very important that you are comfortable with whatever calculator you use, as the SAT is no place to learn the functions of a new calculator.

The SAT is not the appropriate time to experiment with a new calculator!

Despite this recommendation, we urge you to complete as much of the test as possible without your calculator. You should not be reaching for it to multiply 4 and 8. This is a basic math fact, and the time it takes to turn on your calculator and punch in the problem is time you lose on a future question. Plus, many SAT questions and answer choices involve variables, and therefore are impossible to solve on a calculator. Some of the questions on the SAT are intentionally designed to test your reliance on a calculator; students who reach for their math machines are often sent on a "wild goose chase," because the solution involves simple multiplication or division. When taking timed practice sections, try taking a couple without a calculator. You will be surprised at how much of the section you can complete, and how many questions become easier when you are forced to think about them a little bit longer without a calculator crutch.

Becoming less dependent on your calculator can boost your SAT score.

Now for the bad news: scratch paper is not permitted on the SAT. However, you are allowed to write in your test booklet.

Each math page of the SAT contains 2 to 4 questions. Below each question is space for calculating. The amount of space provided can help you determine the number of calculations required for a specific question. If a question has only a quarter page of space, it's a relatively short solution. But if the question uses the whole column, be prepared for longer calculations. The only time this might not be true is for the last two to four questions in a section. Because they are at the end of the section, the test designers may have had an extra page, and thus spread out the questions. So extra space for these Hard questions may have been given for aesthetic purposes rather than for calculations.

You should always use the space below each question to jot down notes or rewrite the mathematical equation. Figuring the problem in your head is a bad habit to practice on the SAT, as questions are sometimes designed to test your short-term memory. Writing down the numbers from the problem will prevent you from forgetting important information.

Write down all numbers and equations in the space provided below the question. This prevents short-term memory loss.

Multiple-Choice Questions

There are 44 multiple-choice questions on the SAT, and they appear in all three Math sections. The multiple-choice questions contain five separate answer choices, lettered (A) through (E), each of which offers a different solution to the problem. The answer choices may be expressions, equations, inequalities, items, statements, or actual values. Consider an example:

13. If x is an integer and 2 is the remainder when $3x + 4$ is divided by 4, then which of the following could be a value of x?

    (A) 0
    (B) 3
    (C) 4
    (D) 5
    (E) 6

In this book, multiple-choice example questions will be numbered as if they appear in a 20-question section.

All multiple-choice questions progress in difficulty from Easy to Hard. Notice that the question above is #13. In a section with 20 questions, this question would fall just over halfway through the section, indicating Medium to Medium-Hard difficulty level. Throughout this book we will number multiple-choice questions according to their difficulty level in the 20-question section.

When the answers are numbers or values in integer, fraction, or decimal form, multiple-choice answers are almost always arranged from smallest to largest, or from largest to smallest. You will rarely see answers out of ascending or descending order. This is also true for answers containing *pi* or square roots that have been left in simplified terms. Look at five answer choices from a problem about right triangles and then compare their decimal values:

	If the answers appeared in decimal form:
(A) $5\sqrt{2}$	(A) 7.07
(B) $6\sqrt{2}$	(B) 8.49
(C) $5\sqrt{3}$	(C) 8.66
(D) $7\sqrt{2}$	(D) 9.90
(E) $6\sqrt{3}$	(E) 10.39

Even though this set of answers may appear orderless, as you can see from the decimal version, the answers are indeed in ascending order. It is important to understand this ordering concept in the event that you cannot solve a problem and choose to estimate an answer instead.

☠ CAUTION: SAT TRAP! The College Board has anticipated the most common calculation errors for each question and used these errors in answer choices. This tactic gives students a false sense of confidence and prevents many of them from checking their work.

The test makers have determined the most common wrong answers for each multiple-choice question and use them as the four wrong answer choices. For example, if a solution requires finding the *square root* of 9 (which is 3), the test makers anticipate that some students will mistakenly *square* 9 instead (resulting in 81), which is reflected in an answer choice. This is why it is so important that you double-check your work.

Be aware that the answers are often cleverly selected and arranged to deceive the unsuspecting test taker. The "best" wrong answer is usually situated above the right answer. Students who do not read or test all of the answer choices may select the wrong answer choice simply because it occurs first in the list. Let's look at an example:

16. If a is a positive, even integer, which of the following is the least value of a for which $\dfrac{\frac{1}{a}}{\frac{1}{24}}$ is an integer?

    (A) 9
    (B) 8
    (C) 7
    (D) 5
    (E) 3

☠ CAUTION: SAT TRAP!
Popular wrong answers are often placed above the right answer in order to trap students who do not read all five answer choices.

To solve this question, turn the complex fraction into a simple fraction division problem:

$$\frac{\frac{1}{a}}{\frac{1}{24}} = \frac{1}{a} \div \frac{1}{24} = \frac{1}{a} \times \frac{24}{1} = \frac{24}{a}$$

Did you forget how to simplify complex fractions? It's okay! We'll cover the process in detail in a later chapter, as well as all of the other math discussed in these introductory chapters.

Now go through the answer choices, plugging each answer in for a:

(A) 9	24/9 = 2.67	Not an integer
(B) 8	24/8 = 3	Choice (B) produces an integer!

At this point, many students will pick answer (B) and get the question wrong. The question asked for the *least value* of a. There is another answer choice that creates an integer and has a lesser value:

(C) 7	24/7 = 3.43	Not an integer
(D) 5	24/5 = 4.8	Not an integer
(E) 3	24/3 = 8	Choice (E) produces the smallest integer!

The correct answer is (E). However, if you do not look at all of the answer choices, you can easily miss this question. For questions that have more than one possible answer, you must look at every answer choice—especially in the Medium and Hard questions—as an attractive wrong answer is usually placed above the correct answer.

Sometimes the wording of multiple-choice questions can cause confusion. A common question type asks students to solve for one variable in terms of another. Review an example:

6. If x is 25 percent of z and y is 55 percent of z, what is $x + y$ in terms of z ?

(A) $0.5z$
(B) $0.6z$
(C) $0.7z$
(D) $0.8z$
(E) $0.9z$

"In terms of z" simply means that z will appear in all of the answer choices. If this confuses you, cross out "in terms of z:"

☠ CAUTION: SAT TRAP!
If the words "in terms of ?" appear in a question, cross them out to find the variable requested.

6. If x is 25 percent of z and y is 55 percent of z, what is $x + y$ ~~in terms of z~~ ?

Now it is clear that you must find $x + y$:

x is 25% of z and y is 55% of z
$x = 0.25z$ $y = 0.55z$

$$x + y = 0.25z + 0.55z = 0.8z$$

For questions that ask you to find a variable "in terms of x" or in terms of any other variable, crossing out the phrase "in terms of" makes your required task more clear.

Some multiple-choice questions attempt to complicate the problem with the use of Roman numeral statements. These questions are quite simple if you attack each Roman numeral separately:

Roman numeral problems are like three separate questions about the same equation.

19. If r is an odd negative integer less than -1 and if $s = r + \frac{r}{r}$, which of the following must be true?

I. s is an integer.
II. $r + s$ is odd.
III. $r^2 < s^2$

(A) I only
(B) II only
(C) I and II only
(D) II and III only
(E) I, II, and III

Treat each Roman numeral as a separate question and set out to prove or disprove each statement.

To begin, assign a number to r based on the rule (more about this strategy in Chapter 4). The variable r can be any number as long as you follow the rule that r is an odd negative integer less than -1. So r can be -3, -5, -7, etc.

Once you determine r, you can solve for s (because $s = r + \frac{r}{r}$). Because the first Roman numeral asks whether it *must* be true that s is an integer, look at several possible values of r and s to make sure that the statement is ALWAYS true:

		Is s an integer?
If $r = -3$, then $s = -3 + \frac{-3}{-3}$	$s = -3 + 1 = -2$	Yes
If $r = -5$, then $s = -5 + \frac{-5}{-5}$	$s = -5 + 1 = -4$	Yes
If $r = -99$, then $s = -99 + \frac{-99}{-99}$	$s = -99 + 1 = -98$	Yes

As you can see, a pattern developed. If r is a negative, odd integer less than –1, then s is always going to be the negative, even integer immediately preceding r. So Roman numeral I is proven true. This eliminates answer choices (B) and (D), because I is not included in their possible list of true statements. Cross these answer choices out as you eliminate them.

When a Roman numeral is proven true, eliminate all answer choices that do not include it.

If the Roman numeral is proven untrue, eliminate all answer choices that include it.

Now look at Roman numeral II. Is it ALWAYS true that $r + s$ is odd? Use your results from the previous statement to test this new statement:

		Is $r + s$ odd?
If $r = -3$, then $s = -2$	$r + s = -3 + -2 = -5$	Yes
If $r = -5$, then $s = -4$	$r + s = -5 + -4 = -9$	Yes
If $r = -99$, then $s = -98$	$r + s = -99 + -98 = -197$	Yes

We have proven that Roman numeral II is always true, and thus we can further eliminate answer choice (A) as it does not account for the second statement.

Finally, test the third Roman numeral using your previous values for r and s:

		Is $r^2 < s^2$?
If $r = -3$, then $s = -2$	$r^2 < s^2 = -3^2 < -2^2 = 9 < 4$	No

We do not have to test the other values of r and s; if just one set of values is false, then the statement is not always true.

The correct answer is (C), I and II only.

As you can see, these questions are not always that difficult to solve. However, the use of Roman numerals makes understanding the question itself a little more difficult, which is why most Roman numeral questions are higher-numbered Medium and Hard questions. Treat each statement as its own question, and the entire question is suddenly much easier to solve.

Throughout this book are problem sets, the majority of which are multiple-choice questions that progress in order of difficulty. When you come to these multiple-choice questions, follow the directions given previously in this section to select the best answer.

Grid-In Questions

Grid-ins only occur in one math section.

Grid-ins, officially referred to as Student-Produced Response questions, occur in only one of the math sections on the test. The first 8 questions of the section are multiple-choice. The last ten questions in the section are grid-ins.

Grid-ins do not have answer choices. Instead, students must produce their own answer and then transfer that answer to a box like this one:

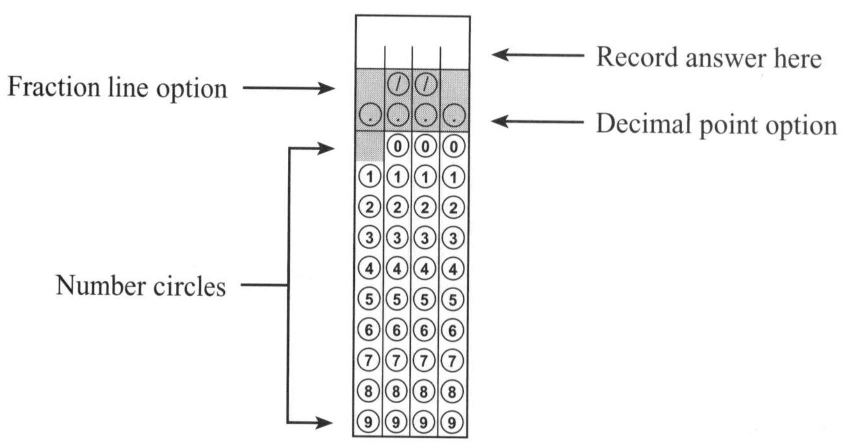

You must bubble in the ovals to receive points for grid-in questions!

The four half-boxes at the top are for recording the answer, but note that answers will not be scored if the ovals are not completed. Fractional answers can be submitted as fractions or decimals:

Answer: 12/5 or 2.4

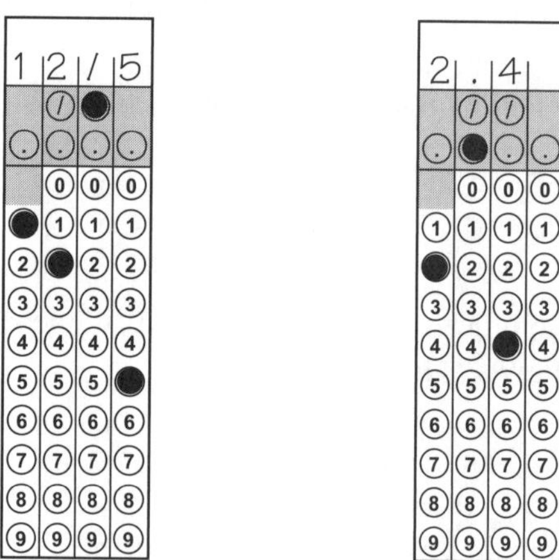

Some students are intimidated by grid-ins and prefer multiple-choice questions. While it is true for grid-ins that the one right answer is not given, it is more beneficial to the test taker that the four wrong answers are not given either, especially when those wrong answers contain common traps and tricks.

You should never take time to read the directions on the SAT; these directions should be studied before taking the actual test.

Grid-ins have some unique rules that may take up to two or three minutes to read during the actual test. Therefore, you should learn and memorize those rules here so that can spend your test time solving problems rather than reading directions.

Grid-In Rules and Tips

1. Do not place a "0" in front of decimal points. Notice that there is no "0" bubble in the first column of the answer grid:

Example answer: $\frac{1}{8}$ or 0.125

Correct

Correct

Incorrect

Unlike classroom math, NEVER put a "0" in front of a decimal point on a grid-in answer.

2. Decimal answers with more digits than grids can be truncated (cut short) or rounded. However, you must fill all four grids with the answer.

Example answer: $\frac{1}{6}$ or $0.1666\overline{6}$

Correct — *Correct* — *Correct* — ***Incorrect*** — ***Incorrect***

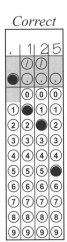

Truncated — Rounded

To avoid rounding mistakes, PowerScore recommends that you truncate all answers that have more decimals than grid spaces.

Rounding can create careless errors, so truncating long decimals is a better strategy.

3. For questions with more than one possible answer, only grid-in one of those possibilities. For example, if 4 and 6 are both right answers, choose 4 or 6 but not both.

4. While answers may start in any of the empty grids, we highly recommend that you start in the far left box when entering an answer. This creates consistency and eliminates the possibility that you accidentally omit part of a decimal answer.

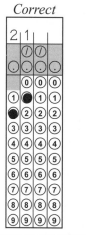

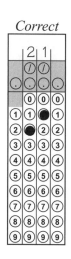

Correct — 2 1 — RECOMMENDED

Correct — 2 1

Correct — 2 1

5. You do not have to reduce fractions to their most basic form unless the fraction does not fit into the four spaces.

Example answer: $\frac{8}{20}$ or $\frac{16}{40}$

Correct — 8/20

Correct — 4/10

Correct — 2/5

The largest fraction does not fit: ***Incorrect***

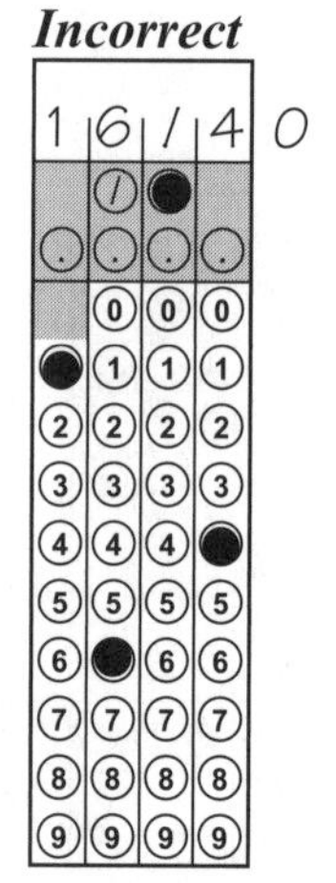

6. You will never have a negative answer for a grid-in question. There are no negative signs available in the grid. If your only solution is negative, you have made a mistake. Go back and double-check your work.

If forced to guess on a grid-in question, remember that common answers are 2 and 3.

7. There are no penalties for wrong answers on grid-ins, so you must always guess on these questions.

8. Mixed numbers must be written in decimal or regular fraction from. The machine that scores the SAT cannot tell the difference between mixed fractions and regular fractions, so it would read 1 1/3 (one and one third or 1.334) as 11/3 (eleven thirds or 3.667).

   Example answer: $1\frac{1}{3}$

Correct

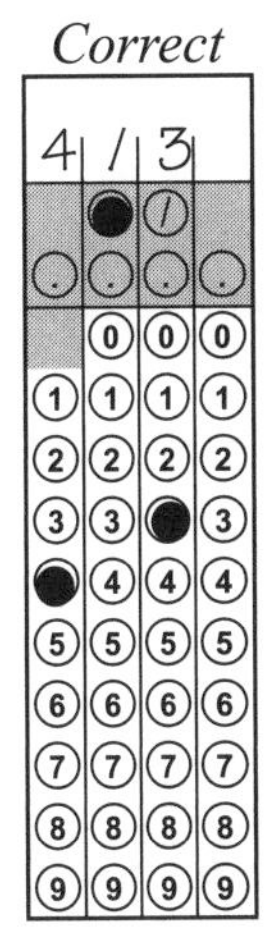

Correct

Incorrect

9. If the requested value is in absolute value brackets, you should expect a negative result prior to inserting it into the absolute value.

   14. If $r < 0$ and if $r^2 + 4r = 32$, what is the value of $|r|$?

The problem states that r is less than 0, so therefore r is negative. But according to the grid-in rules, you can never have a negative answer for a grid-in question. The test makers compensate for this by using absolute value brackets. Plus, the phrasing adds difficulty to an otherwise simple quadratic equation.

In this question, look for a negative answer for r:

$r^2 + 4r = 32$
$r^2 + 4r - 32 = 0$
$(r + \quad)(r - \quad) \rightarrow (r + 8)(r - 4)$

If $(r + 8)(r - 4)$, then $r = -8$ or $r = 4$

Since $r < 0$, then $r = -8$

If $r = -8$, then $|r| = 8$

What do you think the most common wrong answer is for this question? If you said "4", you are correct! Students who do not carefully read and reread often miss the phrase "r < 0".

You will find some grid-in questions in the problem sets in later chapters. Follow the rules found in this section to correctly enter your answers into the grids.

Classroom Math vs. SAT Math

Has anyone ever told you to "think outside the box?" It's an old cliche used to encourage people to think creatively. To think inside the box is to follow traditional methods or use unoriginal processes, while to think outside the box is to find clever solutions or apply innovative strategies.

A math classroom is like the metaphorical box. When you are in class, you are in the box, using the same thought processes you have been trained to use for the last ten years. You rely heavily on your calculator and math formulas. You can methodically solve for x, find the area of triangle, compute rate using distance and time, and figure out any other mundane math question that you've seen a dozen times before. You are thinking in the box, which is appropriate for these questions, as they were designed to be given in the box.

SAT questions are written differently and solved differently than the questions in your high school math class.

The makers of the SAT however, have devised a test that is intended to assess your ability to think outside of the box. Most of the questions are written differently than questions from your high school textbooks, and many of the solutions require creative and critical thinking rather than rote formulas and strategies. It is important to understand these two differences between classroom math and SAT math if you want to reach your highest scoring potential on the test:

1. Questions Are Written Differently

Compare an example of a geometry question from a textbook to a geometry question from the SAT:

Classroom Question

9a. If a rectangular room is 8 feet by 10 feet, what is the perimeter of the room?

9b. What is the area of the room?

SAT Question

9. The perimeter of a rectangular room is 36 feet. If the length of one side of the room is 10 feet, what is the area of the room in square feet?

(A) 8
(B) 80
(C) 116
(D) 288
(E) 360

The classroom question is very straightforward; it presents both the length and the width and requests the perimeter and area. All of the information you need for the two formulas (perimeter $= 2\ell + 2w$ and area $= \ell w$) is provided in the question. Plus, the area is presented as a separate problem from the perimeter question. It is easy for students to see that the question has two parts, thus requiring two different formulas.

The SAT question, on the other hand, only gives the perimeter and length. The width is not given. This causes many students to panic, because they are so used to being spoon-fed the length and width when finding perimeter or area. When the width is withheld, some test takers shut down and assume the question is too difficult to attempt. Other students become confused because the question includes both perimeter and area; they are not separate problems as in the classroom question. The makers of the SAT know that this simple question, which is essentially the same question asked in a classroom, can cause confusion when asked in a way that differs from typical textbooks.

To combat these intentional tactics, you must learn to carefully read SAT math questions. As stated earlier in the chapter, the omission of one word can cause you to miss an otherwise easy question. If the problem states that "x and y are negative even consecutive integers," to misinterpret just one word of that phrase will lead to the wrong answer. If the question asks for the value of $x + 3$, you can be certain that one of the answer choices is just the value of x in order to trick test takers who did not carefully read the question. PowerScore recommends that you reread every question after you find the solution to make sure that you correctly answered the question.

Reread every question after finding the answer to make sure that you solved for the right variable.

If the wording of the question is difficult to understand, try to paraphrase the question. By putting the question into your own words, you may find that it is much easier to comprehend. Consider an example:

5. What is the least integer value of x for which $4x - 20 > 0$?

If the wording of this question causes you difficulty, put it into your own words:

What is the smallest integer that x can be in order for $4x - 20 > 0$ to be true?

Most students would find the paraphrased version of this question easier to understand and thus easier to answer.

Understanding that SAT questions are written differently than classroom questions can help you start to read and interpret them differently. This is the first step in changing your approach to SAT math. The second comes in choosing a solution method.

Confidence Quotation
"Success is not the result of spontaneous combustion. You must set yourself on fire."
—Reggie Leach, NHL hockey player and coach

2. Questions Are Solved Differently

On the SAT, many of the Medium and most of the Hard level questions can be solved using strategies not typically covered in high school math courses. The SAT asks you to treat these questions as puzzles rather than number-crunching problems, in order to assess your critical thinking skills instead of your ability to follow a time-tested formula or procedure. Let's compare the classroom solution to the SAT solution for the following question:

20. If $rs = 5$ and $r - s = 3$, then $r^2s - rs^2 =$

(A) 2
(B) 8
(C) 15
(D) 16
(E) 40

Classroom Solution

Most students, trained to find one variable at a time, will use substitution to solve for either r or s:

If $r - s = 3$, then $r = 3 + s$

Since $r = 3 + s$, substitute $3 + s$ for r in $rs = 5$:

$$(3 + s)\,s = 5 \;\rightarrow\; 3s + s^2 = 5 \;\rightarrow\; s^2 + 3s - 5 = 0$$

You are left with a quadratic equation, in which factoring (the reverse of FOIL) will not work to find the roots of the polynomial. You must use the quadratic formula:

$$s^2 + 3s - 5 = 0 \;\rightarrow\; ax^2 + bx + c = 0, \text{ so } a = 1, b = 3, \text{ and } c = 5$$

$$x = \frac{-b \pm \sqrt{b^2 - 4ac}}{2a} \;\rightarrow\; s = \frac{-(3) \pm \sqrt{3^2 - 4(1)(-5)}}{2(1)} \;\rightarrow\; s = \frac{-3 \pm \sqrt{9 + 20}}{2}$$

$$\rightarrow\; s = \frac{-3 \pm 5.4}{2} \;\rightarrow\; s = \frac{-8.4}{2} \text{ or } \frac{2.4}{2} \;\rightarrow\; s = -4.2 \text{ or } 1.2$$

Plug one of these values for s into an original equation to solve for r:

If $r = 3 + s$, and if $s = 1.2$, then $r = 3 + 1.2 = 4.2$

If $s = 1.2$ and $r = 4.2$, solve for $r^2s - rs^2 =$

$$(4.2)^2(1.2) = (4.2)(1.2)^2 \;\rightarrow\; (17.64)(1.2) = (4.2)(1.44) \;\rightarrow\; 21.2 - 6.1 = 15.1$$

Because we rounded, our solution is off by a tenth. Choice (C), 15, is the correct answer.

Confidence Quotation

"The human mind prefers to be spoon-fed with the thoughts of others, but deprived of such nourishment it will, reluctantly, begin to think for itself—and such thinking, remember, is original thinking and may have valuable results." —Agatha Christie, author

We found the correct answer using the classroom solution, but it involved intricate operations where errors are easily made and minutes are eaten up by the clock. Plus, it required a formula that is not provided by the makers of the test. Anytime you find yourself using a formula other than those given in the Formula Box or recommended in this book, you've missed an easier solution. The quadratic formula is never required on the SAT! And when you find yourself in the middle of a long, complicated solution, you should begin to suspect that you have missed a faster way to solve the question.

The SAT does not include long, tedious operations. If you encounter a problem that uses complicated formulas or solutions, look for a logical, more efficient solution instead.

Let's look at how good test takers solve the same math question.

SAT Solution

Because this question is number 20, you know it is the last question in the section, and therefore it has a Hard difficulty level. Simple substitution is a basic algebra skill, so it will not be the main focus of the most difficult question in the section. The question has a trick or logical solution.

Knowing this, examine $r^2s - rs^2$. How was this expression created? In order to obtain an r^2 and an s^2, two r's and two s's would need to be multiplied. You would also need one of the multiplied expressions to have a subtraction sign, in order to create a subtraction sign in the final expression. Examine the other two expressions in the problem: rs and $r - s$. Here we have two r's, two s's, and a subtraction sign. What happens when they are multiplied together?

$$(rs)\,(r - s) = r^2s - rs^2$$

Wow! Since $rs = 5$ and $r - s = 3$, multiplying rs and $r - s$ is the same thing as multiplying 5 and 3:

$$(rs)\,(r - s) = r^2s - rs^2$$
$$(5)\quad(3)\quad = 15$$

The correct answer is 15, Choice (C).

As you can see, spending a few seconds analyzing a question can save you minutes on the solution. You should always look for logical solutions to the Hard questions, and to many of the Medium questions as well. It is extremely important that you learn to analyze a question before choosing a solution strategy, because classroom strategies on which you have relied for your entire academic career may actually be used against you. The College Board knows your habits and your tendency to rely on traditional math, and thus creates some questions that can easily trick you into using old formulas and procedures—such as the quadratic formula in the problem we just examined.

If you find yourself reaching for a calculator to do anything more than add, subtract, multiply, divide, square, or cube large numbers, ask yourself if there is an easier solution method.

Once you begin to rethink the SAT and your approach to the math questions, you will gain confidence in your ability to solve questions, the results of which will be seen in your increased math section score.

Chapter Three:
Operation Mastery

A student's ability to manipulate numbers and variables is an invaluable asset on the SAT. Too often students have become so dependent on their calculators that they no longer remember the basic functions behind certain operations. The College Board knows this, which is why so many of the questions require the manipulation of variables; a calculator cannot simplify expressions that include variables.

A fundamental principle of all math is the order of operations. This rule sets precedence for which operations are preformed first when solving or simplifying expressions and equations. The six operations are addition, subtraction, multiplication, division, exponentiation, and grouping, and their order of precedence is often remembered using the acronym PEMDAS.

Each of the letters in PEMDAS represents an operation and its order of priority:

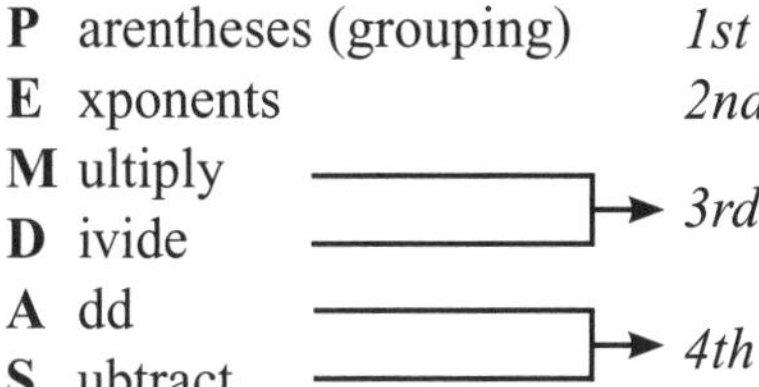

It may help you to remember PEMDAS by memorizing the mnemonic "Please Excuse My Dear Aunt Sally."

Consider the following expression:

$$5\,(3 \times 4)^2 + 6$$

Since the first order of operation is grouping, solve the expression in parentheses first:

$$5\,(3 \times 4)^2 + 6 \quad \rightarrow \quad 5\,(12)^2 + 6$$

Then remove the exponents:

$$5\,(12)^2 + 6 \quad \rightarrow \quad 5\,(144) + 6$$

Now multiply and/or divide:

$$5\,(144) + 6 \quad \rightarrow \quad 720 + 6$$

And finally add and/or subtract:

$$720 + 6 \quad \rightarrow \quad 726$$

If you were to ignore the order of operations and calculate each operation in the order it appeared, the result would be incorrect. The College Board expects all students to perform calculations using the order of operations.

Manipulating values is a skill that many students have lost because calculators do the work for them. The College Board exposes this weakness by using variables, which cannot be plugged into a calculator.

The following pages contain many practice problems to help you re-hone your basic operation skills. You have most likely seen similar questions before in your math courses in middle school and high school. However, many students report that they have forgotten some of the basic skills needed to manipulate variables.

You may choose to work through these problems now, or to come back to them after reviewing some of the concepts in this book, such as fractions and exponents. But it is imperative that you master these skills before taking the SAT. Do not use a calculator. If you miss a question, you must review it in order to avoid making the same mistake again. These operations are the basis of all SAT questions.

Note: Some of the following exercises may be more difficult than they first appear. If you find yourself struggling with any of these questions, do not despair. Later chapters of this book will review these operations. In the meantime, we strongly recommend that you review the solution in the answer key; if you still do not understand the process, return to the question after reviewing the content section in a future chapter.

Simplify the following fractions:

1. $\frac{x}{y} \times \frac{w}{z} =$ ______

2. $\frac{s}{t} \times \frac{x}{y} \times \frac{w}{z} =$ ______

3. $\frac{x}{y} \times \frac{z}{x} =$ ______

4. $\frac{s}{t} \times \frac{t}{s} \times \frac{s}{z} =$ ______

5. $z\left(\frac{x}{y}\right) =$ ______

6. $8\left(\frac{x}{2}\right) =$ ______

7. $8\left(\frac{x}{3}\right)\left(\frac{6}{2}\right) =$ ______

8. $\frac{x}{y} \div \frac{w}{z} =$ ______

9. $\frac{s}{t} \div \frac{x}{y} \div \frac{w}{z} =$ ______

10. $\frac{x}{w} \div \frac{x}{y} \div \frac{y}{w} =$ ______

11. $\frac{x}{w} \div \frac{x}{y} \div \frac{y}{z} =$ ______

12. $x \div \frac{y}{z} =$ ______

13. $\frac{x}{\frac{y}{z}} =$ ______

14. $\frac{\frac{x}{y}}{z} =$ ______

15. $\frac{\frac{x}{y}}{\frac{w}{z}} =$ ______

16. $\frac{1}{\frac{x}{2}} =$ ______

17. $\frac{\frac{1}{x}}{5} =$ ______

18. $\frac{\frac{x}{2}}{x} =$ ______

Answers appear at the end of this chapter.

For further assistance with fractions, see the section "Fractions and Decimals" in Chapter 5.

Simplify the following exponential and radical expressions:

1. $x^6 \times x^2 =$ ________

2. $(x^6)^2 =$ ________

3. $x^6 \bullet y^6 =$ ________

4. $(xy)^6 =$ ________

5. $x^6 \div x^2 =$ ________

6. $\left(\frac{x}{y}\right)^3 =$ ________

7. $x^6 \div y^6 =$ ________

8. $\frac{x^7}{x^2} =$ ________

9. $x^{-3} =$ ________

10. $\frac{x^4}{x^8} =$ ________

11. $4x^{-2} =$ ________

12. $x^0 =$ ________

13. $5x^0 =$ ________

14. $x^{-2}y =$ ________

15. $(xy)^{-5} =$ ________

16. $x^{-4} \times x^4 =$ ________

17. $x^{-7} \times x^4 =$ ________

18. $\frac{x^6}{x^{-3}} =$ ________

19. $\frac{x^{-2}}{y^{-3}} =$ ________

20. $(x^{-4})^3 =$ ________

21. $\left(\frac{x^{-2}}{y^{-3}}\right)^{-2} =$ ________

22. $(x^5 \times x^2) \div (x^6 \times x^3) =$ ________

23. $x^{\frac{2}{3}} =$ ________

24. $x^{-\frac{1}{2}} =$ ________

25. $3x^{-\frac{1}{3}} =$ ________

26. $[(2x^3y^4)^2(3x^2y^3)]^0 =$ ________

27. $5\sqrt{x} + 2\sqrt{x} =$ ________

28. $5\sqrt{x} - 2\sqrt{x} =$ ________

29. $5\sqrt[3]{x} + 2\sqrt[3]{x} =$ ________

30. $5\sqrt[3]{x} - 2\sqrt[3]{x} =$ ________

31. $\sqrt[3]{x^3} =$ ________

32. $\sqrt[4]{x^2} =$ ________

33. $\sqrt{x^4} =$ ________

34. $\sqrt[5]{x^{20}} =$ ________

35. $x\sqrt{12} =$ ________

36. $\sqrt{9x} =$ $3\sqrt{x}$

37. $3\sqrt{12x} =$ ________

38. $2x\sqrt{18} + 5x\sqrt{2} =$ $6x\sqrt{2} + 5x\sqrt{2}$ $11x\sqrt{2}$

39. $\left(3x\sqrt{12}\right) \times \left(4x\sqrt{3}\right) =$ ________

For further assistance with exponents and radicals, see the sections "Exponents" and "Roots and Radical Equations" in Chapter 6.

Expand the following expressions:

1. $2(x + y) =$ $2x + 2y$

2. $3(x - y) =$ ________

3. $-(x + y) =$ $-x - y$

4. $-(x - y) =$ ________

5. $2x(x - y) =$ $2x^2 - 2xy$

6. $3x(x^2 + x - 1) =$ ________

7. $(x + 5)(x + 6) =$ $x^2 + 11x + 30$
   $x^2 + 6x + 5x + 30$

8. $(x + 5)(x - 6) =$ ________

9. $(x - 5)(x + 6) =$ $x^2 + x - 30$
   $x^2 + 6x - 5x - 30$

10. $(x - 5)(x - 6) =$ ________

11. $2(x + 5)(x + 6) =$ $2x^2 + 22x + 60$
    $2(x^2 + 11x + 30)$

12. $(x + 5)(y + 6) =$ ________

13. $(x + 5)^2 =$ $x^2 + 10x + 25$

14. $(x - 5)^2 =$ ________

15. $-(x + 5)(x + 6) =$ $-x^2 - 11x - 30$

16. $-(x - 5)^2 =$ ________

For further assistance with expanding expressions and performing FOIL, see the sections "Expressions" in Chapter 6 and "Quadratic Equations" in Chapter 7.

Factor the following expressions:

1. $x^2 + 11x + 24 =$ ________

2. $x^2 + 5x - 24 =$ ________

3. $x^2 - 5x - 24 =$ ________

4. $x^2 - 11x + 24 =$ ________

5. $-x^2 - 11x - 24 =$ ________

6. $-x^2 + 11x - 24 =$ ________

7. $2x + xy =$ ________

8. $x(2x - xy) =$ ________

9. $14x - 2xy =$ ________

10. $2x^3 - 8x^2 + 6x =$ ________

11. $x^5 + 2x^2 =$ ________

12. $4x^5 + 2x^2 =$ ________

13. $(2x)^{3y} - (2x)^y z =$ ________

14. $(4x)^{3y} - (4x)^{2y} z =$ ________

For further assistance with factoring, see the section "Quadratic Equations" in Chapter 7.

Write using scientific notation:

1. $123 =$ ________

2. $123{,}456 =$ ________

3. $0.0123 =$ ________

4. $0.000123 =$ ________

Write using ordinary decimal notation:

5. $1.34 \times 10^3 =$ ________

6. $5.4321 \times 10^2 =$ ________

7. $2.667 \times 10^{-4} =$ ________

8. $1.5 \times 10^{-6} =$ ________

Solve using scientific notation:

9. $(1.2 \times 10^5)(2.3 \times 10^3) =$ ________

10. $(1.2 \times 10^5)(2.3 \times 10^{-3}) =$ ________

11. $(4.5 \times 10^3)(8.9 \times 10^{-8}) =$ ________

12. $(1.2 \times 10^6) \div (3.2 \times 10^{-2}) =$ ________

For further assistance with scientific notation, see the section "Scientific Notation" in Chapter 6.

Now combine these operation skills to solve for x:

1. $\frac{x+y}{2}=6$ $\quad x =$ $12-y$

$x+y=12$

$x=12-y$

2. $\frac{x}{x+1}=x$ $\quad x =$ __________

3. $\frac{x}{30}\left(y^2-2\right)=z$ $\quad x =$ $\frac{30z}{y^2-2}$

$\frac{x}{30}=\frac{z}{y^2-2}$

$x=\frac{30z}{y^2-2}$

4. $\frac{1}{x^2-x}+\frac{1}{x}=3$ $\quad x =$ __________

5. $\frac{(x-y)^2}{y}-\frac{x^2+y^2}{y}=1$ $\quad x =$ $-\frac{1}{2}$

$\frac{x^2-xy-xy+y^2-x^2-y^2}{y}=1$

$-\frac{2xy}{y}=1$

$-2x=1$

$x=-\frac{1}{2}$

6. $(x+3)(y-2)=2x+xy$ $\quad x =$ __________

7. $\frac{(x+6)(x-8)}{2}=3(x-8)$ $\quad x =$ 0

$(x+6)(x-8)=6(x-8)$

$\frac{(x+6)(x-8)}{(x-8)}=6$

$x+6=6$

$x=0$

8. $\frac{x+4}{x-3}=\frac{x+2}{x+5}$ $\quad x =$ __________

9. $\frac{1}{\frac{1}{x}}=\frac{\frac{1}{x}}{4}$ $\quad x =$ $\pm\frac{1}{2}$

$4=\frac{1}{x^2}$

$4x^2=1$

$x^2=\frac{1}{4}$

$x=\pm\sqrt{\frac{1}{4}}=\pm\frac{1}{2}$

10. $x+2=\frac{1}{3}x+8$ $\quad x =$ __________

11. $3x - 9 = \dfrac{x^2}{4}$ $x =$ 6

$12x - 36 = x^2$
$-x^2 + 12x - 36 = 0$
$x^2 - 12x + 36 = 0$
$(x-6)(x-6) = 0$

12. $17 = \sqrt{(20-x)^2 + (13-x)^2}$ $x =$ ____________

13. $\sqrt{3x - 11} = x - 3$ $x =$ ____________

14. $y \times \dfrac{y-2}{y} \times \dfrac{y+2}{y} = \dfrac{x}{3}$ $x =$ ____________

15. $\dfrac{x}{x^{\frac{1}{2}}} = 2 - x$ $x =$ ____________

16. $(x+4)^{\frac{1}{2}} = (x-4)^{-\frac{1}{2}}$ $x =$ ____________

17. $\left(x^{-\frac{1}{3}} y^{\frac{1}{2}}\right)^6 = \dfrac{9}{4}$ $x =$ ____________

18. $x^{-\frac{1}{2}} = y^{\frac{3}{2}}$ $x =$ ____________

Notes

• 12/21/10 – review exponents, roots and radical equations (Chp. 6)
(esp. fraction, negative exponents)

Fractions Answer Key

1. $\frac{x}{y} \times \frac{w}{z} = \frac{x \times w}{y \times z} = \frac{xw}{yz}$

2. $\frac{s}{t} \times \frac{x}{y} \times \frac{w}{z} = \frac{s \times x \times w}{t \times y \times z} = \frac{sxw}{tyz}$

3. $\frac{x}{y} \times \frac{z}{x} = \frac{\cancel{x} \times z}{y \times \cancel{x}} = \frac{z}{y}$

4. $\frac{s}{t} \times \frac{t}{s} \times \frac{s}{z} = \frac{\cancel{s} \times \cancel{t} \times s}{\cancel{t} \times \cancel{s} \times z} = \frac{s}{z}$

5. $z\left(\frac{x}{y}\right) = \frac{z}{1} \times \frac{x}{y} = \frac{zx}{y}$

6. $8\left(\frac{x}{2}\right) = \frac{8}{1} \times \frac{x}{2} = \frac{8x}{2} = 4x$

7. $8\left(\frac{x}{3}\right)\left(\frac{6}{2}\right) = \frac{8}{1} \times \frac{x}{\cancel{3}} \times \frac{\cancel{6}}{\cancel{2}} = \frac{8x}{1} = 8x$

8. $\frac{x}{y} \div \frac{w}{z} = \frac{x}{y} \times \frac{z}{w} = \frac{xz}{wy}$

9. $\frac{s}{t} \div \frac{x}{y} \div \frac{w}{z} = \frac{s}{t} \times \frac{y}{x} \times \frac{z}{w} = \frac{syz}{txw}$

10. $\frac{x}{w} \div \frac{x}{y} \div \frac{y}{w} = \frac{\cancel{x}}{\cancel{w}} \times \frac{\cancel{y}}{\cancel{x}} \times \frac{\cancel{w}}{\cancel{y}} = \frac{1}{1} \times \frac{1}{1} \times \frac{1}{1} = 1$

11. $\frac{x}{w} \div \frac{x}{y} \div \frac{y}{z} = \frac{\cancel{x}}{w} \times \frac{\cancel{y}}{\cancel{x}} \times \frac{z}{\cancel{y}} = \frac{z}{w}$

12. $x \div \frac{y}{z} = \frac{x}{1} \times \frac{z}{y} = \frac{xz}{y}$

13. $\frac{x}{\frac{y}{z}} = \frac{x}{1} \div \frac{y}{z} = \frac{x}{1} \times \frac{z}{y} = \frac{xz}{y}$

14. $\frac{\frac{x}{y}}{z} = \frac{x}{y} \div \frac{z}{1} = \frac{x}{y} \times \frac{1}{z} = \frac{x}{yz}$

15. $\frac{\frac{x}{y}}{\frac{w}{z}} = \frac{x}{y} \div \frac{w}{z} = \frac{x}{y} \times \frac{z}{w} = \frac{xz}{yw}$

16. $\frac{1}{\frac{x}{2}} = 1 \div \frac{x}{2} = 1 \times \frac{2}{x} = \frac{2}{x}$

17. $\frac{\frac{1}{x}}{5} = \frac{1}{x} \div \frac{5}{1} = \frac{1}{x} \times \frac{1}{5} = \frac{1}{5x}$

18. $\frac{\frac{x}{2}}{x} = \frac{x}{2} \div \frac{x}{1} = \frac{x}{2} \times \frac{1}{x} = \frac{\cancel{x}}{2} \times \frac{1}{\cancel{x}} = \frac{1}{2}$

Exponents and Radicals Answer Key

1. $x^6 \times x^2 = x^{6+2} = x^8$

2. $(x^6)^2 = = x^{6 \cdot 2} = x^{12}$

3. $x^6 \bullet y^6 = (xy)^6$

4. $(xy)^6 = x^6y^6$

5. $x^6 \div x^2 = x^{6-2} = x^4$

6. $\left(\frac{x}{y}\right)^3 = \frac{x^3}{y^3}$

7. $x^6 \div y^6 = \frac{x^6}{y^6}$ or $(x \div y)^6$

8. $\frac{x^7}{x^2} = x^{7-2} = x^5$

9. $x^{-3} = \frac{1}{x^3}$

10. $\frac{x^4}{x^8} = x^{4-8} = x^{-4} = \frac{1}{x^4}$

11. $4x^{-2} = 4\frac{1}{x^2} = \frac{4}{x^2} = \frac{2}{x}$

12. $x^0 = 1$ (Any base raised to the power of 0 is 1).

13. $5x^0 = 5 \times 1 = 5$

14. $x^{-2}y = \frac{1}{x^2} \times \frac{y}{1} = \frac{y}{x^2}$

15. $(xy)^{-5} = \frac{1}{(xy)^5} = \frac{1}{x^5y^5}$

16. $x^{-4} \times x^4 = x^{-4+4} = x^0 = 1$

17. $x^{-7} \times x^4 = x^{-7+4} = x^{-3} = \frac{1}{x^3}$

18. $\frac{x^6}{x^{-3}} = x^{6--3} = x^{6+3} = x^9$

19. $\frac{x^{-2}}{y^{-3}} = \frac{\frac{1}{x^2}}{\frac{1}{y^3}} = \frac{1}{x^2} \div \frac{1}{y^3} = \frac{1}{x^2} \times \frac{y^3}{1} = \frac{y^3}{x^2}$

20. $(x^{-4})^3 = x^{-4 \cdot 3} = x^{-12} = \frac{1}{x^{12}}$

21. $\left(\frac{x^{-2}}{y^{-3}}\right)^{-2} = \frac{(x^{-2})^{-2}}{(y^{-3})^{-2}} = \frac{x^{-2\times-2}}{y^{-3\times-2}} = \frac{x^4}{y^6}$

22. $(x^5 \times x^2) \div (x^6 \times x^3) = (x^{5+2}) \div (x^{6+3}) = x^7 \div x^9 =$

$x^{7-9} = x^{-2} = \frac{1}{x^2}$

23. $x^{\frac{2}{3}} = \sqrt[3]{x^2}$

24. $x^{-\frac{1}{2}} = \frac{1}{\sqrt[2]{x^1}} = \frac{1}{\sqrt{x}}$

25. $3x^{-\frac{1}{3}} = 3\frac{1}{\sqrt[3]{x^1}} = \frac{3}{\sqrt[3]{x}}$

26. $[(2x^3y^4)^2(3x^2y^3)]^0 = 1$

27. $5\sqrt{x} + 2\sqrt{x} =$ $7\sqrt{x}$

28. $5\sqrt{x} - 2\sqrt{x} =$ $3\sqrt{x}$

29. $5\sqrt[3]{x} + 2\sqrt[3]{x} =$ $7\sqrt[3]{x}$

30. $5\sqrt[3]{x} - 2\sqrt[3]{x} =$ $3\sqrt[3]{x}$

31. $\sqrt[3]{x^3} = (x^3)^{\frac{1}{3}} = \left(x^{3\times\frac{1}{3}}\right) = x^1 = x$

32. $\sqrt[4]{x^2} = (x^2)^{\frac{1}{4}} = \left(x^{2\times\frac{1}{4}}\right) = x^{\frac{1}{2}} = \sqrt{x}$

33. $\sqrt{x^4} = (x^4)^{\frac{1}{2}} = \left(x^{4\times\frac{1}{2}}\right) = x^2$

34. $\sqrt[5]{x^{20}} = (x^{20})^{\frac{1}{5}} = \left(x^{20\times\frac{1}{5}}\right) = x^4$

35. $x\sqrt{12} = x \times \sqrt{4} \times \sqrt{3} = x \times 2 \times \sqrt{3} = 2x\sqrt{3}$

36. $\sqrt{9x} = \sqrt{9} \times \sqrt{x} = 3\sqrt{x}$

37. $3\sqrt{12x} = 3 \times \sqrt{4} \times \sqrt{3} \times \sqrt{x} = 3 \times 2 \times \sqrt{3} \times \sqrt{x} = 6\sqrt{3x}$

38. $2x\sqrt{18} + 5x\sqrt{2} = \left(2x\sqrt{9}\sqrt{2}\right) + 5x\sqrt{2} =$

$\left(2x3\sqrt{2}\right) + 5x\sqrt{2} = 6x\sqrt{2} + 5x\sqrt{2} = 11x\sqrt{2}$

39. $\left(3x\sqrt{12}\right) \times \left(4x\sqrt{3}\right) = \left(3x\sqrt{4}\sqrt{3}\right) \times \left(4x\sqrt{3}\right) =$

$\left(3x2\sqrt{3}\right) \times \left(4x\sqrt{3}\right) = \left(6x\sqrt{3}\right) \times \left(4x\sqrt{3}\right) = 24x^2 \times 3 = 72x^2$

Expansion Answer Key

1. $2(x + y) = 2x + 2y$
2. $3(x - y) = 3x - 3y$
3. $-(x + y) = -x - y$
4. $-(x - y) = -x + y$
5. $2x(x - y) = 2x^2 - 2xy$
6. $3x(x^2 + x - 1) = 3x^3 + 3x^2 - 3x$
7. $(x + 5)(x + 6) = x^2 + 6x + 5x + 30 =$
   $x^2 + 11x + 30$
8. $(x + 5)(x - 6) = x^2 - 6x + 5x - 30 =$
   $x^2 - x - 30$
9. $(x - 5)(x + 6) = x^2 + 6x - 5x - 30 =$
   $x^2 + x - 30$
10. $(x - 5)(x - 6) = x^2 - 6x - 5x + 30 =$
    $x^2 - 11x + 30$

11. $2(x+5)(x+6) = 2(x^2+6x+5x+30) =$
$2(x^2+11x+30) = 2x^2+22x+60$

12. $(x+5)(y+6) = xy+6x+5y+30$

13. $(x+5)^2 = (x+5)(x+5) = x^2+5x+5x+25 =$
$x^2+10x+25$

14. $(x-5)^2 = (x-5)(x-5) = x^2-5x-5x+25 =$
$x^2-10x+25$

15. $-(x+5)(x+6) = -(x^2+6x+5x+30) =$
$-(x^2+11x+30) = -x^2-11x-30$

16. $-(x-5)^2 = -(x-5)(x-5) = -(x^2-5x-5x+25)$
$= -(x^2-10x+25) = -x^2+10x-25$

Factoring Answer Key

1. $x^2+11x+24 = (x+\ \ )(x+\ \ ) = (x+3)(x+8)$
2. $x^2+5x-24 = (x-\ \ )(x+\ \ ) = (x-3)(x+8)$
3. $x^2-5x-24 = (x-\ \ )(x+\ \ ) = (x-8)(x+3)$
4. $x^2-11x+24 = (x-\ \ )(x-\ \ ) = (x-3)(x-8)$
5. $-x^2-11x-24 = -(x^2+11x+24) =$
$-(x+\ \ )(x+\ \ ) = -(x+3)(x+8)$
6. $-x^2+11x-24 = -(x^2-11x+24) =$
$-(x-\ \ )(x-\ \ ) = -(x-3)(x-8)$
7. $2x+xy = x(2+y)$
8. $x(2x-xy) = x^2(2-y)$
9. $14x-2xy = 2x(7-y)$
10. $2x^3-8x^2+6x = 2x(x^2-4x+3)$
11. $x^5+2x^2 = x^2(x^3+2)$
12. $4x^5+2x^2 = 2x^2(2x^3+1)$
13. $(2x)^{3y}-(2x)^y z = 2x^y(x^{2y}-z)$
14. $(4x)^{3y}-(4x)^{2y} z = 4x^{2y}(x^y-z)$

Scientific Notiation Answer Key

1. $123 = 1.23 \times 10^2$
2. $123{,}456 = 1.23456 \times 10^5$
3. $0.0123 = 1.23 \times 10^{-2}$
4. $0.000123 = 1.23 \times 10^{-4}$
5. $1.34 \times 10^3 = 1340$
6. $5.4321 \times 10^2 = 543.21$
7. $2.667 \times 10^{-4} = 0.0002667$
8. $1.5 \times 10^{-6} = 0.0000015$
9. $(1.2 \times 10^5)(2.3 \times 10^3) = (1.2 \times 2.3)(10^5 \times 10^3) =$
$(2.76)(10^{5+3}) = 2.76 \times 10^8$
10. $(1.2 \times 10^5)(2.3 \times 10^{-3}) = (1.2 \times 2.3)(10^5 \times 10^{-3})$
$= (2.76)(10^{5+-3}) = 2.76 \times 10^2$
11. $(4.5 \times 10^3)(8.9 \times 10^{-8}) = (4.5 \times 8.9)(10^3 \times 10^{-8})$
$= (40.05)(10^{3+-8}) = 40.05 \times 10^{-5}$
$= 4.005 \times 10^{-4}$
12. $(1.2 \times 10^6) \div (3.2 \times 10^{-2}) =$
$(1.2 \div 3.2)(10^6 \div 10^{-2}) = (0.375)(10^{6--2}) =$
$0.375 \times 10^8 = 3.75 \times 10^7$

Solve for *x* Answer Key

1. $\frac{x+y}{2} = 6$

$x+y = (6)(2) \rightarrow x+y = 12 \rightarrow \boxed{x = 12-y}$

2. $\frac{x}{x+1} = x$

$x = x(x+1) \rightarrow \frac{x}{x} = \frac{x(x+1)}{x} \rightarrow \frac{\cancel{x}}{\cancel{x}} = \frac{\cancel{x}(x+1)}{\cancel{x}}$

$\rightarrow 1 = (x+1) \rightarrow \boxed{0 = x}$

3. $\frac{x}{30}\left(y^2-2\right)=z$

$x(y^2 - 2) = z(30) \rightarrow$ $\boxed{x=\frac{30z}{\left(y^2-2\right)}}$

4. $\frac{1}{x^2-x}+\frac{1}{x}=3$

$\frac{(x)(1)}{(x)(x^2-x)}+\frac{(1)(x^2-x)}{(x)(x^2-x)}=3 \rightarrow$

$\frac{(x)+(x^2-x)}{x^3-x^2}=3 \rightarrow \frac{x^2}{x^3-x^2}=3 \rightarrow$

$\frac{x^2(1)}{x^2(x^1-1)}=3 \rightarrow \frac{x^2(1)}{x^2(x^1-1)}=3 \rightarrow \frac{1}{x-1}=3$

$1 = 3(x - 1) \rightarrow 1 = 3x - 3 \rightarrow 4 = 3x \rightarrow \boxed{\frac{4}{3}=x}$

5. $\frac{(x-y)^2}{y}-\frac{x^2+y^2}{y}=1$

$\frac{(x-y)(x-y)-(x^2+y^2)}{y}=1 \rightarrow$

$\frac{(x^2-xy-xy+y^2)-x^2-y^2}{y}=1 \rightarrow \frac{-2xy}{y}=1$

$\frac{-2xy}{y}=1 \rightarrow -2x = 1 \rightarrow \boxed{x=-\frac{1}{2}}$

6. $(x + 3)(y - 2) = 2x + xy$

$xy - 2x + 3y - 6 = 2x + xy$

$3y - 6 = 4x$

$\boxed{\frac{3y-6}{4}=x}$

7. $\frac{(x+6)(x-8)}{2}=3(x-8)$

$(x + 6)(x - 8) = 6(x - 8)$

$\frac{(x+6)(x-8)}{(x-8)}=\frac{6(x-8)}{(x-8)} \rightarrow x + 6 = 6 \rightarrow \boxed{x=0}$

8. $\frac{x+4}{x-3}=\frac{x+2}{x+5} \rightarrow (x + 4)(x + 5) = (x - 3)(x + 2)$

$x^2 + 9x + 20 = x^2 - x - 6 \rightarrow 9x + 20 = -x - 6$

$10x = -26 \rightarrow \boxed{x=-\frac{26}{10}=-\frac{13}{5}}$

9. $\frac{1}{\frac{1}{x}}=\frac{\frac{1}{x}}{4}$

$4\times1=\frac{1}{x}\times\frac{1}{x} \rightarrow 4=\frac{1}{x^2} \rightarrow 4x^2 = 1 \rightarrow$

$x^2=\frac{1}{4} \rightarrow \sqrt{x^2}=\sqrt{\frac{1}{4}} \rightarrow \boxed{x=\frac{1}{2}}$

10. $x+2=\frac{1}{3}x+8$

$x-6=\frac{1}{3}x \rightarrow 3(x - 6) = x \rightarrow 3x - 18 = x$

$-18 = -2x \rightarrow \boxed{9 = x}$

11. $3x-9=\frac{x^2}{4} \rightarrow 4(3x - 9) = x^2 \rightarrow$

$12x - 36 = x^2 \rightarrow 0 = x^2 - 12x + 36 \rightarrow$

$(x - \ )(x - \ ) = 0 \rightarrow (x - 6)(x - 6) = 0 \rightarrow \boxed{x = 6}$

12. $17=\sqrt{(20-x)^2+(13-x)^2}$

$17^2=\left(\sqrt{(20-x)^2+(13-x)^2}\right)^2$

$289 = (20 - x)^2 + (13 - x)^2$
$289 = (20 - x)(20 - x) + (13 - x)(13 - x)$
$289 = (400-20x-20x+x^2) + (169-13x-13x+x^2)$
$289 = (400 - 40x + x^2) + (169 - 26x + x^2)$
$0 = 280 - 66x + 2x^2$
$0 = 2(140 - 33x + x^2)$
$0 = x^2 - 33x + 140 \rightarrow 0 = (x - \ )(x - \ )$
$0 = (x - 5)(x - 28) \rightarrow$ $\boxed{x = 5 \text{ or } x = 28}$

13. $\sqrt{3x-11}=x-3 \rightarrow \left(\sqrt{3x-11}\right)^2=(x-3)^2$

$3x - 11 = (- 3)(x - 3)$
$3x - 11 = x^2 - 3x - 3x + 9$
$0 = x^2 - 9x + 20 \rightarrow 0 = (x - \ )(x - \ )$
$0 = (x - 4)(x - 5) \rightarrow$ $\boxed{x = 4 \text{ or } x = 5}$

14. $y\times\frac{y-2}{y}\times\frac{y+2}{y}=\frac{x}{3} \rightarrow \cancel{y}\times\frac{y-2}{\cancel{y}}\times\frac{y+2}{y}=\frac{x}{3}$

$\frac{y-2}{1}\times\frac{y+2}{y}=\frac{x}{3} \rightarrow \frac{y^2+2y-2y-4}{y}=\frac{x}{3}$

$\frac{y^2-4}{y}=\frac{x}{3} \rightarrow 3(y^2-4)=xy \rightarrow 3y^2-12=xy$

$\boxed{\frac{3y^2-12}{y}=x}$

15. $\frac{x}{x^{\frac{1}{2}}}=2-x \rightarrow \frac{x}{\sqrt[2]{x^1}}=2-x \rightarrow$

$\left(\frac{x}{\sqrt{x}}\right)^2=(2-x)^2 \rightarrow$

$\frac{x^2}{x}=(2-x)(2-x) \rightarrow$

$x=4-2x-2x+x^2$

$0=x^2-5x+4 \rightarrow 0=(x-\ )(x-\ )$

$0=(x-1)(x-4) \rightarrow \boxed{x=1 \text{ or } x=4}$

16. $(x+4)^{\frac{1}{2}}=(x-4)^{-\frac{1}{2}} \rightarrow \sqrt{x+4}=\frac{1}{\sqrt{x-4}}$

$\left(\sqrt{x+4}\right)^2=\left(\frac{1}{\sqrt{x-4}}\right)^2 \rightarrow x+4=\frac{1}{x-4}$

$(x-4)(x+4)=1 \rightarrow x^2+4x-4x-16=1$

$x^2-16=1 \rightarrow x^2=17 \rightarrow \boxed{x=\sqrt{17}}$

17. $\left(x^{-\frac{1}{3}}y^{\frac{1}{2}}\right)^6=\frac{9}{4} \rightarrow \left(x^{-\frac{1}{3}\times 6}\right)\left(y^{\frac{1}{2}\times 6}\right)=\frac{9}{4} \rightarrow$

$\left(x^{-2}\right)\left(y^3\right)=\frac{9}{4} \rightarrow \frac{1}{x^2}\left(y^3\right)=\frac{9}{4} \rightarrow \frac{y^3}{x^2}=\frac{9}{4}$

$\sqrt{\frac{y^3}{x^2}}=\sqrt{\frac{9}{4}} \rightarrow \frac{\sqrt{y\times y\times y}}{\sqrt{x\times x}}=\frac{\sqrt{9}}{\sqrt{4}} \rightarrow$

$\frac{y\sqrt{y}}{x}=\frac{3}{2} \rightarrow \frac{1}{x}=\frac{3}{2y\sqrt{y}} \rightarrow \boxed{x=\frac{2y\sqrt{y}}{3}}$

18. $x^{-\frac{1}{2}}=y^{\frac{3}{2}} \rightarrow \frac{1}{x^{\frac{1}{2}}}=y^{\frac{3}{2}} \rightarrow \frac{1}{\sqrt[2]{x^1}}=\sqrt[2]{y^3} \rightarrow$

$\frac{1}{\sqrt{x}}=\sqrt[2]{y^2\times y} \rightarrow \frac{1}{\sqrt{x}}=y\sqrt{y} \rightarrow$

$\left(\frac{1}{\sqrt{x}}\right)^2=\left(y\sqrt{y}\right)^2 \rightarrow \frac{1}{x}=y^2y \rightarrow$

$\frac{1}{x}=y^3 \rightarrow 1=y^3x \rightarrow \boxed{\frac{1}{y^3}=x}$

Notes

CHAPTER FOUR:
SOLUTION STRATEGIES

Thomas Edison once said that "Good fortune is what happens when opportunity meets with planning." You will soon have an opportunity in the form of the SAT; your results can open doors to some of the most respected colleges in the country. And it is clear that you are serious about planning, or else it is unlikely that you'd be reading this book. A good plan consists of several strategies that can be adapted and applied to many different situations.

Anyone who has ever played a competitive sport and served under a good coach knows about the need for multiple strategies. For example, take a high school volleyball coach. Before each game, she reviews taped footage of the next opponent and devises several strategies for attacking the opponent's weaknesses. If she notices a weak block on the far side of the net, she maneuvers her best hitters into that position. If she finds a hesitant player in the backcourt, she might direct her team to consistently place the ball in that player's vicinity. And then on game day, the coach may need to make adjustments based on the action on the court. If the opponent's hesitant player is not in the game, the coach needs to determine another weak area on the court, and redirect the ball. A good coach must be willing to use multiple strategies on game day if she wants to win.

Similarly, you must learn several strategies for the SAT math section, and be able to adapt them to many different types of questions if you want to score well. PowerScore teaches eight math strategies that are commonly used on every SAT:

1. ANALYZE the Answer Choices
2. BACKPLUG the Answer Choices
3. SUPPLY Numbers
4. TRANSLATE from English to Math
5. RECORD What You Know
6. SPLIT the Question into Parts
7. DIAGRAM the Question
8. SIZE UP the Figure

The first word of each strategy is capitalized to make it more easily recognized in other chapters of this book.

Most of these strategies can be applied to many different types of questions, but a few are quite specialized. For example, you can BACKPLUG questions from arithmetic, algebra, geometry, statistics, and more, but you can only SIZE UP geometry questions. Each strategy is explained in detail in this chapter.

Plan on rereading this chapter after finishing the book.

NOTE: The problem sets in this chapter contain content that you may not remember. It would be wise to review this chapter again after completing the book, as some of the material might be easier to understand after reading specific chapters on that content. After all, sufficient planning is the key to your good fortune on test day.

1. ANALYZE the Answer Choices

For multiple-choice questions, always preview the answer choices before attempting a solution.

Before attempting any SAT multiple-choice math question, look at the answer choices. They can reveal information about the solution for the question. For example, consider the following group of answer choices for a question on the SAT:

(A) 3π
(B) 6π
(C) 8π
(D) 9π
(E) 12π

The presence of *pi* indicates that this question involves a circle or a portion of a circle. The only formulas that require *pi* on the SAT involve a circle, so you will undoubtedly need to find and use the radius in the solution of the problem. This is true even if the question appears to be about a square or triangle. The circle is hidden somewhere, or else *pi* would not be required.

Similarly, if a geometry question has five answer choices in square root form, you can expect to use the Pythagorean Theorem:

(A) $\sqrt{15}$
(B) $\sqrt{17}$
(C) $3\sqrt{2}$
(D) $\sqrt{19}$
(E) $\sqrt{21}$

These answers are computed using $a^2 + b^2 = c^2$, where the question asks for the length of a specific leg or the hypotenuse. For example, if the legs of the triangle have lengths of 4 and 1, then you can find the length of the hypotenuse using the Pythagorean Theorem:

As you will learn later in this book, the College Board uses questions with hidden triangles. The presence of square root answers can tip you off to the "secret" solution.

$$a^2 + b^2 = c^2$$
$$4^2 + 1^2 = c^2$$
$$16 + 1 = c^2$$
$$17 = c^2$$
$$\sqrt{17} = c$$

Again, the question itself may appear to be about circles, rectangles, or other figures, but the use of square roots in the answer choices should indicate a hidden right triangle.

If a variable is present in all answer choices, you can expect to solve for a different variable. Notice that x is used in all of the following answer choices:

(A) $2x$
(B) $\frac{x}{2}$
(C) $2\sqrt{x}$
(D) $2+x$
(E) $2-x$

From these answer choices you can deduce that you will NOT be solving for x, but rather for another variable. You can also confidently assume that the number 2 will be a part of the solution.

Once you read the problem and begin working on a solution, you may be able to solve the question simply by thinking about the "nature" of the answer choices. You might be able to quickly recognize that the correct answer must be negative or that the solution is greater than or less than a certain number. This allows you to eliminate wrong answers that do not meet these requirements without having to do a lot of work. You may even find the right answer this way! Let's consider an example:

12. Bob and Ted start in different cities and drive towards each other. Bob drives at a constant rate of 50 miles per hour, and Ted drives at a constant rate of 70 miles per hour. If they began 300 miles apart, how many miles had Bob driven when he passed Ted on the road?

(A) 125
(B) 150
(C) 175
(D) 180
(E) 210

You need very little math to solve this question. Instead, draw a picture and then analyze the answer choices:

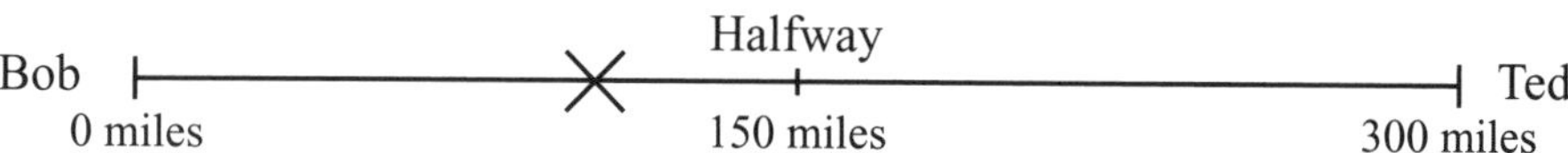

We will address drawing pictures later in this chapter.

If both Bob and Ted were driving the same speed, they would meet at the halfway mark. But Bob is driving slower than Ted, so they will meet on Bob's side of the halfway mark. Therefore, answer choices (B), (C), (D), and (E) are impossible. The correct answer must be (A).

(A) 125
~~(B)~~ 150
~~(C)~~ 175
~~(D)~~ 180
~~(E)~~ 210

There is one more answer choice that is important to understand. Consider the following question:

19. If $4a + c = 3b$ and if $4a + 3b + c = 24$, then $b = ?$

    (A) 3
    (B) 4
    (C) 7
    (D) 8
    (E) It cannot be determined from the information given.

Answer choice (E), "It cannot be determined from the information given," is often used in Hard level questions that contain three variables and only two equations. Many students assume that you cannot solve these questions without a third equation. However, in PowerScore's analysis of hundreds of tests and thousands of questions, we have never found a math question in which the answer is "It cannot be determined from the information given." This isn't to say that it will never be the correct answer, but in our extensive experience, this answer choice has always been used to intentionally trick unsuspecting test takers. The value has always been found, as two of the variables usually cancel each other out when the equations are added together:

☠ CAUTION: SAT TRAP!
Watch for other versions of this bogus answer choice, such as "x can never equal y."

Equation 1: $4a + c = 3b \rightarrow 4a - 3b + c = 0$

Equation 1: $4a - 3b + c = 0$
Equation 2: $4a + 3b + c = 24$

Multiply the second equation by –1 and then add the two equations:

Equation 1: $4a - 3b + c = 0$
Equation 2: $-(4a + 3b + c = 24)$

Equation 1: $4a - 3b + c = 0$
Equation 2: $-4a - 3b - c = -24$

Equation 1: $\cancel{4a} - 3b + \cancel{c} = 0$
Equation 2: $\cancel{-4a} - 3b - \cancel{c} = -24$
$-6b = -24$

$$b = 4$$

As you can see, this question clearly does have a solution. Be wary of any answer choice that implies otherwise.

We have included a short problem set to illustrate the techniques used in each of the Solution Strategies in this chapter. You can choose to attempt these problem sets now, or return to them after reviewing the content chapters. Many students find that the problems are easier to solve once they have reviewed the entire book.

ANALYZING the Answer Choices Problem Set

Solve the following multiple-choice questions by selecting the best answer from the five answer choices. For grid-in questions, write your answer in the grids and completely mark the corresponding ovals. Answers are given on page 107.

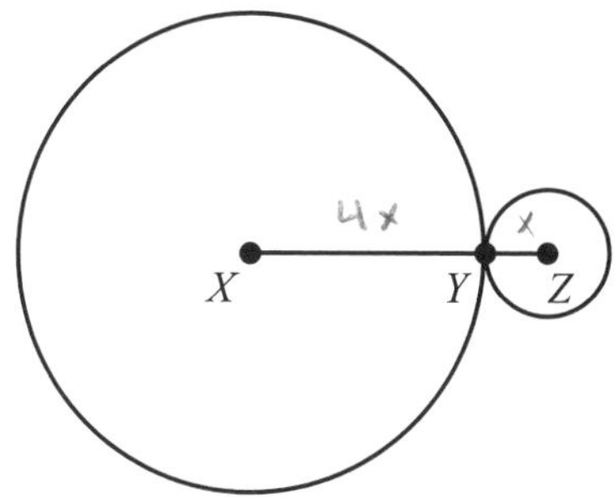

1. The two circles shown above are tangent at point Y and the circumference of the circle with center X is four times greater than the circumference of the circle with center Z. If $XZ = 20$, what is the length of XY?

(A) 4
(B) 5
(C) 8
(D) 10
(E) 16

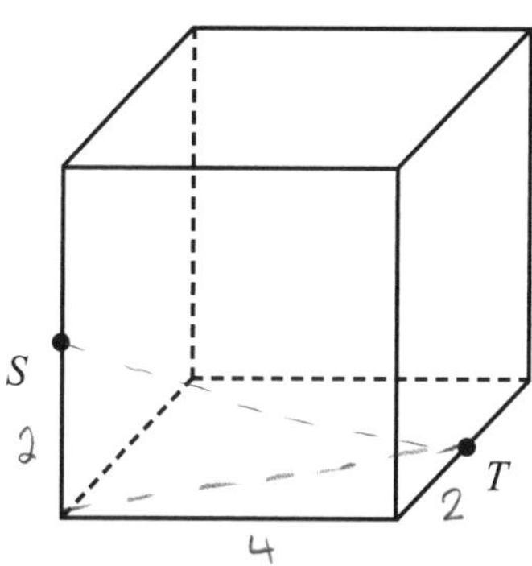

2. In the diagram above, points S and T are the midpoints of two edges and each of the edges have a length of 4. What is the length of $\overline{ST}$ (not shown)?

(A) $\sqrt{12}$
(B) 4
(C) $\sqrt{18}$
(D) $\sqrt{20}$
(E) $\sqrt{24}$

2. BACKPLUG the Answer Choices

One great characteristic of multiple-choice questions is that the answer is given to you! Granted, four wrong answer choices are given as well, but you know that one of the answers provided is the right one. This often allows you to "backplug" the answer choices into the equation in the question. Consider an example:

When backplugging, look for the answer choice that results in a true mathematical statement.

6. If a is a positive integer and $\frac{a}{2^a+4}=\frac{1}{5}$, what is the value of a?

   (A) 3
   (B) 4
   (C) 5
   (D) 6
   (E) 7

While some students may be tempted to isolate a, they will quickly see that this is a complicated process given the exponent. It is much easier to backplug each of the answer choices into the equation to find the one that makes a true mathematical statement. Start with the value in (A). Wherever there is an a in the original equation, insert a 3:

(A) 3 $\quad \frac{3}{2^3+4}=\frac{1}{5} \rightarrow \frac{3}{8+4}=\frac{1}{5} \rightarrow \frac{3}{12}=\frac{1}{5} \rightarrow \frac{1}{4}\neq\frac{1}{5}$

When (A) does not work, use the value in (B).

(B) 4 $\quad \frac{4}{2^4+4}=\frac{1}{5} \rightarrow \frac{4}{16+4}=\frac{1}{5} \rightarrow \frac{4}{20}=\frac{1}{5} \rightarrow \frac{1}{5}=\frac{1}{5}$ ✓

We can stop with answer choice (B), as the question indicates there is only one value for a. Because we found that value in (B), there is no need to backplug (C), (D), or (E).

Best bets for backplugging:
- exponents
- radicals
- absolute value
- inequalities

Backplugging is an efficient strategy that helps test takers avoid calculation errors. It is especially helpful when calculating exponents, radicals (square roots and cube roots), absolute value, and inequalities, as these operations are known to produce calculation mistakes on the SAT. To identify an SAT math problem that can be solved using backplugging, look for answer choices that are numbers and any of the following phrases in the question itself:

"which of the following values..."
"what is the value of x?"
"for what value of x..."
"...then x could be"
"...then x could equal"
"...then the value of x is"
"...then x is"
"...then x must equal"
"...then x = "
"...then one possible value of x is"

For many backplugging problems, it is important to analyze the answer choices in order to choose the most efficient starting point. Consider an example:

10. If $|5 - t| > 4$, then t could be

(A) 6
(B) 7
(C) 8
(D) 9
(E) 10

Answer choice (C) is the median value, so let's start there:

(C) 8 $\quad |5 - t| > 4 \rightarrow |5 - 8| > 4 \rightarrow |-3| > 4 \rightarrow 3 > 4$ False

The value in (C) was too small. We need a larger number to produce a larger result. Therefore we can eliminate answer choices (A) and (B). The answer is either (D) or (E):

(D) 9 $\quad |5 - t| > 4 \rightarrow |5 - 9| > 4 \rightarrow |-4| > 4 \rightarrow 4 > 4$ False

(E) 10 $\quad |5 - t| > 4 \rightarrow |5 - 10| > 4 \rightarrow |-5| > 4 \rightarrow 5 > 4$ ✓

The correct answer is (E). For most backplugging problems, you should start with answer choice (C) as it provides the median value. If the question does not work like the one above, you've lost no time and can continue to backplug through answer choices until the right answer is revealed. However, if answer choice (C) does reveal the estimated size of the correct answer, then you've gained valuable seconds by avoiding the two largest or two smallest answer choices.

Unless a question asks for the greatest or least value, it is best to start backplugging with answer choice (C).

There are other times, however, when you need to analyze the question to decide which answer choice makes the most logical starting point. Take an example:

12. If k is a positive integer, and if $\sqrt{\frac{8k}{5}}$ is an integer, what is the least possible value of k?

(A) 5
(B) 10
(C) 16
(D) 25
(E) 40

In this question, we are looking for the *smallest* possible value of k. Therefore, it makes sense to start with the smallest answer choice!

(A) 5 $\quad \sqrt{\frac{8k}{5}} \rightarrow \sqrt{\frac{(8)(5)}{5}} \rightarrow \sqrt{\frac{40}{5}} \rightarrow \sqrt{8} \rightarrow 2.83$ No

Continue with the next smallest answer choice:

$$\text{(B) } 10 \quad \sqrt{\frac{8k}{5}} \rightarrow \sqrt{\frac{(8)(10)}{5}} \rightarrow \sqrt{\frac{80}{5}} \rightarrow \sqrt{16} \rightarrow 4 \checkmark$$

This is the smallest answer choice that works, thus it is the correct answer.

For questions that require the "least value," start by backplugging the lowest number in the answer choices. If the question requires the "greatest value," begin with the largest number in the list.

It is important to note that one other answer choice produced an integer as well:

$$\text{(E) } 40 \quad \sqrt{\frac{8k}{5}} \rightarrow \sqrt{\frac{(8)(40)}{5}} \rightarrow \sqrt{\frac{320}{5}} \rightarrow \sqrt{64} \rightarrow 8$$

This answer choice, though, is not the *least value*, and is therefore wrong. Unfortunately, many unsuspecting test takers who do not carefully read the question will pick this answer or become confused by the presence of two integer-producing answers and leave the question blank.

One last note about analyzing the answer choices prior to backplugging: if the question requires multiplication and if one of the answer choices is 0, you should suspect this answer and try backplugging it first. The College Board uses many questions that test a student's understanding of the zero property of multiplication. Consider the following example:

13. If $\frac{1}{4}x^2 < (4x)^2$, for what value of x is the statement FALSE?

(A) –4

(B) 0

(C) $\frac{1}{4}$

(D) 1

(E) 4

Remember, questions that use emphasized text are the College Board's way of warning you that the phrasing may be confusing. In this question, the capitalized FALSE warns us that four of the answer choices create TRUE statements. We want to select the one that creates a FALSE mathematical statement. Because 0 is one of the answer choices, we should start backplugging with answer choice (B):

$$\text{(B) } 0 \quad \frac{1}{4}x^2 < (4x)^2 \rightarrow \frac{1}{4}(0)^2 < (4^2)(0^2) \rightarrow 0 < 0 \quad \text{False} \checkmark$$

If zero is an answer choice and the question involves multiplication, start by backplugging this answer choice.

This question assesses your ability to recognize the zero property of multiplication. By selecting the answer choice containing 0, we quickly found the correct answer and eliminated the need to backplug any other answer. On any multiplication question that can be solved using backplugging, always backplug 0 first if it appears as one of the answer choices.

BACKPLUG the Answer Choices Problem Set

Solve the following multiple-choice questions by selecting the best answer from the five answer choices. For grid-in questions, write your answer in the grids and completely mark the corresponding ovals. Answers are given on page 108.

1. If $4^{3x} = 2^{x+5}$, what is the value of x?

   (A) 0
   (B) 1
   (C) 2
   (D) 3
   (E) 4

2. If x is an integer and 2 is the remainder when $4x + 5$ is divided by 3, then x could equal

   (A) 1
   (B) 2
   (C) 3
   (D) 4
   (E) 5

3. If $a = 5b$, for which of the following values of b is $a = b$?

   (A) –5
   (B) 0
   (C) $\frac{1}{5}$
   (D) 1
   (E) a can never equal b

3. SUPPLY Numbers

Many students have a hard time conceptualizing math problems that use variables in place of numbers. Therefore, one of the most effective and most common strategies on the SAT is to supply numbers for variables. By following any conditions or rules set by the text in the question, you can pick numbers for variables and then run the numbers through the expressions or equations to find an answer. Look at an example:

Supplying numbers is an undervalued strategy on the SAT. You can use any value as long as the value satisfies the "rules" provided by the text.

5. If x is an even integer and y is an odd integer, then which of the following must be an even integer?

(A) $\frac{y}{x}$

(B) $2x + y$

(C) $3x - y$

(D) $2x + 2y$

(E) $x^2 + y$

To supply numbers for x and y, follow the rules: x is an even integer and y is an odd integer. Pick two numbers to satisfy these rules:

$$x = 2 \quad y = 3$$

When supplying numbers, choose low numbers that are easy to run through the answer choices. However, be sure to avoid 0, 1, and any numbers between 0 and 1 for your first attempt.

Notice that we chose small values. If larger numbers are used, such as $x = 100$ and $y = 101$, the calculations become much more difficult. You should also avoid 0, 1, and fractions when first supplying numbers, as these values have special properties we will discuss on the following page.

Plug your values for x and y into each of the answer choices:

(A) $\frac{y}{x}$ $\quad \frac{3}{2} \rightarrow 1.5$ Not an integer

(B) $2x + y$ $\quad 2(2) + 3 \rightarrow 4 + 3 \rightarrow 7$ Not an *even* integer

(C) $3x - y$ $\quad 3(2) - 3 \rightarrow 6 - 3 \rightarrow 3$ Not an *even* integer

(D) $2x + 2y$ $\quad 2(2) + 2(3) \rightarrow 4 + 6 = 10$ ✓

(E) $x^2 + y$ $\quad 2^2 + 3 \rightarrow 4 + 3 = 7$ Not an *even* integer

As you can see, answer (D) is the only choice that produces an even integer.

After completing this type of problem, it is extremely important to reread the question to make sure that the numbers you supplied satisfied the rules of the question. Too often, students misread the question and use the wrong numbers. For example, in the previous problem, make sure that you chose an *even* number for x and an *odd* number for y, and not the other way around. A ten second check-up can save you valuable points.

Let's consider another example:

12. If $p < q < r < s$, and if p, q, r, and s are negative consecutive even integers, then which of the following must be true?

    (A) $pq < rs$
    (B) $(pq)^2 < (rs)^2$
    (C) $p + s = q + r$
    (D) $p + r > q + s$
    (E) $2p = q + r$

The question has two rules: $p < q < r < s$ and p, q, r, and s are negative consecutive even integers. Choose four values to satisfy these rules:

$p = -8$ $q = -6$ $r = -4$ $s = -2$

Now, plug your values into all of the answer choices to find one that must be true:

(A)	$pq < rs$	$(-8)(-6) < (-4)(-2)$ →	$48 < 8$ False
(B)	$(pr)^2 < (rs)^2$	$(-8 \bullet -4)^2 < (-6 \bullet -2)^2$ →	$1024 < 144$ False
(C)	$p + s = q + r$	$-8 + -2 = -6 + -4$ →	$-10 = -10$ ✓
(D)	$p + r > q + s$	$-8 + -4 > -6 + -2$ →	$-12 > -8$ False
(E)	$2p = q + r$	$2(-8) = -6 + -4$ →	$16 = -10$ False

Notice that we completed answer choices (D) and (E) even though the supplied numbers worked in answer choice (C). When a problem can be solved by supplying numbers, and the problem asks which answer choice "*must* be" correct, then it is important to supply numbers for all five answer choices. This is because of a trap often set by more difficult questions. Examine the following problem:

19. If $p < q < r < s$, and if p, q, r, and s are consecutive even integers, then which of the following must be true?

    (A) $pq < rs$
    (B) $(pq)^2 < (rs)^2$
    (C) $p + s = q + r$
    (D) $p + r > q + s$
    (E) $2p = q + r$

The original question on this page was #12 in the set, signifying a Medium difficulty level. The second question is #19—the second to last question in the section, indicating a Hard level question. Yet the only difference between the two problems is the omission of the word "negative" in the text of the question. Without this word, most students will supply positive numbers for the variables in the question. This creates a trap during the solution:

$p = 2$ $q = 4$ $r = 6$ $s = 8$

(A) $pq < rs$ $(2)(4) < (6)(8)$ → $8 < 48$ ✓

CAUTION: SAT TRAP!
When a question uses the word "must," as in "which of the following must be true," you would be wise to check every answer choices. Often more than one answer choice will work for your first set of supplied numbers.

An inexperienced test taker would select choice (A) without looking at the other answer choices, as (A) clearly creates a true mathematical statement. A PowerScore student, however, would test all five answer choices and realize that three of the answer choices work:

$p = 2$ $\quad q = 4$ $\quad r = 6$ $\quad s = 8$

(A)	$pq < rs$	$(2)(4) < (6)(8)$ $\rightarrow$	$8 < 48$ ✓
(B)	$(pr)^2 < (rs)^2$	$(2 \bullet 6)^2 < (4 \bullet 8)^2$ $\rightarrow$	$144 < 1024$ ✓
(C)	$p + s = q + r$	$2 + 8 = 4 + 6$ $\rightarrow$	$10 = 10$ ✓
(D)	$p + r > q + s$	$2 + 6 > 4 + 8$ $\rightarrow$	$8 > 12$ False
(E)	$2p = q + r$	$2(2) = 4 + 6$ $\rightarrow$	$4 = 10$ False

When the words "must be true" or "must be false" are included in an SAT math question, you must test all five answer choices. This is especially true when the question is a Hard level problem. The inexperienced test taker probably thought to himself, "That was easy!" and it was easy—too easy, in fact.

If one of the last three or four questions of a section asks which answer choice "must be true," be sure to supply numbers for all five answer choices, as more than one choice is likely to work with positive integers.

To eliminate two of the answer choices, you must try different numbers. If you find more than one answer that works, there are a group of special numbers with special properties that you should always test for a Hard level "Must Be True" question to help eliminate answer choices. These numbers have special properties that cause them to behave differently than positive integers when multiplied, and thus they are likely options for eliminating answer choices when more than one is correct. Study the following table to learn how certain numbers behave when multiplied:

Number	Behavior when multiplied	Example
Positive numbers (greater than 1)	The product of two positive numbers (greater than 1) is greater than either of the numbers being multiplied	$4 \times 7 = 28$ $2 \times 5 \times 8 = 80$
0	Zero times a number equals zero	$0 \times 6 = 0$ $34 \times 2 \times 0 = 0$
1	One times a number equals that number	$1 \times 3 = 3$ $49 \times 1 = 49$
Positive Fractions/ Positive Decimals	The product of any two positive fractions is less than either fraction being multiplied	$\frac{1}{4} \times \frac{1}{3} = \frac{1}{12}$ $0.5 \times 0.4 = 0.2$
Negative Integers	Negative × Positive = Negative (The product is less than either multiplier) Negative × Negative = Positive (Product is greater than either multiplier)	$-5 \times 6 = -30$ $-4 \times -9 = 36$
Negative Fractions/ Negative Decimals	Negative × Positive = Negative (Product is less than the positive multiplier but greater than the negative multiplier) Negative × Negative = Positive (Product is greater than either multiplier)	$-\frac{1}{2} \times \frac{1}{3} = -\frac{1}{6}$ $-0.3 \times -0.2 = 0.6$

As you can see, 0, 1, fractions, and negative numbers create unique outcomes when multiplied. For this reason, these numbers should be used to supply numbers when faced with more than one correct answer in a question that "must be" true.

Try the previous problem again, this time using negative integers and 0 as your supplied numbers:

$p = -6 \qquad q = -4 \qquad r = -2 \qquad s = 0$

Then re-run the numbers through the three answer choices that worked previously:

(A)	$pq < rs$	$(-6)(-4) < (-2)(0)$	$\rightarrow$ $24 < 0$ False
(B)	$(pr)^2 < (rs)^2$	$(-6 \bullet -2)^2 < (-4 \bullet 0)^2$	$\rightarrow$ $144 < 0$ False
(C)	$p + s = q + r$	$-6 + 0 = -4 + -2$	$\rightarrow$ $-6 = -6$ ✓

As we learned in the original problem, and as we see again here, negative numbers rule out the first two answer choices. We have proven that even though (A) and (B) are *sometimes* true, only (C) *must* be true.

Percentages are another type of SAT question that can be solved by supplying numbers. Because percents are based out of 100, they can be easily solved by supplying the number 100. Consider an example:

11. The price of a certain book is discounted by 40 percent. If the reduced price is then discounted by another 10 percent, then the series of successive discounts is equivalent to a single discount of how much?

    (A) 40
    (B) 46
    (C) 50
    (D) 54
    (E) 60

Can you guess the most common wrong answer? Some students erroneously choose (C) because they fail to look at each discount individually.

While \$100 is an unlikely price for a book, it is the best number to supply for a percentage problem:

$$\$100 \xrightarrow[-\$40]{-40\%} \$60$$

Forty percent of \$100 is \$40. The reduced price is now \$60. This price is further reduced 10%:

$$\$100 \xrightarrow[-\$40]{-40\%} \$60 \xrightarrow[-\$6]{-10\%} \$54$$

Even if you are supplying a value for the price of a candy bar, 100 is the best number to use when the question involves percents.

Ten percent of \$60 is \$6. The new price is \$54. For the price to go from \$100 to \$54, there was a single discount of 46%.

$$\$100 \xrightarrow[-\$46]{-46\%} \$54 \qquad \text{Answer choice (B)}$$

You can supply numbers in many different types of SAT math questions, including geometry problems. Examine the following model:

Confidence Quotation
"A man is but a product of his thoughts; what he thinks, he becomes."
—Mahatma Gandhi, political and spiritual leader

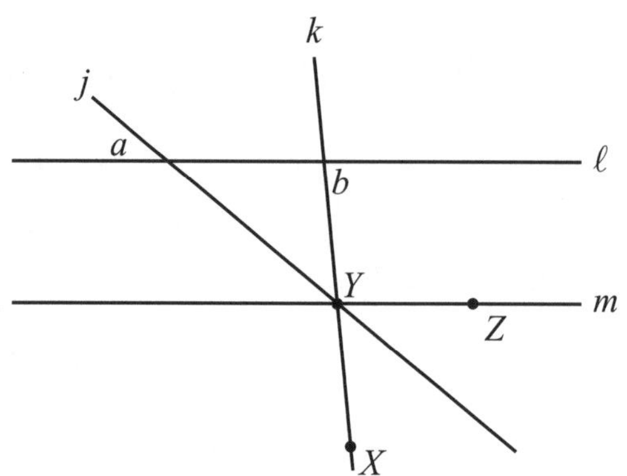

Note: Figure not drawn to scale.

11. In the figure above, $\ell \parallel m$ and j bisects $\angle XYZ$. If $30 < a < 40$, what is one possible value for b ?

Even though the question involves geometry, a rule is still provided: $30 < a < 40$. Satisfy this rule by making $a = 35$. If $a = 35$, then three other angles around line j also equal 35:

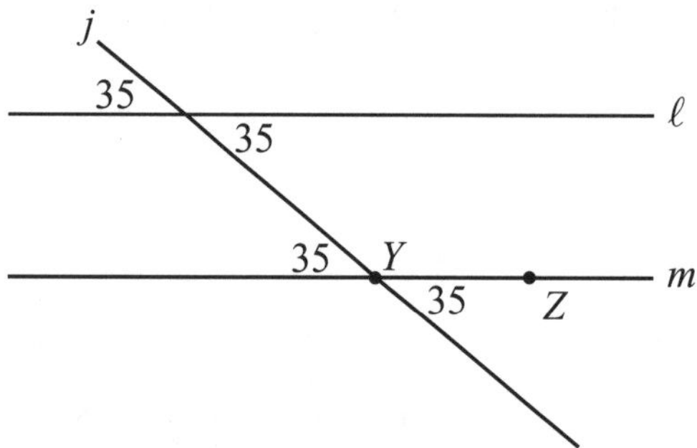

To learn more about vertical and supplementary angles, see Chapter 8.

Now add line k back into the picture. Since line j bisects $\angle XYZ$, both angles are 35:

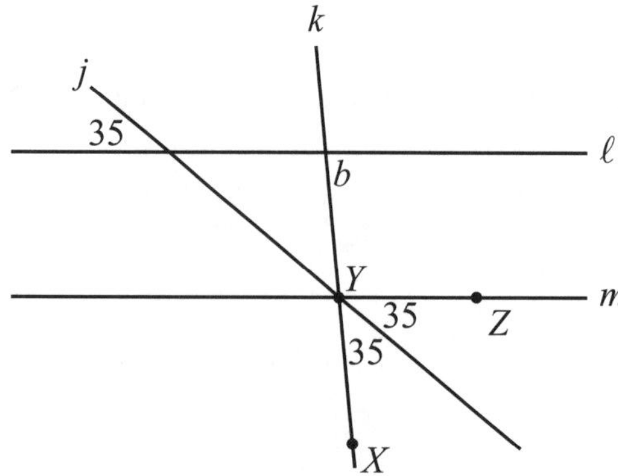

Therefore, $\angle XYZ = 35 + 35 = 70$.

Angle $\angle XYZ$ is the same measurement as b:

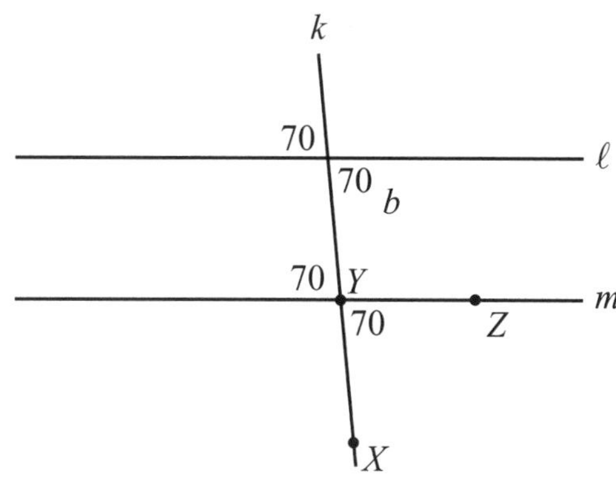

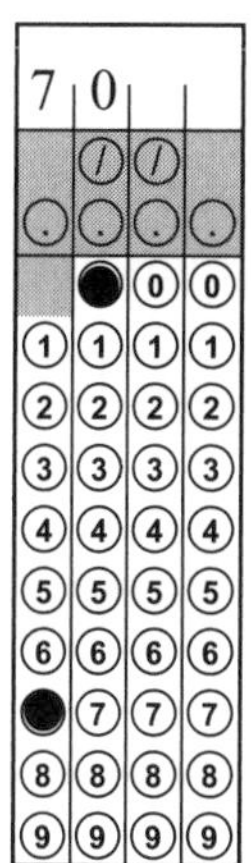

Remember that darkened ovals are the only scored portion of the grid-in questions.

Thus, $b = 70$.

There are many possible values for b, depending on the value assigned to a. The value of a could have been smaller, such as $a = 31$, or larger, such as $a = 39$, as long as it was greater than 30 and less than 40. For example, if $a = 32$, then:

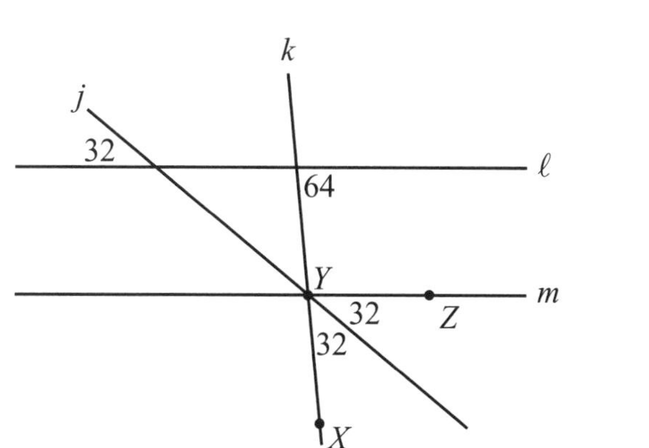

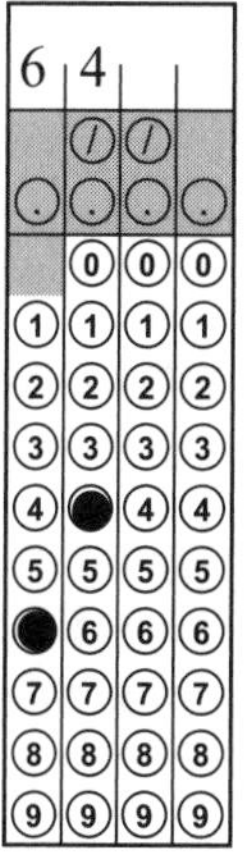

Our value for b has changed, but the answer is still correct. The phrase from the question, "what is one possible value for b?" indicates that there is more than one right answer. In fact, the value of b can be anything from 60 to 80 ($60 < b < 80$) when rounded to fit the student response box.

Supplying numbers is an undervalued strategy for many geometry questions. If a value is not provided for the length of a line or measurement of an angle, but a rule or condition is present (such as $\overline{AB} > 10$ or $\angle STU < 90$), supplying numbers may be the only way to find a solution.

A final use for supplying numbers comes with Hard level questions that ask you to choose a numberless expression to represent a value. These are usually the last question in a section, and they can often be solved more quickly by writing the expression. However, many students find that while they cannot understand the problem in its natural state, they can understand it when using real numbers. Consider the following:

20. To travel to an away football game, n friends agree to ride together and equally split the cost of gas money, which is a total of d dollars. If x of the friends forget their money and fail to pay for their share of the gas, which of the following represents the additional amount, in dollars, that each of the paying friends must contribute to cover the cost of the gas?

    (A) $\frac{d}{x}$

    (B) $\frac{n}{d(n-x)}$

    (C) $\frac{nd}{n-x}$

    (D) $\frac{xd}{n-x}$

    (E) $\frac{xd}{n(n-x)}$

The most efficient way to solve this question is to simply record what you know and create the expression:

This is the most efficient way to solve this question.

n = number of passengers
d = cost of gasoline
x = non-paying passengers

$n-x$ = paying passengers

$\frac{d}{n}\times x$ = total cost not covered by non-paying passengers

$\frac{\frac{dx}{n}}{n-x}$ = additional amount each paying passenger must cover

Since none of the answer choices represent this complex fraction, simplify:

$$\frac{\frac{dx}{n}}{n-x} \rightarrow \frac{dx}{n}\div(n-x) \rightarrow \frac{dx}{n}\times\frac{1}{(n-x)} \rightarrow \frac{dx}{n(n-x)}$$

For some students, this explanation makes perfect sense. If so, skip the next page, as you are already using the most efficient strategy. But if you have read through it twice and are still scratching your head and thinking, "Huh?," then supplying numbers is a better strategy for you.

To begin, supply numbers for each of the variables. Choose numbers that are easily divisible by the other numbers, even if they don't necessarily make sense in the problem. For example, it's unlikely that 10 passengers rode in the vehicle and that the gasoline cost $100, but these round numbers make for easy calculations:

$10 = n$ = number of passengers
$\$100 = d$ = cost of gasoline
$2 = x$ = non-paying passengers

Now, use the numbers to calculate costs:

$100 ÷ 10 passengers = $10 = cost per passenger
10 passengers – 2 non-paying passengers = 8 paying passengers
$10 × 2 = $20 = total cost not paid by non-paying passengers

So the 8 paying passengers must cover the $20 left from the non-paying passengers:

$20 ÷ 8 = $2.50 = additional amount each paying passenger must cover

From this point, there are two ways to find the correct answer.

Method #1

Some students may choose to review the previous calculations and insert the variables back into the calculation for each value:

$\$100 \div 10 = \$10 \qquad d \div n = \frac{d}{n}$ = cost per passenger

$10 - 2 = 8 \qquad n - x$ = paying passengers

$\$10 \times 2 = \$20 \qquad \frac{d}{n} \times x = \frac{dx}{n}$ = total cost not paid by non-paying passengers

So the 8 paying passengers must cover the $20 left from the non-paying passengers:

$\$20 \div 8 = \$2.50$

$\frac{dx}{n} \div (n - x)$ = additional amount each paying passenger must cover

$$\frac{dx}{n} \div (n - x) \quad \rightarrow \quad \frac{dx}{n} \times \frac{1}{(n-x)} \quad \rightarrow \quad \frac{dx}{n(n-x)}$$

This is the second most efficient way to solve this question.

As you can see, by supplying the variables back in for the values, the correct answer is revealed. This method is second in efficiency to simply writing the expression, but some students may still struggle to understand the placement of the variables. If this is the case, try the next method on the following page.

Method #2

As we calculated on the previous page, each of the 8 paying passengers would have to pay an additional $2.50 to cover the cost of gasoline for the 2 non-paying passengers. The numbers we supplied for the three variables were:

$10 = n =$ number of passengers
$\$100 = d =$ cost of gasoline
$2 = x =$ non-paying passengers

Supply these values into each of the answer choices to find the one that results in $2.50:

(A) $\frac{d}{x} \rightarrow \frac{10}{2} = 5$ No

(B) $\frac{n}{d(n-x)} \rightarrow \frac{10}{100(10-2)} \rightarrow \frac{10}{100(8)} \rightarrow \frac{10}{800} \rightarrow \frac{1}{80}$ No

(C) $\frac{nd}{n-x} \rightarrow \frac{(10)(100)}{(10-2)} \rightarrow \frac{1000}{8} \rightarrow 125$ No

(D) $\frac{xd}{n-x} \rightarrow \frac{(2)(100)}{(10-2)} \rightarrow \frac{200}{8} \rightarrow 25$ No

(E) $\frac{xd}{n(n-x)} \rightarrow \frac{(2)(100)}{10(10-2)} \rightarrow \frac{200}{(10)(8)}\ \frac{200}{80} \rightarrow 2.5$ ✓

This is the least efficient way to solve this question, but it makes a difficult question very easy to solve.

We recommend this method if the solutions on the previous pages prove too difficult.

As you can see, choice (E) produced $2.50 as the amount that each of the paying friends must contribute to cover the cost of the gas.

While this method is not the most efficient strategy for solving the problem, it is extremely useful for the student who struggles with variables. Using this method still produces the correct answer in about a minute, and allows the student to gain credit for a question he might otherwise omit.

Supplying numbers is an effective strategy for eliminating confusing variables. Expect to use it on many different types of SAT math questions, from the easiest arithmetic problem to the most difficult geometry question.

SUPPLY Numbers Problem Set

Solve the following multiple-choice questions by selecting the best answer from the five answer choices. For grid-in questions, write your answer in the grids and completely mark the corresponding ovals. Answers begin on page 109.

1. If $\frac{t+4}{2}$ is an integer, then t must be

   (A) a multiple of 4
   (B) a positive integer
   (C) a negative integer
   (D) an even integer
   (E) an odd integer

2. If $a^3 < b < 0$, which of the following has the least value?

   (A) 0
   (B) a^3
   (C) $-a^3$
   (D) $a^3 - b$
   (E) $-(a^3 - b)$

3. At a feed store, the price of bird seed is d dollars for 10 pounds. Each pound of bird seed can fill b bird feeders. In terms of b and d, what is the dollar cost of the bird seed required to fill 1 bird feeder?

   (A) $10bd$
   (B) $\frac{b}{10d}$
   (C) $\frac{d}{10b}$
   (D) $\frac{10b}{d}$
   (E) $\frac{10d}{b}$

4. TRANSLATE from English to Math

Many unruly SAT word problems can be tamed by translating complicated English sentences into easily understood math equations. Certain words have assigned math symbols to take their place in an equation. The following English words and their math symbol equivalents are commonly used on the SAT:

MEMORY MARKER: Throughout this book, we will draw attention to formulas or shortcuts that you should memorize before taking the SAT. To help you with this process, PowerScore has provided free flash cards available on the SAT Math Bible website at www.powerscore.com/satmathbible. English to math translations, which should be memorized to save time on the test, are included in these flash cards.

English	Math	English Example	Math Translation
what, what number	x, n, ? or other variable	What is 15 + 4?	? = 15 + 4
what percent	$\frac{x}{100}$, $\frac{?}{100}$	What percent of 70 is 21?	$\frac{x}{100} \times 70 = 21$
plus, more than, added to, increased by, sum	+	a number is three more than x	$n = 3 + x$
minus, less than, subtracted from, decreased by, reduced by, difference	−	10 less than a is 15	$a - 10 = 15$
of, times, product	×	three-fourths of x is two-fifths of y	$\frac{3}{4}x = \frac{2}{5}y$
per, out of, quotient	÷	He works n days out of 7	$\frac{n}{7}$
is, equals, result	=	x is y	$x = y$

To see how translation is used, consider the following example:

9. What percent of 70 is 21?

(A) 3%
(B) $3.3\overline{3}$%
(C) 14.7%
(D) 30%
(E) $30.3\overline{3}$%

Each word in the problem can be represented by a symbol:

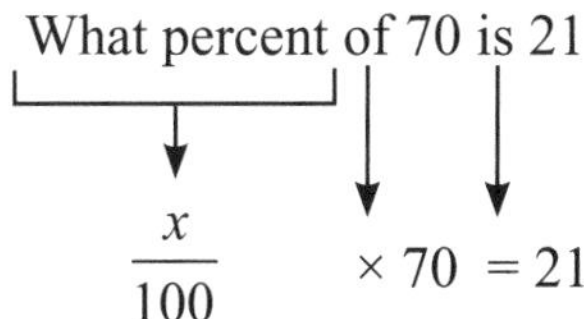

The word "what" represents the value for which we are searching. You can use any variable to represent this value. If x is not used in the question, it is a wise choice since most students are conditioned to find x (however, if x is used in the question, you must choose a different variable, such as n, y, or even ?, to represent the requested value to avoid confusion). Since our value for "what" is a percent, x must be placed over 100. The word "of" is the most commonly-translated word on the SAT; it signifies multiplication. Finally, the word "is" means "equals." The once-difficult question becomes an easily solved equation:

It is imperative that you know "of" means "multiply," as it appears frequently on the SAT.

$$\frac{x}{100} \times 70 = 21 \quad \rightarrow \quad x \times 70 = 2100 \quad \rightarrow \quad x = 30$$

The correct answer is (D), 30%.

Questions involving percentages and fractions can often be solved using translation. You should also expect to see basic arithmetic questions in which you must translate an English sentence into an equation:

7. When a number is subtracted from 27 and the difference is divided by twice that number, the result is 4. What is the value of the number?

   (A) 3
   (B) 4
   (C) 6
   (D) 9
   (E) 16

To solve this question, write an expression for each part of the sentence:

a number is subtracted from 27 $\rightarrow$ $27 - n$

the difference is divided by twice that number $\rightarrow$ $\frac{27-n}{2n}$

the result is 4 $\rightarrow$ $\frac{27-n}{2n} = 4$

Then solve for the number:

$$\frac{27-n}{2n} = 4 \quad \rightarrow \quad 27 - n = 8n \quad \rightarrow \quad 27 = 9n \quad \rightarrow \quad 3 = n$$

Translation is most common in questions with fractions, percents, and other basic arithmetic.

The correct answer is (A).

TRANSLATION Mini-Drill

You should be comfortable creating expressions and equations from English sentences. Translate the following statements from English to math. Check your translations in the answer key on page 110.

1) The product of two numbers equals 15.

$xy = 15$

2) When x is increased by one-third of y, the result is 30.

$x + \frac{1}{3}y = 30$

3) 8 less than 4 times a certain number is 10 more than the number.

$4n - 8 = 10 + n$

Confidence Quotation

"Practice, the master of all things."

—Augustus Octavius, first ruler of the Roman Empire

4) The sum of two numbers is 4 and their difference is 0.

$x + y = 4$

$x - y = 0$

5) What percent of a is 3 less than b ?

$\frac{a}{100} = b - 3$

6) The difference between s and t is one-half more than twice t.

$s - t = \frac{1}{2} + 2t$

7) If two-thirds of x is increased by one-fourth of y, the result is what percent of 10?

$\frac{2}{3}x + \frac{1}{4}y = \frac{?}{100} \times 10$

8) There are 2 lottery tickets per 100 that are winners.

$\frac{2}{100}$

9) What percent of $x - y$ is 12 less than z?

$\frac{x-y}{100} = z - 12$

While translation is most common among problems involving percentages, fractions, and basic arithmetic, you can still apply it to other types of questions as well, such as geometry, algebra, and statistics. Examine the following question:

10. The perimeter of a rectangle is 8 times the width. If the perimeter is 24 inches, what is the length of the rectangle in inches?

    (A) 3
    (B) 4
    (C) 6
    (D) 8
    (E) 9

Begin by translating:

perimeter is 8 times the width $\rightarrow$ $p = 8w$

perimeter is 24 inches $\rightarrow$ $p = 24$

On the SAT, any time there are two values equal to a single variable, set the values equal to each other.

You now have two expressions equal to p. Set them equal to each other to find the width:

$8w = 24 \rightarrow w = 3$

The formula for perimeter will determine the length of the rectangle:

$2l + 2w = \text{perimeter}$

$2(l) + 2(3) = 24 \rightarrow 2(l) + 6 = 24 \rightarrow 2(l) = 18 \rightarrow l = 9$

The length of the rectangle is 9 inches. The correct answer is (E).

Translation is a great tool for all students, but can be especially helpful for the student who is intimidated by word problems.

TRANSLATE from English to Math Problem Set

Solve the following questions. For a multiple-choice question, select the best answer from the five answer choices. For a grid-in question, write your answer in the grid and completely mark the corresponding ovals.

1. The sum of Dan's age and his granddaughter Susan's age is 104. If Dan's age is 8 more years than 3 times Susan's, what is Susan's age in years?

(A) 21
(B) 22
(C) 23
(D) 24
(E) 25

2. If 40 percent of s equals 60 percent of t, which of the following expresses t in terms of s ?

(A) $t = 20\%$ of s
(B) $t = 33\%$ of s
(C) $t = 66\%$ of s
(D) $t = 100\%$ of s
(E) $t = 150\%$ of s

TRANSLATE from English to Math Problem Set

Solve the following multiple-choice questions by selecting the best answer from the five answer choices. For grid-in questions, write your answer in the grids and completely mark the corresponding ovals. Answers are on page 111.

3. When $\frac{1}{5}$ of $\frac{5}{4}$ is subtracted from 6, what is the result?

5. RECORD What You Know

This strategy can be implemented two ways. The first is to take an unfamiliar problem and apply familiar values to the variables or supplied information. Study an example:

7. How old will a person be exactly 1 year from now if exactly r years ago the person was s years old?

    (A) $s + 1$
    (B) $s - r - 1$
    (C) $r - s + 1$
    (D) $r + s - 1$
    (E) $r + s + 1$

Some students may be able to write the expression without applying personal information to the question. This strategy is specifically for students who are unable to recognize the solution using only variables.

To make this problem more familiar, use your own age as a base. If you are now 17 years old, then:

Age now: 17
Age one year from now: 18
Let $r = 5$ years
Age r years ago $(s) = 17 - 5 = 12$

If $r = 5$ and $s = 12$, which of the answer choices equals 18, your age one year from now?

(A) $s + 1$	$12 + 1 = 13$	No
(B) $s - r - 1$	$12 - 5 - 1 = 6$	No
(C) $r - s + 1$	$12 - 5 + 1 = 8$	No
(D) $r + s - 1$	$12 + 5 - 1 = 16$	No
(E) $r + s + 1$	$12 + 5 + 1 = 18$	✓

By applying your own information to a problem, you take some of the mystery out of the question and thus make it more easily solvable.

This strategy can also be helpful for logic problems:

9. If all of the members of the tennis team are right-handed, which of the following statements must be true?

    (A) All left-handed people are members of the tennis team.
    (B) All people who are not members of the tennis team are left-handed.
    (C) No left-handed people are members of the tennis team.
    (D) Every member of the tennis team who is right-handed is a senior.
    (E) There is one left-handed person on the tennis team.

Think about your own school's tennis team. Maybe you know three people who play tennis: Alex, Britney, and Chris. Pretend that all three of these people are right-handed, and then run through the answer choices to prove or disprove each one:

(A) All left-handed people are members of the tennis team. *What about your friend Matt, who is left-handed but is not a member of the tennis team?*

(B) All people who are not members of the tennis team are left-handed. *False. Your mom is not a member of the tennis team, and she's right-handed.*

(C) No left-handed people are members of the tennis team. *This is true, because Alex, Britney, and Chris are all right-handed.*

(D) Every member of the tennis team who is right-handed is a senior. *Britney, who is on the tennis team, is a junior so this is not true.*

(E) There is one left-handed person on the tennis team. *False. Alex, Britney, and Chris are all right-handed.*

By using real people, this question is easier to understand and therefore easier to answer. Only choice (C) *must* be true based on the statement made in the question.

The second, and more common way to apply "Record What You Know" occurs when you read a question, but do not know how to solve it. At that point, record any formulas you know that are mentioned in the problem, as well as any information that is provided in the question. Examine a question that can be solved by recording what you know:

Every test taker can benefit from RECORDING what they know when faced with a problem they do not know how to solve.

12. If the area of a circle is 49π, what is the circumference?

(A) 2π
(B) 7π
(C) 9π
(D) 14π
(E) 36π

Let's say that this question has stumped you (and maybe it really has). What do you know from the problem?

Area = 49π

There are two formulas mentioned in the problem: area and circumference. Record these formulas:

Area = πr^2
Circumference = $2\pi r$

When you have exhausted the information in the problem, review your notes. Notice that there are two expressions equal to area: 49π and πr^2. When there are two values or expressions that equal the same component, set them equal to each other:

$49\pi = \pi r^2$

The only unknown in the equation is r, the radius. Solve for r:

$$49\pi = \pi r^2 \rightarrow 49\cancel{\pi} = \cancel{\pi} r^2 \rightarrow 49 = r^2 \rightarrow 7 = r$$

Return to the question and your notes to see what else can be solved now that you have the length of the radius. Since you are looking for the circumference, you can use the radius to complete the circumference formula:

$$\text{Circumference} = 2\pi r \rightarrow 2\pi(7) \rightarrow 14\pi$$

Answer choice (D) is correct. Let's try one more:

$$x, x + 4, y$$

13. If the average (arithmetic mean) of the 3 numbers above is $x + 4$, what is y in terms of x ?

(A) $2x + 4$
(B) $x + 8$
(C) $2x + 8$
(D) $x^2 + 4x$
(E) $\frac{3}{x+4}$

Recording formulas and information from the problem is a great strategy for attacking the questions you skipped on your first pass through a section.

Begin with recording information from the problem:

$$\text{Average} = x + 4$$

What formula do you know for averages? What information from the problem can be applied to that formula?

$$\text{Average} = \frac{\text{sum of numbers}}{\text{number of numbers}} = \frac{x+(x+4)+y}{3}$$

Two expressions now equal the average; set them equal to each other:

$$x + 4 = \frac{x+(x+4)+y}{3}$$

Now solve for y ("what is y in terms of x ?"):

$$x + 4 = \frac{x+(x+4)+y}{3} \rightarrow 3(x + 4) = x + (x + 4) + y \rightarrow$$

$$3x + 12 = 2x + 4 + y \rightarrow x + 8 = y$$

The correct answer is (B).

While rewriting information from the question is always a safe strategy, recording what you know might not always be the most efficient method for solving questions. This strategy is best applied when you do not know how to solve a question, or when the wording in the question confuses you.

RECORD What You Know Problem Set

Solve the following multiple-choice questions by selecting the best answer from the five answer choices. For grid-in questions, write your answer in the grids and completely mark the corresponding ovals. Answers begin on page 112.

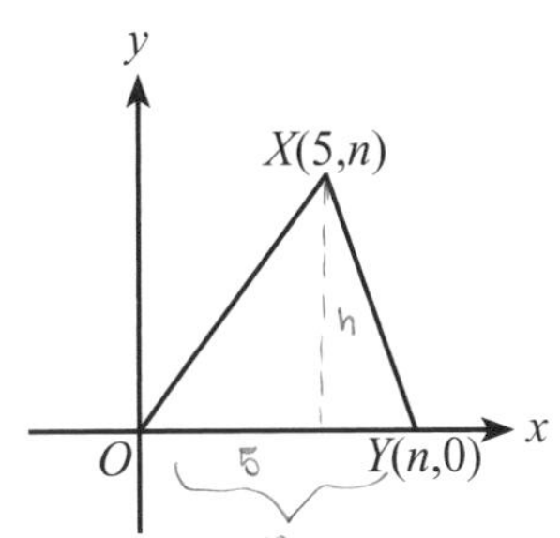

Note: Figure is not drawn to scale.

1. In the xy-plane above, the area of $\triangle OXY$ is 32. What is the value of n ?

2. What was the original weight, in pounds, of a person who now weighs h pounds and who gained 10 pounds and then lost j pounds ?

   (A) $h - j + 10$
   (B) $h + j - 10$
   (C) $h + j + 10$
   (D) $j - h - 10$
   (E) $j - h + 10$

6. SPLIT the Question into Parts

If a question is made up of more than one equation, split the question into parts for easier management.

Many SAT math questions require more than one step to solve. However, test takers can become overwhelmed by the breadth of the question and forget to consider the individual parts, and thus fail to find a solution. Sometimes you must split the question into parts and work on each section separately. This is often true of questions involving fractions, percentages, functions, symbolic functions, geometry, and coordinate geometry, especially if the questions are a higher Medium or Hard difficulty level. Let's look at a function question as an example:

16. Let the function h be defined by $h(x) = 3x + 5$. If $6h\left(\sqrt{n}\right) = 42$, what is the value of n ?

(A) $\frac{1}{4}$

(B) $\frac{4}{9}$

(C) $\frac{2}{3}$

(D) $\frac{3}{2}$

(E) $\frac{9}{2}$

Most function questions on the SAT are two- and three-part questions.

The presence of two functions often indicates at least a two-part solution. Start with the second function, as it can be manipulated:

Part 1:

$$6h\left(\sqrt{n}\right) = 42 \quad \rightarrow \quad \text{divide both sides by 6} \quad \rightarrow \quad h\left(\sqrt{n}\right) = 7$$

Now return to the first function. It is in the form of $h(x)$, but to solve for n, it must be in the form $h\left(\sqrt{n}\right)$:

Part 2:

$$h(x) = 3x + 5$$

$$h\left(\sqrt{n}\right) = 3\left(\sqrt{n}\right) + 5$$

There are now two results equal to $h\left(\sqrt{n}\right)$. Set them equal to each other to solve for n:

Part 3:

$$h\left(\sqrt{n}\right) = 7$$

$$h\left(\sqrt{n}\right) = 3\left(\sqrt{n}\right) + 5$$

$$7 = 3\left(\sqrt{n}\right) + 5 \quad \rightarrow \quad 2 = 3\left(\sqrt{n}\right) \quad \rightarrow \quad \frac{2}{3} = \sqrt{n} \quad \rightarrow \quad \frac{4}{9} = n$$

The correct answer is (B).

A common two-part question occurs when the test makers ask you to solve for something other than a single variable. Students are conditioned to find x, so the test makers might instead require $x + 3$. If $x = 2$, then $x + 3 = 5$. But you can be sure that 2 will be one of the answer choices to try to trick the forgetful test taker. For this reason, it is important to reread the question after finding the solution, to make sure you found the requested value. Consider an example:

15. The area of a square is $2x^2 - 10x$. If the length of one side of the square is $x - 5$, what is the value of $4x$?

    (A) 4
    (B) 5
    (C) 10
    (D) 20
    (E) 25

Notice that the requested value is $4x$, rather than x. However, to find $4x$, we must first find x:

Part 1:
Area = $2x^2 - 10x$
Area = s^2

Area = $(x - 5)^2 \rightarrow (x - 5)(x - 5) \rightarrow x^2 - 10x + 25$

Now set the two expressions equal and solve for x:

$$2x^2 - 10x = x^2 - 10x + 25 \rightarrow x^2 = 25 \rightarrow x = 5$$

Many students would circle (B) and move on to the next question. However, by skipping the final part of the question, they would have lost an easy point. If you reread the question, you will see that you must find $4x$:

Part 2:
$4x \rightarrow 4(5) = 20$

Answer choice (D) is the correct answer.

☠ CAUTION: SAT TRAP!
When solving for x, be sure to reread the problem after solving it to make sure that x is the requested value.

You may want to circle 4x in the question so that you remember to come back and reread the problem after finding x.

Another type of question that must be tackled in parts requires you to compute a value in the question and then find the same value in the answer choices. We have seen a couple examples of this previously in this chapter, but it bears studying as an individual strategy. Functions are very vulnerable to this type of modified backplugging question:

19. Let the function $g(x)$ be defined by $g(x) = x^2 - 2x$. Which of the following is equivalent to $g(9) - g(8)$?

    (A) $g(-1)$
    (B) $g(1)$
    (C) $g(5)$
    (D) $g(7)$
    (E) $g(17)$

This question has two parts: first, the question, which requires you to find the value of $g(9) - g(8)$, and second, the answer choices, which require you to find the value of each one.

Remember, you do not yet have to know how to solve the problems in the examples and the problem sets. We will review functions in Chapter 7, just as we will review all of the content in this chapter in later chapters of the book.

Part 1
$g(x) = x^2 - 2x$
$g(9) = 9^2 - 2(9) \quad \rightarrow \quad 81 - 18 = 63$
$g(8) = 8^2 - 2(8) \quad \rightarrow \quad 64 - 16 = 48$

$g(9) - g(8) = 63 - 48 = 15$

Now, calculate the answer choice to find the one that also results in 15:

Part 2 $\quad g(x) = x^2 - 2x$
(A) $g(-1)$ $\quad g(-1) = (-1)^2 - 2(-1) \quad \rightarrow \quad 1 + 2 = 3$ No
(B) $g(1)$ $\quad g(1) = 1^2 - 2(1) \quad \rightarrow \quad 1 - 2 = -1$ No
(C) $g(5)$ $\quad g(5) = 5^2 - 2(5) \quad \rightarrow \quad 25 - 10 = 15$ ✓

There is no need to check (D) or (E) as choice (C) is equivalent to $g(9) - g(8)$.

Dividing a question into parts and working each part separately is a key strategy for difficult questions, especially those that involve functions, fractions and percentages, and geometry and coordinate geometry. By splitting a question, you can concentrate on a single task and avoid any confusion presented by the problem as a whole.

SPLIT the Question into Parts Problem Set

Solve the following multiple-choice questions by selecting the best answer from the five answer choices. For grid-in questions, write your answer in the grids and completely mark the corresponding ovals. Answers are on page 113.

1. Let the function f be defined by $f(x) = 3(x^2 - 6)$. When $f(x) = 30$, what is the value of $3x - 6$?

   (A) 2
   (B) 3
   (C) 4
   (D) 5
   (E) 6

2. Let the function g be defined by $g(t) = t^2 - 4$. Which of the following is equivalent to $g(x - 2)$?

   (A) $g(x)$
   (B) $g(x^2)$
   (C) $g(2 - x)$
   (D) $g(x^2 - 2)$
   (E) $g(x + 4)$

7. DIAGRAM the Question

To diagram the question, add information from the text to the existing figure or use that information to create a figure.

Many students find that SAT math questions are easier to understand if they are represented graphically. Some questions, especially those assessing geometry skills, already have an existing picture, while others only contain the text of the question. In either scenario, diagramming the question can help you to understand the information presented in the text.

Add Information to an Existing Figure

Most geometry and coordinate geometry questions will have an existing figure. However, the figure rarely contains all of the information provided in the text. Look at an example:

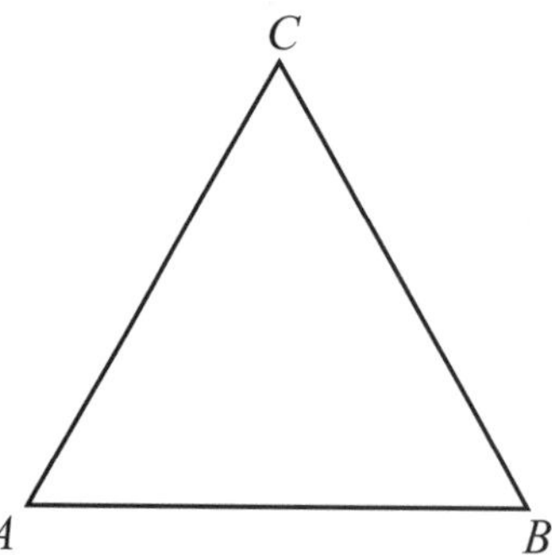

14. In the figure above, $\triangle ABC$ is equilateral. If $\overline{AC}$ is 8, what is the area of the triangle?

The existing figure does not contain any information about side length or angle measurement. This information is provided in the text of the question as another obstacle for inexperienced test takers. Some students will not be able to retain the values in their short term memory, and others will make careless mistakes when transferring the information to the figure. For this reason, you should carefully complete the diagram with the information in the text:

The test makers expect you to know the meaning of words like "equilateral" and "bisect." It is important to illustrate these words in the diagram.

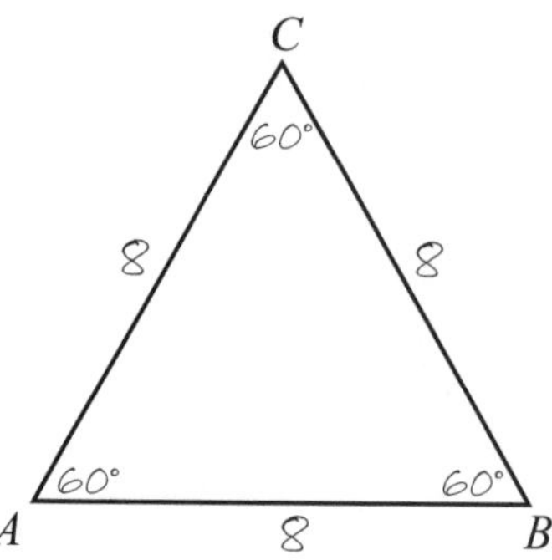

14. In the figure above, $\triangle ABC$ is equilateral. If $\overline{AC}$ is 8, what is the area of the triangle?

After adding the information from the text to the figure, apply your own knowledge and further complete the diagram:

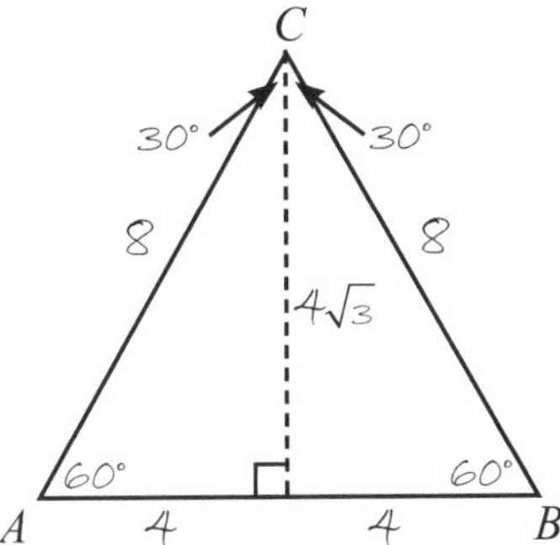

14. In the figure above, $\triangle ABC$ is equilateral. If $\overline{AC}$ is 8, what is the area of the triangle?

For more information on how to find the height of an equilateral triangle, see Chapter 8.

A completed diagram will more easily lead to a solution. The height and the base were found in the diagram above, and thus the area can be calculated.

Scale presents another potential pitfall with existing diagrams. A figure that is drawn to scale is a visually accurate diagram with proportional measurements. Unless a figure explicitly states "not drawn to scale," you can assume that it is in fact drawn to scale. The figure in the previous problem is drawn to scale, because there is nothing indicating that it is not proportionate.

A figure is drawn to scale unless the problem indicates otherwise.

When a figure is not proportionate, a note will be placed just below the figure that says "Figure not drawn to scale." If you see this note, take a few seconds to evaluate the figure. It will either be extremely close to scale, or grossly disproportionate:

CLOSE TO SCALE

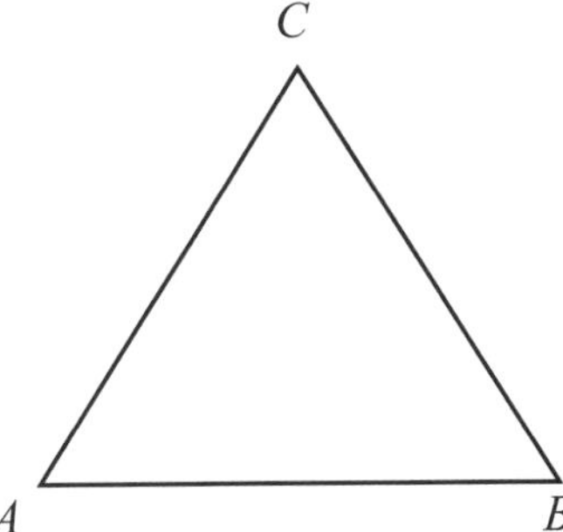

Note: Figure not drawn to scale

14. In the figure above, $\triangle ABC$ is equilateral. If $\overline{AC}$ is 8, what is the area of the triangle?

DISPROPORTIONATE

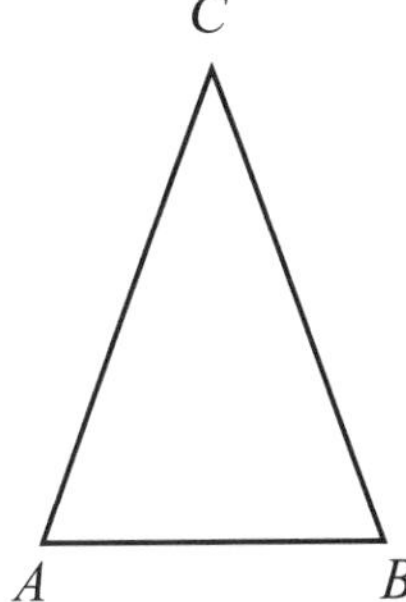

Note: Figure not drawn to scale

14. In the figure above, $\triangle ABC$ is equilateral. If $\overline{AC}$ is 8, what is the area of the triangle?

If you find that a figure is highly disproportionate, suspect a trap. The test makers hope that by drawing the figure so far off scale, you will miss a key piece of information. To avoid this common trap, redraw the figure closer to scale.

CAUTION: SAT TRAP! Figures that are grossly disproportionate are intentionally designed to cause students to miss important information.

Draw a Figure

Some geometry and coordinate geometry questions will not have a figure. Depending on the question, you may want to draw your own:

11. Mary leaves her house and walks 6 miles due south, then 3 miles due west, and finally 2 miles due north, where she arrives at a park. What is the straight-line distance, in miles, between Mary's house and the park?

This question is very difficult to solve without a sketch. To make the problem easier, draw a simple figure:

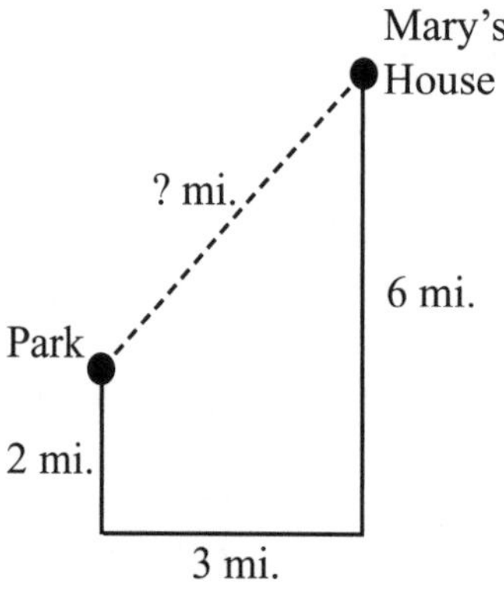

Confidence Quotation
"Imagination is everything. It's a preview of life's coming attractions."
—Albert Einstein, theoretical physicist

Once you establish the basic figure, use your prior knowledge to add more information and solve the problem:

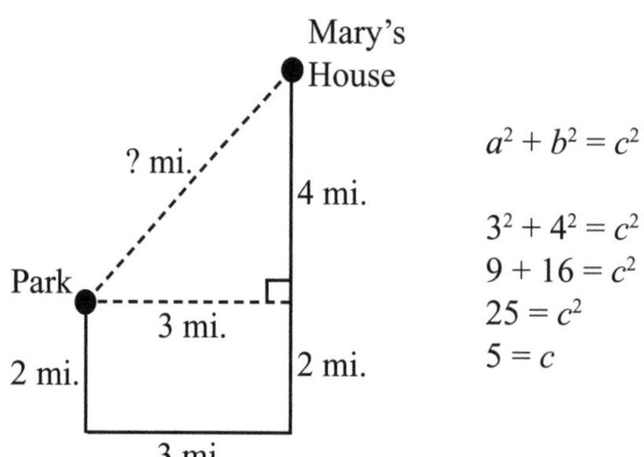

$a^2 + b^2 = c^2$

$3^2 + 4^2 = c^2$
$9 + 16 = c^2$
$25 = c^2$
$5 = c$

There may be times when you do not need to represent the problem graphically. If you can solve the question without drawing a picture, and you are confident about your answer, then you should not waste time drawing figures. But if you are at all unsure, then a graphic representation can make a problem more solvable.

While this strategy is most often used with geometry questions, it can also be used with coordinate geometry questions. A problem may be more clear if it is represented on the coordinate plane.

Consider the following coordinate geometry question:

13. In the xy coordinate plane, line k is a reflection of line j about the y-axis. If the slope of line k is 2, what is the slope of line j ?

(A) -2

(B) $-\frac{1}{2}$

(C) $\frac{1}{2}$

(D) 1

(E) 2

If you are confident in your knowledge of reflections and can solve this without a figure, then select an answer and move on. But if you are like most students, you will need to verify the reflection with a sketch. Start by drawing a simple coordinate plane, placing four or five tick marks as evenly as possible in each direction.

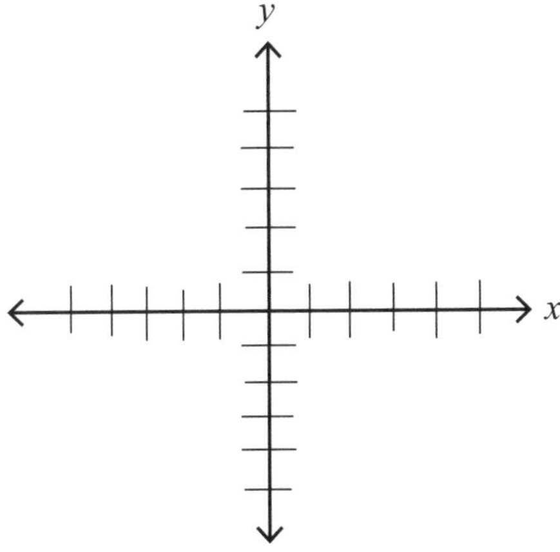

Line j could be anywhere on the coordinate plane. No matter where you graph it, the slope of j and k will always be the same. Choose the origin as a point on line j, and graph it with a slope of 2 (Figure A).

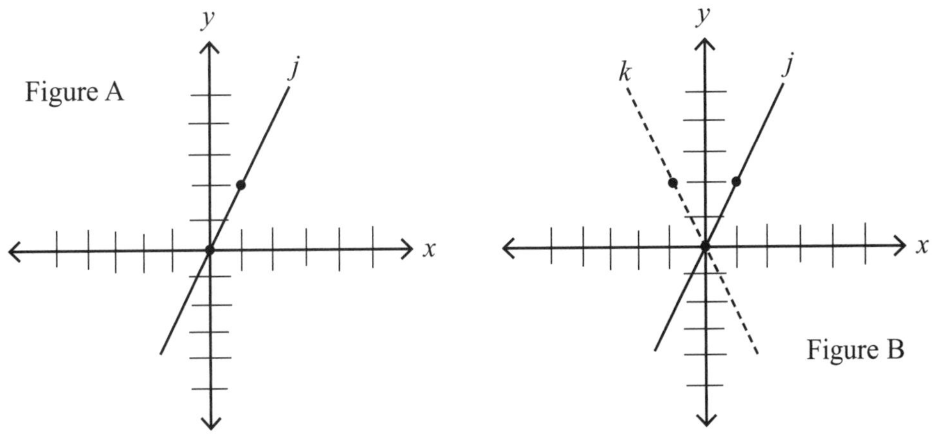

Then, draw a reflection of this line over the y-axis (Figure B). From the picture, it is clear that the slope of k is -2.

To learn more about reflections in the coordinate plane, visit Chapter 9.

Diagramming questions is a main strategy for geometry and coordinate geometry problems, but you can sometimes graphically represent arithmetic, algebra, and statistics questions as well. This is particularly helpful for students who tend to learn best through spatial or visual representations. Let's examine a grid-in problem that can be solved graphically:

Another arithmetic question that was represented graphically was the Bob and Ted driving problem used to illustrate ANALYZING the Answer Choices.

> 15. One-third of a bucket is filled with sand. It is then filled to the top with a mixture that contains equal parts soil, fertilizer, and sand. What fraction of the final mixture is sand?

Start by drawing a picture of the bucket with one-third sand:

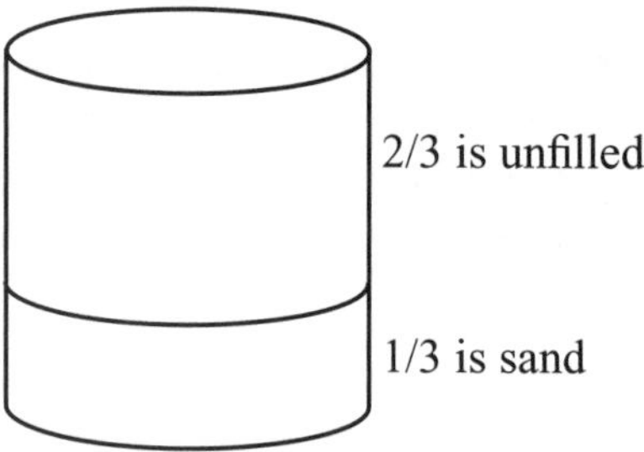

Then visualize the bucket being filled with equal parts soil, fertilizer, and sand:

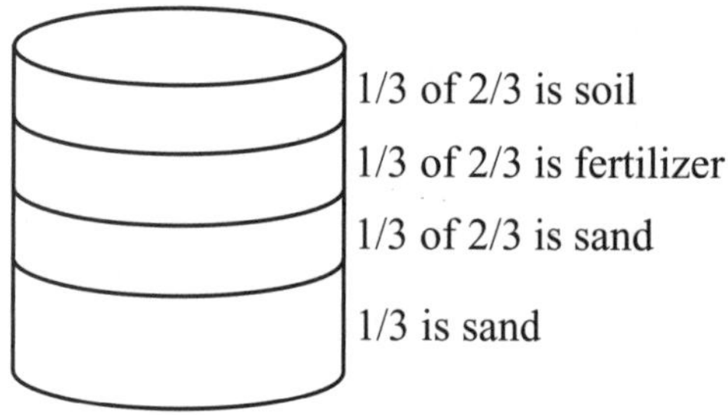

Now, write a simple equation using your picture as a reference:

This question requires TRANSLATION, too. Many questions require multiple strategies.

$$\text{The fraction of sand} = \frac{1}{3} + \left(\frac{1}{3} \text{ of } \frac{2}{3}\right)$$

$$\text{The fraction of sand} = \frac{1}{3} + \left(\frac{1}{3} \times \frac{2}{3}\right) \rightarrow \frac{1}{3} + \frac{2}{9} \rightarrow \frac{3}{9} + \frac{2}{9} = \frac{5}{9}$$

Creating quick, simple pictures can make a question less intimidating for many students. These figures do take a little time to set up, but not so much time that you should avoid the question. This method is not for every student every time, but certainly every student could benefit from it at some point in their SAT preparation.

DIAGRAM the Question Drill

Diagram each of the following questions. Compare your diagram to the figures beginning on page 114. Note that you do not need to solve these questions, although answers are provided should you choose to solve them.

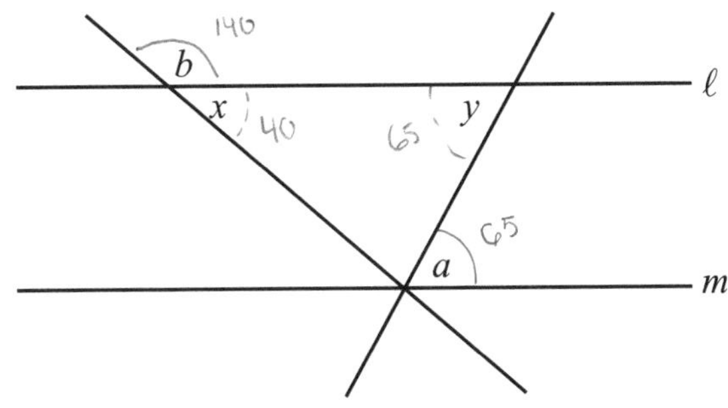

1. In the figure above, line ℓ is parallel to line m, $a = 65°$, and $b = 140°$. What is the value of $x + y$?

   $65 + 40 = 105°$

2. An isosceles triangle has a perimeter of 18. One of the sides is 5 units long. What is one possible value of the area of the triangle?

   $A = \frac{1}{2}bh$

   $= \frac{1}{2}(8)(3)$

   $= 12$ sq. units

3. In the xy-coordinate plane, a line passes through point $(0, -3)$, and between points $(2, 2)$ and $(4, 2)$, but does not contain either of them. What is one possible value of the slope of the line?

   $m = \frac{5}{3}$

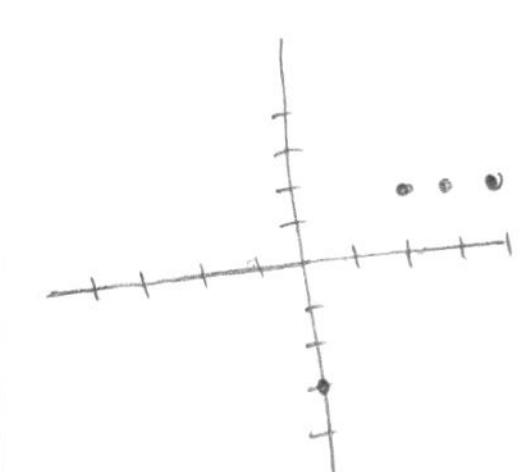

B

A

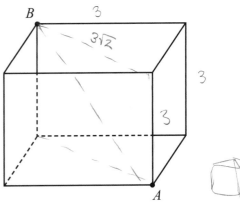

Note: Figure not drawn to scale

4. The figure above is a cube with side length of 3. What is the length of $\overline{AB}$ (not shown)?

   $3^2 + (3\sqrt{2})^2 = c^2$

   $9 + 18 = 27$

   $c = \sqrt{27}$

8. SIZE UP the Figure

This strategy, which uses estimation to size up a figure, should only be used when time is running out, or when you do not know how to solve the question mathematically.

The final PowerScore solution strategy is reserved primarily for geometry and coordinate geometry questions, and should only be used when you do not know how to solve the question or do not have time to solve the question.

For figures that are drawn to scale, you can size up the figure and estimate measurements. Estimation often eliminates two or three answer choices, thus giving you the opportunity to guess and earn points. Remember though, this strategy should not be used if you can work through the problem to produce an exact answer.

Triangles present many opportunities to size up the figures. You can estimate any angle based on your knowledge of benchmark measurements, which are common measurements that are easily recognized. You should memorize the following benchmark measurements if you do not already know them:

MEMORY MARKER: Memorize these benchmark measurements for more accurate estimates.

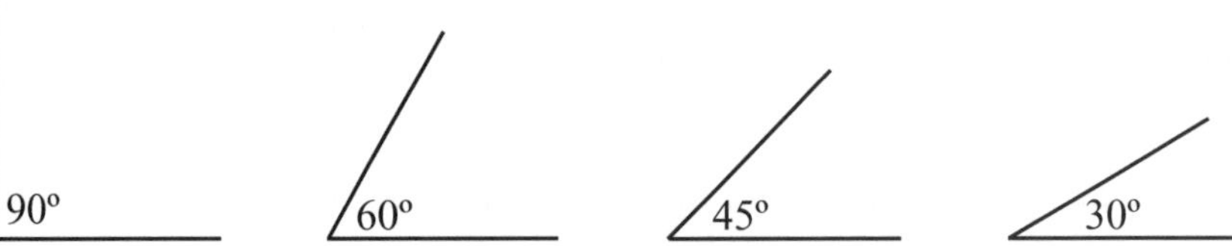

The following figure is drawn to scale. Consider $\angle x$ in $\triangle XYZ$:

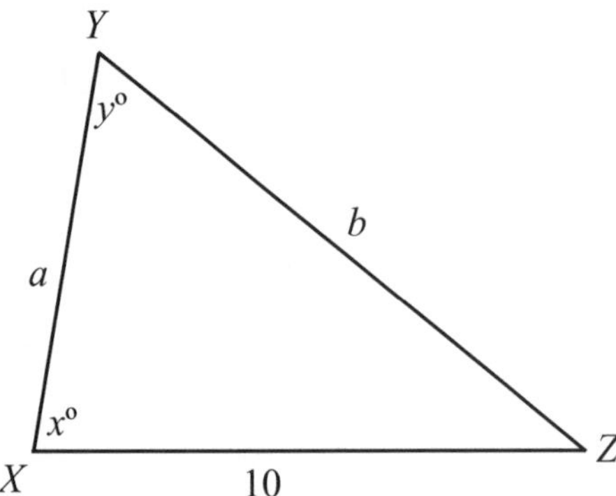

Draw benchmarks around the angle in question. It is somewhere between 60° and 90°:

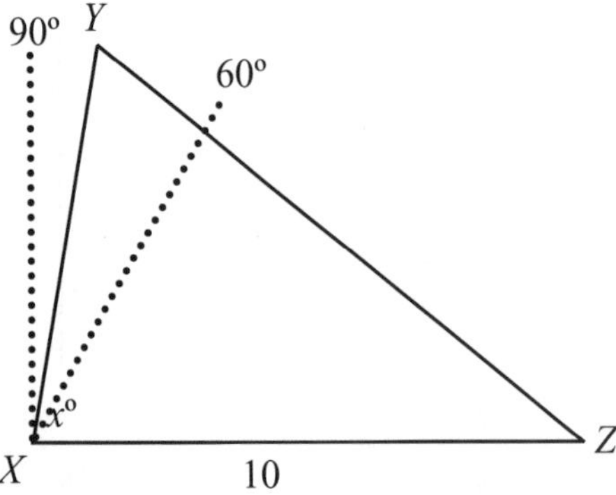

Angle x is nearer to 90° than to 60°. Since the midpoint between 60° and 90° is 75°, $\angle x$ must be greater than 75°. A good estimate is 80°.

Now look at $\angle y$ in $\triangle XYZ$. How can we estimate its measurement? Simply turn your test book so that $\overline{XY}$ becomes the base of the triangle. Then draw in benchmark angles around $\angle y$:

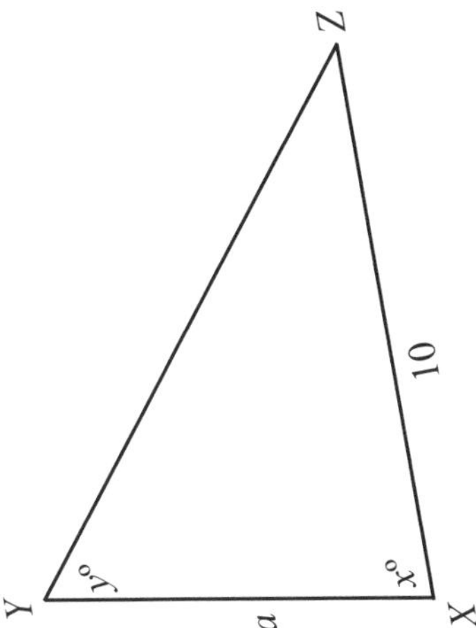

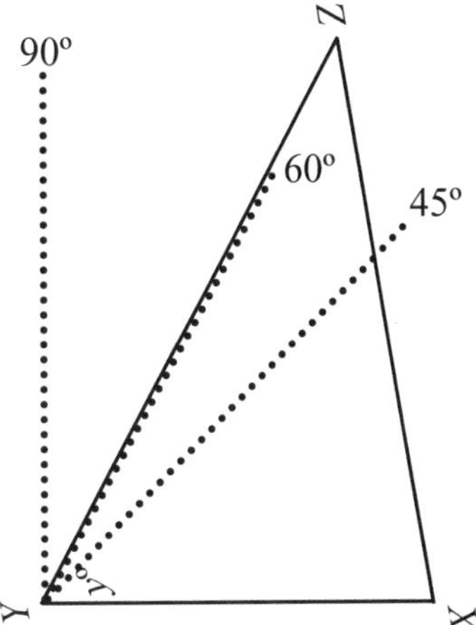

As you can see, angle y is about 60°.

If you are given the length of just one side of a triangle, you can estimate the remaining two side lengths. Return to $\triangle XYZ$ and estimate the length of side a. Imagine that side b was removed, and a fell onto $\overline{XZ}$. Draw the arc that a would make as it falls:

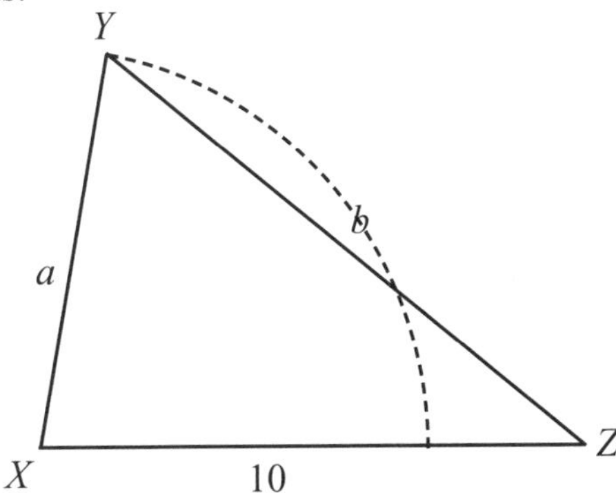

You may wonder why $\overline{XY}$ makes an arc as it falls. Imagine that $\overline{XY}$ is the radius of a circle. When point X is anchored, point Y creates a circle as it turns.

Now you can compare side a to $\overline{XZ}$, which is 10 units. If you need further reference points, divide $\overline{XZ}$ into halves and quarters:

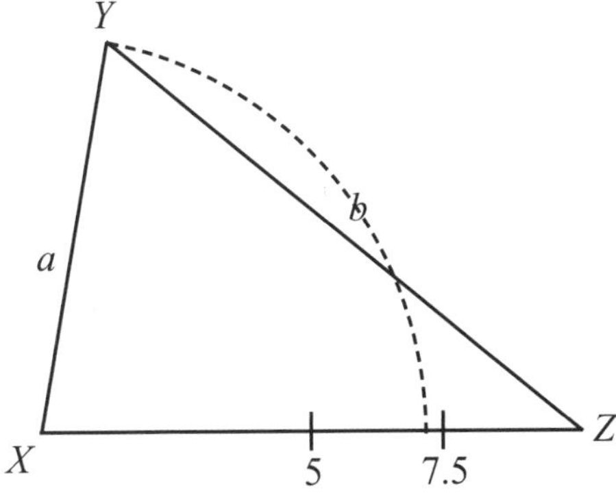

Side a is closer to 7.5 than to 5. A good estimate is 7.

Even though side b is longer than $\overline{XZ}$, you can still make an estimate:

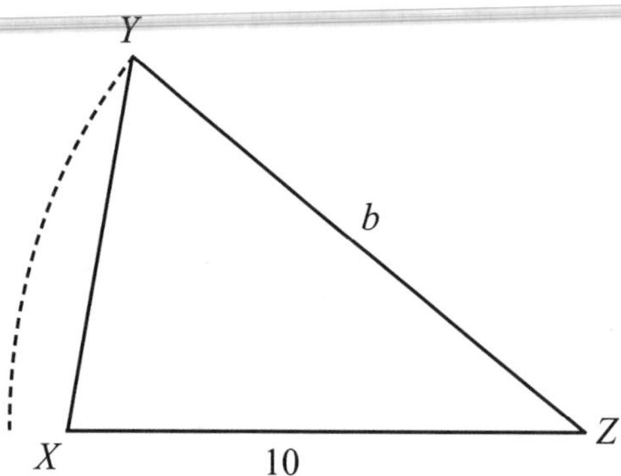

You should only rely on sizing up figures when the figures are drawn to scale.

Side b is between 11 and 12 units. If you need further reference, divide $\overline{XZ}$ using tick marks:

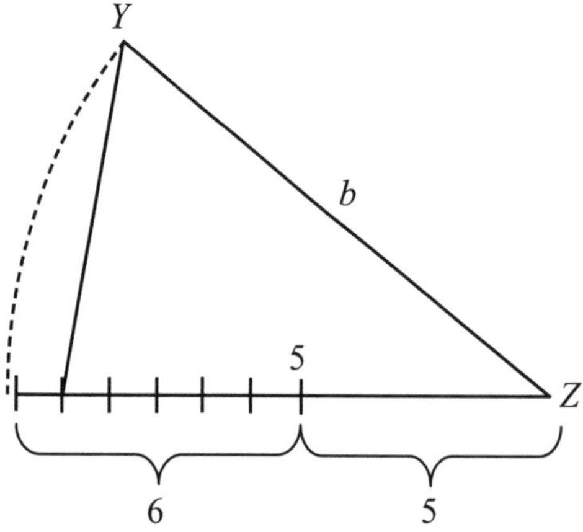

This diagram shows that our estimate should be closer to 11 than to 12. A good estimate is 11 or 11.2, depending on the answer choices given.

You can SIZE UP any diagram or figure—lines, circles, squares, cylinders, and more. Using the techniques demonstrated here, you can estimate any line length or angle measurement. You can also estimate area, given the entire area of a figure or the area of a similar figure. Let's see how it applies to an SAT problem:

Remember, only size up a figure when time is running out or you do not know how to solve the question.

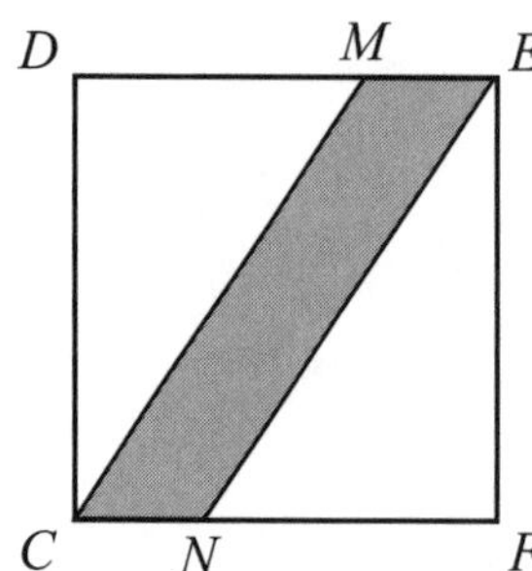

16. In square $CDEF$ to the left, $\overline{ME} = \frac{1}{3}\,\overline{DE}$ and $\overline{CN} = \frac{1}{3}\,\overline{CF}$. If $\overline{CD}$ is 6, what is the area of the shaded region?

(A) 9
(B) 12
(C) 20
(D) 24
(E) 36

The question is drawn to scale (you know this because it does not specifically note that it is not drawn to scale) so you can estimate the area if we do not know how to find it using calculations.

Begin by DIAGRAMMING the question. Then find the area of the whole figure:

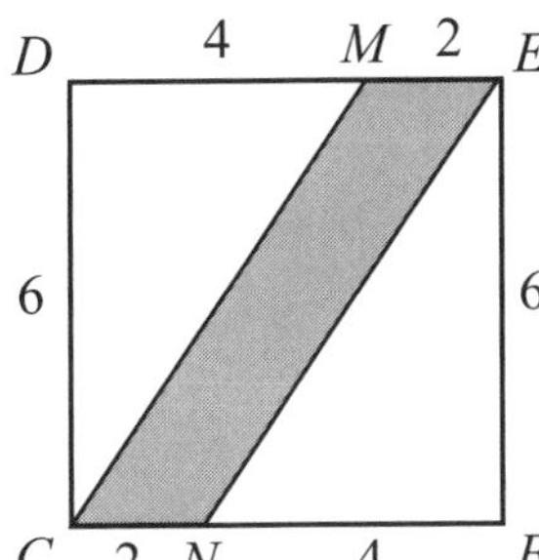

Area of a square = side^2

Area = 6^2
Area = 36

If the area of the whole square is 36, then we can eliminate answer choice (E). We can also eliminate (C) and (D), as the shaded region is smaller than half the square.

What if we juxtaposed the unshaded spaces:

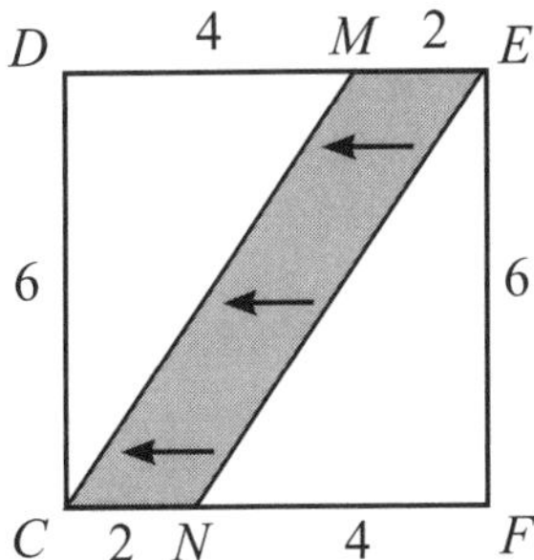

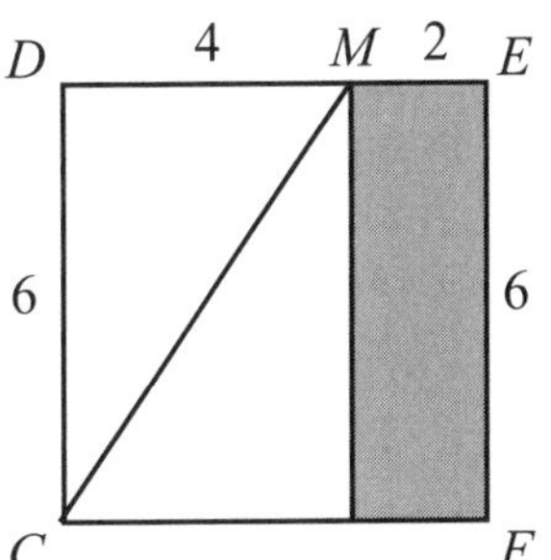

Just as visualizing a great SAT score takes practice, visualizing a rearranged figure may take some practice, too.

If you can visualize this shift of $\triangle EFN$, then you can see that the area of the shaded region is about one-third of the entire area (one-third of 36 = 12). To confirm, you can multiply length times width:

$$6 \times 2 = 12$$

Remember that sizing up figures only works when the figures are drawn to scale. And this method should never be used in place of simply calculating side length or angle measurement if you know how to set up the calculations.

You can also size up figures in the coordinate plane. Look at how estimation can help with the following midpoint problem:

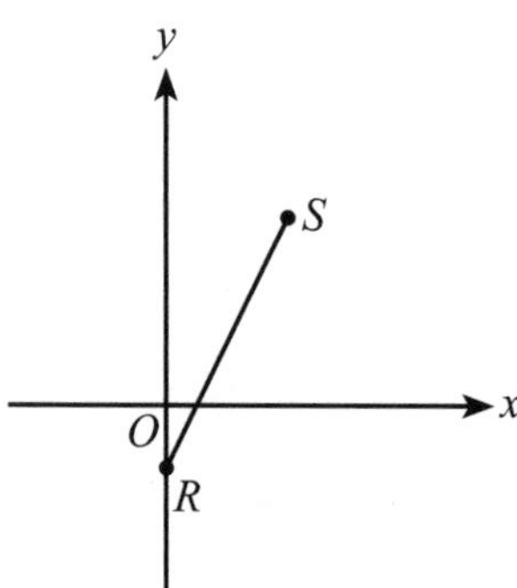

9. In the figure above, line $\overline{RS}$ has endpoints at $R(0, -2)$ and $S(4, 6)$. What is the midpoint of $\overline{RS}$?

(A) (1, 1)
(B) (1, 2)
(C) (2, 1)
(D) (2, 2)
(E) (2, 4)

This question is relatively easy if you know the midpoint formula. But what if you forget it? You can size up the graph and estimate. Begin by adding tick marks that satisfy the coordinates of the exiting points:

The midpoint formula is a much more efficient tool to solve this question. However, if you freeze up and forget formulas on test day, sizing up figures increases your odds of guessing correctly.

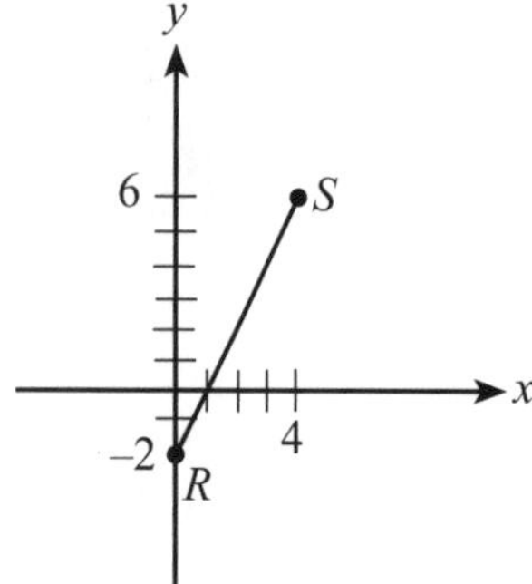

Then estimate the midpoint:

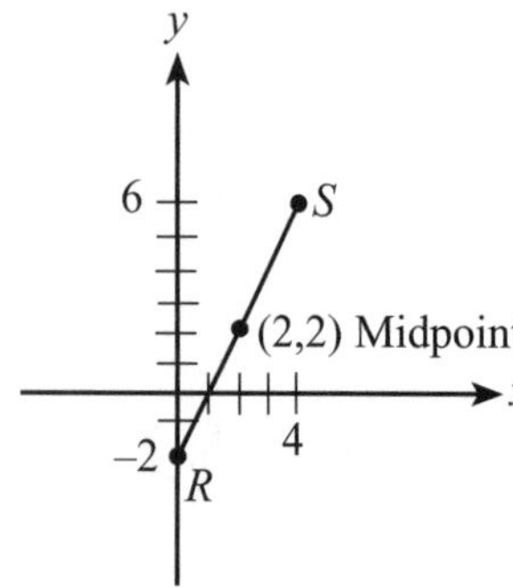

As you can see, by applying estimation techniques when other strategies fail, you can size up a geometry or coordinate geometry figure and make an educated guess.

SIZE UP the Figure Drill

Estimate the answer for each of the following questions. Compare your estimates to those starting on page 116.

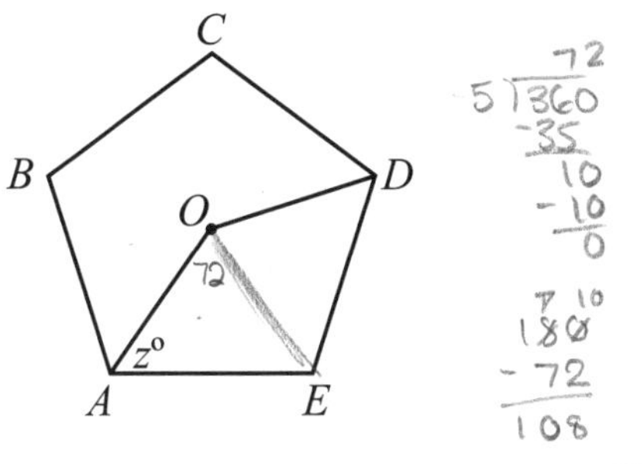

1. In the figure above, O is the center of regular pentagon $ABCDE$. What is the value of z ?

(A) 70
(B) 65
(C) 54
(D) 36
(E) 30

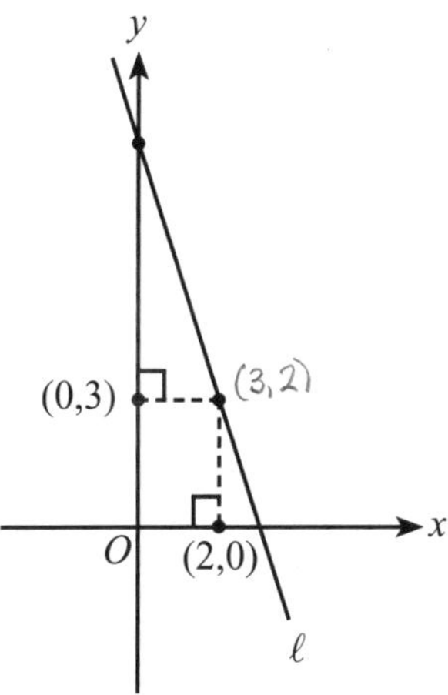

2. In the figure above, line ℓ has a slope of –3. What is the y-intercept of line ℓ ?

(A) 5
(B) 6
(C) 9
(D) 12
(E) 15

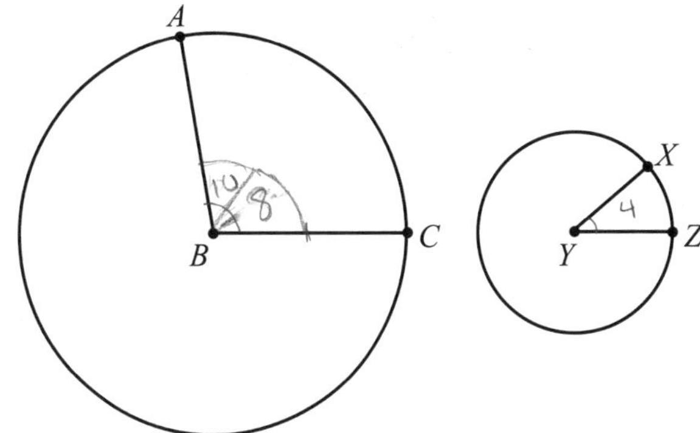

3. In the figure above, the area of sector XYZ is 4. The measure of $\angle ABC$ is two and a half times that of $\angle XYZ$, and the radius of the circle with center B is twice the radius of the circle with center Y. What is the area of sector ABC ?

(A) 40
(B) 20
(C) 10
(D) 5
(E) 2.5

Solution Strategies Summary

Remember, many questions will require a combination of these solution strategies.

You have eight strategies to employ when tackling SAT math questions:

1. ANALYZE the Answer Choices
- Recognize answer choices that include special characters (like *pi*) as a signal for a specific solution
- Eliminate some answers due to their value being too large or too small to fit the nature of the question
- Be wary of the answer choice "It cannot be determined from the information given" as it has never been a correct answer on any test analyzed by PowerScore

2. BACKPLUG the Answer Choices
- Plug answer choices into the variables in the question
- Decide which answer choice makes the most logical starting point

3. SUPPLY Numbers
- Choose numbers to satisfy the rules set up by the question
- Select small, positive integers to start
- Try fractions and negative numbers if more than one answer choice works with your original values
- Supply 100 for percentage problems

4. TRANSLATE from English to Math
- Turn complicated English into simple math by substituting symbols for words
- Know that the word "of" is a common translation meaning "multiply"

5. RECORD What You Know
- Use information from your own life to make a question more personal, or
- Write information from the problem and add relevant formulas

6. SPLIT the Question into Parts
- Tackle some questions, especially functions and geometry problems, in sections
- Beware of questions that ask you to find an expression $(x + 3)$ rather than x

7. DIAGRAM the Question
- Add information to an existing diagram
- Redraw the figure if it is not drawn to scale and is greatly disproportionate
- Draw a figure if a geometry or coordinate geometry question does not have one

8. SIZE UP the Figure
- Estimate side lengths, angle measurements, and area of geometry figures that are drawn to scale
- Estimate line lengths and midpoints for coordinate geometry questions
- Use this strategy only if you cannot solve the question mathematically or if time is running out

Confidence Quotation
"Success comes from within, not from without."
—Ralph Waldo Emerson, philosopher and poet

SOLUTION STRATEGIES ANSWER KEY

ANALYZE the Answer Choices Answer Key—Page 65

1. (E) Medium

Notice that the figure is drawn to scale (we know this because of the absence of the phrase "Figure not drawn to scale"). Therefore, we can estimate the halfway point of $\overline{XZ}$:

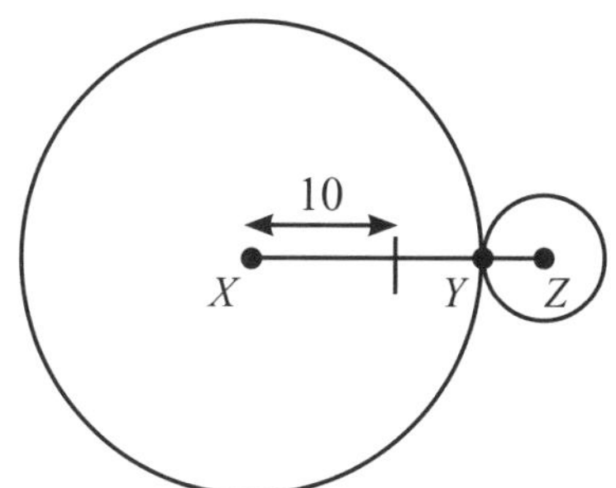

If $\overline{XZ} = 20$, then $\overline{XY}$ must be greater than 10.

Analyze the answer choices. The only answer choice greater than 10 is (E) 16.

$\overline{XY} = 16$

2. (E) Hard

Look at the answer choices. Four of the five answers are in square root form. This most likely indicates that there is a hidden triangle in the problem, as the square roots are the result of the Pythagorean Theorem. Find the triangle in the original figure:

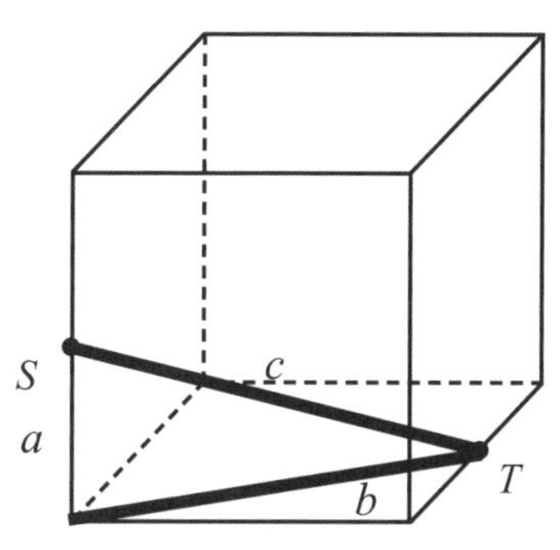

FIGURE 1

The thick lines in Figure 1 represent a right triangle. Side c, the hypotenuse, is the side length for which we are searching. Since each of the edges of this cube are 4 units long, side a is 2. In order to find the length of side b, we need to use another right triangle, as seen in Figure 2: →

S
?
T
2
4

FIGURE 2

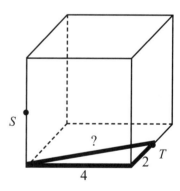

To find the side represented by ? in Figure 2, use the Pythagorean Theorem:

$$a^2 + b^2 = c^2$$
$$4^2 + 2^2 = ?^2$$
$$16 + 4 = ?^2$$
$$20 = ?^2$$
$$\sqrt{20} = ?$$

Since ? = $\sqrt{20}$, then b in Figure 1 = $\sqrt{20}$. Now use the Pythagorean Theorem to find c:

$$a^2 + b^2 = c^2$$
$$2^2 + (\sqrt{20})^2 = c^2$$
$$4 + 20 = c^2$$
$$24 = c^2$$
$$\sqrt{24} = c$$

BACKPLUG the Answer Choices Answer Key—Page 69

1. (B) Medium

Start with the median value answer choice, (C):

(C) 2 $\quad 4^{3x} = 2^{x+5} \rightarrow 4^{3(2)} = 2^{2+5} \rightarrow 4^6 = 2^7 \rightarrow 4096 \neq 128$

We need a smaller value for 4^{3x} in order to make the product smaller:

(B) 1 $\quad 4^{3x} = 2^{x+5} \rightarrow 4^{3(1)} = 2^{1+5} \rightarrow 4^3 = 2^6 \rightarrow 64 = 64$ ✓

2. (C) Medium

Begin by setting up the problem:

$$3\overline{)4x+5}\ \text{R}\,2$$

Then backplug the answer choices in for x. Since it is difficult to determine the nature of the answer choice with remainder problems, you can start with (A):

(A) 1 $\quad 3\overline{)4(1)+5} \rightarrow 3\overline{)4+5} \rightarrow 3\overline{)9} \rightarrow 3\overline{)9}$ gives 3 R 0 ($9 - 9 = 0$)

(B) 2 $\quad 3\overline{)4(2)+5} \rightarrow 3\overline{)8+5} \rightarrow 3\overline{)13} \rightarrow 3\overline{)13}$ gives 4 R 1 ($13 - 12 = 1$)

(C) 3 $\quad 3\overline{)4(3)+5} \rightarrow 3\overline{)12+5} \rightarrow 3\overline{)17} \rightarrow 3\overline{)17}$ gives 5 R 2 ✓ ($17 - 15 = 2$)

3. (B) Medium

Notice that one of the answer choices is 0. Since this is a multiplication problem, and 0 is one of the answer choices, we should suspect the zero property of multiplication as the solution. Start with choice (B) and backplug:

(B) 0 $\quad a = 5b \rightarrow a = 5(0) \rightarrow 0 = 0$

When $b = 0, a = 0 \rightarrow a = b$ ✓

SUPPLY Numbers Answer Key—Page 79

1. (D) Easy

Supply a series of positive and negative integers for t:

$t = -2$	$(t + 4) \div 2$	$\rightarrow$	$(-2 + 4) \div 2$	$\rightarrow$	$2 \div 2 = 1$	✓
$t = -1$	$(t + 4) \div 2$	$\rightarrow$	$(-1 + 4) \div 2$	$\rightarrow$	$3 \div 2 = 1.5$	Not an integer
$t = 2$	$(t + 4) \div 2$	$\rightarrow$	$(2 + 4) \div 2$	$\rightarrow$	$6 \div 2 = 3$	✓
$t = 3$	$(t + 4) \div 2$	$\rightarrow$	$(3 + 4) \div 2$	$\rightarrow$	$7 \div 2 = 3.5$	Not an integer

Then look at the answer choices to find one that MUST be true:

(A) a multiple of 4	When $t = 2$, the result was an integer. However, 2 is not a multiple of 4.
(B) a positive integer	When $t = -2$, the result was an integer. However, –2 is not a positive integer.
(C) a negative integer	When $t = 2$, the result was an integer. However, 2 is not a negative integer.
(D) an even integer	When $t = 2$ or –2, the result was an integer. Both are even integers. ✓
(E) an odd integer	When $t = 3$, the result was not an integer.

2. (B) Medium

Supply numbers for a and b so that the condition ($a^3 < b < 0$) is satisfied:

$a = -2$
$b = -1$

$a^3 < b < 0 \rightarrow -8 < -1 < 0$

Now run your supplied numbers through the answer choices to reveal the one with the least value:

(A) 0	
(B) a^3	$(-2)^3 = -8$
(C) $-a^3$	$-(-2)^3 = -(-8) = 8$
(D) $a^3 - b$	$(-2)^3 - (-1) = -8 + 1 = -7$
(E) $-(a^3 - b)$	$-[(-2)^3 - (-1)] = -(-8 + 1) = -(-7) = 7$

Choice (B) has the least value at –8.

3. (C) Hard

For both supply methods, begin by supplying numbers for the variables:

$d = \$100$
$c = 2$ feeders

Using these supplied numbers, compute basic information:

$100 ÷ 10 pounds = $10 per pound
1 pound is $10 dollars
1 pound fills 2 feeders
$10 ÷ 2 feeders = $5 per feeder

Method #1

Return to your calculations and supply the variables back in for the values:

$100 ÷ 10 pounds = $10 per pound $\qquad d \div 10 = \frac{d}{10}$
1 pound is $10 dollars
1 pound fills 2 feeders
$10 ÷ 2 feeders = $5 per feeder $\qquad \frac{d}{10} \div b \rightarrow \frac{d}{10} \times \frac{1}{b} \rightarrow \frac{d}{10b}$

Method #2

As we learned when we supplied $100 for d and 2 for b, the cost to fill one feeder is $5. Supply these same values into the answer choices to find one that equals $5.

(A) $10bd$ $\quad$ (10)(2)($100) = $2000 $\quad$ No

(B) $\frac{b}{10d}$ $\quad$ 2 ÷ 10($100) → 2 ÷ $1000 → $0.002 $\quad$ No

(C) $\frac{d}{10b}$ $\quad$ $100 ÷ 10(2) → $100 ÷ 20 → $5 ✓

(D) $\frac{10b}{d}$ $\quad$ (10)(2) ÷ $100 → 20 ÷ $100 → $0.20 $\quad$ No

(E) $\frac{10d}{b}$ $\quad$ (10)($100) ÷ 2 → $1000 ÷ 2 → $1000 $\quad$ No

TRANSLATION Mini-Drill Answer Key—Page 82

1) The product of two numbers equals 15. → $xy = 15$

2) When x is increased by one-third of y, the result is 30. → $x + \frac{1}{3}y = 30$

3) 8 less than 4 times a certain number is 10 more than the number. → $4n - 8 = n + 10$

4) The sum of two numbers is 4 and their difference is 0. → $x + y = 4$ and $x - y = 0$

5) What percent of a is 3 less than b? $\quad \frac{x}{100} \times a = b - 3$

6) The difference between s and t is one-half more than twice t. → $s - t = 2t + \frac{1}{2}$

7) If two-thirds of x is increased by one-fourth of y, the result is what percent of 10? → $\frac{2}{3}x + \frac{1}{4}y = \frac{?}{100} \times 10$

8) There are 2 lottery tickets per 100 that are winners. → $\frac{2}{100}$

9) What percent of $x - y$ is 12 less than z? → $\frac{?}{100}(x - y) = z - 12$

TRANSLATE from English to Math Answer Key—Pages 84-85

1. (D) Medium

Translate the two statements from the problem:

The sum of Dan's age and his granddaughter Susan's age is 104 → $D + S = 104$
Dan's age is 8 more years than 3 times Susan's → $D = 3S + 8$

Rework the first equation so that Dan's age is isolated:

$D + S = 104 \rightarrow D = 104 - S$

Now we have two equations that equal Dan's age:

$D = 104 - S$
$D = 8 + 3S$

Set them equal and solve for S, Susan's age:

$104 - S = 8 + 3S \rightarrow 96 = 4S \rightarrow 24 = S$

2. (C) Medium

Translate: 40 percent of s equals 60 percent of t → $0.40s = 0.60t$

Now find t: $\frac{0.40}{0.60}s = t \rightarrow 0.6\overline{6}s = t$ Answer choice (C) is correct.

3. $23/4$ or 5.75 Hard

Translate the statement from the question:

what is the result when $\frac{1}{5}$ of $\frac{5}{4}$ is subtracted from 6 → $x = 6 - \left(\frac{1}{5} \times \frac{5}{4}\right)$

Then solve for x:

$$x = 6 - \left(\frac{1}{5} \times \frac{5}{4}\right) \rightarrow x = 6 - \left(\frac{1}{\cancel{5}} \times \frac{\cancel{5}}{4}\right) \rightarrow x = 6 - \frac{1}{4} \rightarrow 5\frac{3}{4}$$

Remember, you must put mixed-fractions into regular fraction form or decimal form for all grid-in questions:

$$5\frac{3}{4} \rightarrow (5 \times 4) + 3 = \frac{23}{4} \text{ or } 5.75$$

RECORD What You Know Answer Key—Page 89

1. 8 Hard

Start by recording the information provided in the question:

Area = 32

What is the formula for area of a triangle?

$$\text{Area} = \frac{1}{2}bh$$

Set the two parts equal:

$$32 = \frac{1}{2}bh$$

Now use the figure to find the base (b) and height (h). The base is n units. The height is also n units:

$$32 = \frac{1}{2}(n)(n)$$

Now you can solve for n:

$$32 = \frac{1}{2}(n)(n) \quad \rightarrow \quad 32 = \frac{1}{2}n^2 \quad \rightarrow \quad 64 = n^2 \quad \rightarrow \quad 8 = n$$

2. (B) Hard

There are two methods to solve this. The most efficient way is also the most elusive method. It simply requires you to create an equation using the variables in the question:

O = Original weight
10 = pounds gained
j = pounds lost
h = weight now

Original weight + 10 pounds gained – j pounds lost = h weight now

$$O + 10 - j = h$$

Now solve for the original weight ("what was the original weight?"):

$$O + 10 - j = h \quad \rightarrow \quad O = h + j - 10$$

For students who struggle with writing equations, an easier method is to supply their own information.

For example, use your own weight. Say that you originally weighed 150 pounds.

O = Original weight		150 pounds
10 = pounds gained		+10 pounds
j = pounds lost	Supply for j: 40 lbs	–40 pounds
h = weight now		120 pounds

Now, plug your values for h and j into the answer choices to find one that produces an original weight of 150 pounds:

(A) $h - j + 10$	$120 - 40 + 10 = 90$	No
(B) $h + j - 10$	$120 + 40 - 10 = 150$	✓
(C) $h + j + 10$	$120 + 40 + 10 = 170$	No
(D) $j - h - 10$	$40 - 120 - 10 = -90$	No
(E) $j - h + 10$	$40 - 120 + 10 = -70$	No

SPLIT the Question into Parts Answer Key—Page 93

1. (E) Medium

Instead of solving for x, we are searching for $3x - 6$. Still, you must find x first:

Part 1
$f(x) = 3(x^2 - 6)$
$f(x) = 30$

Since both expressions equal $f(x)$, set them equal to each other to solve for x:

$3(x^2 - 6) = 30 \rightarrow 3x^2 - 18 = 30 \rightarrow 3x^2 = 48 \rightarrow x^2 = 16 \rightarrow x = 4$

Notice that 4 is an answer choice. However, the question requires one more step:

Part 2
$3x - 6 \rightarrow 3(4) - 6 \rightarrow 12 - 6 = 6$

Choice (E) is the correct answer.

2. (C) Hard

Begin by solving the function in the question:

Part 1
$g(t) = t^2 - 4$
$g(x - 2) = (x - 2)^2 - 4 \rightarrow (x - 2)(x - 2) - 4 \rightarrow x^2 - 4x + 4 - 4 \rightarrow x^2 - 4x$

Now calculate the answer choice to find the one that equals $x^2 - 4x$:

Part 2	$g(t) = t^2 - 4$
(A) $g(x)$	$g(x) = x^2 - 4$ No
(B) $g(x^2)$	$g(x^2) = (x^2)^2 - 4 \rightarrow x^4 - 4$ No
(C) $g(2 - x)$	$g(2 - x) = (2 - x)^2 - 4 \rightarrow (2 - x)(2 - x) - 4 \rightarrow (4 - 4x + x^2) - 4 \rightarrow x^2 - 4x$ ✓

DIAGRAM the Question Drill Answer Key—Page 99

1.

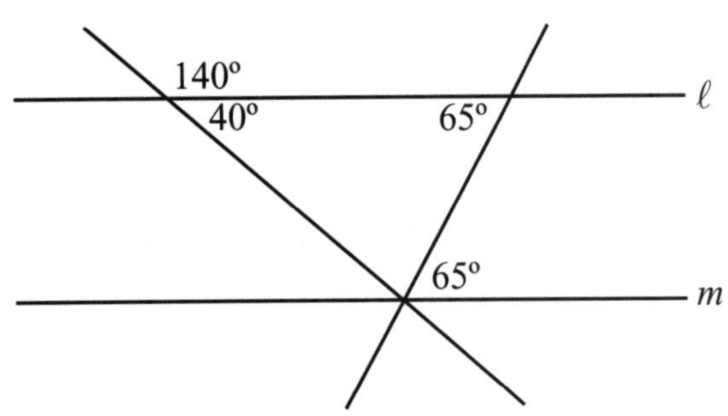

Answer: $x + y = 105$

2. Option 1:

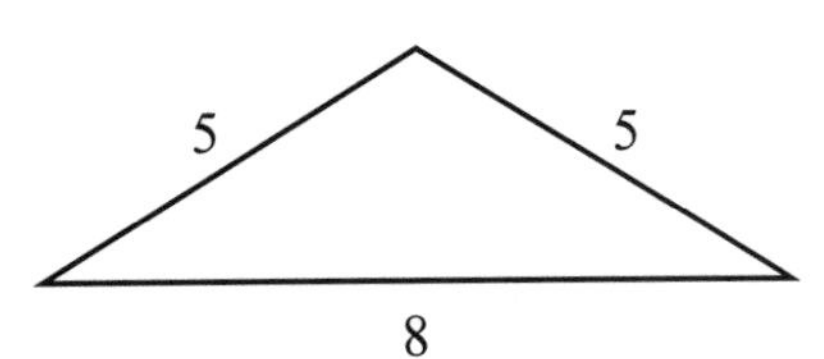

→

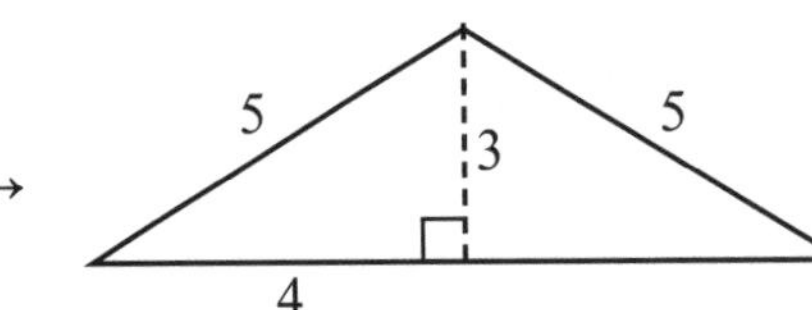

$a^2 + b^2 = c^2$

$a^2 + 4^2 = 5^2$
$a^2 + 16 = 25$
$a^2 = 9$
$a = 3$

Answer: Area = 12

Option 2:

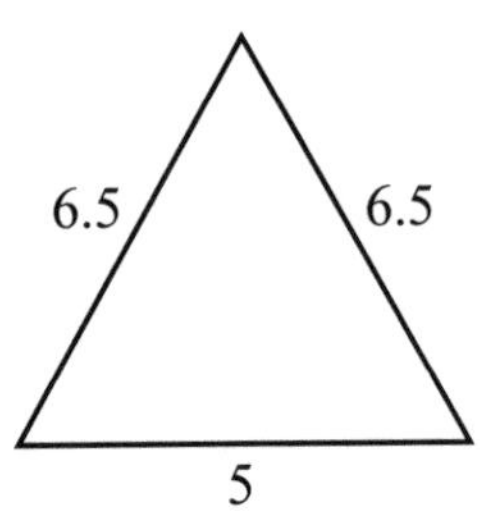

→

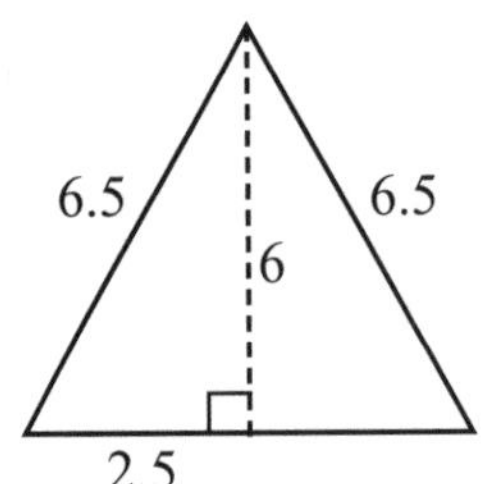

$a^2 + b^2 = c^2$

$a^2 + 2.5^2 = 6.5^2$
$a^2 + 6.25 = 42.25$
$a^2 = 36$
$a = 6$

Answer: Area = 15

3.

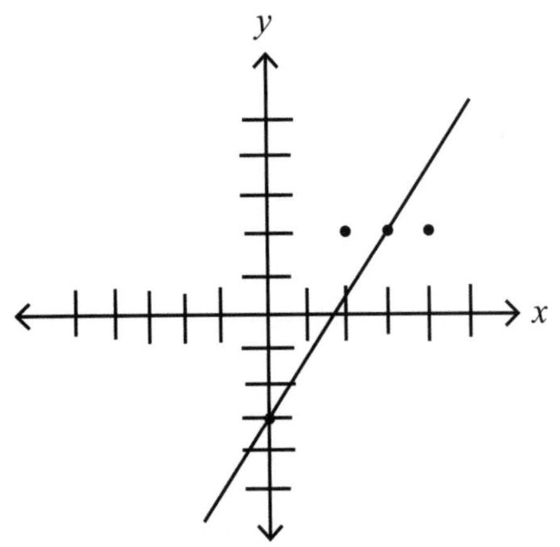

Answer: $\frac{5}{4} < \text{slope} < \frac{5}{2}$

4. Redraw the cube closer to scale:

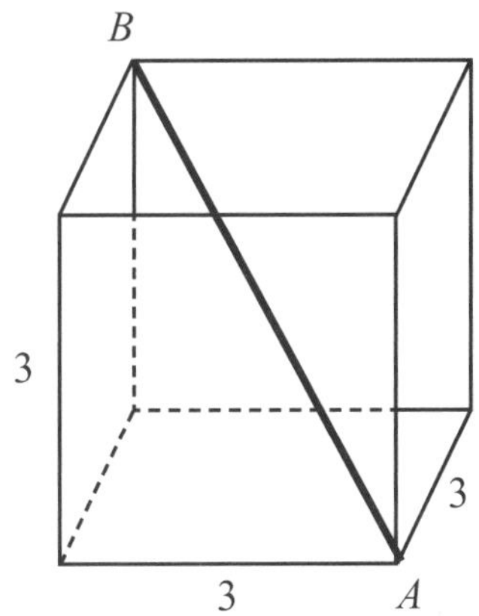

→

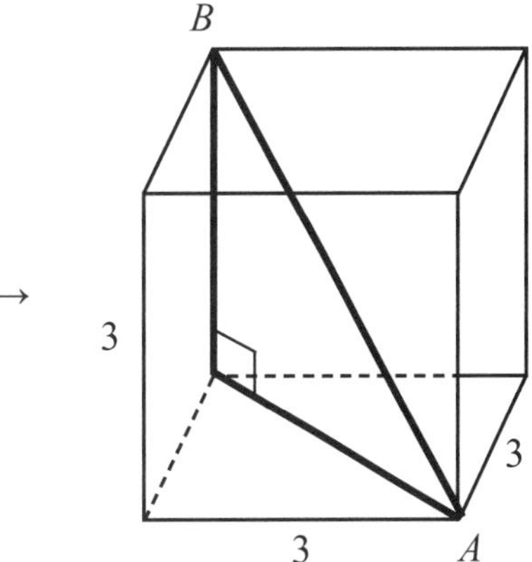

To find the base of the outlined right triangle, you must first find the hypotenuse of the 45:45:90 triangle on the base of the cube:

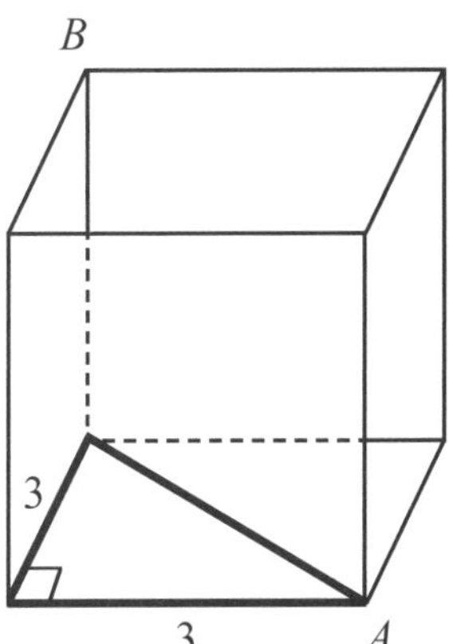

→

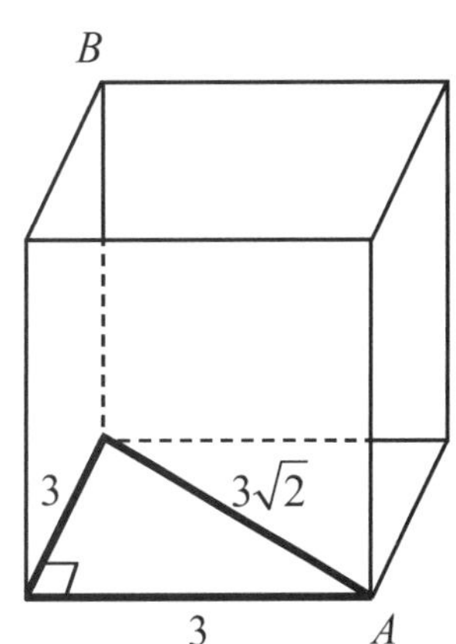

Now turn back to the original right triangle:

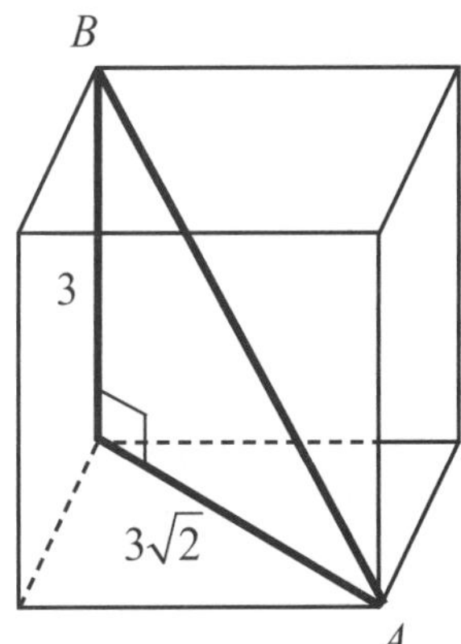

$a^2 + b^2 = c^2$

$3^2 + (3\sqrt{2})^2 = c^2$
$9 + (9)(2) = c^2$
$9 + 18 = c^2$
$27 = c^2$
$\sqrt{27} = c$

Answer = $\sqrt{27}$

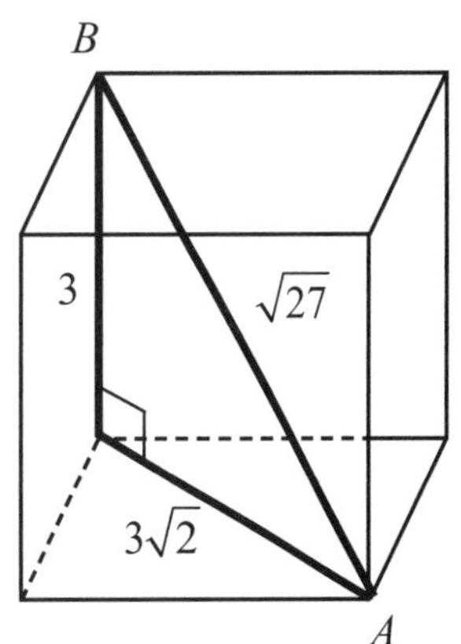

SIZE UP the Figure Drill Answer Key—Page 105

1. (C) Medium

Use benchmark angles to estimate z:

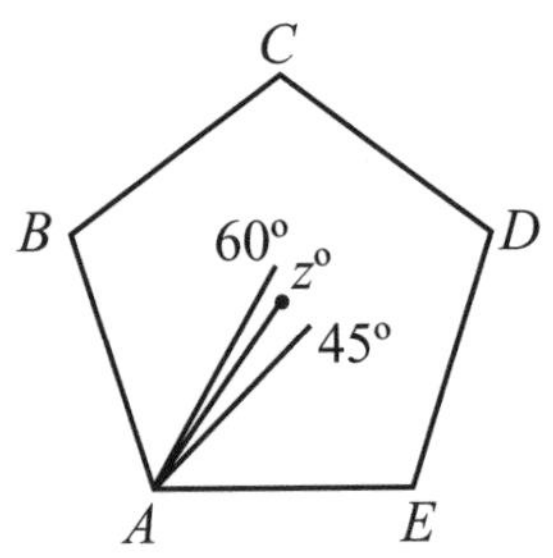

As you can see, z is between 60 and 45 (but closer to 60). Choice (C), 54, is the best estimate.

2. (C) Hard

The first reference point on the y-axis is 3. Use evenly-spaced tick marks to show every three digits on the y-axis:

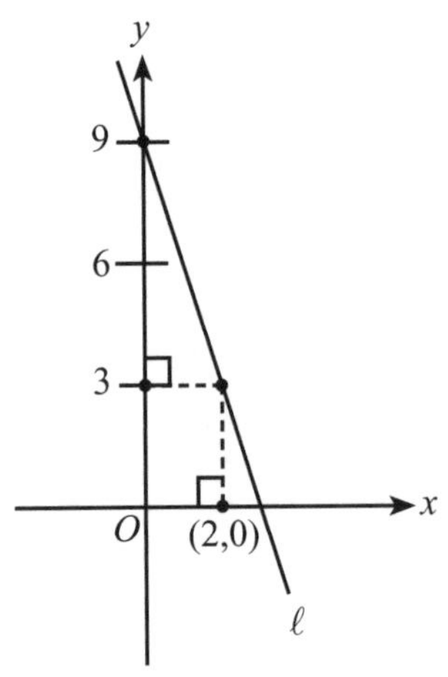

Based on our estimates, the y-intercept is 9.

3. (A) Hard

Imagine placing multiple sectors XYZ into sector ABC:

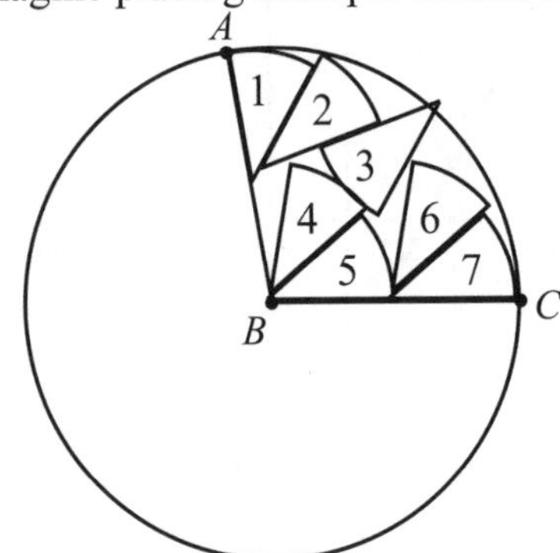

We can place 7 little sectors into the big sector without much overlap. If a single little sector has an area of 4, then the big sector has an area of at least 28 (4×7) = 28. Since there is some unfilled space in the big sector, the area will actually be larger than 28. This eliminates answer choices (B), (C), (D), and (E), as they are all too small. Only choice (A) is greater than 28.

Notes

Chapter Five:
Arithmetic Mastery

Now that you are armed with the eight strategies used to tackle SAT math questions, you must study and understand specific content on the test. Unlike other standardized admissions tests, you do not have to remember a copious number of rules and endless amounts of material for the SAT. The College Board uses a finite number of skills to test your ability to reason critically. Therefore, the SAT math sections can be conquered given intense study and repeated practice. The more exposure you have to real SAT questions, the fewer surprises await you on test day.

Arithmetic, which deals with the computation of numbers, is the most basic branch of mathematics.

As we discussed earlier, much of the test content and many of the required skills are quite basic—such as remainders or fractions. You may find that some of the content review in the following chapters seems elementary and inappropriate to high school students. We urge you to read through it, however, as tips, tricks, and shortcuts are mentioned in the content area discussion. We often advise you to solve problems much differently than the ways you were previously taught, so skipping content can cause you to miss important information.

It is important to read all of the material in this book, no matter how basic. We will show many examples of how basic concepts hold the key to difficult questions.

Each content section (such as Number Properties or Percents) contains two parts: the "Required Knowledge and Skill Set" and "Application on the SAT." The "Required Knowledge and Skill Set" is a review of basic concepts and applications involving the content area. The "Application on the SAT" section examines how this basic content may be presented on the test.

At the beginning of each content area, in the notes column near the edge of the page, is a frequency guide where the content is assigned a number. This number indicates the likelihood that this content will be tested on your SAT. Based on PowerScore's extensive analysis of real tests, you can use the following key to predict the general frequency of each question type:

Rating	Frequency
5	Extremely High: 3 or more questions typically appear on every SAT
4	High: at least 2 questions typically appear on every SAT
3	Moderate: at least one question typically appears on every SAT
2	Low: one question typically appears on every two or three SATs
1	Extremely Low: one question appears infrequently and without a pattern

Frequency Guide: 2

This is an example of the Frequency Guide that appears with each content area. If a content area, such as "Ratios," has a frequency of 2, you can generally expect to see one ratio question on every second or third SAT.

Following each individual content review is a short problem set. An answer key is provided at the end of the chapter, where each question is assigned a degree of difficulty. If you struggle with any Easy or Medium difficulty level questions, reread the content section and then use the map to find additional practice questions in the "Blue Book Database," located on the book owners' website (www.powerscore.com/satmathbible). It is only through repeated practice that you can become confident with a specific question type.

Number Properties

Frequency Guide: 5

The properties of numbers, and more specifically, the properties of integers, are tested on every single SAT.

Required Knowledge and Skill Set

There are a few terms that the College Board assumes all students know:

MEMORY MARKER: Each of these terms are defined in the free flash cards available at www.powerscore.com/satmathbible.

integer: Any number in the set of positive or negative whole numbers and 0 {...–4, –3, –2, –1, 0, 1, 2, 3, 4...}
- Integers do not include fractions or decimals
- Integers are the most commonly used numbers on the SAT
- It is important to remember that 0 is an integer

set: A collection of numbers marked by brackets {4, 6, 9, 13}
- Sets can contain any amount of numbers
- Sets may have rules, such as "all even integers"

digit: The numbers 0 through 9 {0, 1, 2, 3, 4, 5, 6, 7, 8, 9}
- *Place* is used to represent where in a number a digit occurs
- The ones place or units digit in 3748 is 8
- The tens digit in 3748 is 4
- The hundreds digit in 3748 is 7

consecutive: Describes integers following in order without interruption {1, 2, 3, 4}
- Example of a set of consecutive positive *even* integers: {6, 8, 10, 12, 14, 16}
- Example of a set of consecutive negative *odd* integers: {–105, –103, –101, –99, –97}

For questions that use the word "consecutive," it is extremely important to read and reread the question carefully. Test takers often miss the inclusion of the word "even" or "odd," thus leading them to a wrong answer.

sum: The amount obtained by adding numbers
- The sum of 2, 3, and 4 is 9: $(2 + 3 + 4 = 9)$

product: The amount obtained by multiplying numbers
- The product of 2, 3, and 4 is 24: $(2 \times 3 \times 4 = 24)$

multiple: An integer that is divisible by another integer without a remainder
- Multiples of 3 include {–3, 3, 6, 9, 12...}
- Multiples of 4 include {–8, –4, 4, 8, 12, 16...}

divisible: Describes a number capable of being divided without a remainder. A number that is divisible by x is also said to be a multiple of x.
- 18 is divisible by 1, 2, 3, 6, 9, and 18
- xy is divisible by 1, x, y, and xy

factor: One of two or more numbers that divides into a larger number without a remainder
- Factors of 18 are 1 and 18, 2 and 9, and 3 and 6
- Factors of xy include 1 and xy, as well as x and y

prime: An integer that does not have any factors besides itself and 1

- Memorizing the first 10 to 15 prime numbers can save time: {2, 3, 5, 7, 11, 13, 17, 19, 23, 29, 31, 37, 41, 43, 47}
- One (1) is not a prime number
- When two prime numbers are multiplied together, the product's factors are limited to itself, one, and the prime numbers themselves.

←This relatively simple concept is often forgotten and thus used for hard level questions. If 3 and 7 are multiplied, the product's factors will only be 1, 3, 7, and 21. For further exploration, what happens when three prime numbers are multiplied together?

prime factor: Prime number that divides into a larger number without a remainder

- Factors of 18 are 1 and 18, 2 and 9, and 3 and 6; the prime factors are 2 and 3

common factor: Factor shared by two numbers

- Factors of 18 are 1 and 18, 2 and 9, and 3 and 6.
- Factors of 15 are 1 and 15 and 3 and 5.
- The common factors of 15 and 18 are 1 and 3.

Some properties of numbers provide shortcuts on the SAT. There are rules of divisibility that can quickly help you to determine the factors or multiples of larger numbers. PowerScore recommends that you memorize the following rules:

Rules of Divisibility

2	If the last digit of a number is even, it is a multiple of 2. Some numbers that are divisible by 2: 28, 104, 132642
3	If the sum of the digits is divisible by 3, the entire integer is divisible by 3. The number 132 is divisible by 3 because $1 + 3 + 2 = 6$, which is divisible by 3.
4	If the last two digits are a multiple of 4, the entire number is a multiple of 4. Numbers that are divisible by 4: 136, 4664, 17520
5	If the last digit ends in 0 or 5, the entire number is divisible by 5. Numbers that are divisible by 5: 70, 175, 4805, 21680
6	If the number is both divisible by 2 and 3, it is divisible by 6. The number 342 is divisible by both 2 and 3, so it is also divisible by 6.
9	If the sum of the digits is divisible by 9, the entire integer is divisible by 9. The number 2106 is divisible by 9 because $2 + 1 + 0 + 6 = 9$, which is divisible by 9.

MEMORY
The rules c
should be
save time

Divisibility Mini Drill

Place a mark in the box if the number on the left is divisible by the numbers in the top row. For example, the number 372 is divisible by 2. Answers are on page 162.

	2	3	4	5	6	9
372	✓					
5940						
14022						

Memorizing the rules of addition and multiplication of integers can also save you time on the test:

M. RKER:
The involved when ..g or multiplying integers should be memorized for optimal performance.

Addition of Integers

even + even = even	$2 + 4 = 6$
odd + odd = even	$3 + 7 = 10$
odd + even = odd	$5 + 6 = 11$
positive + positive = positive	$2 + 3 = 5$
negative + negative = negative	$-2 + -3 = -5$
positive + negative = can be either	$5 + -4 = 1$ or $3 + -4 = -1$

Multiplication of Integers

even × even = even	$2 \times 4 = 8$
odd × odd = odd	$3 \times 7 = 21$
odd × even = even	$5 \times 6 = 30$
positive × positive = positive	$2 \times 3 = 6$
negative × negative = negative	$-2 \times -3 = 6$
positive × negative = negative	$5 \times -4 = -20$

Application on the SAT

Now let's look at an example of how number properties are applied on the SAT:

If this question is number 11 in a 20-question section, what is the likely level of difficulty?

11. If n is an integer and $\frac{n+8}{2}$ is an integer, which of the following must be true?

    I. n is even

    II. n is a factor of 8

    III. $\frac{n+2}{2}$ is an integer

    (A) I only
    (B) III only
    (C) I and II
    (D) I and III
    (E) I, II, and III

To answer this question, you must be able to define *integer* and *factor*. While it may appear to be an easy question, it is complicated by the Roman numeral set-up and the "must be true" requirement. Each Roman numeral must *always* true, and not just *sometimes* true.

If a condition "must be true," then it is ALWAYS true, given any value for the variable.

Most questions involving number properties can be solved when you SUPPLY numbers. Because this question sets forth two rules, (n is an integer and $\frac{n+8}{2}$ is an integer), you can supply numbers to satisfy the rules. But due to the "must be true" clause, you will need to try several numbers for n in order to prove or disprove that a particular case is *always* true.

Start with Roman numeral I, n is even. Try some odd and even numbers:

If $n = 2$, then $\frac{2+8}{2} \rightarrow \frac{10}{2} \rightarrow 5$ ✓

If $n = 3$, then $\frac{3+8}{2} \rightarrow \frac{11}{2} \rightarrow 5.5$ (5.5 is not an integer)

If $n = 4$, then $\frac{4+8}{2} \rightarrow \frac{12}{2} \rightarrow 6$ ✓

If $n = 7$, then $\frac{7+8}{2} \rightarrow \frac{15}{2} \rightarrow 7.5$ (7.5 is not an integer)

As you can see, a pattern is developing. Only even numbers create integers. Therefore, the first Roman numeral is true. This eliminates answer choice (B).

When you eliminate an answer, use your pencil to cross out the corresponding letter in the test booklet. This helps ensure accuracy and makes guessing easier, if needed.

Now attempt to prove or disprove Roman numeral II, n is a factor of 8. In our previous calculations, when $n = 2$ and $n = 4$, we created integers, and 2 and 4 are factors of 8. But to see if n must *always* be a factor of 8, try some even numbers that are not factors of 8:

If $n = 10$, then $\frac{10+8}{2} \rightarrow \frac{28}{2} \rightarrow 14$ ✓

Since 10 is not a factor of 8, and we created an integer, than n is only *sometimes* a factor of 8. Roman numeral II is not true; this eliminates choices (C) and (E).

Now cross out (C) and (E) in your test booklet.

We have narrowed the answers to (A) and (D). Examine the last Roman numeral to find the correct answer.

So far we have found three possible values for n (2, 4, and 10). Try them in the expression in Roman numeral III ($\frac{n+2}{2}$):

If $n = 2$, then $\frac{2+2}{2} \rightarrow \frac{4}{2} \rightarrow 2$ ✓

If $n = 4$, then $\frac{4+2}{2} \rightarrow \frac{6}{2} \rightarrow 3$ ✓

If $n = 10$, then $\frac{10+2}{2} \rightarrow \frac{12}{2} \rightarrow 6$ ✓

It appears that all even numbers will make the statement work in Roman numeral III so it is also true. Therefore, the answer is (D).

Number Properties Problem Set

Solve the following multiple-choice questions by selecting the best answer from the five answer choices. For grid-in questions, write your answer in the grids and completely mark the corresponding ovals. Answers begin on page 162.

1. If x and y are positive integers and $x + 3y = 19$, what is the sum of all possible values of y?

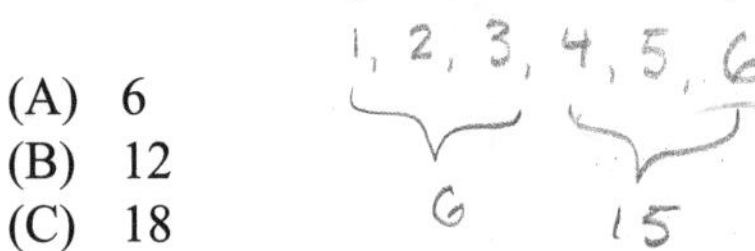

(A) 6
(B) 12
(C) 18
(D) 20
(E) 21

2. If w, x, y, and z are consecutive even integers, and if $w < x < y < z$, then $w + x$ is how much less than $y + z$?

(A) 4
(B) 5
(C) 6
(D) 8
(E) 10

3. If X is the set of two-digit prime numbers and Y is the set of two-digit numbers whose units digit is 2, how many numbers are common to both sets?

(A) None
(B) One
(C) Four
(D) Seven
(E) Fourteen

Number Properties Problem Set

Solve the following multiple-choice questions by selecting the best answer from the five answer choices. For grid-in questions, write your answer in the grids and completely mark the corresponding ovals. Answers begin on page 162.

4. If the sum of two numbers is 1 and their difference is $\frac{1}{2}$, what is their product?

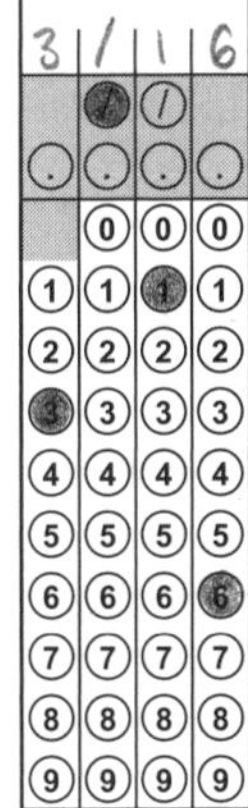

5. The numbers *a, b,* and *c* are consecutive integers and $0 < a < b < c$. If the units digit of ac is 4, what is the units digit of b?

(A) 4
(B) 5
(C) 6
(D) 7
(E) 8

Remainders

Frequency Guide: 2

Remainders are the best example of the basic content that is tested on the SAT. The concept seems too simple to be included on a college admissions exam. After all, remainders were introduced in first grade and concluded by fourth grade. But it is this long recess from elementary school to high school that makes remainders so difficult for the typical teenage student. Trained to find decimals to the nearest thousandth using a high-tech graphing calculator, a high school math student is out of practice with hand-written long division. Plus, the College Board makes the problems more difficult by substituting variables for values in the dividend and divisor. For these reasons, questions involving remainders usually have a Medium-Hard or Hard level of difficulty.

Required Knowledge and Skill Set

To begin your study of remainders, you must review the basics. Look at the following division problem:

$$4\overline{)7}$$

Four is the *divisor*, and seven is the *dividend*. Four goes into seven one time with three left over:

$$\begin{array}{r} 1\text{ R }3 \\ 4\overline{)7} \\ -4 \\ \hline 3 \end{array}$$

Using elementary school notation to set up remainder questions can make the solution more clear. Using fraction notation or plugging the problem into your calculator can cause confusion.

It is important to use this "old school" long division style when setting up an SAT question involving a remainder.

Application on the SAT

Consider an example of how remainders are applied on the exam:

17. If the integer p is divided by 7, the remainder is 5. What is the remainder if $3p$ is divided by 7?

(A) 0
(B) 1
(C) 2
(D) 3
(E) 4

To begin, set up the problem using elementary notation:

$$\begin{array}{l} \;\;\,?\text{ R }5 \\ 7\overline{)p} \end{array}$$

Now comes the critical reasoning part. You can SUPPLY numbers for p until you find one that creates a remainder of 5 when plugged into the long division problem, or you can think about the nature of the problem:

$$\begin{array}{r} ? \\ 7\overline{)p} \\ \underline{-7} \\ 5 \end{array}$$

If $p - 7 = 5$, then $p = 12$. This is one possible value of p. There are others: p could also be 19, 26, 33, and more. The value of p can be 5 plus any multiple of 7:

Remainder + Multiple of Divisor = Dividend

5	+	7	=	12
5	+	14	=	19
5	+	21	=	26
5	+	28	=	33

This formula creates an efficient solution for questions that require you to find the dividend, especially when the dividend is a higher multiple.

While you have found four possible values for p, you only need one. The smallest possible value, 12, is the easiest to manipulate.

Now, using $p = 12$, find the remainder when $3p$ is divided by 7:

If $p = 12$, then $3p = 36$

$$\begin{array}{r} 5 \text{ R } 1 \\ 7\overline{)36} \\ \underline{-35} \\ 1 \end{array}$$

The remainder is 1. This is true for any value of p. Test $p = 19$ or $p = 26$ to prove the remainder is always 1.

Successfully solving remainder questions is as much about carefully reading the problem as it is about understanding remainders. By setting up the problem using long division notation, you are more clearly able see what the question is asking.

ARITHMETRICK

This formula will work when the variable is the sole dividend.

When more than one number can be supplied for an unknown, choose the least value, as it makes the calculations easier.

Remainders Problem Set

Solve the following multiple-choice questions by selecting the best answer from the five answer choices. For grid-in questions, write your answer in the grids and completely mark the corresponding ovals. Answers begin on page 164.

1. If the integer s is divided by 9, the remainder is 1. If s is divided by 4, the remainder is 2. If s is between 30 and 90, what is one possible value of s ?

2. When $7x + 3$ is divided by 6, the remainder is 2. If x is an integer, then x could be

    (A) 4
    (B) 5
    (C) 6
    (D) 7
    (E) 8

Fractions and Decimals

Frequency Guide: 4

Fractions are tested on nearly every SAT administration. To make operations with fractions more complicated, the College Board uses fractions with variables; these questions cannot be solved using a calculator.

Decimals are not tested as often. You can expect to see them, however, in any word problems involving money.

Required Knowledge and Skill Set

There are several key points about fractions and decimals that are covered on the SAT:

1. You must know the terms *numerator* and the *denominator*, and where each of these appear in a fraction:

$$\frac{3}{4} \quad \begin{matrix} \longleftarrow \text{Numerator} \\ \longleftarrow \text{Denominator} \end{matrix}$$

2. Having a knowledge of basic fraction and decimal equivalents can save valuable time when estimating. You should memorize the following equivalents:

$$\frac{1}{8} = 0.125 \qquad \frac{1}{6} = 0.16\overline{6} \qquad \frac{1}{5} = 0.2$$

$$\frac{1}{4} = 0.25 \qquad \frac{1}{3} = 0.3\overline{3} \qquad \frac{1}{2} = 0.5$$

$$\frac{2}{5} = 0.4 \qquad \frac{2}{3} = 0.6\overline{6} \qquad \frac{3}{4} = 0.75$$

MEMORY
These equi
included in
flash card
powerscore.com/
satmathbible.

3. You may be required to reduce or factor fractions:

$$\frac{6t}{16} \rightarrow \frac{\cancel{6}t^{\div 2}}{\cancel{16}_{\div 2}} \rightarrow \frac{3t}{8}$$

$$\frac{2x-4}{12} \rightarrow \left(\frac{2}{2}\right)\left(\frac{x-2}{6}\right) \rightarrow (1)\left(\frac{x-2}{6}\right) \rightarrow \frac{x-2}{6}$$

$$\frac{w^3+w^2}{3w^2} \rightarrow \left(\frac{w^2}{w^2}\right)\left(\frac{w+1}{3}\right) \rightarrow (1)\left(\frac{w+1}{3}\right) \rightarrow \frac{w+1}{3}$$

CAUTION: SAT TRAP!
If you factor out an entire quantity, a 1 must be left in the quantity's place, as in the third example to the left. Many students forget to leave 1 in place of w^2. This factoring situation is a common SAT trap.

3. You must be able to manually add, subtract, multiply, and divide fractions.

To add fractions, find a common denominator:

$$\frac{z}{6}+\frac{3}{4} \rightarrow \frac{z^{\times 2}}{6_{\times 2}}+\frac{3^{\times 3}}{4_{\times 3}} \rightarrow \frac{2z}{12}+\frac{9}{12} \rightarrow \frac{2z+9}{12}$$

Confidence Quotation
"Once you replace negative thoughts with positive ones, you'll start having positive results."
—Willie Nelson, country music artist

To subtract fractions, find a common denominator:

$$\frac{2}{5}-\frac{4a}{7} \rightarrow \frac{2^{\times 7}}{5_{\times 7}}-\frac{4a^{\times 5}}{7_{\times 5}} \rightarrow \frac{14}{35}-\frac{20a}{35} \rightarrow \frac{14-20a}{35}$$

To add or subtract multiple fractions, you still must find a common denominator:

$$\frac{x}{4}+\frac{2x}{3}+\frac{3x}{8} \rightarrow \frac{x^{\times 6}}{4_{\times 6}}+\frac{2x^{\times 8}}{3_{\times 8}}+\frac{3x^{\times 3}}{8_{\times 3}} \rightarrow \frac{6x}{24}+\frac{16x}{24}+\frac{9x}{24} \rightarrow \frac{31x}{24}$$

To multiply fractions, multiply the numerators to create a new numerator. Then multiply the denominators to create a new denominator:

$$\frac{3}{5}\times\frac{4x}{9} \rightarrow \frac{3\times 4x}{5\times 9} \rightarrow \frac{12x}{45}$$

To divide fractions, multiply the dividend (the fraction being divided) by the reciprocal of the divisor (the fraction that is dividing the dividend):

$$\frac{2y}{9}\div\frac{3}{8} \rightarrow \frac{2y}{9}\times\frac{8}{3} \rightarrow \frac{2y\times 8}{9\times 3} \rightarrow \frac{16y}{27}$$

Remember to flip any fraction that is a divisor!

If you divide by more than one fraction, be sure to multiply by the reciprocal of all divisors:

$$\frac{2y}{9}\div\frac{3}{8}\div\frac{2}{y} \rightarrow \frac{2y}{9}\times\frac{8}{3}\times\frac{y}{2} \rightarrow \frac{2y\times 8\times y}{9\times 3\times 2} \rightarrow \frac{16y^2}{54}$$

4. Expect to square and cube fractions, and to find the square root and cube root of fractions.

To square or cube a fraction, simply multiply the fraction by itself the number of times indicated by the exponent:

$$\left(\frac{2}{3}\right)^2 \rightarrow \frac{2}{3}\times\frac{2}{3} \rightarrow \frac{4}{9}$$

$$\left(\frac{3}{5}\right)^3 \rightarrow \frac{3}{5}\times\frac{3}{5}\times\frac{3}{5} \rightarrow \frac{27}{125}$$

To find the square root or cube root of a fraction, you must find the square or cube root of both the numerator and denominator:

$$\sqrt{\left(\frac{1}{4}\right)} \rightarrow \frac{\sqrt{1}}{\sqrt{4}} \rightarrow \frac{1}{2}$$

$$\sqrt[3]{\left(\frac{8}{27}\right)} \rightarrow \frac{\sqrt[3]{8}}{\sqrt[3]{27}} \rightarrow \frac{2}{3}$$

If you are required to find the square or cube root of a fraction, it will likely be in an equation where the requested variable is squared or cubed. For example, in the following equation, find x:

$$3x^2 = \frac{25}{12} \rightarrow x^2 = \frac{25}{12} \div 3 \rightarrow x^2 = \frac{25}{12} \times \frac{1}{3} \rightarrow x^2 = \frac{25}{36} \rightarrow$$

$$\sqrt{x} = \sqrt{\frac{25}{36}} \rightarrow \sqrt{x} = \frac{\sqrt{25}}{\sqrt{36}} \rightarrow x = \frac{5}{6}$$

When solving an SAT equation with a square or cube root, isolate the root. Then square or cube both sides of the equation. This skill is tested frequently on the SAT.

5. Nearly every SAT will assess your ability to manipulate complex fractions. Complex fractions contain a fraction in the numerator, denominator, or both:

$$\frac{\frac{1}{5}}{n} \qquad \frac{4}{\frac{x}{8}} \qquad \frac{\frac{a}{3}}{\frac{b}{4}}$$

To simplify complex fractions, treat them as division problems:

$$\frac{\frac{1}{5}}{n} \rightarrow \frac{1}{5} \div \frac{n}{1} \rightarrow \frac{1}{5} \times \frac{1}{n} \rightarrow \frac{1}{5n}$$

$$\frac{4}{\frac{x}{8}} \rightarrow \frac{4}{1} \div \frac{x}{8} \rightarrow \frac{4}{1} \times \frac{8}{x} \rightarrow \frac{32}{x}$$

$$\frac{\frac{a}{3}}{\frac{b}{4}} \rightarrow \frac{a}{3} \div \frac{b}{4} \rightarrow \frac{a}{3} \times \frac{4}{b} \rightarrow \frac{4a}{3b}$$

For more practice manipulating fractions, see Chapter 3.

Application on the SAT

Some fraction questions are straightforward mathematical sentences that test your ability to perform operations on fractions. For example, study the following grid-in question:

This question is easy to solve with your calculator, but could you manually solve it if one of the integers was replaced with a variable? You must be able to manipulate fractions!

14. If $\frac{1}{3}$ of $\frac{3}{4}$ is subtracted from 2, what is the resulting value?

Many questions involving fractions can be solved by TRANSLATING from English to Math. This question is a good example:

$\frac{1}{3}$ of $\frac{3}{4}$ is subtracted from 2

$$2-\left(\frac{1}{3}\times\frac{3}{4}\right)=?$$

This question requires multiplication and subtraction of fractions, following the order of operations:

Use the abbreviation PEMDAS to remember the order of operations: Parenthesis, Exponents, Multiply and Divide, and Add and Subtract.

$$\frac{1}{3}\times\frac{3}{4} \rightarrow \frac{1\times 3}{3\times 4} \rightarrow \frac{3}{12} \rightarrow \frac{1}{4}$$

$$2-\frac{1}{4} \rightarrow \frac{2}{1}-\frac{1}{4} \rightarrow \frac{2^{\times 4}}{\cancel{1}_{\times 4}}-\frac{1}{4} \rightarrow \frac{8}{4}-\frac{1}{4} \rightarrow \frac{7}{4} \text{ or } 1.85$$

Results of multiple choice questions are usually in fraction form, but occasionally occur in decimal form. You should be prepared for either type of answer.

Other fraction questions are more complicated. The SAT often uses word problems with fractions, and these word problems provide a combination of fractional parts and whole parts to complete a total:

15. In a jar of jelly beans, $\frac{1}{6}$ of the jelly beans are green, $\frac{1}{5}$ of the jelly beans are purple, $\frac{1}{3}$ of the jelly beans are red, and the remaining 18 jelly beans are yellow. How many total jelly beans are in the jar?

(A) 21
(B) 30
(C) 54
(D) 60
(E) 90

Clearly Green + Purple + Red + Yellow = Total. However, the first three colors of jelly beans are presented as fractions of the total. The fourth color is given as a whole portion of the total. This can make the solution seem complicated for many students.

The question can be solved several ways, but the easiest and most efficient method is to write an equation that accommodates both the fractional and whole parts. You may want to create an English sentence first :

Green	+	Purple	+	Red	+ Yellow = Total
$\frac{1}{6}$ of the total	+	$\frac{1}{5}$ of the total	+	$\frac{1}{3}$ of the total	+ 18 = total

Now TRANSLATE from English to Math. Since you are searching for the total number of jelly beans in the jar, call the total t:

$$\frac{1}{6}t + \frac{1}{5}t + \frac{1}{3}t + 18 = t$$

Subtract 18 from both sides to isolate the fractions:

$$\frac{1}{6}t + \frac{1}{5}t + \frac{1}{3}t = t - 18$$

Find a common denominator in order to add the fractions. Then solve:

$$\frac{5}{30}t + \frac{6}{30}t + \frac{10}{30}t = t - 18 \quad \rightarrow \quad \frac{21t}{30} = t - 18 \quad \rightarrow \quad 21t = (t - 18)30 \quad \rightarrow$$

$$21t = 30t - 540 \quad \rightarrow \quad -9t = -540 \quad \rightarrow \quad t = 60 \qquad \text{Answer choice (D)}$$

If writing equations intimidates you, or if you have a hard time setting up the equation, you can also use critical reasoning to solve this question. Think about the equation if all four colors were fractional parts:

$$G + P + R + Y = \text{whole jar}$$

$$\frac{1}{6} + \frac{1}{5} + \frac{1}{3} + \frac{?}{?} = 1 \text{ whole jar}$$

We know that the common denominator for the first three fractions is 30. Make this the denominator of the fourth fraction as well:

$$\frac{5}{30} + \frac{6}{30} + \frac{10}{30} + \frac{?}{30} = 1 \text{ whole jar}$$

If $\frac{30}{30} = 1$, what must the numerator of the last fraction be?

$$\frac{21 + ?}{30} = 1 \quad \rightarrow \quad ? = 9$$

The yellow jelly beans make up $\frac{9}{30}$ of the whole jar. Using this fraction and the actual number of yellow jelly beans, TRANSLATE to find the total:

$$\frac{9}{30} \text{ of the total is 18 (yellow)} \quad \rightarrow \quad \frac{9}{30}t = 18 \quad \rightarrow \quad 9t = 540 \quad \rightarrow \quad t = 60$$

Expect to see at least two or three fraction questions on your SAT.

All students learn differently and with different parts of their brain. It's why we present two solutions here and elsewhere throughout this book, to make sure that all students understand the question. It's usually more efficient to use the first solution presented. However, if you do not understand the first solution, it is completely acceptable to rely on the second.

Fractions and Decimals Problem Set

Solve the following multiple-choice questions by selecting the best answer from the five answer choices. For grid-in questions, write your answer in the grids and completely mark the corresponding ovals. Answers begin on page 166.

1. If $\frac{4}{9}$ of x is 36, what is $\frac{7}{9}$ of x?

   (A) 28
   (B) 30
   (C) 42
   (D) 63
   (E) 81

2. If $x = \frac{2}{3}$, what is the value of $\frac{4}{x} - \frac{x}{4}$?

3. The school store ordered a box of t-shirts. By Tuesday, $\frac{1}{5}$ of the t-shirts were sold. On Wednesday, 20 more t-shirts were sold, raising sales to $\frac{1}{3}$ of the original number of t-shirts in the box. How many t-shirts were originally in the box?

   (A) 20
   (B) 80
   (C) 150
   (D) 180
   (E) 200

Fractions and Decimals Problem Set

Solve the following multiple-choice questions by selecting the best answer from the five answer choices. For grid-in questions, write your answer in the grids and completely mark the corresponding ovals. Answers begin on page 166.

4. If r is $\frac{1}{4}$ of s and s is $\frac{2}{3}$ of t, what is the value of $\frac{r}{t}$?

   (A) $\frac{1}{6}$

   (B) $\frac{1}{4}$

   (C) $\frac{1}{3}$

   (D) $\frac{2}{5}$

   (E) $\frac{3}{7}$

5. If the positive difference between t and $\frac{1}{6}$ is the same as the positive difference between $\frac{2}{3}$ and $\frac{1}{4}$, which of the following could be the value of t ?

   (A) $\frac{1}{3}$

   (B) $\frac{5}{12}$

   (C) $\frac{1}{2}$

   (D) $\frac{7}{12}$

   (E) $\frac{13}{12}$

6. A pizza is being divided by 2 girls and 1 boy, each of whom receive an equal share. If one of the girls gives $\frac{1}{3}$ of her share to the boy, and the other girl keeps $\frac{3}{4}$ of her share but gives the rest to the boy, what fraction of the pizza will the boy have?

   (A) $\frac{1}{3}$

   (B) $\frac{19}{36}$

   (C) $\frac{7}{12}$

   (D) $\frac{2}{3}$

   (E) $\frac{25}{36}$

Percentages

Frequency Guide: 3

Percentages also appear often on the SAT. Because of their frequent use with data, they are popular in SAT word problems.

Required Knowledge and Skill Set

1. Decimals can be expressed as fractions or decimals:

   $$19\% = \frac{19}{100} = 0.19$$

   While the answer choices may all be in percentage form, some students choose to convert to fractions or decimals, depending on their comfort level with manipulating percents. They must convert their final answer back to a percentage.

2. The most common solution strategy for questions involving percents is to TRANSLATE from English to Math:

   What is 35% of x ?

   $? = 0.35 \times x$

   TRANSLATION can help you solve almost every SAT percentage question. However, the College Board knows that most students have a difficult time translating the phrase "what percent." The phrase "what percent" should always be expressed as $\frac{x}{100}$, $\frac{?}{100}$, or other variable over 100:

   MEMORY MARKER: You must know that "what percent" translates into the unknown variable over 100.

   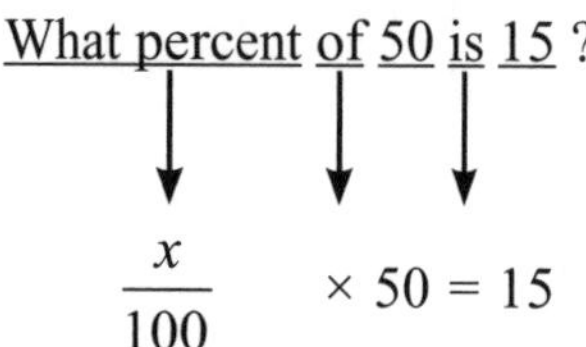

   Failure to place the variable over 100 will lead to the wrong answer, especially when the question uses several variables in place of values. For this reason, "what percent" questions are almost always Hard difficulty level, even though they are relatively easy to solve using proper translation.

3. If you choose to SUPPLY numbers to solve a percentage question, use the number 100. This is a good strategy when percentages are used to calculate discounts on merchandise. It may seem absurd to assign $100 to the price of a candy bar, but since percentages are based out of 100, the use of 100 as a supplied number makes for easy calculations.

   Use 100 when supplying for a percentage question.

Application on the SAT

Percentage questions on the SAT may be simple, straightforward math questions, or presented as more cryptic word problems. The simple math questions usually require TRANSLATION:

8. 65 percent of 140 is the same as 70 percent of what number?

(A) 63
(B) 91
(C) 130
(D) 145
(E) 152

TRANSLATION is the most common solution strategy for percentage questions.

TRANSLATE from English to math:

65 percent of 140 is the same as 70 percent of what number
$0.65 \times 140 = 0.70 \times x$

Then solve for x:

$0.65 \times 140 = 0.70 \times x \quad \rightarrow \quad 91 = 0.7x \quad \rightarrow \quad 130 = x$

Most percentage questions will be combined with word problems. Some may require you to find a value:

10. A total of 80 pets were adopted from the humane society this week. Of the 50 pets adopted on Monday, 30% were cats. Of the 20 pets adopted on Wednesday, 80% were cats. And of the 10 pets adopted on Friday, 50% were cats. What percent of the 80 pets adopted were cats?

(A) 35%
(B) 45%
(C) 50%
(D) 65%
(E) 70%

Word problems should be solved using a combination of SPLITTING the question into parts and TRANSLATING from English to math. This question has four parts:

Part 1: Monday
30% of 50 were cats
$0.30 \times 50 = \text{cats}$
$15 = \text{cats}$

Part 2: Wednesday
80% of 20 were cats
$0.80 \times 20 = \text{cats}$
$16 = \text{cats}$

Part 3: Friday
50% of 10 were cats
$0.50 \times 10 = \text{cats}$
$5 = \text{cats}$

Part 4:
Total cats $= 15 + 16 + 5 = 36$
What percent of the 80 pets adopted were cats?
What percent of 80 is 36?

$\frac{?}{100} \times 80 = 36 \quad \rightarrow \quad ? \times 80 = 3600 \quad \rightarrow \quad ? = 45$

Answer choice (B) is correct.

Don't forget to correctly translate "what percent."

The most difficult percentage questions will ask you to find an expression, rather than a value:

If this is #20 in a 20-question section, it is likely considered the most difficult question in the section.

20. A surf shop held a going-out-of-business sale. On Sunday, a surfboard was k dollars. Each day after that, the surfboard was 20 percent less than the previous day. In terms of k, what was the price of the surfboard on Wednesday?

(A) $0.4k$
(B) $0.512k$
(C) $0.6k$
(D) $0.7592k$
(E) $0.8k$

This question can be more easily understood and solved if you SUPPLY a number for k. Remember to supply 100 for questions involving percents:

$k = 100$

However, by using PowerScore's solution strategies, it's really not that hard!

Sunday = \$100
Monday = \$100 – (20% of \$100) → \$100 – \$20 → \$80
Tuesday = \$80 – (20% of \$80) → \$80 – \$16 → \$64
Wednesday = \$64 – (20% of \$64) → \$64 – \$12.80 → \$51.20

Answer choice (B) should immediately appear attractive. Substitute 100 for k to show that (B) = \$51.20:

(B) $0.512k \rightarrow 0.512(100) \rightarrow 51.20$ ✓

By supplying 100, the answer choice is immediately clear. You could have supplied any number for k, but you probably would have had to plug that number into every answer choice to find the one that was equivalent to your final result. Since percentages are based out of 100, the answer is more easily found when you supply 100 for the base price of the surfboard.

Percentages Problem Set

Solve the following multiple-choice questions by selecting the best answer from the five answer choices. For grid-in questions, write your answer in the grids and completely mark the corresponding ovals. Answers begin on page 169.

1. A student won a \$20,000 scholarship. After 76 percent of the scholarship is deducted for tuition and room and board, the student receives the balance of the scholarship in annual payments of equal amounts over a 4-year period. How many dollars will the student receive each year of the 4 years? (Ignore the dollar sign when completing the grid-in).

2. What is the greatest possible integer for which 40 percent of that integer is less than 1.5 ?

   (A) 7
   (B) 6
   (C) 5
   (D) 4
   (E) 3

3. The price of a textbook is discounted by 45 percent, and then the reduced price is raised by 20 percent. This series of price changes is equivalent to a single discount of

   (A) 32.5%
   (B) 34%
   (C) 55%
   (D) 65%
   (E) 66%

4. If a and b are positive integers, what percent of $(9 - a^3)$ is b ?

   (A) $100b(3 - a)$

   (B) $b(9 - a^3)$

   (C) $\frac{b}{9-a^3}$

   (D) $\frac{9-a^3}{100b}$

   (E) $\frac{100b}{9-a^3}$

Ratios

Frequency Guide: 2

A ratio is a comparison of two quantities. Although a simple concept, ratios are briefly discussed in many classroom curriculums, thus making them intimidating to test takers and earning them at least a Medium level difficulty.

Required Knowledge and Skill Set

1. Ratios are often expressed using a colon (:), but can appear other ways. In a problem where there are 5 trees for every 12 flowers, the ratio could be expressed any of the following ways:

   $5:12$ $\qquad \frac{5}{12}$ $\qquad$ A ratio of 5 to 12

2. Just as fractions can be reduced, ratios can be reduced:

   $$100:150 \rightarrow \frac{100}{150} \rightarrow \frac{10}{15} \rightarrow \frac{1}{3} \rightarrow 1:3$$

3. To solve most SAT ratio questions, you must know how to find the fraction of each quantity. To do this, add the parts of the ratio together to create the denominator. For example, say a classroom has 7 boys for every 2 girls:

   Boys : Girls
   $7 : 2$ $\qquad$ Boys + Girls = Denominator $\rightarrow$ $7+2=9$

   $\frac{7}{9}, \frac{2}{9}$ $\qquad$ $\frac{7}{9}$ of the class is boys, $\frac{2}{9}$ of the class is girls

Finding the sum of the parts of a ratio is the key to solving most ratio questions.

   Labeling the ratio is extremely important. It helps you avoid careless errors and makes the ratio much easier to understand. Look at another:

   Dogs : Cats
   $1 : 4$ $\qquad$ Dogs + Cats = Denominator $\rightarrow$ $1+4=5$

   $\frac{1}{5}, \frac{4}{5}$ $\qquad$ $\frac{1}{5}$ of the animals are dogs, $\frac{4}{5}$ of the animals are cats

   Some ratios are made up of more than 2 quantities:

   Dogs : Cats : Birds
   $1 : 4 : 3$ $\qquad$ Dogs + Cats + Birds = Denominator $\rightarrow$ $1+4+3=8$

   $\frac{1}{8}, \frac{4}{8}, \frac{3}{8}$ $\qquad$ $\frac{1}{8}$ of the animals are dogs, $\frac{4}{8}$ of the animals are cats, and $\frac{3}{8}$ of the animals are birds

Application on the SAT

One type of ratio question requires you to find the amount of a portion, given the amount of the total. These questions are often used in word problems involving recipes or dollar amounts:

11. A recipe for dough calls for cups of sugar, flour, and salt in a ratio of 2:6:1, respectively. In order to make 21 cups of this dough, how many cups of sugar are required?

   (A) $\frac{2}{9}$

   (B) 3

   (C) $4\frac{2}{3}$

   (D) $5\frac{1}{4}$

   (E) 7

Notice that three of the answer choices are in fraction form. This means that using fractions rather than decimals to solve the question eliminates two extra steps: converting the fractions to decimals and then reverting the decimal answer back to a fraction. The makers of the test know that most high school students are more comfortable with decimals, which is why you should become fluent with fractions.

To solve this question (and to solve most ratio questions), find the fractional part of the required ingredient.

Sugar : Flour : Salt

2 : 6 : 1 Sugar + Flour + Salt = Denominator $\rightarrow 2 + 6 + 1 = 9$

$\frac{2}{9}$, $\frac{6}{9}$, $\frac{1}{9}$

Since you need to find the amount of sugar in 21 cups of the recipe, you need only use the fraction of the sugar (two-ninths). Use TRANSLATION to find the required number of cups of sugar:

$\frac{2}{9}$ of 21 cups of dough is sugar

$$\frac{2}{9} \times 21 = \text{sugar} \rightarrow \frac{2}{9} \times 21 = \frac{42}{9} \rightarrow 4\frac{6}{9} \rightarrow 4\frac{2}{3}$$

Answer choice (C) is correct.

If the question had asked for the number of cups of flour or salt, then you would have multiplied the fraction of flour (six-ninths) or salt (one-ninth) by 21 cups.

$\frac{6}{9}$ of 21 cups of dough is flour $\rightarrow \frac{6}{9} \times 21 = \frac{126}{9} \rightarrow 14$

$\frac{1}{9}$ of 21 cups of dough is salt $\rightarrow \frac{1}{9} \times 21 = \frac{21}{9} \rightarrow 2\frac{1}{3}$

Notice that the number of cups of sugar, flour, and salt combined equal the 21 cups in the dough mixture:

Sugar + Flour + Salt = 21 cups of dough $\rightarrow 4\frac{2}{3} + 14 + 2\frac{1}{3} = 21$

The previous problem is indicative of a popular type of ratio question. Another common question requires you to find a possible total number of items given a ratio of parts:

9. In a dresser drawer, there are only white and brown socks. If the ratio of white socks to brown socks is 3 to 8, which of the following could be the total number of socks in the drawer?

(A) 40
(B) 41
(C) 42
(D) 43
(E) 44

The answer for these questions must always be a multiple of the sum of the ratio:

White : Brown
3 : 8 White + Brown = Sum $\rightarrow$ $3 + 8 = 11$

The only multiple of 11 is 44, answer choice (E).

To see why this is true, return to the fractional representation:

These types of ratio questions will always use items that <u>can't</u> be divided, such as socks, marbles, or people. You cannot have one half of a person!

The type of question on the previous page will use items that <u>can</u> be divided by weight, such as sugar, bird seed, or sand.

White : Brown
3 : 8 White + Brown = Denominator $\rightarrow$ $3 + 8 = 11$

$\frac{3}{11}$, $\frac{8}{11}$ $\frac{3}{11}$ of the socks are white, $\frac{8}{11}$ of the socks are brown

If the total number of socks is 44, the number of white and brown socks is an integer:

Number of white socks = $\frac{3}{11}$ of 44 $\rightarrow$ $\frac{3}{11} \times 44$ $\rightarrow$ 12 white socks

Number of brown socks = $\frac{8}{11}$ of 44 $\rightarrow$ $\frac{8}{11} \times 44$ $\rightarrow$ 32 white socks

But if the total number of socks is not a multiple of 11, then the number of white and brown socks is not an integer. Say the number of socks is 40:

Number of white socks = $\frac{3}{11}$ of 40 $\rightarrow$ $\frac{3}{11} \times 40$ $\rightarrow$ 10.9 white socks

Number of brown socks = $\frac{8}{11}$ of 40 $\rightarrow$ $\frac{8}{11} \times 40$ $\rightarrow$ 29.1 white socks

You can not have 10.9 white socks! The number of socks must be a whole number.

Remember, to solve this question, you simply need to find the answer that is a multiple of the sum of the ratio. You do not need to find the fractional representation. We showed the parts of the ratio as fractions to help you see why the shortcut works.

Because ratios add difficulty to any problem, be prepared to see them combined with other types of math questions, including geometry, probability, and sequencing.

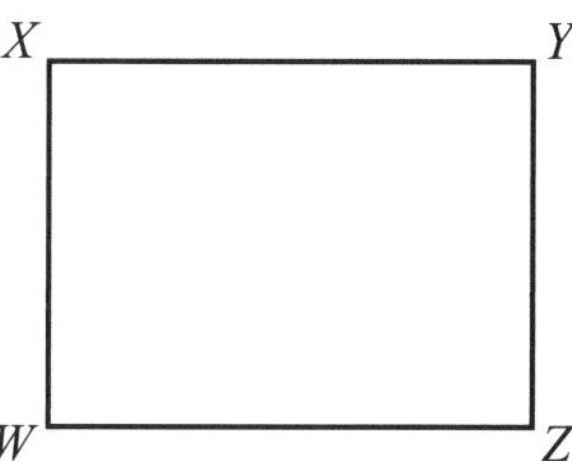

15. In the figure above, $WXYZ$ is a rectangle. If $\overline{XY}$ is 12 and $\overline{YZ}$ is 9, what is the ratio of $\overline{XW}$ to $\overline{XZ}$ (not shown)?

    (A) 1 to 2
    (B) 2 to 3
    (C) 2 to 5
    (D) 3 to 4
    (E) 3 to 5

Does Geometry make you nervous? Have no fear! We will cover SAT Geometry thoroughly in Chapter 7.

Although we have not covered geometry yet, it is still useful to study the solution to this question in order to see how ratios can influence other areas of math. To start, DIAGRAM the question. You can find the length of $\overline{XZ}$ using the Pythagorean Theorem:

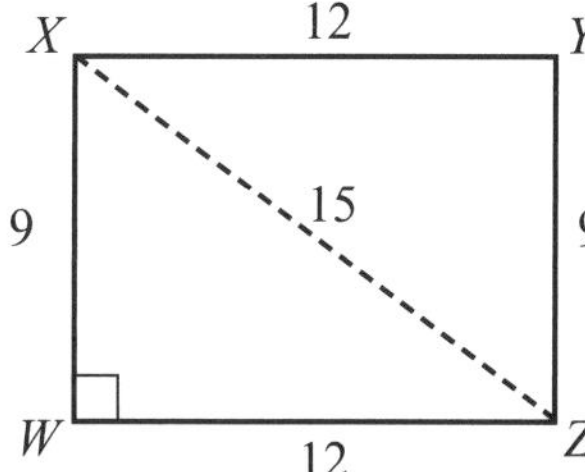

$a^2 + b^2 = c^2$

$9^2 + 12^2 = c^2$
$81 + 144 = c^2$
$225 = c^2$
$\sqrt{225} = \sqrt{c}$
$15 = c$

The ratio of $\overline{XW}$ to $\overline{XZ}$ is 9 to 15. Remember, though, that ratios can be reduced:

$$\frac{9}{15} = \frac{3}{5}$$

The ratio of $\overline{XW}$ to $\overline{XZ}$ is 3 to 5, answer choice (E).

Although ratio questions may seem intimidating at first, take your time to understand what the question is asking. The solution is usually very simple, allowing you to earn quick points when other test takers falter or fail.

Ratios Problem Set

Solve the following multiple-choice questions by selecting the best answer from the five answer choices. For grid-in questions, write your answer in the grids and completely mark the corresponding ovals. Answers begin on page 144.

1. In a box of fruit, there are only oranges, lemons, and kiwis in a ratio of 2:3:5, respectively. Which of the following could be the total number of pieces of fruit in the box?

2+3+5 = 10 fruits

(A) 77
(B) 78
(C) 79
(D) 80
(E) 81

2. Three charities are sharing a donation of \$48,000 in the ratio of 8:4:3. What is the amount of the greatest share?

(A) \$3,200
(B) \$9,600
(C) \$12,800
(D) \$16,000
(E) \$25,600

3. There are 15 girls and 15 boys in a first grade classroom. No girls are to be added or removed from the classroom. What is the least number of boys that can be removed from of the class so that the ratio of girls to boys is 5 to 3?

15/15 → 5/3

15/9 6 boys

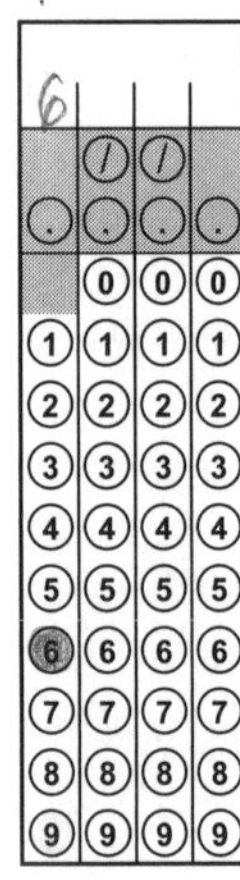

4. If the ratio of a to b is 6 to 7, which of the following could be true?

6:7

(A) $a = 36,\ b = 49$

(B) $a = 14,\ b = 12$

(C) $a = 7,\ b = 8$

(D) $a = 3,\ b = \frac{7}{2}$

(E) $a = 0,\ b = \frac{6}{7}$

Proportions and Rates

Frequency Guide: 3

Proportions are covered on nearly every single SAT, although sometimes they are disguised as ratios. This is because a proportion is two equal ratios:

$$\frac{1}{2}=\frac{4}{8} \qquad \frac{2}{3}=\frac{8}{12} \qquad \frac{3}{5}=\frac{9}{15}$$

Required Knowledge and Skill Set

1. The cross products of a proportion are equal. Therefore, proportions with variables can be solved by cross-multiplying:

$$\frac{2}{3}=\frac{x}{15} \rightarrow \frac{2}{3} \times \frac{x}{15} \rightarrow (2)(15)=(3)(x) \rightarrow 30=3x \rightarrow 10=x$$

Proportions are most easily solved by cross-multiplying.

2. Proportions are generally Easy or Medium difficulty level. The College Board may attempt to make them more difficult by using fractions and/or variables as values in the ratios but you can still solve them by cross-multiplying:

$$\frac{\frac{1}{3}}{6}=\frac{15t}{?} \rightarrow \frac{\frac{1}{3}}{6} \times \frac{15t}{?} \rightarrow (\frac{1}{3})(?)=(6)(15t) \rightarrow \frac{1}{3}?=90t \rightarrow ?=180t$$

3. When setting up a proportion, it is extremely important that you label your diagram to ensure that you place each value in the correct location. Consider an SAT question as an example:

    13. On a map, the straight-line distance between two towns is measured to be k inches. On this map, 3 inches represents an actual distance of 12 miles. In terms of k, what is the actual straight-line distance, in miles, between the two towns?

   Begin by setting up a cross-multiplying diagram in which each of the sections is labeled. The numerators in the proportion are the measurements, in inches, on the map. The denominators are the actual distance in miles. The ratio on the left is based on the map key, while the ratio on the right is based on the distance between the towns:

	Key:		Towns:
Map Measurements:	______	=	______
Real Distance:			

The majority of wrong answers in proportion questions occur because test takers do not place the values in the correct locations in the proportion.

   Then add values in their proper places:

	Key:		Towns:
Map Measurements:	k in.	=	3 in.
Real Distance:	? mi.		12 mi.

Now cross multiply:

$$\frac{k}{?}=\frac{3}{12} \rightarrow (12)(k)=(3)(?) \rightarrow \frac{12k}{3}=? \rightarrow 4k=?$$

The straight-line distance between the two towns is $4k$.

☠ CAUTION: SAT TRAP! Proportion answer choices are designed to trip students who set up a proportion incorrectly.

If you fail to label a proportion, you may place the values from the question into the wrong part of the ratio, thus resulting in a different answer. Often, the test makers anticipate this mistake, and create wrong answers to match these types of errors. For example, say that you incorrectly placed the values like so:

$$\frac{k}{12}=\frac{3}{?} \rightarrow (?)(k)=(3)(12) \rightarrow (?)(k)=36 \rightarrow ?=\frac{36}{k}$$

You can be sure that $\frac{36}{k}$ is one of the incorrect answer choices! Labeling the proportion is the only way to make sure that you correctly place all values into the proportion.

4. A rate is a ratio that measures distance or work over time:

$$\text{Jud can paint 2 rooms per hour} = \frac{2 \text{ rooms}}{1 \text{ hour}}$$

The example above shows Jud's rate of speed in painting rooms. Notice that the word "per" translates into a division symbol.

Most rate problems can be treated as proportions:

5. If Jud can paint at a rate of two rooms per hour, how many rooms can he paint in a 12 hour period?

Most rate questions can be solved by cross-multiplying the proportion. A few, however, must be solved using the formula for finding rate.

Set up a proportion with a variable and solve:

$$\begin{array}{lccc} \text{Rooms} & 2 & & ? \\ & \overline{} & = & \overline{} \\ \text{Hours} & 1 & & 12 \end{array}$$

Cross multiply: $24 = ?$

On rare occasion, however, you may need to use the formulas for finding rate (r), distance (d), or time (t):

MEMORY MARKER: The rate formula is not provided on the test.

$$r=\frac{d}{t} \qquad d=rt \qquad t=\frac{d}{r}$$

Application of this formula will be discussed in the next section.

Application on the SAT

Proportion questions may appear in several forms. Some will conveniently use the word "proportional" in the question:

5. If x is directly proportional to y and $x = 5$ when $y = 8$, what is the value of x when $y = 120$?

Some will be disguised as ratios:

4. The ratio of 9 to 5 is equal to the ratio of 36 to what number?

Others will be disguised as rate problems:

3. A furniture factory produces 200 tables a day. At this rate, how many days will it take to produce 5000 tables?

And some may not use any terms at all to indicate the problem involves a proportion:

6. If 5 cups of flour makes 15 cookies, how many cookies are made from 30 cups of flour?

All of these Easy level questions can be solved with a cross-multiplying diagram. For the answers to each question, look on the next page.

No matter how the question is worded, the solution method is always the same: carefully label a cross-multiplying diagram and place each of the values in the appropriate fields.

Many proportion questions are recognizable by their use of measurements in recipes, on maps, or on blueprints. Recipe questions use cups, quarts, liters, gallons, and other liquid and dry measurements. Map and blueprint problems will compare the scale of the image to actual distances, using inches, feet, kilometers, miles, and other units of length measurement.

To add difficulty to proportion problems, the College Board may require you to convert units of measurement:

15. If 5 cups of sugar weighs 36 ounces, what is the weight, in ounces, of 10 gallons of sugar? (1 gallon = 16 cups)

    (A) 22
    (B) 72
    (C) 800
    (D) 1152
    (E) 1800

Note that the conversion is given. You will not be expected to remember that 1 gallon is 16 cups or that 1 mile is 1.6 kilometers. However, you will be expected to know these time conversions:
60 seconds = 1 minute
60 minutes = 1 hour
24 hours = 1 day

While this question will ultimately be solved using a proportion, you cannot set up the cross-multiplying diagram until you convert 10 gallons to cups:

$$10 \text{ gallons} \times \frac{16 \text{ cups}}{1 \text{ gallon}} \rightarrow 10 \cancel{\text{gallons}} \times \frac{16 \text{ cups}}{1 \cancel{\text{gallon}}} = 160 \text{ cups}$$

Answers from previous page:
5. 75
4. 20
3. 25 days
6. 90 cookies

Now you can create the correct diagram:

Cups of sugar: ——— = ———

Weight in ounces:

And then add values, knowing that 10 gallons is 160 ounces:

$$\text{Cups of sugar: } \frac{5}{36} = \frac{160}{?} \text{ Weight in ounces}$$

And then cross multiply:

$$\frac{5}{36} = \frac{160}{?} \rightarrow (5)(?) = (36)(160) \rightarrow 5(?) = 5760 \rightarrow ? = 1152$$

Answer choice (D) is correct.

As mentioned in the previous section, nearly all rate questions on the SAT can be solved as proportions. However, you should be prepared for more difficult rate questions that require use of the rate formula, where r is rate, d is distance, and t is time:

$$r = \frac{d}{t}$$

Remember, this formula can be rearranged to find distance or time:

$$d = rt \qquad t = \frac{d}{r}$$

One type of question that involves this formula asks that you find an average rate, given a person who takes the same route at two different rates of speed:

18. A car travels from Town A to Town B at a constant speed of 20 miles per hour. It returns along the same route from Town B to Town A at a constant speed of 60 miles per hour. What is the car's average speed, in miles per hour, for the entire round trip?

(A) 30
(B) 35
(C) 40
(D) 45
(E) 50

CAUTION: SAT TRAP!
Be sure to avoid this possible SAT trap.

Many students will mistakenly assume that the answer is (C) because $(20 + 60) \div 2 = 40$. But average speeds are not calculated the same way as regular averages.

The formal for finding the average speed is:

$$\text{average rate of speed} = \frac{\text{total distance}}{\text{total time}}$$

To find the total distance and total time, record each piece of information separately and then add them together. Use subscript 1 to indicate the information for the trip from Town A to Town B and subscript 2 to indicate information for the return trip:

$d_1 = ?$ and $d_2 = ?$ but $d_1 = d_2$ Total distance $= 2d$

$t = \frac{d}{r}$ $t_1 = \frac{d}{20}$ $t_2 = \frac{d}{60}$ Total time $= \frac{d}{20} + \frac{d}{60}$

Now solve for the average rate of speed:

$$\text{average rate of speed} = \frac{\text{total distance}}{\text{total time}}$$

$$\text{average speed} = \frac{2d}{\frac{d}{20}+\frac{d}{60}} \rightarrow \frac{2d}{\frac{3d}{60}+\frac{d}{60}} \rightarrow \frac{2d}{\frac{4d}{60}} \rightarrow \frac{(2d)(60)}{4d} \rightarrow \frac{120}{4} \rightarrow 30$$

The correct answer is (A).

As long as the distance is the same for the two trips, you can use a shortcut formula to find this same result:

$$\text{average rate of speed} = \frac{2 \times \text{rate}_1 \times \text{rate}_2}{\text{rate}_1 + \text{rate}_2}$$

To prove this, return to the previous problem:

$\text{rate}_1 = 20$ mi/hr $\text{rate}_2 = 60$ mi/hr

$$\text{average rate of speed} = \frac{2 \times \text{rate}_1 \times \text{rate}_2}{\text{rate}_1 + \text{rate}_2} \rightarrow \frac{2 \times 20 \times 60}{20 + 60} \rightarrow \frac{2400}{80} \rightarrow 30$$

Remember, though, that this formula only works when the distances of the two routes are equal.

While this question seems difficult enough, the College Board may attempt to make rate questions even more complicated by requiring you to find the total distance or total time, rather than the average rate.

Confidence Quotation
"You create your own universe as you go along."
—Winston Churchill, British Prime Minister

ARITHMETRICK
This shortcut formula works when the distance is the same for both routes.

Consider an example:

20. Tony spent a total of 4 hours driving to and from a college. On the way there he drove at an average speed of 50 miles per hour and on the way home he drove the same route at an average speed of 30 miles per hour. How many miles did he drive to get to the college?

(A) 40
(B) 55
(C) 75
(D) 80
(E) 150

Note that the question requires the one-way distance, rather than the total distance. The distance one way is the total distance divided by two.

Since you are now looking for the distance rather than the average speed, you must rearrange the basic equation:

$$\text{average rate of speed} = \frac{\text{total distance}}{\text{total time}}$$

$$\text{total distance} = (\text{average rate of speed}) \times (\text{total time})$$

You have information about the total distance and the total time:

$d_1 = ?$ and $d_2 = ?$ but $d_1 = d_2$ — Total distance $= 2d$

$t_1 + t_2 = 4$ hours — Total time $= 4$ hours

$2d = (\text{average rate of speed}) \times 4 \text{ hours}$

As you can see, you must find the average rate of speed before returning to the formula. You can use the shortcut formula because the distances to and from the college are equal:

Rate questions like this one are very rare on the SAT, but you would be wise to be prepared.

$\text{rate}_1 = 50$ mi/hr $\qquad \text{rate}_2 = 30$ mi/hr

$$\text{average rate of speed} = \frac{2 \times \text{rate}_1 \times \text{rate}_2}{\text{rate}_1 + \text{rate}_2} \rightarrow \frac{2 \times 50 \times 30}{50 + 30} \rightarrow \frac{3000}{80} \rightarrow 37.5$$

Now plug the average rate of speed into the original equation:

$$\text{total distance} = (\text{average rate of speed}) \times (\text{total time})$$

$$2d = 37.5 \text{ mi/hr} \times 4 \text{ hours} \rightarrow 2d = 150 \text{ mi} \rightarrow d = 75 \text{ mi}$$

The distance to the college is 75 miles (the round trip distance is 150 miles), so the correct answer is (C).

Another complicated rate problem involves work. Work problems require you to find the amount of time it takes people or machines to complete a task. Obviously, the faster people work the faster the task is completed. Similarly, the more people that work, the faster the task is completed. Knowing this, you may be able to use critical reasoning to solve some work problems. For others, you must find individual rates to find the solution.

Work questions are slightly more common than rate questions involving the average speed, distance, or time.

Work questions with Medium difficulty level usually give the rate of a group:

13. If 5 people can build a deck in 4 days, how many days will it take 2 people working at the same rate to build the same deck?

(A) 2.5
(B) 6
(C) 8
(D) 10
(E) 12

If the project is being completed by fewer people, it is going to take more days to finish the deck. More specifically, since the work force has been cut by more than half (from 5 people to 2 people), the number of days will increase by more than twice as much (from 4 days to more than 8 days). You can eliminate answers (A) and (B) and even (C).

To solve this question, you must find the rate of one person. If it takes 5 people 4 days, it will take a single person 5 times as long:

The key to "work questions" is to find the rate of 1 person.

One person rate = 4 days × 5 = 20 days

Now, find the rate of two people. It should take two people half the time it takes one person:

20 days ÷ 2 people = 10 days

Answer choice (D) is correct.

The most difficult work questions involve combined work rates:

20. If Bruce can paint a house in 8 hours and Courtney can paint the house in 6 hours, how long will it take them to paint the house, in hours, working together?

(A) $3\frac{3}{7}$

(B) $3\frac{7}{9}$

(C) $4\frac{2}{5}$

(D) 7

(E) 14

SIZE UP this question! Which answers can you eliminate based on estimation?

Since Bruce, the slowest worker, can paint the house alone in 8 hours, together they will paint it faster than Bruce alone. The correct answer will be faster than 8 hours, thus eliminating choice (E).

Some students worry that Bruce's and Courtney's rates may change now that they are working together. Maybe they will push each other to work faster or take more breaks to talk and work slower. Never apply your personal experiences to the SAT! All work questions are idealized, and each person's rate will be maintained when they work together.

To solve this problem, figure out each person's rate for the same unit of time. For example, how much work can each person do in one hour?

$$\text{Bruce} = \frac{1}{8} \text{ of the house in 1 hour}$$

$$\text{Courtney} = \frac{1}{6} \text{ of the house in 1 hour}$$

Note that this is just a new interpretation of the rate formula, in which the amount of work (w) is substituted for the distance (d):

MEMORY MARKER:
The formula for work is not given on the SAT.

$$r = \frac{d}{t} \qquad\qquad r = \frac{w}{t}$$

$$\text{Bruce's rate} = \frac{\text{1 house}}{\text{8 hours}} \qquad\qquad \text{Courtney's rate} = \frac{\text{1 house}}{\text{6 hours}}$$

Now, find their work rate together by adding their individual rates:

$$\text{Bruce} + \text{Courtney} = \text{Rate Together}$$

$$\frac{1}{8} + \frac{1}{6} \quad \rightarrow \quad \frac{3}{24} + \frac{4}{24} = \frac{7}{24}$$

The time it takes them together is simply the inverse of $\frac{7}{24}$: $\frac{24}{7}$ hours or $3\frac{3}{7}$ hours.

To see why the inverse of their rate is their time, return to the work formula:

$$\text{rate together} = \frac{\text{work together}}{\text{time together}} \quad \rightarrow$$

$$(\text{time together}) \times (\text{rate together}) = \text{work together} \quad \rightarrow$$

$$\text{time together} = \frac{\text{work together}}{\text{rate together}}$$

$$\text{time together} = \frac{\text{1 house}}{\frac{7}{24}} \rightarrow t = \frac{1}{\frac{7}{24}} \rightarrow t = \frac{1 \times \frac{24}{7}}{\frac{7}{24} \times \frac{24}{7}} \rightarrow t = \frac{\frac{24}{7}}{1} \rightarrow t = \frac{24}{7}$$

The correct answer is (A).

It is important to understand the solution to work problems, which is why we thoroughly reviewed it. However, there is a shortcut for work problems that combine the work of each person or machine:

t_1 = time taken by first person

t_2 = time taken by second person

t_T = time together

$$\frac{1}{t_1}+\frac{1}{t_2}=\frac{1}{t_T}$$

ARITHMETRICK
The formula for combined rates makes a difficult question quite simple.

Plug in Bruce's and Cindy's times to show how this formula works:

8 = time taken by Bruce

6 = time taken by Cindy

$$\frac{1}{8}+\frac{1}{6}=\frac{1}{t_T} \rightarrow \frac{3}{24}+\frac{4}{24}=\frac{1}{t_T} \rightarrow \frac{7}{24}=\frac{1}{t_T} \rightarrow t_T=\frac{24}{7}$$

Note that this formula also works when more than two people are contributing. Simply add another rate for each new person:

If 3 people complete the job: $\frac{1}{t_1}+\frac{1}{t_2}+\frac{1}{t_3}=\frac{1}{t_T}$

If 4 people complete the job: $\frac{1}{t_1}+\frac{1}{t_2}+\frac{1}{t_3}+\frac{1}{t_4}=\frac{1}{t_T}$

Confidence Quotation
"Every thought is a seed. If you plant crab apples, don't count on harvesting Golden Delicious." —Bill Meyer

While proportions are quite common on the SAT, including rate questions that can be solved using proportions, actual rate and work questions that require formulas are more rare.

Proportions and Rates Problem Set

Solve the following multiple-choice questions by selecting the best answer from the five answer choices. For grid-in questions, write your answer in the grids and completely mark the corresponding ovals. Answers begin on page 172.

TABLE OF CONVERSIONS

Distance on map in inches	2.4	4.8	9.6
Actual distance in miles	15	30	t

1. In the table above, what is the value of t?

(A) 45
(B) 50
(C) 60
(D) 72
(E) 144

2. A bottling machine can fill 3800 bottles of soda per hour. At this rate, in how many minutes can the bottling machine fill 570 bottles?

(A) $6.6\overline{6}$
(B) 9
(C) 9.5
(D) 18
(E) $63.3\overline{3}$

3. To make 5 cookies, a recipe requires $\frac{1}{4}$ cup of milk. Using this same recipe, how many cookies can be made from 3 gallons of milk? (16 cups = 1 gallon)

(A) 20
(B) 48
(C) 240
(D) 720
(E) 960

Proportions and Rates Problem Set

Solve the following multiple-choice questions by selecting the best answer from the five answer choices. For grid-in questions, write your answer in the grids and completely mark the corresponding ovals. Answers begin on page 172.

4. The ratio of a to b to c to d is 20 to 10 to 5 to 2.5. If $a = 120$, what is the value of c ?

   (A) 4.8
   (B) 12
   (C) 24
   (D) 30
   (E) 48

5. A bicyclist travels up a hill at 10 miles per hour and returns down the same hill at 30 miles per hour. What is the bicyclist's average speed, in miles per hour, for the entire journey up and down the hill?

   (A) 15
   (B) 17
   (C) 20
   (D) 22
   (E) 25

6. Working alone, Navid can mow the field in 2 hours, Janey can do it in 4 hours, and Steven can do it in 12 hours. Working together, how many hours will it take them to mow the field?

   (A) 0.3
   (B) 0.6
   (C) 0.8
   (D) 1
   (E) 1.2

Number Lines

Frequency Guide: 2

Number lines may be introduced in early elementary school, but they are a bit more complicated on the SAT! Still, they can be easily tackled if you understand a few basic principles and strategies.

Required Knowledge and Skill Set

1. A number line is a line in which numbers are placed according to their value. Each number has a corresponding point on a number line, and numbers increase in value from left to right:

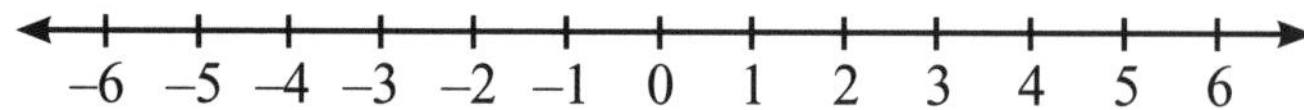

2. Points on a number line are called *coordinates*. The vertical lines on the number line are called *tick marks*, and they are used to evenly mark space between values:

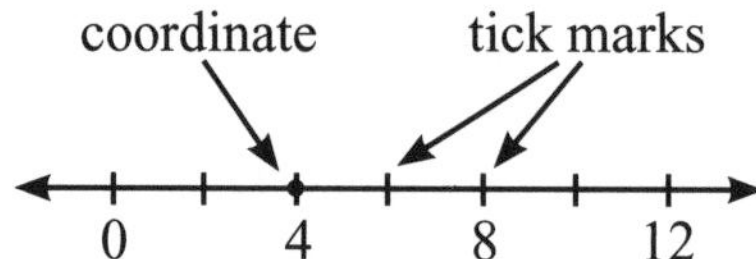

The terms *coordinates* and *tick marks* are used on the SAT.

3. Number lines are drawn to scale on the SAT. If a number line has tick marks with unmarked values, you may be able to confidently assign a value. Consider an example:

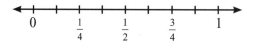

Because number lines are drawn to scale, you can SIZE UP number lines to make an estimate if time is running out or if you do not know how to solve the question.

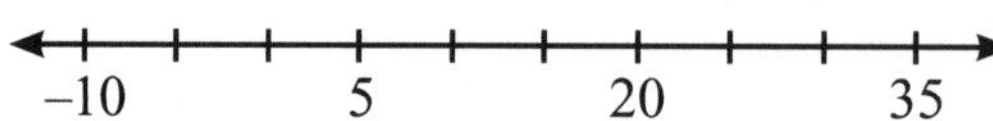

Four tick marks are given values; the space between each of these marked tick marks represent 15 spaces. Therefore, you know the value of each tick mark:

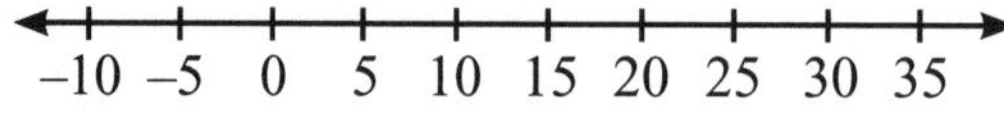

You can even estimate points on the number line that do not have tick marks:

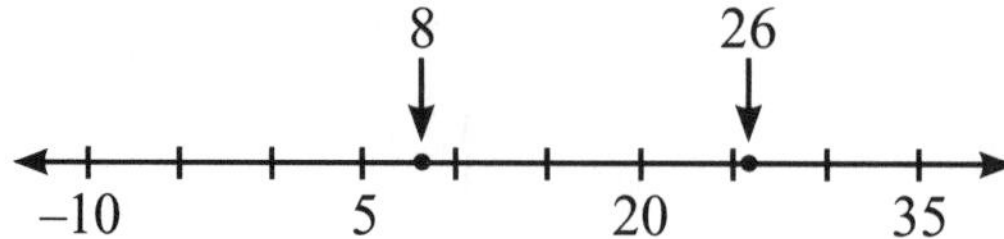

4. Fractions and decimals can also be placed on a number line:

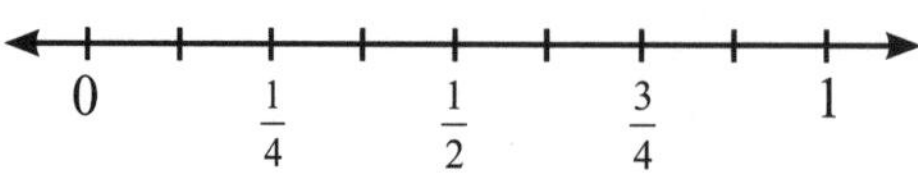

Application on the SAT

Questions involving number lines can occur at any difficulty level. They tend to be Easy and Medium questions, but do occasionally present as Hard level questions.

As you learned in the chapter on Solution Strategies, it is helpful to DIAGRAM a question. If a question contains an illustration of a number line, fill in the figure with as much information as possible. But if a number line question does not contain a figure, you may want to draw one yourself. A number line is a visual concept, and it is often more easily solved when represented graphically.

DIAGRAMMING numbers line problems is essential to finding their solutions.

One number line question type involves determining the length of a line segment. Sometimes these questions are placed on a traditional number line:

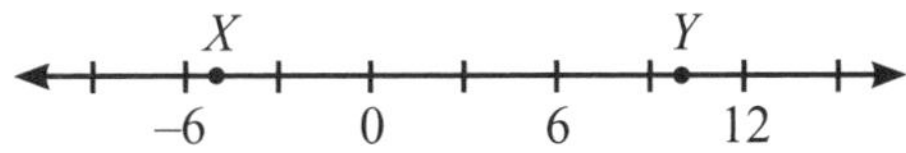

3. On the number line above, the best estimate of the length of line segment $\overline{XY}$ is

   (A) 5
   (B) 6
   (C) 12
   (D) 15
   (E) 18

To solve this question, first fill in the missing values on the number line:

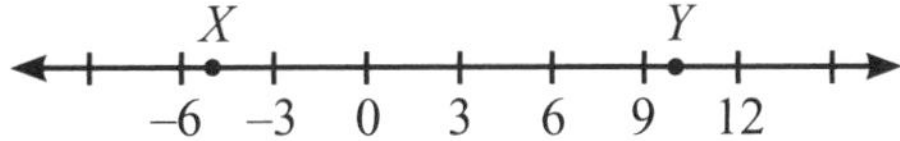

Now, estimate the value of *X*:

X is closer to –6 than to –3, so a good approximation is –5

Then do the same with *Y*:

Y is closer to 9 than to 12, so a good estimate is 10

Now, find the distance between –5 and 10:

–5 10
–6 –3 0 3 6 9 12

–5 to 0 = 5 spaces
0 to 10 = 10 spaces
5 + 10 = 15 spaces between *X* and *Y*

The best estimate is 15, (D).

Other questions that require you to find the length of a line segment may not use a traditional number line. They may instead provide information about a line and ask you to find the distance between two points on that line:

9. Four points, R, S, T, and U lie on a line, but not necessarily in that order. Line segment $\overline{RS}$ has a length of 36, and $\overline{RT} = \overline{TS}$. Point U is the midpoint of $\overline{TS}$. What is the distance between point R and point U?

   (A) 9
   (B) 18
   (C) 27
   (D) 36
   (E) 45

Confidence Quotation
"Getting ahead in a difficult profession requires avid faith in yourself. That is why some people with mediocre talent, but with great inner drive, go much further than people with vastly superior talent."
—Sophia Loren, actress

Without a figure, this question is difficult to solve. Begin by DIAGRAMMING the question. $\overline{RS}$ has a length of 36:

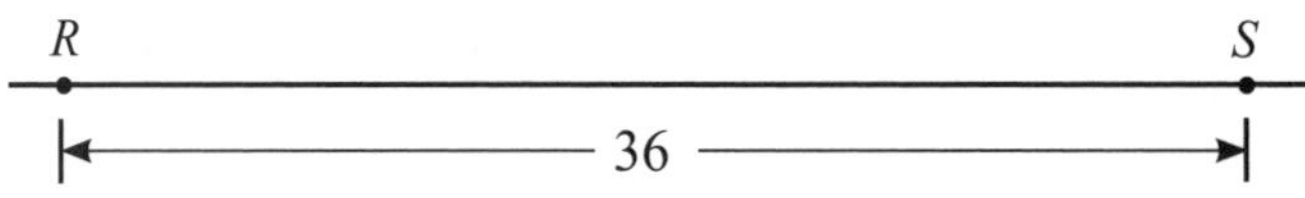

And $\overline{RT} = \overline{TU}$:

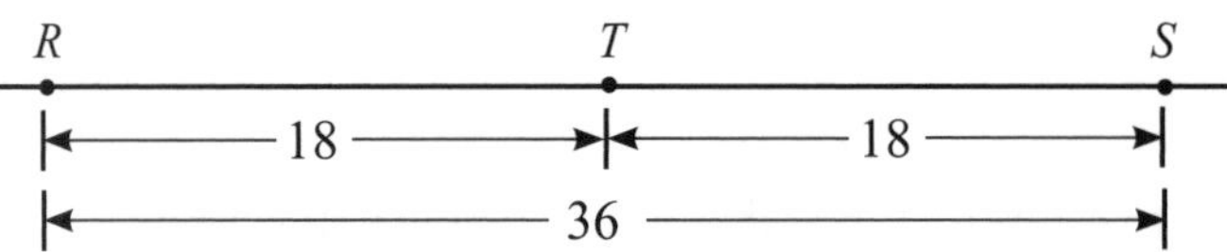

Finally, point U is the midpoint of $\overline{TS}$:

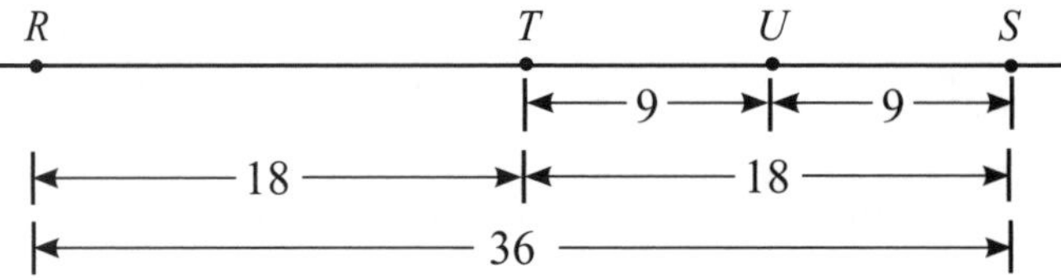

Now you are able to find the distance between R and U:

$$\overline{RT} + \overline{TU} = 18 + 9 = 27$$

Answer choice (C) is correct.

Questions that do not involve a traditional number line can still be solved using your knowledge of number lines. You simply need to sketch the number line, creating your own tick marks and coordinates, and apply the information provided in the question.

A common number line question type asks you perform an operation on the values of two coordinates to find a new coordinate. You may be able to precisely determine the values of the two coordinates, or you may need to estimate the values:

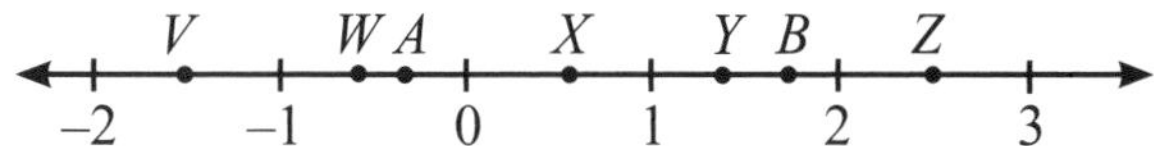

10. On the number line above, which of the coordinates most closely corresponds to the result of $|A \times B|$?

(A) V
(B) W
(C) X
(D) Y
(E) Z

In this question, you will need to estimate the value of point A and point B:

Point A is between –0.5 and 0, but closer to –0.5. A good estimate is –0.4

Point B is between 1.5 and 2, but closer to 1.5. A good estimate is 1.7

Now, plug your estimates into operation:

$$|A \times B| \quad \rightarrow \quad |-0.4 \times 1.7| \quad \rightarrow \quad |-0.68| \quad \rightarrow \quad 0.68$$

This common question type may ask you to add, subtract, multiply, or divide the values of two coordinates.

The coordinate closest to 0.68 is X.

Note that if we had picked different estimates for A and B, point X would still be the best result:

If $A = -0.33$ and $B = 1.8$

$$|A \times B| \quad \rightarrow \quad |-0.33 \times 1.8| \quad \rightarrow \quad |-0.594| \quad \rightarrow \quad 0.594$$

The closest coordinate is still X.

This question required two coordinates to be multiplied and their result placed in absolute value. Expect to apply addition, subtraction, division, exponents, radicals, or other operations to coordinates on a number line.

The most difficult type of number line question involves unmarked values, often which are fractions or decimals. For these questions, assign values. Consider the following number line:

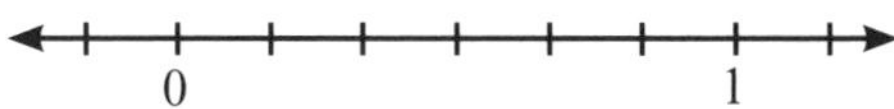

On this number line, there are 6 tick marks from 0 and 1. Therefore, the tick marks indicate increments of one-sixth:

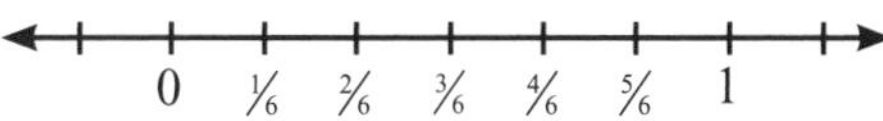

Fractions are often used to make a number line question more difficult.

Number Lines Problem Set

Solve the following multiple-choice questions by selecting the best answer from the five answer choices. For grid-in questions, write your answer in the grids and completely mark the corresponding ovals. Answers begin on page 173.

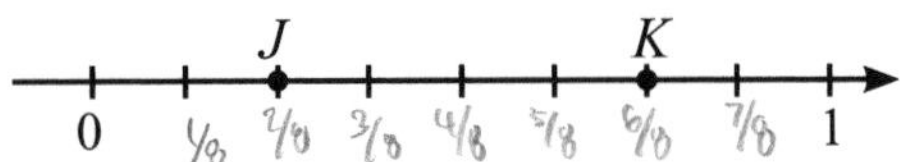

1. The tick marks are equally spaced on the number line above. What is the value of $K \div J$?

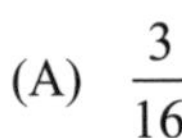

(A) $\frac{3}{16}$

(B) $\frac{1}{3}$

(C) $\frac{1}{2}$

(D) 1

(E) 3

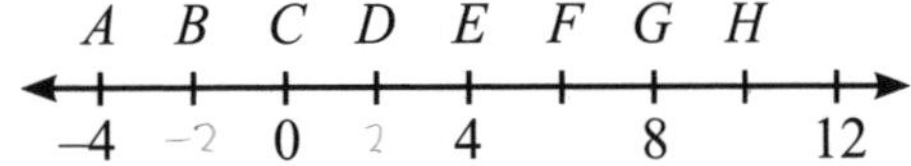

2. The tick marks are equally spaced on the number line above. The ratio of the length of $\overline{BD}$ to the length of $\overline{BH}$ is equivalent to the ratio of the length of $\overline{EF}$ to the length of which of the following?

(A) $\overline{AD}$
(B) $\overline{BC}$
(C) $\overline{BG}$
(D) $\overline{DH}$
(E) $\overline{EG}$

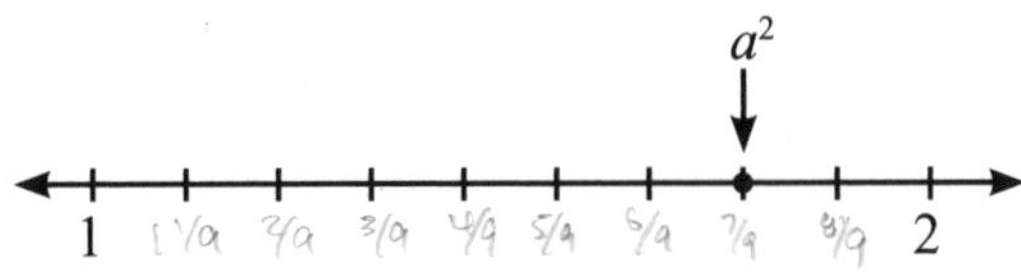

3. On the number line above, the tick marks are equally spaced. What is the value of a?

Notes

ARITHMETIC MASTERY ANSWER KEY

Divisibility Mini Drill—Page 121

	2	3	4	5	6	9
372	✓	✓	✓		✓	
5940	✓	✓	✓	✓	✓	✓
14022	✓	✓			✓	✓

Number Properties Problem Set—Page 124-125

1. (E) Medium

Create a table showing all the possible values of x and y (note that the value of x is irrelevant, but may help some students to understand the problem more clearly):

x	y	$x + 3y = 19$
16	1	$16 + 3(1) = 19$
13	2	$13 + 3(2) = 19$
10	3	$10 + 3(3) = 19$
7	4	$7 + 3(4) = 19$
4	5	$4 + 3(5) = 19$
1	6	$1 + 3(6) = 19$

Then find the sum of all possible values of y:

$1 + 2 + 3 + 4 + 5 + 6 = 21$

2. (D) Medium

Supply four consecutive even integers to satisfy the two rules:

$w = 2$ $\quad x = 4$ $\quad y = 6$ $\quad z = 8$

Then find $w + x$ and $y + z$:

$w + x = 2 + 4 = 6$
$y + z = 6 + 8 = 14$

Subtract $w + x$ from $y + z$:

$14 - 6 = 8$

3. (A) Medium

Write a small sample from each set:

Set X {11, 13, 17, 19, 23...}
Set Y {22, 32, 42, 52, 62...}

Set X will never contain any of the numbers in Set Y. Every number in Set Y is divisible by 2; the numbers in Set X are not divisible by any number besides themselves and 1.

4. .187, .188, or $\frac{3}{16}$ Medium

Start by translating the information from the problem into two equations:

The sum of two numbers is 1: $x + y = 1$
Their difference is $\frac{1}{2}$: $x - y = \frac{1}{2}$

Now use addition (or substitution) to find x:

$$\begin{array}{rl} & x + y = 1 \\ + & x - y = \frac{1}{2} \\ \hline & 2x \quad = 1\frac{1}{2} \end{array}$$

$$\text{or} \quad 2x = \frac{3}{2} \quad \rightarrow \quad x = \frac{3}{2} \div 2 \quad \rightarrow \quad x = \frac{3}{2} \times \frac{1}{2} \quad \rightarrow \quad x = \frac{3}{4}$$

Now find y:

$$x + y = 1 \quad \rightarrow \quad \frac{3}{4} + y = 1 \quad \rightarrow \quad y = 1 - \frac{3}{4} \quad \rightarrow \quad y = \frac{1}{4}$$

Finally, find the product xy:

$x = \frac{3}{4}$ and $y = \frac{1}{4}$

$$xy = \frac{3}{4} \times \frac{1}{4} = \frac{3}{16}$$

5. (B) Hard

It is easiest to understand this question when looking at some examples of *a*, *b*, and *c*:

	a	*b*	*c*	$a \times c$
Example 1:	1	2	3	3
Example 2:	2	3	4	6
Example 3:	3	4	5	15
Example 4:	4	5	6	24
Example 5:	5	6	7	35

Notice that Example 4 produces a product of *a* and *c* that has a 4 in the units digit:

	a	*b*	*c*	$a \times c$
Example 4:	4	$\underline{5}$	6	$2\underline{4}$

The units digit of *b*, then, is 5. To see if this is always true, try some other consecutive numbers where the units digit of *b* is 5:

	a	*b*	*c*	$a \times c$
Example 6:	14	$1\underline{5}$	16	$22\underline{4}$
Example 7:	24	$2\underline{5}$	26	$62\underline{4}$

When the units digit of *ac* is 4, the units digit of *b* is always 5.

Remainders Problem Set—Page 128

1. 46 or 82 Medium

Set up *s* divided by 9.

$$\begin{array}{r} ?\ \text{R } 1 \\ 9\overline{)s} \end{array}$$

Remainder	+	Multiple of Divisor	=	Dividend
1	+	9	=	10
1	+	18	=	19
1	+	27	=	28
1	+	18	=	37
1	+	18	=	46
1	+	18	=	55
1	+	18	=	64
1	+	18	=	73
1	+	18	=	82
1	+	18	=	91

The value of *s* will always be 1 plus any multiple of 9: 19, 28, 37, 46, 55, 64, 73, 82, 91, 100, and so on.

The values between 30 and 90 are: 37, 46, 55, 64, 73, and 82

Now set up *s* divided by 4 and find the lowest possible value for *s*:

$$\begin{array}{r} ?\ \text{R }1 \\ 4\overline{)s} \end{array}$$

Remainder	+	Multiple of Divisor	=	Dividend
2	+	28	=	30
2	+	32	=	34
2	+	36	=	38
2	+	40	=	42
2	+	44	=	46
2	+	48	=	50
2	+	52	=	54
2	+	56	=	58
2	+	60	=	62
2	+	64	=	66
2	+	68	=	70
2	+	72	=	74
2	+	76	=	78
2	+	80	=	82
2	+	84	=	86
2	+	88	=	92

The value of *s* will always be 1 plus any multiple of 4. The values between 30 and 90 are: 34, 38, 42, 46, 50, 54, 58, 62, 66, 70, 74, 78, 82, and 86.

Which two values are shared when 9 and 4 are the divisors? 46 and 82

You would not need to calculate ALL of these values. Most students would have realized that 46 was a shared value and stopped calculating possible dividends.

2. (B) Medium

Because the question says "*x* could be," you should recognize BACKPLUGGING as a solution:

(A) 4

$$6\overline{)7x+3} \qquad 6\overline{)7(4)+3} \rightarrow 6\overline{)28+3} \rightarrow \begin{array}{r} 5\text{ R }1 \\ 6\overline{)31} \\ \underline{-30} \\ 1 \end{array} \qquad \text{No}$$

(B) 5

$$6\overline{)7x+3} \qquad 6\overline{)7(5)+3} \rightarrow 6\overline{)35+3} \rightarrow \begin{array}{r} 6\text{ R }2 \\ 6\overline{)38} \\ \underline{-36} \\ 2 \end{array} \qquad \checkmark$$

Fractions and Decimals Problem Set—Page 134-135

1. (D) Easy

SPLIT the question into parts and then TRANSLATE from English to math:

Part 1:

$\frac{4}{9}$ of x is 36 $\rightarrow$ $\frac{4}{9}\times x = 36$ $\rightarrow$ $4\times x = 36(9)$ $\rightarrow$ $4x = 324$ $\rightarrow$ $x = 81$

Part 2:

what is $\frac{7}{9}$ of x $\rightarrow$ $? = \frac{7}{9}\times x$ $\rightarrow$ $? = \frac{7}{9}\times 81$ $\rightarrow$ $? = \frac{567}{9}$ $\rightarrow$ $? = 63$

2. 35/6 or 5.83 Medium

This question requires manipulation of complex fractions:

$$\frac{4}{\frac{2}{3}} - \frac{\frac{2}{3}}{4} = ?$$

Work on each complex fraction individually first:

$$\frac{4}{\frac{2}{3}} \rightarrow 4\div\frac{2}{3} \rightarrow 4\times\frac{3}{2} \rightarrow \frac{12}{2} \rightarrow 6$$

$$\frac{\frac{2}{3}}{4} \rightarrow \frac{2}{3}\div 4 \rightarrow \frac{2}{3}\times\frac{1}{4} \rightarrow \frac{2}{12} \rightarrow \frac{1}{6}$$

Then complete the subtraction:

$$6-\frac{1}{6} \rightarrow \frac{6}{1}-\frac{1}{6} \rightarrow \frac{36}{6}-\frac{1}{6} \rightarrow \frac{35}{6} \text{ or } 5.83$$

3. (C) Medium

Write an equation in which t is the total number of shirts originally in the box:

$$\frac{1}{5}t + 20 = \frac{1}{3}t$$

Then solve for t:

$$20 = \frac{1}{3}t - \frac{1}{5}t \rightarrow 20 = \frac{5}{15}t - \frac{3}{15}t \rightarrow 20 = \frac{2t}{15} \rightarrow (15)20 = 2t \rightarrow 300 = 2t \rightarrow 150 = t$$

4. (A) Medium

There are two ways to solve this question. The first involves TRANSLATING and manipulating the equations:

$r \text{ is } \frac{1}{4} \text{ of } s \rightarrow r = \frac{1}{4}s$ $\qquad s \text{ is } \frac{2}{3} \text{ of } t \rightarrow s = \frac{2}{3}t \rightarrow \frac{3}{2}t = s$

$$\frac{r}{t} = \frac{\frac{1}{4}s}{\frac{3}{2}s} \rightarrow \frac{\frac{1}{4}}{\frac{3}{2}} \rightarrow \frac{1}{4} \div \frac{3}{2} \rightarrow \frac{1}{4} \times \frac{2}{3} \rightarrow \frac{2}{12} \rightarrow \frac{1}{6}$$

The second method may be easier for students who struggle with variables. SPLIT this question into parts and begin with t. SUPPLY a value for t that is divisible by 3 (because we will be multiplying by $\frac{2}{3}$).

$t = 30$

$$s \text{ is } \frac{2}{3} \text{ of } t \rightarrow s \text{ is } \frac{2}{3} \text{ of } 30 \rightarrow s = \frac{2}{3} \times 30 \rightarrow s = \frac{60}{3} \rightarrow s = 20$$

$$r \text{ is } \frac{1}{4} \text{ of } s \rightarrow r \text{ is } \frac{1}{4} \text{ of } 20 \rightarrow r = \frac{1}{4} \times 20 \rightarrow r = \frac{20}{4} \rightarrow r = 5$$

Now find $\frac{r}{t}$:

$r = 5, t = 30$

$$\frac{r}{t} = \frac{5}{30} = \frac{1}{6}$$

5. (D) Medium

TRANSLATE to create the equation:

the positive difference between t and $\frac{1}{6}$ is the same as the positive difference between $\frac{2}{3}$ and $\frac{1}{4}$

$$t - \frac{1}{6} = \frac{2}{3} - \frac{1}{4}$$

And then solve for t:

$$t - \frac{1}{6} = \frac{2}{3} - \frac{1}{4} \rightarrow t = \frac{2}{3} - \frac{1}{4} + \frac{1}{6} \rightarrow t = \frac{8}{12} - \frac{3}{12} + \frac{2}{12} \rightarrow t = \frac{7}{12}$$

6. (B) Hard

DIAGRAM the question:

$$\text{Girl 1} + \text{Girl 2} + \text{Boy} = \text{Whole Pizza}$$

$$\frac{1}{3} + \frac{1}{3} + \frac{1}{3} = 1$$

Girl 1 gives one-third of her share to the boy:

$$\frac{1}{3} \text{ of } \frac{1}{3} \rightarrow \frac{1}{3} \times \frac{1}{3} = \frac{1}{9}$$

$$\text{Girl 1} + \text{Girl 2} + \text{Boy} = \text{Whole Pizza}$$

$$\frac{1}{3} - \frac{1}{9} + \frac{1}{3} + \frac{1}{3} + \frac{1}{9} = 1$$

$$\frac{3}{9} - \frac{1}{9} + \frac{1}{3} + \frac{3}{9} + \frac{1}{9} = 1$$

$$\frac{2}{9} + \frac{1}{3} + \frac{4}{9} = 1$$

Girl 2 keeps three-fourths of her share but gives the rest to the boy:

If she keeps three-fourths of her share, she gives away one-fourth of her share:

$$\frac{1}{4} \text{ of } \frac{1}{3} \rightarrow \frac{1}{4} \times \frac{1}{3} = \frac{1}{12}$$

$$\text{Girl 1} + \text{Girl 2} + \text{Boy} = \text{Whole Pizza}$$

$$\frac{2}{9} + \frac{1}{3} - \frac{1}{12} + \frac{4}{9} + \frac{1}{12} = 1$$

$$\frac{2}{9} + \frac{4}{12} - \frac{1}{12} + \frac{16}{36} + \frac{3}{36} = 1$$

$$\frac{2}{9} + \frac{1}{4} + \frac{19}{36} = 1$$

The fraction of the pizza owned by the boy is $\frac{19}{36}$.

Note: The total fraction for each of the girls is not significant, but is calculated only to show the parts of the whole.

Percentages Problem Set—Page 139

1. 1200 Easy

This is a two-part question. First, find the balance the student receives after the tuition and room and board are deducted:

Part 1:
76% of \$20,000 is deducted
$0.76 \times \$20{,}000 =$ deducted
\$15,200 = deducted

$\$20{,}000 - \$15{,}200 = \$4800$ (balance)

Then, divide the balance by 4 years to find the amount received each year:

Part 2:
$\$4800 \div 4$ years = \$1200 per year
(Remember to disregard the \$ when gridding the answer).

2. (E) Medium

This TRANSLATION question requires you to BACKPLUG each of the answer choices to find the one that satisfies the question.

40 percent of that integer is less than 1.5
$0.40 \times ? < 1.5$ (in which ? represents each answer choice)

Solve for ?:

$0.40 \times ? < 1.5$
$? < 3.75$

The only answer less than 3.75 is (E) 3.

3. (B) Medium

SUPPLY a number for the price of the textbook: \$100

If a \$100 textbook is discounted 45%, it now costs \$55:

45% of \$100 = \$45
$\$100 - \$45 = \$55$

If the \$55 price of the textbook is increased by 20%, the new price is \$66:

20% of \$55 = \$11
\$55 + \$11 = \$66

The original price was \$100 and the final price is \$66. This is a single discount of 34%:

\$100 – \$66 = \$34

$\frac{34}{100} = 34\%$

4. (E) Hard

This higher level difficulty question becomes an easy math problem if you TRANSLATE and remember that "what percent" means "x/100":

What percent of $(9 - a^3)$ is b

$$\frac{?}{100} \times (9 - a^3) = b$$

$$\frac{?}{100} \times (9 - a^3) = b \quad \rightarrow \quad ? \times (9 - a^3) = b(100) \quad \rightarrow \quad ? = \frac{100b}{9 - a^3}$$

Answer choice (E) is correct. Can you see why answer choice (C) is the most commonly selected incorrect answer? Students who chose (C) improperly translate "what percent."

Ratios Problem Set—Page 144

1. (D) Medium

The answer for these questions must always be a multiple of the sum of the ratio:

O : L : K
2 : 3 : 5 Oranges + Limes + Kiwis = Sum → 2 + 3 + 5 = 10

The only multiple of 10 is 80, answer choice (D).

2. (E) Medium

Find the fractional part of the greatest share:

Greatest : Median : Least
8 : 4 : 3 Greatest + Median + Least = Denominator → 8 + 4 + 3 = 15

$\frac{8}{15}$

Now multiply the fractional amount by the total amount to find the value of the greatest share:

$\frac{8}{15}$ of \$48,000 is the greatest share

$$\frac{8}{15} \times \$48{,}000 = \$25{,}600$$

Answer choice (E) is correct.

3. 6 Medium

This question assesses your ability to reduce a ratio. The current ratio is 15 girls to 15 boys, or $\frac{15}{15}$. Boys must be removed until the ratio can be reduced to $\frac{5}{3}$.

Remove one boy: $\frac{15}{14}$ Cannot be reduced

Remove two boys: $\frac{15}{13}$ Cannot be reduced

Remove three boys: $\frac{15}{12}$ Reduces to $\frac{5}{4}$ (Incorrect ratio)

Remove four boys: $\frac{15}{11}$ Cannot be reduced

Remove five boys: $\frac{15}{10}$ Reduces to $\frac{5}{2}$ (Incorrect ratio)

Remove six boys: $\frac{15}{9}$ Reduces to $\frac{5}{3}$ ✓

The fewest number of boys that can be removed to create a ratio of 5 to 3 is 6.

4. (D) Hard

When you reduce a fraction or a ratio, you must divide the numerator and the denominator by the same value. Similarly, when you create equivalent fractions or ratios, you must multiply the numerator and denominator by the same value: For example:

$$\frac{12}{14} \rightarrow \frac{12 \div 2}{14 \div 2} \rightarrow \frac{6}{7} \quad \text{and} \quad \frac{3}{4} \rightarrow \frac{3 \times 5}{4 \times 5} \rightarrow \frac{15}{20}$$

The only answer choice that multiplies or divides the original ratio ($\frac{6}{7}$) by the same number (2) is (D):

$$\frac{6}{7} \rightarrow \frac{6 \div 2}{7 \div 2} \rightarrow \frac{3}{\frac{7}{2}}$$

Choice (A) multiplies the numerator by 6 and the denominator by 7. Answers (B), (C), and (E) do not have a multiplication or division relationship with the original ratio.

Proportions and Rates Problem Set—Page 154-155

1. (C) Easy

The first two conversions are proportional:

$$\frac{\text{Distance on map in inches:}}{\text{Actual distance in miles:}} \quad \frac{2.4}{15} = \frac{4.8}{30}$$

Therefore, both conversions are also proportional with the third conversion. Either of the first two will work in the proportion. We chose the second one:

$$\frac{\text{Distance on map in inches:}}{\text{Actual distance in miles:}} \quad \frac{4.8}{30} = \frac{9.6}{t} \rightarrow (4.8)(t) = (30)(9.6) \rightarrow 4.8t = 288 \rightarrow t = 60$$

The correct answer is (C), 60.

2. (B) Easy

You must be careful with the conversion from hours to minutes in this problem. Replace 1 hour with 60 minutes in the proportion:

$$\frac{\text{Number of Bottles:}}{\text{Time in } \textit{minutes}\text{:}} \quad \frac{3800}{60} = \frac{570}{?} \rightarrow (3800)(?) = (60)(570) \rightarrow (3800)(?) = 34200 \rightarrow ? = 9$$

3. (E) Medium

This is another question where conversions must be made before setting up the proportion. If 1 gallon is 16 cups, then 3 gallons is $3 \times 16 = 48$ cups.

Key: 1 gallon:

$$\frac{\text{Cups of milk:}}{\text{Number of cookies:}} \quad \frac{\frac{1}{4}}{5} = \frac{48}{?} \rightarrow \frac{1}{4}(?) = (48)(5) \rightarrow \frac{1}{4}? = 240 \rightarrow ? = 960$$

4. (D) Medium

In this proportion, there is extraneous information. The only ratios that matter are *a* and *c*:

a: *c*:

$$\frac{\text{Original ratio values:}}{\text{New ratio values:}} \quad \frac{20}{120} = \frac{5}{?} \rightarrow (20)(?) = (120)(5) \rightarrow 20(?) = 600 \rightarrow ? = 30$$

5. (A) Hard

Because the distances up and down the hill are the same, use the shortcut formula for average speeds:

$$\text{average rate of speed} = \frac{2 \times \text{rate}_1 \times \text{rate}_2}{\text{rate}_1 + \text{rate}_2} \rightarrow \frac{2 \times 10 \times 30}{10 + 30} \rightarrow \frac{600}{40} \rightarrow 15$$

6. (E) Hard

Use the shortcut formula for work problems, where:

t_1 = time taken by first person (Navid)
t_2 = time taken by second person (Janey)
t_3 = time taken by third person (Steven)
t_T = time together

$$\frac{1}{t_1} + \frac{1}{t_2} + \frac{1}{t_3} = \frac{1}{t_T} \rightarrow \frac{1}{2} + \frac{1}{4} + \frac{1}{12} = \frac{1}{t_T} \rightarrow \frac{6}{12} + \frac{3}{12} + \frac{1}{12} = \frac{1}{t_T} \rightarrow \frac{10}{12} = \frac{1}{t_T} \rightarrow t_T = \frac{12}{10} \text{ or } 1.2$$

Number Lines Problem Set—Page 160

1. (E) Medium

There are 8 tick marks from 0 to 1. Therefore, each tick mark represents and increment of one-eighth:

J (at 2/8), K (at 6/8)
0 1/8 2/8 3/8 4/8 5/8 6/8 7/8 1

As you can see, $K = \frac{6}{8}$ and $J = \frac{2}{8}$.

$$K \div J \rightarrow \frac{6}{8} \div \frac{2}{8} \rightarrow \frac{6}{8} \times \frac{8}{2} \rightarrow \frac{6}{\cancel{8}} \times \frac{\cancel{8}}{2} \rightarrow \frac{6}{2} \rightarrow 3$$

2. (A) Medium

Begin by finding the lengths of $\overline{BD}$, $\overline{BH}$, and $\overline{EF}$:

A B C D E F G H
–4 0 4 8 12

$\overline{BD} = 4$, $\overline{BH} = 12$, $\overline{EF} = 2$

Now, set up a proportion:

$\overline{BD}$ to $\overline{BH}$ $\quad$ $\overline{EF}$ to ?

First length: 4 / 2
Second length: 12 / ?

$$\frac{4}{12} = \frac{2}{?} \rightarrow (4)(?) = (12)(2) \rightarrow 4(?) = 24 \rightarrow ? = 6$$

Which of the answer choices has a length of 6? Start with (A), $\overline{AD}$:

A B C D E F G H

–4 0 4 8 12

The distance from A to D is 6. Therefore, (A) is correct and there is no need to check the other answer choices.

3. $\frac{4}{3}$ or 1.33 Hard

In order to find a^2, you must fill in the values of the tick marks. Since there are 9 spaces from 1 to 2, each tick mark represents an increment of one-ninth.

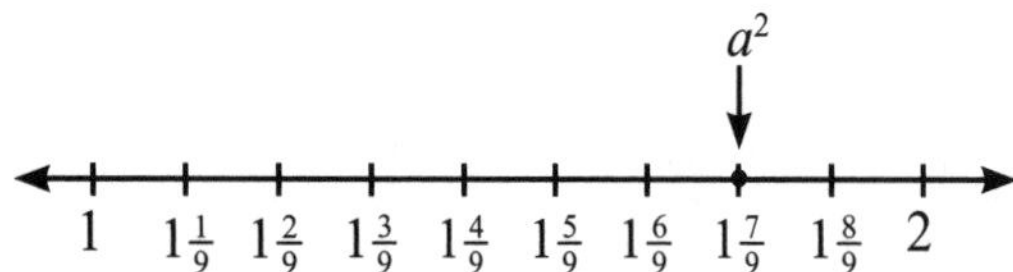

Now you know that $a^2 = 1\frac{7}{9}$. To make your calculations easier, turn the mixed fraction into regular fraction form:

$$1\tfrac{7}{9} = \frac{16}{9}$$

If $a^2 = \frac{16}{9}$, take the square root of both sides to find a:

$$\sqrt{a^2} = \sqrt{\frac{16}{9}} \quad \rightarrow \quad a = \frac{4}{3}$$

If you choose to grid-in the decimal, make sure that your answer is 1.33 (and not 1.3).

Notes

Chapter Six:
Algebra Mastery

There is sometimes a blurred line between arithmetic and algebra questions on the SAT. Often, the content can be introduced in either subject area, or a question will require skills from both arithmetic and algebra. We have divided the subject areas as evenly as possible, and tried to closely model the subject area descriptions used by the College Board.

Algebra is the branch of mathematics that deals with equations, which use variables to represent specific values.

Much of the content in algebra involves writing and solving equations. Just as in the chapter on Arithmetic Mastery, each content area in Algebra Mastery will be assigned a number from the frequency guide. Based on PowerScore's extensive analysis of real tests, you can use the following key to predict the general frequency of each type of question:

Rating	Frequency
5	Extremely High: 3 or more questions typically appear on every SAT
4	High: at least 2 questions typically appear on every SAT
3	Moderate: at least one question typically appears on every SAT
2	Low: one question typically appears on every two or three SATs
1	Extremely Low: one question appears infrequently and without a pattern

Remember that there are eight solution strategies you can employ on SAT math questions:

1. ANALYZE the Answer Choices
2. BACKPLUG the Answer Choices
3. SUPPLY Numbers
4. TRANSLATE from English to Math
5. RECORD What You Know
6. SPLIT the Question into Parts
7. DIAGRAM the Question
8. SIZE UP the Figures

Be sure to check your answers in the answer key following the chapter. It is important to understand why you missed a particular question, in order to avoid making this same mistake on the SAT.

Exponents

Frequency Guide: 4

Exponents appear frequently on the SAT, but you should not count on your calculator for much help with solving them. The makers of the test tend to use variables as the base or exponent to assess your ability to manipulate equations using exponents.

Required Knowledge and Skill Set

Most SAT questions involving exponents use variables to keep you from relying on a calculator.

1. The number being multiplied is called the *base*. The exponent, also called a *power*, indicates the number of times the base is multiplied by itself:

   $3^4 = 3 \times 3 \times 3 \times 3$ The base is 3 and the power is 4.

   $x^2 = x \times x$ The base is x and the power is 2.

2. The SAT is designed so that all problems can be solved without a calculator. For this reason, you will not be asked to calculate the value of a number raised to a power higher than the 5th or 6th power. If you are asked to find a number raised to the 6th power, expect the number to be small, such as 2 or 3 (2^6 or 3^6). If you are required to find a number to a power higher than the 6th (such as 5^8), look for a shortcut using the rules of exponent operation, discussed later in this section.

3. A negative number raised to an even-numbered power will be positive:

   $-3^6 = +729$ $-4^4 = +256$ $-7^2 = +49$

   And a negative number raised to an odd-numbered power will be negative:

   $-3^5 = -243$ $-4^3 = -64$ $-7^1 = -7$

   These rules are also true when a negative variable is raised to even- and odd-numbered powers:

   $-y^2 = y^2$ $-k^4 = k^4$ $-x^{10} = x^{10}$

   $-y^3 = -(y^3)$ $-k^7 = -(k^7)$ $-x^9 = -(x^9)$

4. You may not always have to compute the value of a number with an exponent. Some questions may preserve the exponent in the answer choices:

   (A) 3^8
   (B) 4^8
   (C) 5^8
   (D) 6^8
   (E) 7^8

   This is especially true when the base has a power higher than the 6th, as mentioned above.

Any value raised to the power of 0 equals 1.

5. A base raised to the power of zero results in 1:

   $3^0 = 1$ $12^0 = 1$ $x^0 = 1$

6. An exponent outside of a set of parentheses must be distributed to all numbers and variable within the parentheses:

$$(5x)^2 \rightarrow 5^2 \times x^2 \rightarrow 25x^2$$

The College Board often uses this rule to increase the difficulty level of an exponent question, as most students forget to distribute the exponent.

7. When a fraction is raised to a power, raise both the numerator and the denominator to the power in the exponent:

$$\left(\frac{2}{3}\right)^3 \rightarrow \frac{2^3}{3^3} \rightarrow \frac{2\times2\times2}{3\times3\times3} \rightarrow \frac{8}{27}$$

$$\left(\frac{1}{z}\right)^4 \rightarrow \frac{1^4}{z^4} \rightarrow \frac{1\times1\times1\times1}{z\times z\times z\times z} \rightarrow \frac{1}{z^4}$$

8. A negative exponent results in an inverted base and the removal of the negative sign:

$$2^{-3} \rightarrow \left(\frac{1}{2}\right)^3 \rightarrow \frac{1^3}{2^3} \rightarrow \frac{1}{8}$$

You can use the following formula to help you remember how to simplify negative exponents:

$$x^{-n} = \frac{1}{x^n}$$

9. You should be able to simplify a fraction raised to a negative exponent:

$$\left(\frac{2}{5}\right)^{-2} \rightarrow \frac{1}{\left(\frac{2}{5}\right)^2} \rightarrow \frac{1}{\frac{2^2}{5^2}} \rightarrow \frac{1}{\frac{4}{25}} \rightarrow 1\div\frac{4}{25} \rightarrow$$

$$1\times\frac{25}{4} \rightarrow \frac{25}{4}$$

$$\left(\frac{1}{t}\right)^{-3} \rightarrow \frac{1}{\left(\frac{1}{t}\right)^3} \rightarrow \frac{1}{\frac{1^3}{t^3}} \rightarrow \frac{1}{\frac{1}{t^3}} \rightarrow 1\div\frac{1}{t^3} \rightarrow$$

$$1\times\frac{t^3}{1} \rightarrow t^3$$

CAUTION: SAT TRAP!
Watch for exponents outside of parentheses. The College Board assumes many students will not distribute the exponent.

MEMORY MARKER:
Remember, all formulas are included in the free flash cards at www.powerscore.com/satmathbible.

ARITHMETRICK
As you can see in these two examples, when the number 1 is the numerator of a fraction with a fractional denominator, the simplified result is the inverse of the fraction in the denominator.

10. There are five rules of exponent operation involving multiplication and division:

Rule 1: Multiplication with the Same Base

The base remains unchanged and the exponents are added together:

MEMORY MARKER:
The rules of exponent operation must be memorized before taking the SAT.

$(x^n)(x^m) = x^{n+m}$

Examples:

$5^3 \times 5^4 = 5^{3+4} = 5^7$ $\qquad (a^4)(a^6) = a^{4+6} = a^{10}$

$(6x) \times (6x)^2 = (6x)^{1+2} = (6x)^3 = (6^3)(x)^3 = 216x^3$

Rule 2: Division with the Same Base

The base remains unchanged and the exponent of the denominator is subtracted from the exponent of the numerator:

$x^n \div x^m = x^{n-m}$ or $\dfrac{x^n}{x^m} = x^{n-m}$

Examples:

$5^8 \div 5^3 = 5^{8-3} = 5^5$ $\qquad \dfrac{a^4}{a^6} = a^{4-6} = a^{-2}$

$(6x)^5 \div (6x) = (6x)^{5-1} = (6x)^4 = (6^4)(x)^4 = 1293x^4$

Rule 3: Multiplication with the Same Power

The bases are multiplied and the exponents remain unchanged:

$(x^n)(y^n) = (xy)^n$

Examples:

$5^4 \times 3^4 = (5 \times 3)^4 = 15^4$ $\qquad (a^6)(b^6) = (ab)^6$

$(6x)^{10} \times 2^{10} = (6x \times 2)^{10} = (12x)^{10}$

Rule 4: Division with the Same Power

The bases are divided and the exponents remain unchanged:

$x^n \div y^n = (x \div y)^n$ or $\dfrac{x^n}{y^n} = \left(\dfrac{x}{y}\right)^n$

Examples:

$6^4 \div 3^4 = (6 \div 3)^4 = 2^4$ $\qquad \dfrac{a^6}{b^6} = \left(\dfrac{a}{b}\right)^6$

$(6x)^{10} \div 2^{10} = (6x \div 2)^{10} = (3x)^{10}$

CAUTION: SAT TRAP!
When an expression is raised to an exponent, be sure to distribute the exponent to all parts of the expression, as in the last example. Forgetting to distribute the power of 2 to the 6 is a common error.

Rule 5: Multiplication with a Single Base and Multiple Powers

The base remains unchanged and the exponents are multiplied:

$(x^n)^m = x^{n \times m} = x^{nm}$

Examples:

$(5^3)^4 = 5^{3\times4} = 5^{12}$ $\qquad (a^4)^6 = a^{4\times6} = a^{24}$

$(6x^5)^2 = 6^2x^{5\times2} = 36x^{10}$

11. The rules of exponent operation only work when the bases are multiplied and divided. There are no shortcuts or rules when a base is added or subtracted:

Examples:
$5^3 + 5^4 = 125 + 625 = 750$

$2^6 - 3^2 = 64 - 9 = 55$

$x^4 + y^6 = x^4 + y^6$ (No shortcut)

$r^8 - r^3 = r^8 - r^3$ (No shortcut)

However, you may be able to factor exponential expressions that are added or subtracted:

Examples:
$a^4 + a^6 = a^4(1 + a^2)$

$9x^3 - 6x^2 = 3x^2(3x - 2)$

$(6x)^2 + (2x)^3 = 36x^2 + 8x^3 = 4x^2(9 + 2x)$

There are no shortcuts for adding and subtracting expressions with exponents, although the expressions may be able to be factored.

Application on the SAT

The easiest exponent questions ask you to solve simple equations:

1. If $a^4 = b^2$ and $a = 3$, what is the value of b ?

(A) 1
(B) 3
(C) 9
(D) 27
(E) 81

To solve, plug the value of a into the equation:

$$a^4 = b^2 \quad \rightarrow \quad 3^4 = b^2 \quad \rightarrow \quad 81 = b^2$$

Then, solve for b:

$$81 = b^2 \quad \rightarrow \quad \sqrt{81} = \sqrt{b^2} \quad \rightarrow \quad 9 = b$$

These questions usually have an Easy difficulty level, and a scientific or graphing calculator is handy (although certainly not required).

Some exponent questions require you to "work backwards." In the classroom, these questions are typically introduced in Algebra II, where logarithms are the suggested solution strategy. However, on the SAT, logarithms are not required; all exponent questions can be solved with a little logic and math combined. Consider the example on the following page:

Confidence Quotation
"Whether you think you can or think you can't, either way you are right." —Henry Ford, inventor and automotive industrialist

9. If $3^y = 5$, then what is the value of 3^{2y}?

(A) 2.5
(B) 5
(C) 10
(D) 20
(E) 25

The SAT tests your ability to find patterns and apply those patterns to new situations.

This question is not so much about figuring out how 5 is the result of 3^y, as it is about understanding what happens when an exponent is doubled. Let's supply a couple of numbers for y to examine this relationship:

If $y = 2$, then $3^y = 9$ and $3^{2y} = 3^{(2)(2)} = 3^4 = 81$
The result of 3^{2y} is 3^y squared ($9^2 = 81$). To see if this is an exception or a rule, try another number for y.

If $y = 3$, then $3^y = 27$ and $3^{2y} = 3^{(2)(3)} = 3^6 = 729$
The result of 3^{2y} is 3^y squared ($27^2 = 729$). This appears to be the rule. If you still are not convinced, supply $y = 4$.

Now apply this relationship to the original question:

Since the result of 3^{2y} is 3^y squared, then $5^2 = 25$.

The correct answer is (E). Remember, logarithms are not required to solve any SAT math question.

A third type of exponent question involves the rules of exponents. The College Board may assess your ability to multiply and divide the same base or same exponent, and your ability to simplify a base with a negative exponent:

15. If $a^{20} \div a^x = a^5$ and $(a^y)^4 = a^{12}$, what is the value of $x - y$?

(A) –4
(B) 1
(C) 5
(D) 7
(E) 12

This question tests your ability to divide the same base and multiply two powers with a single base. Start with the first equation:

$a^{20} \div a^x = a^5$

$x^n \div x^m = x^{n-m}$ Therefore, $a^{20-x=5}$ $20 - x = 5$ $x = 15$

Now, find y in the second equation:

$(a^y)^4 = a^{12}$

$(x^n)^m = x^{n \times m}$ Therefore, $a^{y \times 4 = 12}$ $y \times 4 = 12$ $y = 3$

Finally, find $x - y$:

$x = 15$ and $y = 3$ $\quad x - y \rightarrow 15 - 3 \rightarrow 12$

The correct answer is (E).

Let's look at the question again to see how the test makers set several traps in the equations. Start with the first equation:

$$a^{20} \div a^x = a^5$$

You already know that the exponents are subtracted from each other when the base is the same. But imagine a student that did not know this rule or was not confident of the rule. He would see two possible scenarios:

Correct: $20 - x = 5 \quad x = 15$ Incorrect: $20 \div x = 5 \quad x = 4$

Most students who are unsure of the rule would go with the second scenario, which is incorrect. Their reasoning? "Well, the test makers used 20 and since 5 is a factor of 20, they must have wanted me to divide." Yes, they did want you to divide. Because it would lead to a wrong answer. The test makers chose these values to intentionally mislead you.

CAUTION: SAT TRAP!
The test makers choose values that are divisible in order to trick test makers who do not know that subtraction is the proper operation. They will also use answer choices that are the result of multiplying the exponents. If you are not completely confident in manipulating exponents, you could be a victim of these traps.

If a student went with the incorrect second scenario and found that $x = 4$, there are several answer choices set to trap him. Even if he solves the second equation correctly, and finds that $y = 3$, he will still miss the question:

$x = 4$ and $y = 3$ $\quad x - y \rightarrow 4 - 3 \rightarrow 1$ Answer choice (B)

Many students will also stumble with the second equation. They may confuse Rule 1 and Rule 5, and add the exponents instead of multiplying them:

$$(a^y)^4 = a^{12}$$

Correct: $y \times 4 = 12 \quad y = 3$ Incorrect: $y + 4 = 12 \quad y = 8$

Again, an answer choice is designed to trap the test taker who erroneously thinks that $y = 8$:

$x = 4$ and $y = 8$ $\quad x - y \rightarrow 4 - 8 \rightarrow -4$ Answer choice (A)

Or, if they correctly found x but missed y:

$x = 15$ and $y = 8$ $\quad x - y \rightarrow 15 - 8 \rightarrow 7$ Answer choice (C)

The College Board has spent decades studying common student errors and designing questions to take advantage of these mistakes. It is imperative that you memorize these rules of exponents, and be completely confident in your ability to manipulate them. If you have just a fair understanding of these rules, you are likely to be tricked or trapped when faced with the pressure on test day.

☠ CAUTION: SAT TRAP!
Watch for exponent questions in which the bases are added or subtracted. These problems are designed to tempt you with a fake shortcut.

Another reason that it is important you have a firm understanding of the rules of exponents is the presence of "counterfeit" questions. These exponent questions appear to be testing the rules of exponents, but the bases are being added and subtracted. As you learned earlier, there are no rules that govern the addition or subtraction of bases. These questions are used to intentionally trick you. Consider an example:

8. If n is an integer, then $6^n + 6^n =$

(A) 6^{2n}
(B) $(6n)^2$
(C) 12^n
(D) $2(6^n)$
(E) 36^n

If you do not read this question carefully, you may be duped into choosing answer choice (A), because of Rule 1 (Multiplication with the Same Base). As you remember, the base remains unchanged and the exponents are added together when the base is the same in both expressions:

$$(x^n)(x^m) = x^{n+m}$$

If you carefully read the question, though, then you know that the bases are being *added* together rather than *multiplied*. Therefore, there are no rules or shortcuts for solving this question. Answer choice (A) is incorrect.

The most efficient way to solve this question is to simply understand it. Say that we were adding two variables. The result is two times that variable:

$$x + x = 2x$$

The same is true for 6^n, even though it has an exponent:

$$6^n + 6^n = 2(6^n)$$

The correct answer is (D).

You can also SUPPLY numbers to solve this question, although it is a much slower solution method. Begin by supplying a number for n in the original equation:

If $n = 2$, then $6^n + 6^n \rightarrow 6^2 + 6^2 \rightarrow 36 + 36 \rightarrow 72$

SUPPLYING numbers for this problem is inefficient but effective, and should only be used if you forget the best solution method.

Now find the answer choice that equals 72:

(A) $6^{2n} \rightarrow 6^{2(2)} \rightarrow 6^4 \rightarrow 1296$ No
(B) $(6n)^2 \rightarrow (6\times2)^2 \rightarrow 12^2 \rightarrow 144$ No
(C) $12^n \rightarrow 12^2 \rightarrow 144$ No
(D) $2(6^n) \rightarrow 2(6^2) \rightarrow 2(36) \rightarrow 72$ ✓
(E) $36^n \rightarrow 36^2 \rightarrow 1296$ No

The final type of exponent question is often the most difficult. For these problems involving variables, you will need to rely on logic and BACKPLUGGING or SUPPLYING a number to solve. Let's examine one of these questions:

19. If $2^a = b$, then in terms of a which of the following equals $4b$?

(A) 2^{-a}
(B) 2^{2a}
(C) 2^{a+2}
(D) 2^{4a}
(E) 8^a

Begin with the first part of the question: $2^a = b$. SUPPLY a number for a to determine the value of b and $4b$:

If $a = 2$, then $2^a = 2^2 = 4$ $\quad b = 4 \quad 4b = 4(4) = 16$

Now turn your attention to the second part of the question: in terms of a which of the following equals $4b$? You can mentally remove the phrase "in terms of a" so that the question simply asks "which of the following equals $4b$?" Since $4b = 16$, you must find the answer choice that also equals 16. To do this, plug 2 in for a into all of the answer choices:

(A) 2^{-a} $\quad 2^{-2} \rightarrow \frac{1}{2^2} \rightarrow \frac{1}{4}$ No
(B) 2^{2a} $\quad 2^{2(2)} \rightarrow 2^4 \rightarrow 16$ ✓
(C) 2^{a+2} $\quad 2^{2+2} \rightarrow 2^4 \rightarrow 16$ ✓
(D) 2^{4a} $\quad 2^{4(2)} \rightarrow 2^8 \rightarrow 256$ No
(E) 8^a $\quad 8^2 \rightarrow 64$ No

Because two answer choices produced the desired result, you must supply a new number to eliminate one of the answer choices:

If $a = 3$, then $2^a = 2^3 = 8$ $\quad b = 8 \quad 4b = 4(8) = 32$

Now test the two answer choices to find the one that results in 32:

(B) 2^{2a} $\quad 2^{2(3)} \rightarrow 2^6 \rightarrow 64$ No
(C) 2^{a+2} $\quad 2^{3+2} \rightarrow 2^5 \rightarrow 32$ ✓

Answer choice (C), the correct answer, is the only one that worked when we supplied a new number. Answer choice (C) will always equal $4b$, no matter what number you supply for a.

CAUTION: SAT TRAP! Remember to avoid 0 and 1 when supplying numbers, especially to problems involving exponents, because of the special properties of those numbers.

CAUTION: SAT TRAP! Some test takers may be tempted to select answer choice (B) as the correct answer and move on. However, because this is question #19, you should also try the remaining answer choices to see if more than one equals 16. Difficult questions that require supplied numbers often have more than one answer choice that works with the first set of supplied numbers.

Exponents Problem Set

Solve the following multiple-choice questions by selecting the best answer from the five answer choices. For grid-in questions, write your answer in the grids and completely mark the corresponding ovals. Answers begin on page 222.

1. If $8^{x-4} = 8$, then $x =$?

(A) 1
(B) 2
(C) 3
(D) 4
(E) 5

2. What is the value of t if $3^{2t} = 27^{t-1}$?

(A) 0
(B) 1
(C) 2
(D) 3
(E) 4

3. If $x^{-3} = 5y$, what does y equal in terms of x ?

(A) $\frac{1}{5x^3}$

(B) $\frac{x^3}{5}$

(C) $\frac{5}{x^3}$

(D) $5x^2$

(E) $5x^3$

4. If $4^x = 64^{10}$, what is the value of x ?

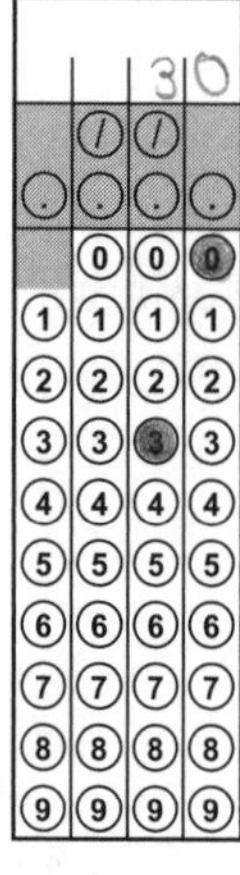

5. If x and y are positive integers, which of the following is equal to $\frac{(9^{2x})^y}{9^x}$?

(A) 1^{xy}
(B) 9^{xy}
(C) 9^{2y}
(D) $9^{x(2y-1)}$
(E) 9^{2xy-y}

Roots and Radical Equations

Frequency Guide: 3

Finding the root of a number is the inverse operation of raising a number to a power. While roots appear a little less frequently than exponents, you can still expect to see at least one square or cube root question on every SAT.

Required Knowledge and Skill Set

Having a firm understanding of radical operations will help you feel more confident on test day. You should be comfortable with all of the following key points:

1. Expect to see square roots ($\sqrt{x}$) and cube roots ($\sqrt[3]{x}$)on the SAT. Larger roots, such as a fourth root ($\sqrt[4]{x}$), are rarely—if ever—tested.

Square roots and cube roots are tested on the SAT.

2. Calculators are valuable tools when dealing with roots on the SAT. However, keep in mind that many radical questions cannot be solved with a calculator. For example, some questions require you to find the square root of a variable. Other questions leave answer choices in root terms:

   (A) $3\sqrt{2}$
   (B) 4
   (C) $4\sqrt{2}$
   (D) $4\sqrt{3}$
   (E) 5

   A calculator that converts square roots to decimals will not help you solve the question when the answers are left in root terms.

3. The root of a negative number (i.e. $\sqrt{-16}$) is not tested.

4. Positive square roots have two possible answers:

   $\sqrt{25} = -5$ or 5 $\qquad\qquad$ $\sqrt{z^2} = -z$ or z

   The test makers know that many students forget about the negative value of a square root, and thus they may use this knowledge to design difficult questions. If a question involving a square root asks "What is one possible value of x?," you should be prepared to find both the positive and negative values of the square root, as the negative value will likely be required to solve the question.

☠ CAUTION: SAT TRAP!
Never forget to find both the negative and the positive values of a square root. For difficult SAT questions, the negative value is often required.

4. Roots can be factored:

   $$\sqrt{75} \rightarrow \sqrt{25}\times\sqrt{3} \rightarrow 5\sqrt{3}$$

   $$\sqrt{24b^3} \rightarrow \sqrt{6}\times\sqrt{4}\times\sqrt{b^3} \rightarrow \sqrt{6}\times 2\times\sqrt{b^2}\times\sqrt{b} \rightarrow 2\sqrt{6}\times b\times\sqrt{b} \rightarrow 2b\sqrt{6b}$$

   $$5\sqrt{72x} \rightarrow 5\times\sqrt{2}\times\sqrt{4}\times\sqrt{9}\times\sqrt{x} \rightarrow 5\times\sqrt{2}\times 2\times 3\times\sqrt{x} \rightarrow 30\sqrt{2x}$$

5. Values with identical roots can be added and subtracted:

Chapter 3, "Operation Mastery," assesses your ability to manipulate roots.

$$5\sqrt{2}+3\sqrt{2}=8\sqrt{2} \qquad 10\sqrt[3]{7a}-\sqrt[3]{7a}=9\sqrt[3]{7a}$$

6. When a fraction is subject to roots, the root applies to both the numerator and the denominator:

$$\sqrt{\frac{1}{4}}=\frac{\sqrt{1}}{\sqrt{4}}=\frac{1}{2} \text{ or } -\frac{1}{2} \qquad \sqrt{\frac{25}{x^2}}=\frac{\sqrt{25}}{\sqrt{x^2}}=\frac{5}{x} \text{ or } -\frac{5}{x}$$

7. Roots can be expressed as a number raised to a fractional exponent:

MEMORY MARKER: You must memorize the formula for fractional exponents.

Formula: $x^{\frac{n}{m}}=\sqrt[m]{x^n}$

$$a^{\frac{1}{2}}=\sqrt[2]{a^1}=\sqrt{a} \qquad (8y)^{\frac{2}{3}}=\sqrt[3]{(8y)^2}=\sqrt[3]{64y^2}=4\sqrt[3]{y^2}$$

To help you remember this formula, think of the fractional exponent as a pesky weed that must be removed from a flower bed. It takes *power* to pull the greens growing above ground. The *root* of the weed is below ground. If you think of the ground as the fraction bar, you will always remember which number in a fractional exponent is the power and which is the root:

ARITHMETRICK
Having a hard time remembering the fractional exponent formula? Start pulling weeds!

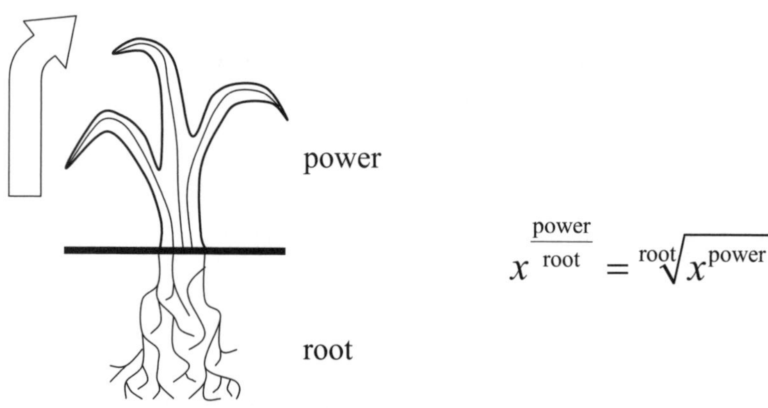

$$x^{\frac{\text{power}}{\text{root}}}=\sqrt[\text{root}]{x^{\text{power}}}$$

If you are faced with a fractional exponent and forget the meaning of the numerator and denominator, remember the weed:

$$x^{\frac{1}{3}} \rightarrow x \text{ or } x^{\frac{\text{power}}{\text{root}}} \rightarrow \sqrt[3]{x^1}$$

Note that if a variable with a fractional exponent is raised to another exponent, you may not have to convert the expression into root form. The exponents are multiplied, and often simplified to a single whole number exponent:

$$\left(x^{\frac{1}{2}}\right)^6 \rightarrow x^{\frac{1}{2}\times 6} \rightarrow x^{\frac{6}{2}} \rightarrow x^3$$

Application on the SAT

Because of the Pythagorean Theorem ($a^2 + b^2 = c^2$) and the formula for the area of a circle ($A = \pi r^2$), geometry questions will require you to manipulate roots on the SAT. In classroom math, students are conditioned to find c in the Pythagorean Theorem and the *area* in the formula for the area of the circle. On the SAT, however, it is more likely that you will be required to find a or b in the Pythagorean Theorem, and the radius in the formula for the area of a circle. You should expect to find square and cube roots in many different types of math problems on the SAT.

Because of the Pythagorean Theorem and the area of a circle, geometry questions often involve square roots.

Your ability to manipulate roots may be tested with a radical equation. A radical equation is any equation that contains a root:

12. If x is a positive integer, and $5\sqrt[3]{x} + 7 = 22$, what is the value of x?

    (A) 3
    (B) 4
    (C) 9
    (D) 27
    (E) 64

The key to solving these questions is to isolate the root:

To solve radical equations, isolate the root before removing it.

$$5\sqrt[3]{x} + 7 = 22 \quad \rightarrow \quad 5\sqrt[3]{x} = 15 \quad \rightarrow \quad \sqrt[3]{x} = 3$$

When the root is left alone on one side of the equation, cube both sides of the equation to eliminate the cube root:

$$\sqrt[3]{x} = 3 \quad \rightarrow \quad \left(\sqrt[3]{x}\right)^3 = 3^3 \quad \rightarrow \quad x = 27$$

The correct answer is (D).

Because this was a cube root, you cubed both sides. If it had been a square root, you would have squared both sides:

$$\sqrt{x} = 7 \quad \rightarrow \quad \left(\sqrt{x}\right)^2 = 7^2 \quad \rightarrow \quad x = 49$$

The final way that the SAT assesses your ability to manipulate roots is by using fractional exponents. These questions usually have a higher difficulty level:

17. If $(x+y)^{\frac{1}{2}} = x^{-\frac{1}{2}}$, what is y in terms of x ?

(A) $1 - x^2$

(B) $\frac{1-x}{x}$

(C) $\frac{1-x^2}{x}$

(D) $x^2 - 1$

(E) $\frac{x^2-1}{x}$

Given the variables, a calculator cannot help you solve this question. You must have memorized the rules of exponents and roots to simplify this equation. Simplify each side of the equation separately:

Left side:

$$(x+y)^{\frac{1}{2}} \rightarrow \sqrt[2]{(x+y)^1} \rightarrow \sqrt{x+y}$$

Remember the weed! The numerator is the power you have to exert to remove the weed. The denominator is the root.

Right side:

$$x^{-\frac{1}{2}} \rightarrow \frac{1}{x^{\frac{1}{2}}} \rightarrow \frac{1}{\sqrt[2]{x^1}} \rightarrow \frac{1}{\sqrt{x}}$$

And then set the two expressions back equal to each other:

$$\sqrt{x+y} = \frac{1}{\sqrt{x}}$$

Square both sides to remove the square roots:

$$\sqrt{x+y} = \frac{1}{\sqrt{x}} \rightarrow \left(\sqrt{x+y}\right)^2 = \left(\frac{1}{\sqrt{x}}\right)^2 \rightarrow x+y = \frac{1}{x}$$

ARITHMETRICK

Remember, you can perform any operation on an equation as long as you do the same thing to both sides of the equation. Multiplying powers may be easier than simplifying fractional exponents.

If you forget how to simplify fractional exponents, for this problem you can also square both sides of the equation to simplify the exponents:

$$\left[(x+y)^{\frac{1}{2}}\right]^2 = \left(x^{-\frac{1}{2}}\right)^2 \rightarrow (x+y)^{\frac{2}{2}} = x^{-\frac{2}{2}} \rightarrow (x+y)^1 = x^{-1} \rightarrow x+y = \frac{1}{x}$$

Now solve for y to learn that the correct answer is (C):

$$x+y = \frac{1}{x} \rightarrow x(x+y) = 1 \rightarrow x^2 + xy = 1 \rightarrow xy = 1 - x^2 \rightarrow \frac{1-x^2}{x}$$

Roots and Radical Equations Problem Set

Solve the following multiple-choice questions by selecting the best answer from the five answer choices. For grid-in questions, write your answer in the grids and completely mark the corresponding ovals. Answers begin on page 224.

1. If $9 + \sqrt[3]{x} = 12$, then what is the value of x ?

(A) 1
(B) 8
(C) 9
(D) 27
(E) 64

2. For what positive number is the number divided by 50 the same as the square root of the number?

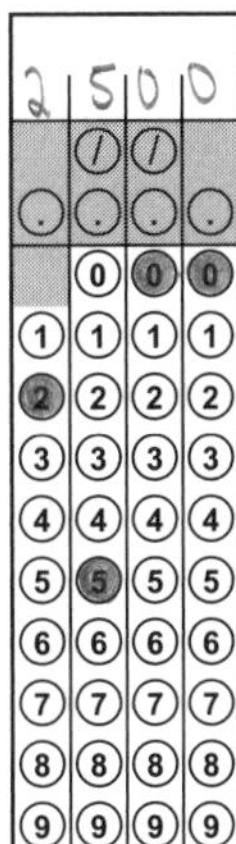

$$12\sqrt{12} = k\sqrt{m}$$

3. In the equation above, if k and m are positive integers and $k > m$, which of the following is a possible value of $k + m$?

(A) 15
(B) 24
(C) 27
(D) 41
(E) 51

4. If $(cd)^{\frac{1}{3}} = (c+d)^{\frac{-1}{3}}$, which of the following must be true?

(A) $c + d = 1$
(B) $cd = 1$
(C) $2c + 2d = 1$
(D) $c^2 + d^2 = 1$
(E) $c^2d + cd^2 = 1$

Scientific Notation

Frequency Guide: 1

Scientific notation was created in order to express very large numbers in as few symbols as possible. The makers of the SAT expect you to understand and manipulate numbers written in scientific notation.

Required Knowledge and Skill Set

Scientific notation occurs infrequently on the SAT.

1. Scientific notation uses two numbers: the coefficient and the base. The coefficient is a number from 1 to 10. The base is always the number 10 with an exponent to indicate the power of 10 that must be multiplied to find the exact number. Consider an example:

   $1240000 = 1.24 \times 10^6$

   Coefficient (1.24) — Base (10^6)

2. To write a number in scientific notation, place a decimal point after the first digit. If the last digits of the number are zeros, drop the zeros. Then, count the number of digits after the decimal point to determine the exponent:

   53000 — $53000 = 5.3 \times 10^4$ (digits counted: 1 2 3 4)

   9342854 — $9342854 = 9.342854 \times 10^6$ (digits counted: 1 2 3 4 5 6)

3. To read numbers in scientific notation, simply move the decimal point the number of spaces indicated by the exponent in the base. Add zeros if there are not enough digits:

   7.42705×10^5 — $7.42705 \times 10^5 = 7.42705.$ (1 2 3 4 5) — 742705

   4.16×10^6 — $4.16 \times 10^6 = 4.160000.$ (1 2 3 4 5 6) — 4160000

4. Long decimals can also be written using scientific notation:

   $0.00004 = 4.0 \times 10^{-5}$

   Again, place the decimal after the first non-zero digit. Count the number of digits that the decimal point moved to determine the negative exponent:

   0.0036 — $0.003.6 = 3.6 \times 10^{-3}$ (1 2 3)

   0.02319 — $0.02.319 = 2.319 \times 10^{-2}$ (1 2)

Confidence Quotation
"It doesn't matter who you are or where you come from. The ability to triumph always begins with you. Always."
—Oprah Winfrey, television host and producer

5. Numbers written in scientific notation can be added and subtracted if they have the same power of 10 base. Simply add or subtract the coefficients and use the same base:

   $(4.5 \times 10^4) + (2.1 \times 10^4) \rightarrow (4.5 + 2.1) \times 10^4 \rightarrow 6.6 \times 10^4$

   $(7.82 \times 10^8) - (3.5 \times 10^8) \rightarrow (7.82 - 3.5) \times 10^8 \rightarrow 4.32 \times 10^8$

   If the power of 10 is not the same in the two bases, you must convert one of the numbers so that it has the same power of 10 as the other:

   $(2.9 \times 10^3) + (5.1 \times 10^5)$

   The second number has a base with the power of 5. How can it have a base with the power of 3? Move the decimal point two places to the right:

   $5.1 \times 10^5 = 510000$
   $510.000 \times 10^3 = 510000$

   Therefore, $5.1 \times 10^5 = 510 \times 10^3$

   Now you can add the two numbers:

   $(2.9 \times 10^3) + (510 \times 10^3) \rightarrow (2.9 + 510) \times 10^3 \rightarrow 512.9 \times 10^3$

   Remember, proper scientific notation has a coefficient between 1 and 9. You must write your result using proper notation:

   $512.9 \times 10^3 \rightarrow 5.129 \times 10^5$

   By moving the decimal point two spaces to the left, you must add 2 to the exponent on the base.

Moving the decimal to the left ADDS to the value of the exponent; moving the decimal to the right SUBTRACTS from the value of the exponent. You can remember this because there are four letters in both LEFT and ADDS.

6. Numbers written in scientific notation can also be multiplied and divided.

   $(2.35 \times 10^7) \times (4.8 \times 10^5)$

   First, multiply the coefficients:

   $2.35 \times 4.8 = 11.28$

   Next, multiply the bases. When multiplying numbers with the same base, simply add the exponents and keep the base:

   $10^7 \times 10^5 = 10^{(7+5)} = 10^{12}$

   The result is the new coefficient and new base:

   $11.28 \times 10^8 \rightarrow 1.128 \times 10^9$

   Remember, you must use proper notation, thus moving the decimal point one place to the left and adding one to the base power.

Application on the SAT

Most SAT questions regarding scientific notation are Medium difficulty level. A variable will be used for a coefficient or base exponent:

12. If $(5 \times 10^8) + (2 \times 10^x) = 5.02 \times 10^8$, what is the value of x ?

    (A) 1
    (B) 2
    (C) 3
    (D) 6
    (E) 8

The resulting coefficient is 5.02, indicating that 5 and 0.02 were added together. You know from the rules of scientific notation that the bases must be the same in order to add the coefficients. Therefore:

Scientific notation questions on the SAT are straightforward and usually ask you to find the value of one of the exponents.

$$(5 \times 10^8) + (0.02 \times 10^8) = 5.02 \times 10^8$$

In order for this new coefficient to equal 2, the decimal place must be moved two places to the right, thus decreasing the exponent by 2:

$$0.02 \times 10^8 = 2 \times 10^6$$

$$2 \times 10^6 = 2 \times 10^x$$

Therefore, $x = 6$.

These questions are fairly rare on the SAT. Still, you should be well-versed in the rules of adding, subtracting, multiplying, and dividing scientific notation in the event that one of these operations appears on your test.

Scientific Notation Problem Set

Solve the following multiple-choice questions by selecting the best answer from the five answer choices. For grid-in questions, write your answer in the grids and completely mark the corresponding ovals. Answers begin on page 225.

1. If $(4 \times 10^4) + (5 \times 10^5) = y \times 10^5$ then what is the value of y ?

2. If $(3 \times 10^3) \times (5 \times 10^4) = 1.5 \times 10^n$, what is the value of n ?

   (A) 4
   (B) 6
   (C) 7
   (D) 8
   (E) 12

Expressions

Frequency Guide: 4

While classroom math focuses mainly on solving equations, the SAT asks you to also set up those equations by creating expressions.

Required Knowledge and Skill Set

1. An expression is a group of two or more terms. Each of the following are examples of expressions:

   $3x$

   $24ab$

   $4(m+n) \quad \rightarrow \quad 4m+4n$

   x^2-5

   $4z^2(y+2) \quad \rightarrow \quad 4yz^2+8z^2$

   $\dfrac{2t+7}{3}$

Confidence Quotation
"He started to sing as
he tackled the thing
That couldn't be done,
and he did it."
—Edgar A. Guest, poet

2. Word problems can be turned into expressions by assigning variables to unknown quantities. Use your knowledge of TRANSLATION terms to determine the operations required by the question:

   An ice cream company can make g gallons per h hours $\rightarrow \dfrac{g}{h}$

   On the SAT, variables are often provided in the problem, as in the example above.

3. You will never be asked to find an expression for a Student-Produced Response question. The grid-ins do not allow for variables as answers.

Application on the SAT

Most SAT questions concerning expressions ask you to select an expression that accurately reflects a word problem:

8. An ice cream company can make ice cream at a rate of g gallons in h hours. If the ice cream earns the company \$10 per gallon, how many dollars worth of ice cream will be made in x hours?

    (A) $\frac{10h}{gx}$

    (B) $\frac{gx}{10h}$

    (C) $\frac{gh}{10x}$

    (D) $\frac{10gh}{x}$

    (E) $\frac{10gx}{h}$

As with so many SAT math questions, there are multiple ways to solve this question. The most efficient solution is to understand what the question is asking, and create an expression based on the information presented in the problem.

The most efficient solution method for expression questions is usually TRANSLATION.

Start with the first statement:

An ice cream company can make ice cream at a rate of g gallons in h hours

TRANSLATE:

$$\frac{g}{h}$$

Now look at the second sentence and translate:

the ice cream earns the company \$10 per gallon $\rightarrow$ $10g$

Combine the information from the two statements:

$$\frac{10g}{h}$$

Finally, consider the question:

how many dollars worth of ice cream will be made in x hours?

$\frac{10g}{h} \times x = \frac{10gx}{h}$ The correct answer is (E).

SUPPLYING numbers is an adequate, although less efficient, solution for those who struggle with TRANSLATING variables.

For some students, creating expressions from variables proves too confusing. If this is true for you, SUPPLY numbers for each of the variables. This solution method is not the most efficient, but it will lead to the correct answer:

g gallons = 2
h hours = 1
x hours = 5

Now TRANSLATE:

make ice cream at a rate of g gallons in h hours $\rightarrow \dfrac{2 \text{ gallons}}{1 \text{ hour}} \rightarrow \dfrac{g}{h}$

the ice cream earns the company \$10 per gallon $\rightarrow \dfrac{\$20}{1 \text{ hour}} \rightarrow \dfrac{10g}{h}$

how many dollars worth of ice cream will be made in x hours?

$$\frac{\$20}{1 \text{ hour}} \times 5 \text{ hours} = \$100 \rightarrow \frac{10g}{h} \times x$$

By resupplying the variable after each numerical expression, it is clear that the correct answer is (E).

But what if resupplying the variable still confuses you? Simply plug your supplied numbers into the answer choices to find the one that results in \$100.

g gallons = 2
h hours = 1
x hours = 5

(A) $\dfrac{10h}{gx} \rightarrow \dfrac{(10)(1)}{(2)(5)} \rightarrow \dfrac{10}{10} \rightarrow 1$ No

(B) $\dfrac{gx}{10h} \rightarrow \dfrac{(2)(5)}{(10)(1)} \rightarrow \dfrac{10}{10} \rightarrow 1$ No

(C) $\dfrac{gh}{10x} \rightarrow \dfrac{(2)(1)}{(10)(5)} \rightarrow \dfrac{2}{50} \rightarrow \dfrac{1}{25}$ No

(D) $\dfrac{10gh}{x} \rightarrow \dfrac{(10)(2)(1)}{5} \rightarrow \dfrac{20}{5} \rightarrow 4$ No

(E) $\dfrac{10gx}{h} \rightarrow \dfrac{(10)(2)(5)}{1} \rightarrow \dfrac{100}{1} \rightarrow 100$ ✓

We show every step of each solution for the benefit of all readers. However, on the SAT, you should be able to do many of these calculations in your head.

While this is not the most efficient solution method, you can do many of the calculations in your head to quickly eliminate wrong answer choices.

Another type of expression question asks you to follow directions to produce an equation:

The product of x and $\frac{1}{2}$ is equal to the sum of $3x$ and 10.

2. For the statement above, which of the following equations could be used to find x ?

(A) $\frac{1}{2}x + 10 = 3x$

(B) $x + \frac{1}{2} = 10(3x)$

(C) $\frac{1}{2}x = 3(x + 10)$

(D) $x \div \frac{1}{2} = 3x + 10$

(E) $\frac{1}{2}x = 3x + 10$

Equations may be made up of many different expressions.

To solve this problem, simple TRANSLATION can be used to see that the correct answer is (E). Sometimes, though, this same question may ask you to find the value of x rather than the expression or equation:

The product of x and $\frac{1}{2}$ is equal to the sum of $3x$ and 10.

8. For the statement above, what is the value of x ?

(A) –4
(B) 0
(C) 2
(D) 4
(E) 12

Notice that the difficulty of the question was raised from an Easy to a Medium just by requesting a value. This is because students are now required to write their own equations:

The product of x and $\frac{1}{2}$ is equal to the sum of $3x$ and 10

$\frac{1}{2}x = 3x + 10$

$x = 2(3x + 10)$
$x = 6x + 20$
$-5x = 20 \rightarrow x = -4$ The correct answer is (A)

When writing expressions or equations based on directions in the problem, watch for

You should know how to represent consecutive numbers (n, n+1, n+2,...), consecutive even numbers (n, n+2, n+4,...), and consecutive odd numbers (also n, n+2, n+4,...).

the use of consecutive odd or even numbers:

The sum of three consecutive even integers is multiplied by 10.

16. If n represents the least of these three integers, which of the following expressions represents the relationship in the statement above?

(A) $30n$
(B) $30n + 20$
(C) $30n + 40$
(D) $30n + 60$
(E) $30n + 90$

As the sixteenth question in the section, this question is Medium-Hard or Hard. Remember, though, that these labels are used to show how many students previously missed the question. It is actually quite easy as long as you remember how to represent consecutive odd or even numbers:

First number: n
Second number: $n + 2$
Third number: $n + 4$
Fourth number: $n + 6$ Etc.

To see why this is true, supply both an even and odd number for n:

	Even	Odd	Relationship:
First number (n):	4	5	n
Second number:	6	7	$n + 2$
Third number:	8	9	$n + 4$
Fourth number:	10	11	$n + 6$

Now use TRANSLATION to solve the question:

The sum of three consecutive even integers is multiplied by 10

(First + Second + Third) × 10

$[(n) + (n + 2) + (n + 4)] \times 10$

$(3n + 6)10$

$30n + 60$

The correct answer is (D).

It is easy to see why so many people miss this question. You must remember the proper representation of consecutive odd and even integers.
One other question style to watch for is a sales problem in which the first item (or

occasionally a time period) is a different price than the remaining items purchased. Consider an example:

18. At the grocery store, Lynne can buy one roll of paper towels for d dollars. Each additional roll costs x dollars less than the first roll. For example, the price of the second roll is $d - x$ dollars. Which of the following expressions represents Lynne's cost, in dollars, for n rolls of paper towels bought at the grocery store?

    (A) $n(d - x)$

    (B) $(d + n)(d - x)$

    (C) $d + (n - 1)(d - x)$

    (D) $\dfrac{d+(d-x)}{n-1}$

    (E) $(d-x)+\dfrac{d-x}{n}$

To understand this problem, look at the cost of four rolls of paper towels:

First roll:	d
Second roll:	$d - x$
Third roll:	$d - x$
Fourth roll:	$d - x$
Total cost:	$d + (d - x) + (d - x) + (d - x)$
Total cost:	$d + 3(d - x)$

Why do we multiply $d - x$ by 3 rolls instead of by 4 rolls? Because the first roll has a different price, and thus must be subtracted from the number multiplied by $d - x$. This is represented by $n - 1$:

Total cost: $d + (n - 1)(d - x)$

The correct answer is (C).

For questions like this, where the first item has a different price than all of the other items, look for answers with a binomial that subtracts 1 from the variable representing the total number of items.

Working with expressions is a highly common procedure on the SAT. TRANSLATION is the most efficient tool for creating an expression, but you can always SUPPLY numbers if concrete values are easier to manipulate.

If you are forced to guess on a question like this, improve your odds by choosing an answer with a binomial that subtracts 1 from the variable representing the total number of objects, such as (n – 1) or (x – 1).

Expressions Problem Set

Solve the following multiple-choice questions by selecting the best answer from the five answer choices. For grid-in questions, write your answer in the grids and completely mark the corresponding ovals. Answers begin on page 226.

1. A clothing store had $4x$ shirts for sale at a price of t dollars each. If a is the number of shirts that did not sell, which of the following represents the total dollar amount earned from the shirts that did sell?

   (A) $t(a - 4x)$
   (B) $t(4x - a)$
   (C) $4x - a$
   (D) $4xt - a$
   (E) $4ax - t$

   $t(4x-a)$

2. A group of x salespeople have visited 200 potential clients. If each salesperson visits c more clients per day for the next d days, which of the following expressions represents the total number of clients visited by all of the salespeople?

   (A) $200 + cdx$ $\quad 200 + cdx$

   (B) $200cdx$

   (C) $200 + \frac{dx}{c}$

   (D) $200 - \frac{dc}{x}$

   (E) $200 + cd + x$

3. The rate for a long distance telephone call is \$1.00 for the first minute and \$0.10 for each additional minute. Which of the following describes the cost, in dollars, of a phone call that lasts for m minutes?

   (A) $1.10m$
   (B) $1.00 + 0.10m$
   (C) $1.00m + 0.10m$
   (D) $1.00 + 0.10(m - 1)$
   (E) $1.00m + 0.10(m - 1)$

   \$1.00 (1)
   \$0.10(m-1)

Expressions Problem Set

Solve the following multiple-choice questions by selecting the best answer from the five answer choices. For grid-in questions, write your answer in the grids and completely mark the corresponding ovals. Answers begin on page 226.

4. The expression $\frac{b}{2}+\frac{3+b}{2}$ is how much greater than b?

5. Four is added to a certain number y. The sum is divided by 5. Then 3 is subtracted from the quotient. Which of the following expressions represents the result of these operations?

   (A) $\frac{y+1}{5}$

   (B) $\frac{y+1}{15}$

   (C) $\frac{y+4}{15}$

   (D) $\frac{y-11}{5}$

   (E) $\frac{4y-15}{5}$

6. An animal shelter houses only dogs and cats. There are 40 more dogs than cats living at the shelter. If there are x cats at the shelter, then, in terms of x, what percent of the animals are cats?

   (A) $\frac{x}{x+40}\%$

   (B) $\frac{x}{2x+40}\%$

   (C) $\frac{2x}{100(x+40)}\%$

   (D) $\frac{100x}{x+40}\%$

   (E) $\frac{100x}{2x+40}\%$

Equations

Frequency Guide: 5

Equations are among the most popular SAT questions. You should expect to see at least three questions involving equations on every SAT.

Required Knowledge and Skill Set

1. An equation contains two or more equal expressions. Thus, an equation is easily identifiable by the presence of an equals sign:

$$1 + 5 = 6$$

$$y = 3z$$

$$2(a + b) = 10$$

$$x^2 + 9 = 16$$

$$\frac{x}{x-7} = 2y + 15$$

Equations may prove true or false. On the SAT, correct answers are typically values that make an equation true.

☠ CAUTION: SAT TRAP! High school students are trained to find the value for 'x' that makes an equation TRUE. Watch for the occasional question that ask which of the values makes an equation FALSE. These questions usually involve the phrasing "all of the following EXCEPT...".

2. Any quantity can be added to, subtracted from, multiplied to, and divided into one side of the equation as long as the same operation is performed with the same quantity on the other side of the equation. You can also square, cube, or find the root of both sides of the equation.

$$1 + 5 = 6 \quad \rightarrow \quad 1 + 5 + 10 = 6 + 10$$

$$2(a + b) = 10 \quad \rightarrow \quad \frac{a+b}{2} = \frac{10}{2}$$

$$y = 3z \quad \rightarrow \quad y^2 = (3z)^2$$

$$x^2 + 9 = 16 \quad \rightarrow \quad \sqrt{x^2 + 9} = \sqrt{16}$$

Application on the SAT

Because equations are such a large part of the SAT, we have already discussed many of the ways they are applied on the test. For example, when '0' is an answer choice, you should always backplug it first into the equations, as the College Board may be testing your knowledge of the zero property law. Or, when a question uses "in terms of," as in "what is x in terms of y," you should cross out the phrase "in terms of y" to make it clear that you are solving for x. But there are a few more pointers we can give on this popular question type.

There is almost always one Easy equation problem on every SAT:

1. If $2a + b = 12$ and $b = 4a$, then $a =$

   (A) 0
   (B) 1
   (C) 2
   (D) 3
   (E) 6

These questions usually occur as the first or second question in a section, and should be solved just as you would solve them in math class:

$$2a + b = 12 \text{ and } b = 4a$$

$$2a + 4a = 12 \quad \rightarrow \quad 6a = 12 \quad \rightarrow \quad a = 2$$

Be sure to read the questions about equations carefully to make sure that you are solving for the required variable. Sometimes there are shortcuts you might miss if you proceed too quickly:

8. If $3x + 2y = 17$ and $x + 4y = 9$, what is the value of $5(3x + 2y)$?

   (A) 25
   (B) 60
   (C) 85
   (D) 100
   (E) 108

If you rush through this problem, you might begin using substitution to find the value of a single variable:

$$x + 4y = 9 \quad \rightarrow \quad x = 9 - 4y$$

$$3x + 2y = 17$$
$$3(9 - 4y) + 2y = 17$$
$$27 - 12y + 2y = 17$$
$$27 - 10y = 17$$
$$-10y = -10$$
$$y = 1$$

It is important to carefully read questions with equations. For example, this question could have asked you to solve for 'a' or 'b,' so it is important to take note when reading the question.

We recommend that you reread every question after finding a solution to make sure you solved for the correct variable or expression.

Then you would plug the value for y into an equation to find x:

$$y = 1 \quad \rightarrow \quad x + 4y = 9 \quad \rightarrow \quad x + 4(1) = 9 \quad \rightarrow \quad x = 5$$

Finally, you would plug these two values back into the required expression:

What is the value of $5(3x + 2y)$? $\quad y = 1 \quad x = 5$

$$5[(3 \times 5) + (2 \times 1)] \quad \rightarrow \quad 5(15 + 2) \quad \rightarrow \quad 85$$

☠ CAUTION: SAT TRAP!
Be wary of using substitution as a solution method on the SAT. There is usually a faster and easier method.

While 85 is the correct answer, a careful review of the question would have resulted in the same answer in a fraction of the time. The second equation, $x + 4y = 9$, is unnecessary. Compare the first equation with the required expression:

$$3x + 2y = 17$$
$$5(3x + 2y) = ?$$

Since one side of the equation is multiplied by 5, simply multiply the other side by 5 to find the answer:

$$5(3x + 2y) = 5(17)$$
$$5(3x + 2y) = 85$$

Just a few seconds of careful study can save you several minutes over the course of a test.

In the previous individual sections on fractions and radicals, we discussed the College Board's tendency to use these concepts to make equations more difficult. It is an important point, and it bears mentioning again. You must be confident in your ability to add, subtract, multiply, and divide fractions. Likewise, you should be comfortable isolating a variable in a square or cube root, and then squaring or cubing both sides of the equation. Let's look at an example:

7. If $\frac{4}{z} + \frac{7}{5} = 1$, what is the value of z?

(A) -10
(B) 2.2
(C) 5
(D) 5.4
(E) 5.6

This question tests your ability to add fractions:

$$\frac{4}{z} + \frac{7}{5} = 1 \quad \rightarrow \quad \frac{(4)5}{(z)5} + \frac{(7)z}{(5)z} = 1 \quad \rightarrow \quad \frac{20}{5z} + \frac{7z}{5z} = 1 \quad \rightarrow \quad 20 + 7z = 5z \quad \rightarrow$$

$$20 = -2z \quad \rightarrow \quad -10 = z$$

The test makers may try to complicate equations with fractions by using multiple variables. You must remember what you are solving for, and then perform the operations on variables the same as you would on numbers. Consider a Hard level example:

19. If $\frac{s+x}{rx}=1$, and r and s are positive, then what is the value of x ?

(A) $\frac{r+1}{s}$

(B) $\frac{r+1}{s-1}$

(C) $\frac{r}{s-1}$

(D) $\frac{s}{r-1}$

(E) $\frac{r-1}{s}$

If 'r' and 's' were actual values, how would you solve for 'x'? Use this same strategy when they are variables instead of values.

If this question provided values for r and s rather than variables, it would be a Medium level question. But students tend to be intimidated by variables, so you must be confident in your ability to manipulate them:

$$\frac{s+x}{rx}=1 \quad \rightarrow \quad s+x=rx \quad \rightarrow \quad s=rx-x \quad \rightarrow \quad s=x(r-1) \quad \rightarrow \quad \frac{s}{r-1}=x$$

The correct answer is (D).

Finally, remember to use the Solution Strategies you have learned. Even though equations have equals signs, you still may need to SUPPLY numbers on occasion. This is especially true when you have a single equation with two variables. Examine a grid-in example:

13. If $y+z=6$ and y and z are different positive integers, what is one possible value of $3yz$?

This question is impossible to solve without SUPPLYING numbers for y and z. When supplying numbers for an equation, you can use any numbers that make the equation true (provided they follow any rules, of course). Therefore, y and z could be:

$y=1, z=5 \qquad y=2, z=4 \qquad y=4, z=2 \qquad y=5, z=1$

Note that y and z are *different* positive integers, so neither can equal 3 because then both would equal 3. You need to pick only one of the pairs of supplied numbers above ($y=1$, $z=5$), and then plug them into the final expression:

$3yz = 3(1)(5) = 15$

The possible answers for this question are 15 and 24.

Equations Problem Set

Solve the following multiple-choice questions by selecting the best answer from the five answer choices. For grid-in questions, write your answer in the grids and completely mark the corresponding ovals. Answers begin on page 228.

1. If $4x + 6 = 30$, then $4x - 6 =$

(A) 0
(B) 6
(C) 18
(D) 20
(E) 24

2. If $5c + 3d = 50$ and $3c + d = 24$, what is the value of $\frac{1}{2}(5c + 3d)$?

(A) 20
(B) 22
(C) 25
(D) 26
(E) 28

3. If n is a positive integer, and $3\left(\sqrt{n+4}\right) - 8 = 7$, what is the value of n ?

(A) 3
(B) 9
(C) 21
(D) 23
(E) 29

Equations Problem Set

Solve the following multiple-choice questions by selecting the best answer from the five answer choices. For grid-in questions, write your answer in the grids and completely mark the corresponding ovals. Answers begin on page 228.

4. If $x + 7y = 40$ and x and y are different positive integers, what is the sum of all possible values of y ?

(A) 5
(B) 10
(C) 12
(D) 15
(E) 21

5. If $c > 0$ and $d = 2 - \frac{1}{c}$, then d could equal which of the following?

(A) $\frac{4}{5}$

(B) $\frac{7}{5}$

(C) $\frac{9}{5}$

(D) $\frac{16}{5}$

(E) $\frac{21}{5}$

$$x = y + \frac{y}{5}$$

6. In the equation above, y is an integer greater than 0. Which of the following must be true?

I. $x > y$
II. $y^2 > xy$
III. $y + 2 > x$

(A) I only
(B) II only
(C) I and III only
(D) II and III only
(E) I, II, and III

Systems of Equations

Frequency Guide: 2

A system of equations question occurs when two or more equations are present. There are two ways to solve a system of equations.

Required Knowledge and Skill Set

1. The first solution method is substitution. Substitution uses one equation to isolate a single variable, the value of which is then plugged into another equation to solve for the second variable. To demonstrate substitution, we will find the value of x and y in the following system of equations:

   $3x + 2y = 16 \qquad 6x - 2y = 20$

   Solve for y in the second equation:

   $$6x - 2y = 20 \rightarrow -2y = 20 - 6x \rightarrow y = \frac{20 - 6x}{-2} \rightarrow y = -10 + 3x$$

   Now plug this value for y into the first equation:

   $y = -10 + 3x$

   $3x + 2y = 16 \rightarrow 3x + 2(-10 + 3x) = 16 \rightarrow 3x + -20 + 6x = 16 \rightarrow$
   $9x = 36 \rightarrow x = 4$

   Finally, to find y, plug 4 into one of the equations for x:

   $3x + 2y = 16$

   $3(4) + 2y = 16 \rightarrow 12 + 2y = 16 \rightarrow 2y = 4 \rightarrow y = 2$

   Most students are quite comfortable with this method and rely on it heavily in classroom math.

Of the two solution methods, adding is most commonly used as the trick solution. As indicated in the previous section, be wary of using substitution.

2. The second solution method is adding equations to each other. Look again at the same system of equations:

   $3x + 2y = 16 \qquad 6x - 2y = 20$

   Write the equations in a column fashion, where all of the variables line up:

   $3x + 2y = 16$
   $6x - 2y = 20$

   Now, add them together to eliminate y:

   $$\begin{array}{r} 3x + 2y = 16 \\ \underline{+\ 6x - 2y = 20} \\ 9x \qquad = 36 \end{array}$$

And then solve for x:

$9x = 36 \quad \rightarrow \quad x = 4$

Then you can solve for y by plugging the value of x into either equation:

$6x - 2y = 20$

$6(4) - 2y = 20 \quad \rightarrow \quad 24 - 2y = 20 \quad \rightarrow \quad -2y = -4 \quad \rightarrow \quad y = 2$

Sometimes you must multiply one of the equations by a negative number to eliminate a set of variables:

$3x + 2y = 16 \qquad\qquad 6x + 2y = 20$

When these equations are written in column form, it is clear that you cannot add them to eliminate y:

$$\begin{array}{r} 3x + 2y = 16 \\ \underline{+\ 6x + 2y = 20} \\ 9x + 4y = 36 \end{array}$$

Multiply the second equation by –1:

$-1(6x + 2y = 20) \quad \rightarrow \quad -6x - 2y = -20$

Now add them together:

$$\begin{array}{r} 3x + 2y = \ 16 \\ \underline{+ -6x - 2y = -20} \\ -3x \qquad = \ -4 \end{array}$$

As you can see, $x = \frac{4}{3}$. When plugged into either equation, you will find that $y = 6$.

This solution method is very important on the SAT, as we will discuss in the next section. Therefore, you must also be able to add or subtract equations when the variables are not equal:

$8x + y = 34 \qquad\qquad 3x - 5y = 59$

Write the equations in a column fashion, where all of the variables line up:

$$\begin{array}{r} 8x + \ \ y = 34 \\ 3x - 5y = 59 \end{array}$$

If we add them together now, no variables are eliminated.

$$\begin{array}{r} 8x + \ \ y = 34 \\ \underline{+\ 3x - 5y = 59} \\ 11x - 4y = 36 \end{array}$$

Confidence Quotation
"Impossible is a word only to be found in the dictionary of fools."
—Napoleon Bonaparte, French general and politician

To eliminate one set of variables, they must be equal in the two equations. To do this, multiple one or both equations by any value that creates equal variables. You can multiply the top equation by 3 and the bottom by –8:

$$3(8x + y = 34) \rightarrow 24x + 3y = 102$$
$$-8(3x - 5y = 59) \rightarrow -24x + 40y = 472$$

Or you can simply multiply the first equation by 5:

$$5(8x + y = 34) \rightarrow 40x + 5y = 170$$
$$3x - 5y = 59$$

Both of these solutions will result in $x = 3$ and $y = 10$.

3. Some systems of equation questions contain three or more equations. You can also solve these systems by using substitution or addition.

4. A system of equations has no solution when all variables can be eliminated:

$$10x + 4y = 16$$
$$-5x - 2y = 20$$

Multiply the second equation by 2:

$$2(-5x - 2y = 20) \rightarrow -10x - 4y = 40$$

Now attempt to add them together:

$$\begin{array}{r} 10x + 4y = 16 \\ \underline{+ -10x - 4y = 40} \\ = 56 \end{array}$$

Because both variables are eliminated, this system of equations does not have a solution.

A system of equations has no solution when all variables are eliminated.

Application on the SAT

Expect to use substitution on Easy level questions and those with three two-term equations, like so:

10. If $a = 4b$, $b = 2c$, $a = xc$, and $x \neq 0$, then $x =$

(A) $\frac{1}{8}$

(B) $\frac{1}{2}$

(C) 2

(D) 4

(E) 8

Substitution is usually only needed for Easy level questions.

Substitute values:

$a = 4b$ and $b = 2c$ $\qquad a = 4(2c) \rightarrow a = 8c$

$a = 8c$ and $a = xc$ $\qquad 8c = xc \rightarrow 8 = x$

For simultaneous equations, or those equations with like variables, try to add the equations together before relying on substitution. Pay careful attention to the requested variables, as you may not need to find the value of a single variable:

12. If $2x + 10y = 81$, and $3x - 5y = -36$, then $5x + 5y =$

(A) –117
(B) –45
(C) 45
(D) 81
(E) 117

Notice the you do not need the individual value of x or y. You need the value of $5x + 5y$. You should be instantly suspicious of substitution for this reason. Try addition first:

$$\begin{array}{r} 2x + 10y = 81 \\ +\ 3x - 5y = -36 \\ \hline 5x + 5y = 45 \end{array}$$

This extremely easy question is often made more difficult by students who try to solve it using substitution. The makers of the test know that most students will rely on substitution to solve these problems, so the questions are written to quickly be solved using the addition method to give an advantage to students who carefully study the question.

CAUTION: SAT TRAP! Many systems of equations questions are designed to trick you into using substitution, when addition or subtraction will more efficiently solve the same system.

Systems of Equations Problem Set

Solve the following multiple-choice questions by selecting the best answer from the five answer choices. For grid-in questions, write your answer in the grids and completely mark the corresponding ovals. Answers begin on page 230.

1. If $5x + 8y + 5z = 50$ and $2x + 4y + 3z = 20$, what is the value of $x - z$?

$$r = 6s$$
$$s = 3t$$
$$2t = p$$

2. If $p \neq 0$, what is the value of $\frac{p}{r}$ for the system of equations above?

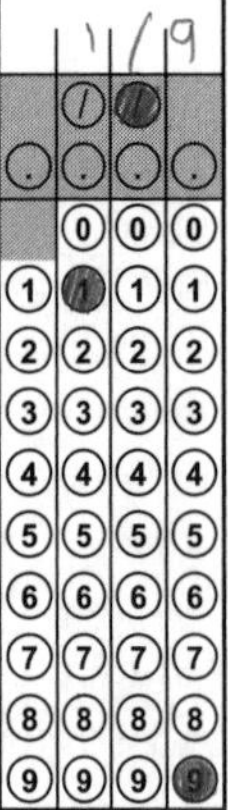

3. If the difference between x and y is -3 and the sum of x and y is 8, what is the product of x and y?

(A) 1.75
(B) 2.5
(C) 13.75
(D) 16.5
(E) 27.5

Systems of Equations Problem Set

Solve the following multiple-choice questions by selecting the best answer from the five answer choices. For grid-in questions, write your answer in the grids and completely mark the corresponding ovals. Answers begin on page 230.

4. If $5x + 6y = 18$ and $10x + dy = 10$, for which of the following values of d will the system of equations have NO solution?

(A) –12
(B) –6
(C) 0
(D) 6
(E) 12

$$d = a - b + 6$$
$$d = c - a - 2$$
$$d = b - c + 5$$

5. What is the value of d in the system of equations above?

(A) 2
(B) 3
(C) 4.5
(D) 6
(E) 9

Inequalities and Absolute Value

Frequency Guide: 3

Inequalities appear on nearly every SAT, while absolute value problems occur less frequently. Absolute value is almost always used with an inequality, which is why the two concepts are grouped together in this book.

Required Knowledge and Skill Set

1. Inequalities function in a manner similar to equations in the sense that you can add, subtract, multiply, divide, or square one side of the inequality as long as you perform the same operation with the same quantity on the other side of the inequality.

2. The inequality sign is reversed if the inequality is multiplied or divided by a negative number:

   Flip the sign!

   $$7-3x>-2 \quad\rightarrow\quad -3x>-9 \quad\rightarrow\quad x<\frac{-9}{-3} \quad\rightarrow\quad x<3$$

The most common error in inequality questions is the failure to flip the sign when the inequality is multiplied or divided by a negative number.

3. Sometimes two inequalities are chained together:

   $$5 \le a-6 \le 12$$

   This chained notation actually represents two inequalities. Separate them:

   $5 \le a-6$ and $a-6 \le 12$

   And work on each inequality separately:

   $5 \le a-6$ and $a-6 \le 12$
   $11 \le a$ and $a \le 18$ or $11 \le a \le 18$

4. The absolute value of a number is its distance from 0 on a number line. The numerical value is expressed as a positive number, since distance cannot be expressed with a negative number.

   $|7|=7$ $|-12|=12$ $|-x|=x$

5. Absolute value equations have two answers because two equations are set up to find the values that satisfy the original equation:

   $$|5-x|=7$$

   Split the equation into two cases, one case for each sign, and then solve:

   $5-x=7$ and $5-x=-7$
   $-x=2$ and $-x=-12$
   $x=-2$ and $x=12$

   Both –2 and 12 make the original absolute value equation true.

Absolute value equations have two possible answers.

6. When absolute values are combined with inequalities, the process is similar but much more tricky:

$$|3 - y| > 5$$

Separate the inequality into two cases with two signs. However, when setting up the case with a negative sign, you must flip the inequality sign as the inequality has been multiplied by –1:

$3 - y > 5$ and $3 - y < -5$
$-y > 2$ and $-y < -8$

Both signs must flip when dividing by –1:

$-y > 2$ and $-y < -8$
$y < -2$ and $y > 8$

The value for y must be less than –2 or greater than 8 in order for the inequality to work.

Application on the SAT

Inequalities are tested in a variety of ways on the SAT. The makers of the SAT are likely to include a question that assesses your ability to reverse the inequality sign when the inequality is multiplied or divided by a negative number.

Absolute value questions are usually combined with inequalities. The best way to solve these questions is to BACKPLUG answer choices or SUPPLY numbers to make the inequality true. These questions present many opportunities for careless mistakes (such as forgetting to flip the sign), but BACKPLUGGING and SUPPLYING eliminate the possibility of these errors occurring.

The best solution strategies for most inequalities and absolute value questions are BACKPLUGGING and SUPPLYING numbers.

Consider the question above one more time, this time written in SAT form:

7. If $|3 - y| > 5$, which of the following is a possible value of y ?

(A) –2
(B) 0
(C) 7
(D) 8
(E) 9

There were many opportunities in the previous solution for careless mistakes, such as forgetting to flip the sign or failing to set the problem up correctly. However, if you BACKPLUG, these possibilities for errors are eliminated:

(A) –2 $|3 - y| > 5 \rightarrow |3 - -2| > 5 \rightarrow |5| > 5 \rightarrow 5 > 5$ False
(B) 0 $|3 - y| > 5 \rightarrow |3 - 0| > 5 \rightarrow |3| > 5 \rightarrow 3 > 5$ False
(C) 7 $|3 - y| > 5 \rightarrow |3 - 7| > 5 \rightarrow |-4| > 5 \rightarrow 4 > 5$ False
(D) 8 $|3 - y| > 5 \rightarrow |3 - 8| > 5 \rightarrow |-5| > 5 \rightarrow 5 > 5$ False
(E) 9 $|3 - y| > 5 \rightarrow |3 - 9| > 5 \rightarrow |-6| > 5 \rightarrow 6 > 5$ ✓

It may seem like backplugging took a lot of effort, but those calculations can be quickly done without writing anything down. You are not only saving time, but also the possibility of calculation errors.

SUPPLYING numbers is an even more common strategy than BACKPLUGGING. This is especially true for inequalities, as demonstrated in the following grid-in question:

There are eight possible answers for this question, but you only have to provide one.

$$20 < a < 25$$
$$5 < b < 8$$

11. If a and b are integers that satisfy the inequalities above, what is one possible value of $\frac{a}{b}$?

To solve this question, you must SUPPLY numbers for a and b that satisfy the inequalities:

$a = 21$
$b = 7$

$$\frac{a}{b} = \frac{21}{7} = 3$$

Some of the more difficult inequality questions involve fractions and their relationships when manipulated, because fractions behave much differently than integers. For example, when a positive integer is squared, the result is larger than the original integer. But when a positive fraction is squared, the result is smaller than the original fraction:

$$2^2 = 4$$

$$\left(\frac{1}{2}\right)^2 = \frac{1}{4}$$

Let's look at a question that tests your ability to recognize fractional relationships:

19. Which of the following inequalities is NEVER true for some values of b ?

(A) $b < b^2 < b^3$
(B) $b^2 < b < b^3$
(C) $b < b^3 < b^2$
(D) $b^3 < b^2 < b$
(E) $b^3 < b < b^2$

When a word is capitalized on the SAT, pay close attention. The test makers are warning you that the question is tricky. In this case, you must look for the answer that is never true. That means that four of the answers are SOMETIMES, but not always, true.

You will need to SUPPLY several types of numbers to test this question. Begin with a positive integer:

$b = 2$ $\quad b^2 = 4$ $\quad b^3 = 8$ $\quad$ (A) $b < b^2 < b^3$

These calculations eliminate choice (A), as you have proven that (A) is sometimes true.

Supply a negative integer:

$b = -2$ $\quad b^2 = 4$ $\quad b^3 = -8$ $\quad$ (E) $b^3 < b < b^2$

As you can see, (E) is sometimes true.

Now we must try fractions (or decimals), because we know that they behave differently when squared and cubed. Try both a positive and negative fraction:

$b = \frac{1}{2}$ $\quad b^2 = \frac{1}{4}$ $\quad b^3 = \frac{1}{8}$ $\quad$ (D) $b^3 < b^2 < b$

$b = -\frac{1}{2}$ $\quad b^2 = \frac{1}{4}$ $\quad b^3 = -\frac{1}{8}$ $\quad$ (C) $b < b^3 < b^2$

With fractions, you have eliminated choices (C) and (D). You have now proven that four of the answer choices are sometimes true, so (B) must be the answer that is never true.

For higher-level Medium and Hard inequality questions, be prepared to supply fractions or decimals, as the College Board may test your knowledge of how these numbers behave when manipulated.

Another type of question to look for the on the SAT involves the representation of an inequality. These word problems describe a relationship, and you must pick the inequality that best represents the relationship:

6. Larry has $25 to spend at the movie theater. He buys 2 tickets at $8.00 each. If z represents the dollar amount he can spend on refreshments at the concession stand, which of the following inequalities could be used to find the possible values of z ?

   (A) $(2 \times 8) - z \leq 25$
   (B) $(2 \times 8) - z \geq 25$
   (C) $(2 \times 8) + z \leq 25$
   (D) $(2 \times 8) + z \geq 25$
   (E) $(2 \times 8) \geq 25$

Larry must spend less than or equal to $25. Therefore, you can eliminate (B), (D), and (E), as these inequalities use a sign indicating greater than or equal to $25. Now, is the price of refreshments added to or subtracted from the cost of the tickets? The correct answer is (C), as they are added to the cost of tickets.

Confidence Quotation

"Formulate and stamp indelibly on your mind a mental picture of yourself succeeding. Hold this picture tenaciously. Never permit it to fade. Your mind will seek to develop the picture."
—Norman Vincent Peale, author of "The Power of Positive Thinking"

Inequalities and Absolute Value Problem Set

Solve the following multiple-choice questions by selecting the best answer from the five answer choices. For grid-in questions, write your answer in the grids and completely mark the corresponding ovals. Answers begin on page 231.

1. The cost of the red car is less than the cost of the blue car but greater than the cost of the green car. If r, b, and g represent the cost of each car, respectively, which of the following is true?

(A) $r < b < g$
(B) $r < g < b$
(C) $b < g < r$
(D) $g < r < b$
(E) $b < r < g$

$$|r - 6| = 2$$

2. For how many values of r is the equation above true?

(A) None
(B) One
(C) Two
(D) Three
(E) More than three

3. If $b < -8$ or $b > 8$, which of the following must be true?

I. $b^3 > 8$
II. $b^4 > 8$
III. $|b| + b > 8$

(A) II only
(B) I and II only
(C) I and III only
(D) II and III only
(E) I, II, and III

Inequalities and Absolute Value Problem Set

Solve the following multiple-choice questions by selecting the best answer from the five answer choices. For grid-in questions, write your answer in the grids and completely mark the corresponding ovals. Answers begin on page 231.

$$a < 3 < \frac{1}{a}$$

4. For the inequality above, what is one possible value of a ?

5. If $y < 0$ and $8 < |y - 6| < 9$, what is one possible value of $|y|$?

6. If r and s are positive integers that satisfy $r + s < 18$, and $r > 8$, what is the least possible value of $r - s$?

ALGEBRA MASTERY ANSWER KEY

Exponents Problem Set—Page 186

1. (E) Easy

If $8^{x-4} = 8$, then $8^{x-4} = 8^1$ because $8^1 = 8$.

$x - 4 = 1$
$x = 5$

2. (D) Medium

There are two ways to solve this question. The first, and most efficient, method is to convert 27^{t-1} to a base of 3:

$27^{t-1} = (3^3)^{t-1} = 3^{3t-3}$

Now, we have two expressions with 3 as the base, so we can solve for t:

$3^{2t} = 3^{3t-3}$

$2t = 3t - 3 \quad \rightarrow \quad -t = -3 \quad \rightarrow \quad t = 3$

If you forget the rules of exponents, you can solve this question by BACKPLUGGING. Plug each answer choice in for t to find the answer that makes a true mathematical statement:

(A) 0 $\quad 3^{2t} = 27^{t-1} \rightarrow 3^{2(0)} = 27^{0-1} \rightarrow 3^0 = 27^{-1} \rightarrow 1 = \frac{1}{27}$ False
(B) 1 $\quad 3^{2t} = 27^{t-1} \rightarrow 3^{2(1)} = 27^{1-1} \rightarrow 3^2 = 27^0 \rightarrow 9 = 1$ False
(C) 2 $\quad 3^{2t} = 27^{t-1} \rightarrow 3^{2(2)} = 27^{2-1} \rightarrow 3^4 = 27^1 \rightarrow 81 = 27$ False
(D) 3 $\quad 3^{2t} = 27^{t-1} \rightarrow 3^{2(3)} = 27^{3-1} \rightarrow 3^6 = 27^2 \rightarrow 729 = 729$ ✓

There is no need to test answer choice (E) since (D) is correct.

3. (A) Medium

If you cross out "in terms of x" it becomes clear that you must isolate y ("what is y ?"):

$$x^{-3} = 5y \quad \rightarrow \quad \frac{1}{x^3} = 5y \quad \rightarrow \quad \frac{1}{5} \times \frac{1}{x^3} = y \quad \rightarrow \quad \frac{1}{5x^3} = y$$

Answer choice (A) is correct.

4. 30 Medium

Remember, whenever a number is raised to a power higher than the sixth, there is a shortcut in the solution. You could conceivably find the value of 64^{10}, but the number is too large to work with on the SAT (especially without a calculator!).

Instead, find the shortcut by making the bases equivalent:

$4^x = 64^{10}$

$64^{10} = (4^3)^{10}$

Therefore, $4^x = (4^3)^{10} \rightarrow 4^x = 4^{30} \rightarrow x = 30$

5. (D) Hard

There are two ways to solve this question. The most efficient is by following the rules of exponents:

$$\frac{(9^{2x})^y}{9^x} \rightarrow (9^{2x})^y \div 9^x \rightarrow 9^{2xy} \div 9^x$$

$x^n \div x^m = x^{n-m} \rightarrow 9^{2xy} \div 9^x = 9^{2xy-x}$

Factor the final expression: $9^{2xy-x} = 9^{x(2y-1)}$

The correct answer is (D).

If you forget the rules of exponents, or are not sure how to apply them, you can still solve the question by SUPPLYING numbers.

If $x = 1$ and $y = 2$, then $\frac{(9^{2x})^y}{9^x} = \frac{(9^{(2)(1)})^2}{9^1} \rightarrow \frac{9^4}{9^1} \rightarrow 9^3 \rightarrow 729$

Now find the answer choice that produces 729 when $x = 1$ and $y = 2$:

(A) 1^{xy} $1^{(1)(2)} \rightarrow 1^2 \rightarrow 1$ No
(B) 9^{xy} $9^{(1)(2)} \rightarrow 9^2 \rightarrow 81$ No
(C) 9^{2y} $9^{(2)(2)} \rightarrow 9^4 \rightarrow 6561$ No
(D) $9^{x(2y-1)}$ $9^{(1)(2\times2-1)} \rightarrow 9^{4-1} \rightarrow 9^3 \rightarrow 729$ ✓
(E) 9^{2xy-y} $9^{(2\times1\times2-2)} \rightarrow 9^{4-2} \rightarrow 9^2 \rightarrow 81$ No

Because this question is the last one in the set, it is likely a Hard level question. Therefore, it is important to check answer choice (E) even though answer choice (D) produced the desired result. If answer choice (E) had worked, you would need to supply different numbers and check both (D) and (E) again.

Roots and Radical Equations Problem Set—Page 191

1. (D) Easy

Isolate the root:

$$9 + \sqrt[3]{x} = 12 \quad \rightarrow \quad \sqrt[3]{x} = 3$$

Cube both sides:

$$\left(\sqrt[3]{x}\right)^3 = 3^3 \quad \rightarrow \quad x = 27$$

2. 2500 Medium

TRANSLATE from English to math:

the number divided by 50 is the same as the square root of the number

$$\frac{x}{50} = \sqrt{x}$$

Some students might instantly realize that the number is 50^2. However, if you do not see this, you must solve for x. Since the root is already isolated, simply square both sides:

$$\left(\frac{x}{50}\right)^2 = \left(\sqrt{x}\right)^2 \quad \rightarrow \quad \frac{x^2}{2500} = x$$

Then solve for x:

$$\frac{x^2}{2500} = x \quad \rightarrow \quad x^2 = (x)(2500) \quad \rightarrow \quad \frac{x^2}{x} = \frac{(x)(2500)}{x} \quad \rightarrow \quad x = 2500$$

3. (C) Medium

We are given two rules:

k and m are positive integers
$k > m$

Currently, $k = m$, because both k and m equal 12:

$$12\sqrt{12} = k\sqrt{m}$$

Some students want to argue and say that while $k = 12$, $m = \sqrt{12}$ or 3.464 But this is not true. The variable m is the value under the square root symbol, rather than the value that results when m is placed under the square root symbol. It is important to understand this concept on the SAT.

So if k and m are currently equal, but k must be greater than m, how can we increase k or decrease m? By factoring the square root:

$$12\sqrt{12} \rightarrow 12\times\sqrt{12} \rightarrow 12\times\sqrt{4}\times\sqrt{3} \rightarrow 12\times 2\times\sqrt{3} \rightarrow 24\times\sqrt{3} \rightarrow 24\sqrt{3}$$

Now $k = 24$ and $m = 3$. Find $k + m$: $24 + 3 = 27$. The correct answer is (C).

4. (E) Hard

There are two ways to solve this question. The first is by converting the fractional exponents to roots:

$$(cd)^{\frac{1}{3}} = (c+d)^{-\frac{1}{3}} \rightarrow \sqrt[3]{(cd)^{1}} = \frac{1}{\sqrt[3]{(c+d)^{1}}}$$

Then cube both sides and simplify:

$$\left(\sqrt[3]{cd}\right)^{3} = \left(\frac{1}{\sqrt[3]{c+d}}\right)^{3} \rightarrow cd = \frac{1}{c+d} \rightarrow (cd)(c+d) = 1 \rightarrow c^2d + cd^2 = 1$$

The correct answer is (E).

If you forget the rules of fractional exponents, you can cube both sides of the original equation:

$$(cd)^{\frac{1}{3}} = (c+d)^{-\frac{1}{3}} \rightarrow \left[(cd)^{\frac{1}{3}}\right]^{3} = \left[(c+d)^{-\frac{1}{3}}\right]^{3} \rightarrow (cd)^{1} = (c+d)^{-1} \rightarrow cd = \frac{1}{c+d}$$

Note that this only works because both sides of the original equation have a fractional exponent of one-third. If the right side had a fractional exponent of one-half, cubing both sides would result in a fractional exponent of three-halves on the right side.

Scientific Notation Problem Set—Page 195

1. 5.4 Easy

When adding scientific notation, the bases must be the same. Therefore, the first addend must be converted to a base of 10^5:

$$4 \times 10^4 = 0.4 \times 10^5$$

Now add the coefficients and transfer the base:

$$(0.4 \times 10^5) + (5 \times 10^5) = 5.4 \times 10^5$$

$$y = 5.4$$

2. (C) Medium

First multiply the coefficients:

$3 \times 5 = 15$

Then multiply the bases:

$10^3 \times 10^4 = 10^{3+4} = 10^7$

Some students would select answer choice (D) at this point, but they would be wrong. The coefficient cannot be a two-digit number:

15×10^7

The correct scientific notation is:

1.5×10^8

The correct answer is (D).

Expressions Problem Set—Pages 202-203

1. (B) Easy

Number of shirts: $4x$
Number of shirts that sold: $4x - a$
Price of each shirt: t

Amount earned from shirts that sold: $t\,(4x - a)$

The correct answer is (B).

2. (A) Medium

Total number of clients visited = total number visited as group + total number visited individually

Total number of clients visited = 200 + total number visited individually

Total number of clients visited = 200 + (number of salespeople)(number of clients)(number of days)

Total number of clients visited = $200 + xcd$

The correct answer is (A).

3. (D) Medium

Expression problems for sales that have a different price for the first item (or time period) than for the remaining items have a solution with a binomial where 1 is subtracted from the variable representing the total number of items (or time periods) purchased. The variable representing the total number of minutes is m. Only (D) and (E) have a binomial in which 1 is subtracted from m. If you had to guess, you would have even odds of guessing correctly.

Total cost = cost of first minute + cost of all other minutes
Total cost = \$1.00 + 0.10(all other minutes)
Total cost = \$1.00 + 0.10(m – 1)

4. 1.5 or $\frac{3}{2}$ Medium

SUPPLY a number for b: $b = 1$

Plug $b = 1$ into the expression:

$$\frac{b}{2}+\frac{3+b}{2} \rightarrow \frac{1}{2}+\frac{3+1}{2} \rightarrow \frac{1}{2}+\frac{4}{2} \rightarrow \frac{5}{2} \rightarrow 2.5$$

How much greater is 2.5 than 1? $2.5 - 1 = 1.5$ 1.5 or $\frac{3}{2}$

5. (D) Hard

TRANSLATE:

4 is added to a certain number y $\rightarrow$ $y+4$

The sum is divided by 5 $\rightarrow$ $\frac{y+4}{5}$

3 is subtracted from the quotient $\rightarrow$ $\frac{y+4}{5}-3$

Now simplify:

$$\frac{y+4}{5}-3 \rightarrow \frac{y+4}{5}-\frac{15}{5} \rightarrow \frac{y+4-15}{5} \rightarrow \frac{y-11}{5}$$

The correct answer is (D).

6. (E) Hard

RECORD what you know:

Number of cats: x

Number of dogs: $x + 40$

Total animals = number of cats + number of dogs = $x + (x + 40) = 2x + 40$

And then TRANSLATE:

what percent of the animals are cats?

$$\frac{?}{100} \quad \times (2x + 40) \quad = \quad x$$

$$\frac{?}{100} \times (2x + 40) = x \quad \rightarrow \quad ? \times (2x + 40) = 100x \quad \rightarrow \quad ? = \frac{100x}{2x + 40}\%$$

The correct answer is (E).

Equations Problem Set—Pages 208-209

1. (C) Easy

You do no have to find x to solve this equation; you only need to solve for $4x$:

$$4x + 6 = 30$$
$$4x = 24$$

Therefore, $4x - 6 = 24 - 6 = 18$

2. (C) Easy

Carefully study this question before you begin substituting. If $5c + 3d = 50$, then $\frac{1}{2}(5c + 3d) = 25$.

3. (C) Medium

Isolate the root:

$$3\left(\sqrt{n+4}\right) - 8 = 7 \quad \rightarrow \quad 3\left(\sqrt{n+4}\right) = 15 \quad \rightarrow \quad \left(\sqrt{n+4}\right) = 5$$

Square both sides and then solve for n:

$$\left(\sqrt{n+4}\right)^2 = 5^2 \quad \rightarrow \quad n + 4 = 25 \quad \rightarrow \quad n = 21$$

4. (D) Medium

You must SUPPLY numbers for x and y to make a true equation. It is important to supply for both variables to make sure that they do not share the same integer, as the rule states they must be different integers:

If $y = 1, x = 33$	$x + 7y = 40$	$33 + 7(1) = 40$	✓
If $y = 2, x = 26$	$x + 7y = 40$	$26 + 7(2) = 40$	✓
If $y = 3, x = 19$	$x + 7y = 40$	$19 + 7(3) = 40$	✓
If $y = 4, x = 12$	$x + 7y = 40$	$12 + 7(4) = 40$	✓
If $y = 5, x = 5$	$x + 7y = 40$	$5 + 7(5) = 40$	✓
If $y = 6, x = -2$	$x + 7y = 40$	$-2 + 7(6) = 40$	No (x is not positive)

The sum of the y values is $1 + 2 + 3 + 4 + 5 = 15$. The correct answer is (D).

5. (C) Hard

The value of d will be at least 1 but less than 2. Therefore, you can eliminate (A), (D), and (E). This leaves (B) and (C). SUPPLY numbers for c. Since all of the answer choices have a denominator of 5, SUPPLY 5:

$$c = 5 \quad 2-\frac{1}{5} \rightarrow 1\frac{4}{5} \rightarrow \frac{9}{5}$$

The correct answer is (C).

6. (C) Hard

Supply a few numbers for y to compute x:

$y = 1, x = 1.2$ $\qquad x = 1 + \frac{1}{5} = 1.2$

$y = 5, x = 6$ $\qquad x = 5 + \frac{5}{5} = 6$

$y = 20, x = 24$ $\qquad x = 20 + \frac{20}{5} = 24$

Now run through the Roman numerals:

I. $x > y$ — TRUE. This is true for all three numbers you supplied. Even if you were to supply $y = 100$, x will still be larger.

II. $y^2 > xy$ — FALSE. Because x is larger than y, xy is larger than y^2. To prove this, supply a set of numbers:

$$y = 1, x = 1.2 \rightarrow y^2 = 1, xy = 1.44$$

III. $y + 2 > x$ — FALSE. The last set of supplied numbers proves this false.

$$y = 20, x = 24 \qquad 20 + 2 \text{ is not greater than } 24$$

Systems of Equations Problem Set—Pages 214-215

1. 10 Medium

Since these are simultaneous equations, try adding them. Multiply the second equation by –2 to eliminate the y variable, since the required value does not include y:

$$\begin{aligned} 5x + 8y + 5z &= 50 \\ 2x + 4y + 3z &= 20 \end{aligned} \qquad \begin{aligned} 5x + 8y + 5z &= 50 \\ + \ -4x - 8y - 6z &= -40 \\ \hline x \qquad\quad - z &= 10 \end{aligned}$$

As you can see, $x - z = 10$.

2. $\frac{1}{9}$ Medium

Substitute:

$$r = 6s \quad \text{and} \quad s = 3t$$
$$r = 6s \quad \rightarrow \quad r = 6(3t) \quad \rightarrow \quad r = 18t$$

$$r = 18t \quad \text{and} \quad 2t = p$$

$$\frac{p}{r} = \frac{2t}{18t} = \frac{1}{9}$$

3. (C) Medium

TRANSLATE the English to math:

the difference between x and y is –3 $\rightarrow$ $x - y = -3$
the sum of x and y is 8 $\rightarrow$ $x + y = 8$
what is the product of x and y $\rightarrow$ $xy = ?$

Add the system of equations to eliminate y and solve for x:

$$\begin{aligned} x - y &= -3 \\ + \ x + y &= \ \ 8 \\ \hline 2x \qquad &= \ \ 5 \end{aligned} \qquad\qquad 2x = 5 \quad \rightarrow \quad x = \frac{5}{2}$$

Now plug five-halves in for x into either equation and solve for y:

$$x + y = 8 \quad \rightarrow \quad \frac{5}{2} + y = 8 \quad \rightarrow \quad y = 8 - \frac{5}{2} \quad \rightarrow \quad y = \frac{16}{2} - \frac{5}{2} \quad \rightarrow \quad y = \frac{11}{2}$$

Finally, find the product of x and y:

$$xy = \frac{5}{2} \times \frac{11}{2} = \frac{55}{4} \text{ or } 13.75$$

4. (E) Hard

Remember, a system of equations has no solution when both variables cancel out. Look at the variable x. To eliminate x, we would need to multiply the first equation by –2:

$$5x + 6y = 18 \quad \rightarrow \quad -2(5x + 6y = 18) \quad \rightarrow \quad -10x - 12y = -36$$

Now look at the two equations as if you were going to add them:

$$\begin{aligned} -10x - 12y &= -36 \\ 10x + dy &= 10 \end{aligned}$$

It is clear that x cancels out. What number for d would cancel out y? Positive 12:

$$\begin{array}{r} -10x - 12y = -36 \\ \underline{+10x + 12y = 10} \\ 0 + 0 = -26 \end{array}$$

The correct answer is (E).

5. (B) Hard

When you add the equations together, notice that all of the variables except d cancel each other out:

$$\begin{array}{rl} d &= \cancel{a} - \cancel{b} + 6 \\ d &= \cancel{c} - \cancel{a} - 2 \\ \underline{+ \quad d} &\underline{= \cancel{b} - \cancel{c} + 5} \\ 3d &= \qquad\quad 9 \end{array}$$

$d = 3$ The correct answer is (B).

Inequalities and Absolute Value Problem Set—Pages 220-221

1. (D) Easy

TRANSLATE:

The cost of the red car is less than the cost of the blue car:

$r < b$

But greater than the cost of the green car:

$g < r$

Then create the inequality using chained notation:

$g < r < b$

The correct answer is (D).

2. (C) Easy

All absolute value questions have two solutions; however, they may only have one value that makes the question true if the two solutions produce the same value. Therefore, you must verify:

$$|r-6|=2$$

Set up two cases and solve:

$r-6=2$	and	$r-6=-2$
$r=8$	and	$r=4$

The equation is true for two values: 4 and 8.

3. (A) Medium

SUPPLY two values to satisfy the two inequalities:

b = –10 and b = 10

Then test both supplied numbers in each Roman numeral:

	$b=-10$	b = 10
I. $b^3>8$	$(-10)^3=-1000$ False	
II. $b^4>8$	$(-10)^4=10000$ ✓	$(10)^4=10000$ ✓
III. $|b|+b>8$	$|-10|+-10=0$ False	

Only Roman numeral II is true in both cases. Therefore, (A) is the correct answer.

4. $0<a<\frac{1}{3}$ Medium

The variable *a* cannot be a negative number, because then the inequality would not be true:

$a=-4$ $\qquad -4<3<-\frac{1}{4}$ False

Similarly, *a* cannot be a positive integer, as the inequality is not true:

$a=4$ $\qquad 4<3<\frac{1}{4}$

The only type of number that can make $\frac{1}{a}$ greater than 3 is a fraction. Try $\frac{1}{4}$:

$a=\frac{1}{4}$ $\qquad \frac{1}{4}<3<\frac{1}{\frac{1}{4}} \rightarrow \frac{1}{4}<3<4$ ✓

Any fraction between zero and one-third will make the inequality true.

5. $2 < y < 3$ Medium

The best way to solve this problem is to SUPPLY numbers.

Notice that $|y - 6|$ must be between 8 and 9. Therefore, SUPPLY 8.5:

$|y - 6| = 8.5$

Now set up two cases:

$y - 6 = 8.5$	and	$y - 6 = -8.5$
$y = 14.5$	and	$y = -2.5$

Since $y < 0$, the second case is the correct answer. When –2.5 is plugged into $|y|$, it becomes 2.5. The correct answer is any number between 2 and 3.

6. 1 Hard

SUPPLY several numbers for r and s, including the highest and lowest numbers for r:

$r + s < 18$ and $r > 8$

If $r = 9, s = 8$	$9 + 8 < 18$	$r - s = 9 - 8 = 1$
If $r = 12, s = 5$	$9 + 8 < 18$	
If $r = 16, s = 1$	$9 + 8 < 18$	

Then, find $r - s$

If $r = 9, s = 8$	$9 - 8 = 1$
If $r = 12, s = 5$	$12 - 5 = 7$
If $r = 16, s = 1$	$16 - 1 = 15$

The least possible value is 1.

Chapter Seven:
Algebra II Mastery

When the SAT was reformatted in 2005, the College Board added concepts from Algebra II to the tested math content. This inclusion worried many students, as they had not yet taken this higher-level math course. However, the Algebra II concepts tested on the SAT are quite basic and include functions, quadratic equations, and variation. Many of the questions that appear to be using concepts from Algebra II are actually basic algebra in disguise. Plus, there are only a few questions from Algebra II included on each test.

Algebra II involves intermediate concepts from algebra, including quadratic equations and functions.

As always, we have indicated the frequency of each type of question at the beginning of each concept section. The key is included below for your convenience:

Rating	Frequency
5	Extremely High: 3 or more questions typically appear on every SAT
4	High: at least 2 questions typically appear on every SAT
3	Moderate: at least one question typically appears on every SAT
2	Low: one question typically appears on every two or three SATs
1	Extremely Low: one question appears infrequently and without a pattern

Remember that there are eight solution strategies you can employ on SAT math questions:

1. ANALYZE the Answer Choices
2. BACKPLUG the Answer Choices
3. SUPPLY Numbers
4. TRANSLATE from English to Math
5. RECORD What You Know
6. SPLIT the Question into Parts
7. DIAGRAM the Question
8. SIZE UP the Figures

Be sure to check your answers in the answer key following the chapter. It is important to understand why you missed a particular question in order to avoid making the same mistake on the SAT.

Quadratic Equations

Frequency Guide: 2

True quadratic equations only occur on every two or three SATs. However, some equations masquerade as quadratics, which we will discuss in later in this section, and these appear a little more often.

Required Knowledge and Skill Set

1. Quadratic equations follow a specific form:

   $ax^2 + bx + c = 0$

   The variables a, b, and c represent values, or coefficients:

   $5x^2 + 7x + 6 = 0$ $\quad a = 5, b = 7, c = 6$

   $y^2 - y + 4 = 0$ $\quad a = 1, b = -1, c = 4$

☠ CAUTION: SAT TRAP! Students who use the quadratic formula on the SAT have missed a valuable time-saving shortcut.

2. The quadratic formula is not required on the SAT. If you find yourself using it, you have missed a shortcut.

3. Two binomials can be multiplied together using FOIL (First, Outer, Inner, Last) to create a quadratic form:

   $$(x - 5)(x + 3) = 0$$

   First: Multiply the first terms in each binomial $\quad x \times x = x^2$
   Outer: Multiply the outer terms in the two binomials $\quad x \times 3 = 3x$
   Inner: Multiply the inner terms in the two binomials $\quad -5 \times x = -5x$
   Last: Multiply the last terms in each binomial $\quad -5 \times 3 = -15$

   Then add the results (x^2, $3x$, $-5x$, -15):

   $$x^2 + 3x + -5x + -15 \quad \rightarrow \quad x^2 - 2x - 15 = 0$$

4. Some quadratic equations may be disguised. You must manipulate these equations to get them in quadratic form or FOIL form:

   $x^2 + 8x = -15 \quad \rightarrow \quad x^2 + 8x + 15 = 0$

   $24 - 2z = z^2 \quad \rightarrow \quad z^2 + 2z - 24 = 0$

   $$x - 4 = \frac{1}{x+5} \quad \rightarrow \quad (x - 4)(x + 5) = 1 \quad \rightarrow \quad x^2 + 5x - 4x - 20 = 1 \quad \rightarrow$$

   $$x^2 + x - 20 = 1 \quad \rightarrow \quad x^2 + x - 19 = 0$$

5. Quadratic equations can be solved by factoring. To factor a quadratic equation, you must perform reverse FOIL:

$$x^2 + 5x - 14 = 0$$

Step 1: Set up the binomials.

$x^2 + 5x - 14 = 0$ $\quad (x - \quad)(x + \quad)$

Because the last coefficient (14) in the quadratic form is negative, the binomials have opposite signs. The polarity of the second and third values ($5x$ and 14) determine the signs in the binomials:

$x^2 + 5x + 14 = 0$ $\quad (x + \quad)(x + \quad)$
$x^2 - 5x - 14 = 0$ $\quad (x - \quad)(x + \quad)$
$x^2 - 5x + 14 = 0$ $\quad (x - \quad)(x - \quad)$

Step 2: Find factors for the term (–14):

1 and –14
–1 and 14
2 and –7
–2 and 7

Step 3: Which set of factors, when added, will result in the coefficient for the middle term (+5)?

1 and –14	$1 + -14 = -13$	No
–1 and 14	$-1 + 14 = 13$	No
2 and –7	$2 + -7 = -5$	No
–2 and 7	$-2 + 7 = 5$	✓

Step 4: Complete the binomials using the two factors:

$x^2 + 5x - 14 = 0$ $\quad (x - \quad)(x + \quad)$ $\quad (x - 2)(x + 7) = 0$

Step 5: Find the roots. If $(x - 2)(x + 7) = 0$, then either $x - 2 = 0$ or $x + 7 = 0$.

$x - 2 = 0$ $\quad x + 7 = 0$
$x = 2$ $\quad x = -7$

Therefore, x must equal 2 or –7.

You must be prepared to expand binomials (FOIL) and factor quadratic forms (reverse FOIL).

Confidence Quotation
"Always bear in mind that your own resolution to succeed is more important than any one thing."
—Abraham Lincoln, 16th President of the United States

6. There are three classic SAT quadratic forms used frequently on the SAT:

MEMORY MARKER:
It is a good idea to have these classic forms memorized so that they are easily recognized.

Classic Form #1

$(x+y)(x-y)=x^2-y^2$

Examples:
$(t-5)(t+5) \rightarrow t^2-5^2 \rightarrow t^2-25$
$(3a+b)(3a-b) \rightarrow (3a)^2-b^2 \rightarrow 9a^2-b^2$
$y^2-64 \rightarrow y^2-8^2 \rightarrow (y+8)(y-8)$
$36-n^2 \rightarrow 36^2-n^2 \rightarrow (6+n)(6-n)$

Classic Form #2

$(x+y)(x+y)=x^2+2xy+y^2$
$(x+y)^2=x^2+2xy+y^2$

Examples:
$(t+5)^2 \rightarrow t^2+2(t)(5)+5^2 \rightarrow t^2+10t+25$
$(3a+b)(3a+b) \rightarrow (3a)^2+2(3a)(b)+b^2 \rightarrow 9a^2+6ab+b^2$
$y^2+16y+64 \rightarrow y^2+2(y)(8)+8^2 \rightarrow (y+8)^2$
$36+12n+n^2 \rightarrow 6^2+2(n)(6)+n^2 \rightarrow (6+n)^2$

Classic Form #3

$(x-y)(x-y)=x^2-2xy+y^2$
$(x-y)^2=x^2-2xy+y^2$

Examples:
$(t-5)^2 \rightarrow t^2-2(t)(5)+5^2 \rightarrow t^2-10t+25$
$(3a-b)(3a-b) \rightarrow (3a)^2-2(3a)(b)+b^2 \rightarrow 9a^2-6ab+b^2$
$y^2-16y+64 \rightarrow y^2-2(y)(8)+8^2 \rightarrow (y-8)^2$
$36-12n+n^2 \rightarrow 6^2-2(n)(6)+n^2 \rightarrow (6-n)^2$

By recognizing these forms and understanding their patterns, you can save valuable time by avoiding FOIL or reverse FOIL.

Application on the SAT

The most common way that the quadratic form is tested is through FOIL or reverse FOIL:

13. If $x^2 - x = 56$, a possible value of $x^2 + x$ is

(A) –72
(B) –56
(C) 28
(D) 72
(E) 3080

☠ CAUTION: SAT TRAP! The most common wrong answer is –56 because students mistakenly assume that the change from a subtraction sign to an addition sign results in the reciprocal of the sum.

The phrase "a possible value of" suggests that there is more than one true answer. The question has a quadratic form in disguise:

$$x^2 - x = 56 \quad \rightarrow \quad x^2 - x - 56 = 0$$

As you know, quadratic equations have two roots. Find them by performing reverse FOIL:

$$x^2 - x - 56 = 0$$
$$(x - 8)(x + 7)$$
$$x = 8 \text{ or } x = -7$$

Now use the roots of x to find the possible values of $x^2 + x$:

$$x^2 + x$$
$$x = 8 \quad \rightarrow \quad 8^2 + 8 = 64 + 8 = 72$$
$$x = -7 \quad \rightarrow \quad -7^2 + -7 = 49 - 7 = 42$$

Only 72 appears as an answer choice, so the correct answer is (D).

Some questions will rely on your knowledge of the classic quadratic forms. Consider the following grid-in question:

15. If $r - p = 7$, what is the value of $r^2 + 2rp + p^2$?

Students unfamiliar with the classic quadratic forms may use substitution, which involves a lengthy solution and opportunities for calculation errors:

$$r - p = 7 \quad \rightarrow \quad r = 7 + p$$
$$r^2 - 2rp + p^2 \quad \rightarrow \quad (7 + p)^2 - [2(7 + p)p] + p^2$$

Remember, the College Board uses questions that have quick, clean solutions; lengthy answers with messy calculations are often an indication of an unseen shortcut.

But a student with a knowledge of the classic quadratic forms would recognize that:

$$r^2 - 2rp + p^2 \quad \rightarrow \quad (r - p)^2$$

Since $r - p = 7$, $(r - p)^2 = 7^2$. The correct answer is 49.

See the next chapter for more on right triangles and the Pythagorean Theorem.

Watch for quadratic equations used in Geometry, especially in right triangle questions using the Pythagorean Theorem:

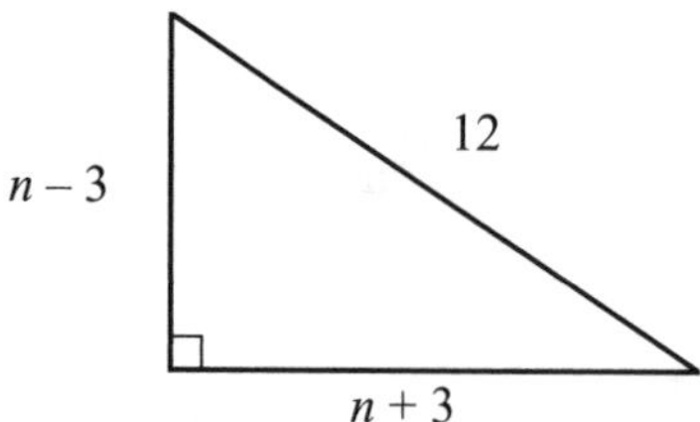

Note: Figure is not drawn to scale.

16. What is the value of $n^2 - 9$?

(A) 54
(B) 63
(C) 72
(D) 126
(E) 144

The Pythagorean Theorem states that $a^2 + b^2 = c^2$, where c is the hypotenuse of a right triangle, and a and b are the legs of the triangle. So to solve for n^2, you must plug in the lengths of the triangle into the Pythagorean Theorem:

$a = n - 3 \qquad b = n + 3 \qquad c = 12$

$$a^2 + b^2 = c^2$$
$$(n - 3)^2 + (n + 3)^2 = 12^2$$
$$(n - 3)(n - 3) + (n + 3)(n + 3) = 144$$
$$(n^2 - 6n + 9) + (n^2 + 6n + 9) = 144$$
$$2n^2 + 18 = 144$$
$$2n^2 = 126$$
$$n^2 = 63$$

Some students would go one step further and solve for n. However, this is not necessary as the required value is $n^2 - 9$, so you can just plug in 63 for n^2:

$$n^2 - 9 \quad \rightarrow \quad 63 - 9 = 54$$

The correct answer is (A).

CAUTION: SAT TRAP! A minus sign followed by a binomial is an SAT trap in which the test makers hope you forget to distribute the negative.

One caution about quadratic equations: watch for questions which require you to distribute a negative. Consider an example of an equation where a negative sign must be distributed:

$$x - (x - 1)(x + 1)$$

Perform FOIL on the two binomials, but place them in brackets to remind yourself of the negative sign:

$$x - [x^2 - 1x + 1x - 1]$$
$$x - [x^2 - 1]$$

Then distribute the negative sign:

$$x - x^2 + 1$$

The makers of the SAT look for opportunities where students will forget the negative outside of the binomials while performing FOIL.

As you can see, true quadratic equations are used to assess your ability to perform FOIL or reverse FOIL. But you should be prepared for questions involving “fake” quadratic equations, or those equations that may appear to involve the quadratic form, but can be solved using simple algebra.

Let’s look at an example to understand how these questions fool many test takers:

10. If $(x + 2)(y - 4) = 0$ and $x > 0$, what is the value of y ?

    (A) –4
    (B) –1
    (C) 0
    (D) 1
    (E) 4

Some students would see the two binomials and automatically begin FOILing. Notice, however, that the variables in the two binomials are different. FOILing would not assist in finding y. Instead, you must study the problem and evaluate the relationships in it:

☠ CAUTION: SAT TRAP!
Binomials with two different variables will not be solved using FOIL.

Since $(x + 2)(y - 4) = 0$, either $(x + 2) = 0$ or $(y - 4) = 0$

But if $x > 0$, then $(x + 2) > 0$

Therefore, $y - 4 = 0 \quad \rightarrow \quad y = 4$

The correct answer is (E).

Beware of questions that appear to have an equation in quadratic form. You should completely evaluate the equation before beginning FOIL or reverse FOIL, as the majority of questions that appear to be quadratic can be solved in a much easier method. In the following problem set, you will find both true quadratics and imposters, so carefully study each question.

Quadratic Equations Problem Set

Solve the following multiple-choice questions by selecting the best answer from the five answer choices. For grid-in questions, write your answer in the grids and completely mark the corresponding ovals. Answers begin on page 262.

1. If y and z are integers, and $y^2 - z^2 = 24$, which of the following could be the value of $y - z$?

   (A) 0
   (B) 4
   (C) 9
   (D) 14
   (E) 16

2. If $(k - 5)^2 = 0$, what is the value of $(k + 3)(k - 3)$?

   (A) −9
   (B) −1
   (C) 9
   (D) 16
   (E) It is impossible to determine.

3. If $(x + 6)(x + 5) - (x - 3)(x - 2) = 0$, what is the value of x ?

   (A) −6
   (B) −2.25
   (C) −1.5
   (D) 1.5
   (E) 6

4. If a and b are integers and $(a + 6)(b - 6) = 0$, what is the least possible value of $a^2 + b^2$?

   (A) 0
   (B) 6
   (C) 12
   (D) 36
   (E) 64

Direct and Inverse Variation

Frequency Guide: 2

Variation is most often tested as a proportion on the SAT. However, there are occasional problems that assess your ability to manipulate the formulas for variation.

Required Knowledge and Skill Set

1. If the high school principal insists that there are 2 adults present for every 25 students attending the senior class trip, the number of adults varies directly with the number of students. This is direct variation. For every change to x's value, y undergoes a proportionate change. So, when x increases, y increases by a constant c. And when x decreases, y decreases by a constant c. The following formula may be required to solve direct variation questions:

   Direct variation: $y = cx$

In direct variation, when 'x' increases, 'y' increases. When 'x' decreases, 'y' decreases.

2. Direct variation problems are the same as proportion problems. These questions are often easier to solve as proportions, as no formulas are required.

3. If the price of the senior trip decreases \$20 for every additional group of 50 students that sign up, the number of students varies inversely with the price. This is inverse variation (also sometimes called indirect variation). When x increases, y decreases by a constant c. And when x decreases, y increases by a constant c. The following formula is required to solve inverse variation questions:

   Inverse variation: $c = xy$

In inverse variation, when 'x' increases, 'y' decreases. When 'x' decreases, 'y' increases.

4. The wording of a variation problem dictates how the formula is set up. You can adapt the two basic formulas to several situations:

y varies directly with x	$y = cx$
r varies inversely with p	$c = rp$
m varies directly with the square root of n	$m = c\sqrt{n}$
j varies inversely with the square of k	$c = jk^2$
the cube of x varies directly with one half of y	$x^3 = c\left(\frac{y}{2}\right)$

MEMORY MARKER: You should memorize the two basic formulas for direct and inverse variation, as well as how to adapt these formulas to special situations.

Application on the SAT

Direct variation problems use the terms "proportion" and "variation" interchangeably. Despite the different terminology, these two questions are the same:

8. If h varies directly with g, and if $h = 30$ when $g = 8$, what is the value of h when $g = 12$?

8. If h is directly proportional with g, and if $h = 30$ when $g = 8$, what is the value of h when $g = 12$?

Direct variation problems are best solved as proportions.

The easiest way to solve this problem is to treat the question as a proportion:

$$\begin{matrix} g: \\ h: \end{matrix} \quad \frac{8}{30} = \frac{12}{?} \quad \rightarrow \quad (8)(?) = (12)(30) \quad \rightarrow \quad (8)(?) = 360 \quad \rightarrow \quad ? = 45$$

But the question can also be solved using the direct variation formula ($y = cx$). Set up the formula as dictated by the question to find the constant c:

h varies directly with g

$h = cg$

$30 = c(8) \quad \rightarrow \quad c = \frac{30}{8} = \frac{15}{4}$

Once you solve for c, find the missing variable when $g = 12$:

$$h = \frac{15}{4}(12) \quad \rightarrow \quad \frac{180}{4} \quad \rightarrow \quad 45$$

It may be more convenient to use the direct variation formula when the variables' relationship involves squares or roots. It would prove difficult for many students to correctly solve the following by just using a proportion:

Q: How would you set up #15 using direct variation?
A: $b = c\sqrt{d}$ (If you set it up correctly, the answer is 256.)

15. If b varies directly with $\sqrt{d}$, and if $b = \frac{1}{8}$ when $d = \frac{1}{25}$, what is the value of d when $b = 10$?

Similarly, questions about inverse variation cannot be solved with proportions and must be set up using the inverse variation formula ($c = xy$):

13. If y varies inversely with the square of x, and if $x = 6$ when $y = \frac{1}{4}$, what is the value of y when $x = 3$?

(A) 0
(B) 1
(C) 6
(D) 9
(E) 36

To solve this question, you must correctly TRANSLATE the inverse relationship:

y varies inversely with the square of x $\rightarrow$ $c = x^2y$

Then plug in values for x and y to solve for c:

$$c = x^2y$$
$$c = 6^2\frac{1}{4} \quad \rightarrow \quad c = (36)\frac{1}{4} \quad \rightarrow \quad c = 9$$

Now use the value for c to find the missing value of y when $x = 3$:

$$c = x^2y$$
$$9 = 3^2y \quad \rightarrow \quad 9 = 9y \quad \rightarrow \quad 1 = y$$

The correct answer is (B).

Direct and inverse variation problems are easy to SIZE UP. If you are running out of time or do not know how to solve a variation question, you can estimate an answer and eliminate answer choices. Consider an example:

ARITHMETRICK
With a little critical thinking, direct and inverse variation questions can be SIZED UP rather easily.

11. If x varies inversely with y, and if $x = 15$ when $y = 10$, what is the value of y when $x = 25$?

(A) 5
(B) 6
(C) $16\frac{2}{3}$
(D) 20
(E) 30

Because this is an inverse variation problem, when x increases, y decreases. The variable x increases from 15 to 25. Therefore, the y variable must decrease. Since $y = 10$ initially, answers (C), (D), and (E) are eliminated because they are greater than y.

Now look at x again. It increased from 15 to 25, an increase which is slightly less than twice the original number ($15 \times 2 = 30$). Therefore, when y decreases, it must create a decrease that is slightly less than half of the original number ($10 \div 2 = 5$). Since answer choice (A) is half of the original number (and we are looking for a decrease slightly less than half), the answer must be (B).

Direct and Inverse Variation Problem Set

Solve the following multiple-choice questions by selecting the best answer from the five answer choices. For grid-in questions, write your answer in the grids and completely mark the corresponding ovals. Answers begin on page 263.

1. Which of the following tables shows a relationship in which p is inversely proportional to r?

(A)

p	r
2	18
3	12
4	9

$p = \frac{k}{r}$

(B)

p	r
4	20
5	25
6	30

(C)

p	r
2	5
3	6
4	7

(D)

p	r
4	8
8	16
12	36

(E)

p	r
2	2
4	8
8	16

2. If y varies directly with $\sqrt{x}$ and $y = \frac{1}{3}$ when $x = \frac{4}{9}$, what is the value of x when $y = 2$?

(A) $\frac{8}{3}$

(B) $\frac{64}{9}$

(C) 2

(D) 4

(E) 16

$y = k\sqrt{x}$

$\frac{1}{3} = k\sqrt{4/9}$

$k = 0.5$

Symbolic Functions

A function is a substitution puzzle, where variables or symbols stand for values. While many high school students are quite familiar with regular functions, most have not experienced symbolic functions, which is why the College Board typically uses at least one on each test.

Frequency Guide: 3

Required Knowledge and Skill Set

1. Symbolic functions use a symbol to represent the function. These symbols do not signify a specific mathematical process or operation; they are used only to alert the reader that the question is a function. Symbols that have been used on past SATs include circles, squares, triangles, number signs, stars, diamonds, and many other shapes and figures. Consider an example:

   10. If $x \text{ ✪ } y = 2x^2 - 3y$, what is the value of $4 \text{ ✪ } 8$?

   (A) 2
   (B) 4
   (C) 8
   (D) 12
   (E) 16

Few students have ever been exposed to symbolic functions, so don't despair if these problems are completely alien to you. We will show you everything you need to know to easily solve them.

2. It may help to think of all function questions as having a 'puzzle' and a 'key.' The puzzle involves the original equation:

   Puzzle: $x \text{ ✪ } y = 2x^2 - 3y$

   The key is the portion of the question that places new expressions or terms into the function:

   Key: $4 \text{ ✪ } 8$

3. To solve these questions, align the key directly below the puzzle:

   Puzzle: $x \text{ ✪ } y = 2x^2 - 3y$
   Key: $4 \text{ ✪ } 8$

   The key reveals that $x = 4$ and $y = 8$. In the puzzle, substitute a 4 for every x and an 8 for every y:

   $$x \text{ ✪ } y = 2\ x^2\ - 3\ y$$
   $$\downarrow \quad \downarrow \quad \downarrow \quad \downarrow$$
   $$4 \text{ ✪ } 8 = 2(4)^2 - 3(8)$$

   Then solve the equation:

   $$4 \text{ ✪ } 8 = 2(4)^2 - 3(8) \rightarrow 2(16) - 24 \rightarrow 32 - 24 \rightarrow 8$$

   The correct answer is (C).

Proper alignment of the puzzle and key is extremely important in the more difficult symbolic function questions.

4. Alignment is extremely important, especially when the key involves expressions:

Puzzle: $g \pentagon h = 5g - 5h$
Key: $x + 3 \pentagon x - 4$

For every g, substitute the expression $x + 3$. For every h, substitute $x - 4$. Use parentheses for the expressions so that you remember to distribute any negatives or exponents:

Always use parentheses when setting up a symbolic function. Failure to do so may result in an undistributed negative or exponent.

$$g \pentagon h = 5g - 5h$$
$$x + 3 \pentagon x - 4 = 5(x + 3) - 5(x - 4)$$

$$x + 3 \pentagon x - 4 = 5x + 15 - (5x - 20)$$
$$x + 3 \pentagon x - 4 = 5x + 15 - 5x + 20$$
$$x + 3 \pentagon x - 4 = 35$$

Application on the SAT

Many of the Easy and Medium level symbolic function questions simply ask you to find the value of a key, as we saw in the previous section. The more difficult symbolic function problems may use confusing mathematical terms or symbols.

One of these types of questions equates the key with a value or a second key. Let's begin by looking at a key equated to a value:

14. For all numbers a and b, where $a \neq b$, $a \diamond b$ is defined as $\frac{\sqrt{a}+2}{b^2-5}$. If $x \diamond 5 = \frac{1}{2}$, what is the value of x ?

(A) 5
(B) 8
(C) 10
(D) 64
(E) 81

This question appears more confusing because of two reasons:

1. The key uses a third variable (x).
2. The key is equal to one-half.

Even though the key uses a variable, you still set the puzzle and key up the same way:

Puzzle: $a \diamond b = \frac{\sqrt{a}+2}{b^2-5}$

Key: $x \diamond 5$

Wherever an a appears in the puzzle, substitute an x. And for every b, substitute 5. Then simplify:

Puzzle: $a \diamond b = \dfrac{\sqrt{a}+2}{b^2-5}$

Key: $x \diamond 5 = \dfrac{\sqrt{x}+2}{5^2-5} \rightarrow \dfrac{\sqrt{x}+2}{25-5} \rightarrow \dfrac{\sqrt{x}+2}{20}$

Now return to the question. It states that $x \diamond 5 = \dfrac{1}{2}$. Therefore, $\dfrac{\sqrt{x}+2}{20} = \dfrac{1}{2}$:

$$\frac{\sqrt{x}+2}{20} = \frac{1}{2} \rightarrow 2(\sqrt{x}+2) = 1(20) \rightarrow 2\sqrt{x}+4 = 20 \rightarrow$$

$$2\sqrt{x} = 16 \rightarrow \sqrt{x} = 8 \rightarrow (\sqrt{x})^2 = 8^2 \rightarrow x = 64$$

The correct answer is (D). When a key is equivalent to a value, run the key through the puzzle and then set the result equal to the value. If the key is equivalent to another key, run both keys through the puzzle and set their results equal:

16. Let the operations ◆ and ■ be defined for all numbers r and s as

r ◆ $s = 6r + s$
r ■ $s = 2r + s$

If $(6x)$ ◆ $8 = 34$ ■ $6x$, what is the value of x ?

(A) 2
(B) 4
(C) 6
(D) 7
(E) 8

Set up each puzzle and key and solve:

Puzzle 1: r ◆ $s = 6r + s$ r ◆ $s =$ 6 r $+ s$
Key 1: $6x$ ◆ 8 $6x$ ◆ $8 = 6(6x) + 8 \rightarrow 36x + 8$

Puzzle 2: r ■ $s = 2r + s$ r ■ $s =$ 2 r $+ s$
Key 2: 34 ■ $6x$ 34 ■ $6x = 2(34) + 6x \rightarrow 68 + 6x$

Then set the two results equal and solve for x:

$36x + 8 = 68 + 6x$
$30x = 60$
$x = 2$

The correct answer is (A).

When a function equals two quantities, set those quantities equal.

Confidence Quotation
"If you have no confidence in self, you are twice defeated in the race of life. With confidence, you have won even before you have started."
—Marcus Garvey, orator and journalist

Another added symbol that tends to confuse students is parentheses. In the order of operations, items in parentheses are solved first. This is also true for symbolic functions in parentheses:

Just as in the order of operations, solve keys in parentheses before solving keys outside of parentheses.

12. For all numbers x and y, let the operation ⁂ be defined as x ⁂ $y = x^2 + x - 3y$. What is the value of (3 ⁂ 2) ⁂ 3 ?

 (A) 25
 (B) 33
 (C) 47
 (D) 63
 (E) 81

To begin, work on the key inside of the parentheses (3 ⁂ 2):

Puzzle: x ⁂ $y = x^2 + x - 3y$
Key: 3 ⁂ 2 $= 3^2 + 3 - 3(2) \quad \rightarrow \quad 9 + 3 - 6 \quad \rightarrow \quad 6$

As you can see, 3 ⁂ 2 = 6, so the new key is (6) ⁂ 3:

Puzzle: x ⁂ $y = x^2 + x - 3y$
Key: 6 ⁂ 3 $= 6^2 + 6 - 3(3) \quad \rightarrow \quad 36 + 6 - 9 \quad \rightarrow \quad 33$

Answer choice (B) is correct.

Some symbolic functions are placed in shapes:

13. Let $\underset{x\quad z}{\overset{y}{\triangle}}$ be defined for all numbers x, y, and z

as $\underset{x\quad z}{\overset{y}{\triangle}} = xy - z$. If $n = \underset{2\quad 8}{\overset{6}{\triangle}}$, what is the

value of $\underset{n\quad 10}{\overset{5}{\triangle}}$?

 (A) 0
 (B) 1
 (C) 4
 (D) 10
 (E) 15

This symbolic function adds another dimension to the key, but the process is still the same.

Begin by solving for n. For every x, substitute 2. For every y, use a 6. And for every z, substitute an 8:

Puzzle: 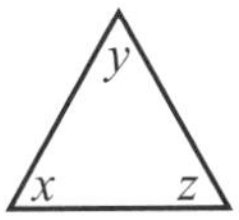

$= xy - z$

Key: 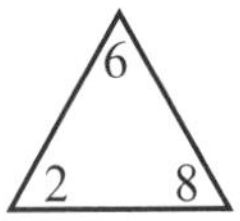

$= (2)(6) - 8 \rightarrow 12 - 8 \rightarrow 4$

Now you know that $n = 4$. Plug this value into the key containing n:

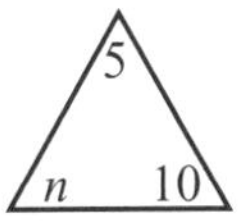

$\rightarrow$
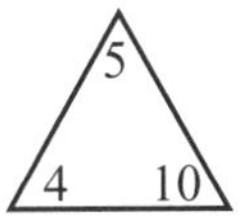

Then solve using the new key:

Puzzle: 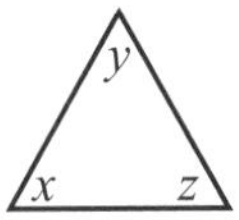

$= xy - z$

Key: 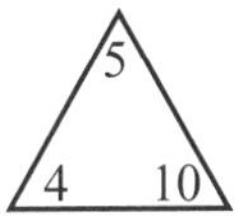

$= (4)(5) - 10 \rightarrow 20 - 10 \rightarrow 10$

The correct answer is 10, answer choice (D).

Symbolic functions are not as difficult as they first appear; by separating the information presented and working with one puzzle and key at a time, they are usually quite easy to solve.

Confidence Quotation
"The person who makes a success of living is the one who sees his goal steadily and aims for it unswervingly. That is dedication."
—Cecil B. DeMille, director and producer

Symbolic Functions Problem Set

Solve the following multiple-choice questions by selecting the best answer from the five answer choices. For grid-in questions, write your answer in the grids and completely mark the corresponding ovals. Answers begin on page 264.

1. If $\begin{array}{c|c} w & x \\ \hline y & z \end{array}$ is defined by $\begin{array}{c|c} w & x \\ \hline y & z \end{array} = \dfrac{w-z}{y-x}$, what is the value of $\begin{array}{c|c} 22 & 5 \\ \hline 9 & 6 \end{array}$?

2. Let the operations ♠ and ♣ be defined for all real numbers x and y as

$$x \spadesuit y = \frac{x+5}{y-5}$$

$$x \clubsuit y = xy + 5$$

What is the value of $(13 \spadesuit 11) \clubsuit 4$?

(A) 3
(B) 5
(C) 12
(D) 15
(E) 17

3. Let $d\triangle$ be defined as $\dfrac{d^2}{2}$ for all values of d. Which of the following is equivalent to $(d\triangle)\triangle$?

(A) $\dfrac{d^4}{16}$

(B) $\dfrac{d^4}{8}$

(C) $\dfrac{d^4}{4}$

(D) $\dfrac{d^4}{2}$

(E) d^4

Functions

Functions are a relatively new addition to the SAT. Their recent inclusion concerned many students, especially those who had not yet taken Algebra II, but the functions that have appeared on the SAT are quite basic and easy to solve.

Frequency Guide: 4

Required Knowledge and Skill Set

1. A function is the relation between two given sets of elements—the domain and the range. Each element in the domain is associated with exactly one element in the domain. Therefore, in a $f(x)$ function, for every input of x, there is one single output for $f(x)$.

On the SAT, most functions use the notation f(x), g(x), or h(x).

2. The most common function notation is $f(x)$, although any variables—$g(x)$, $h(x)$, $f(y)$—can be used to indicate a function.

3. Like symbolic functions, you can think of regular functions as substitution puzzles with keys. Whatever value or expression appears in the parentheses of the function is plugged into the puzzle wherever x (or the assigned variable) appears:

   Puzzle: $f(x) = x^2 - x$
   Key: $f(3)$

   To solve this function, substitute the key (3) for every x in the puzzle:

   $f(x) = x^2 - x$
   $f(3) = 3^2 - 3 \quad \rightarrow \quad 9 - 3 \quad \rightarrow \quad 6$

   The same puzzle can be used with different keys:

   Puzzle: $f(x) = x^2 - x$
   Key: $f(5)$
   $f(x) = x^2 - x$
   $f(5) = 5^2 - 5 \quad \rightarrow \quad 25 - 5 \quad \rightarrow \quad 20$

   Puzzle: $f(x) = x^2 - x$
   Key: $f(-10)$
   $f(x) = x^2 - x$
   $f(-10) = -10^2 - -10 \quad \rightarrow \quad 100 + 10 \quad \rightarrow \quad 110$

   Puzzle: $f(x) = x^2 - x$
   Key: $f\left(\frac{1}{2}\right)$
   $f(x) = x^2 - x$
   $f\left(\frac{1}{2}\right) = \left(\frac{1}{2}\right)^2 - \frac{1}{2} \quad \rightarrow \quad \frac{1}{4} - \frac{1}{2} \quad \rightarrow \quad \frac{1}{4} - \frac{2}{4} \quad \rightarrow \quad -\frac{1}{4}$

4. Sometimes the key is an expression rather than a value. To solve these functions, simply plug the entire expression into the puzzle for every occurrence of x. Be sure to use parentheses when inserting the expression:

Puzzle: $f(x) = x^2$
Key: $f(y + 1)$
$f(x) = x^2$
$f(y + 1) = (y + 1)^2 \rightarrow (y + 1)(y + 1) \rightarrow y^2 + 2y + 1$

Puzzle: $f(x) = x^2 + 4x + 7$
Key: $f(x^2)$
$f(x) = x^2 + 4x + 7$
$f(x^2) = (x^2)^2 + 4(x^2) + 7 \rightarrow x^4 + 4x^2 + 7$

Puzzle: $h(x) = 4x^2 + 6$
Key: $h\left(\frac{a}{2}\right)$
$h(x) = 4x^2 + 6$
$$h\left(\frac{a}{2}\right) = 4\left(\frac{a}{2}\right)^2 + 6 \rightarrow 4\left(\frac{a^2}{4}\right) + 6 \rightarrow a^2 + 6$$

5. It is extremely important to use parentheses when inserting an expression into a function. The makers of the SAT are likely to use functions that cause students to forget to distribute a negative or an exponent:

☠ CAUTION: SAT TRAP!
Beware of functions that require distributing a negative or an exponent!

Puzzle: $g(x) = 5 - x$
Key: $g(x + 1)$

Correct Solution:
$g(x + 1) = 5 - (x + 1) \rightarrow 5 - x - 1 \rightarrow 4 - x$

Incorrect Solution:
$g(x + 1) = 5 - x + 1 \rightarrow 6 - x$

If you fail to use parentheses, you will fail to distribute the negative sign. The same mistake can happen with exponents:

Puzzle: $f(x) = 5x^2$
Key: $f(3x)$

Correct Solution:
$f(3x) = 5(3x)^2 \rightarrow 5(9x^2) \rightarrow 45x^2$

Incorrect Solution:
$f(3x) = 5 \times 3x^2 \rightarrow 15x^2$

6. If everything that occurs *inside* the parentheses is plugged into the puzzle, what happens to values and expressions that occur *outside* of the parentheses? A function can be added, subtracted, multiplied, and divided just like any other value or expression. Therefore, anything that occurs outside of the parentheses should be treated as a normal quantity in a normal equation. To solve these functions, move these quantities to the other side of the equation.

$$4f(x) + 2 = 14 \rightarrow 4f(x) = 12 \rightarrow f(x) = 3$$

$$\sqrt{f(x)} = 4 \rightarrow \left(\sqrt{f(x)}\right)^2 = 4^2 \rightarrow f(x) = 16$$

Values outside of the parentheses in f(x) can be manipulated like values in any equation.

7. Compound functions contain a function within a function:

$f(g(x)) = 10$ $\qquad$ $j(k(x)) = 9 + 2x$ $\qquad$ $f(g(h(x))) = 3x + 25$

To solve compound functions, work from the inside out. Consider an example from the SAT:

> 11. If $f(x) = 2x^2 - x$ and $g(x) = |2 - x|$, what is $f(g(6))$?
>
> (A) 6
> (B) 28
> (C) 36
> (D) 64
> (E) 96

Begin by solving the innermost function, $g(6)$:

Puzzle: $g(x) = |2 - x|$
Key: $g(6)$
$g(x) = |2 - x|$
$g(6) = |2 - 6| \rightarrow |-4| \rightarrow 4$

For compound functions, remember to work from the inside out.

Since $g(6) = 4$, then $f(g(6)) = f(4)$. Now solve for $f(4)$:

Puzzle: $f(x) = 2x^2 - x$
Key: $f(4)$
$f(x) = 2x^2 - x$
$f(4) = 2(4)^2 - 4 \rightarrow 2(16) - 4 \rightarrow 32 - 4 \rightarrow 28$

The correct answer is (B).

What happens if you incorrectly work from the outside in? You get the wrong answer, which almost always appears as an answer choice! Be sure to work compound functions from the inside out.

Application on the SAT

The easiest function questions ask you to solve for f(x). Most function questions, however, have a Medium or Hard difficulty level.

Some function questions simply ask you to evaluate a function:

3. For all numbers x, let the function f be defined by $f(x) = (x-4)^2 + x$. What is the value of $f(9)$?

(A) 25
(B) 34
(C) 45
(D) 90
(E) 178

To solve these Easy function questions, substitute the key for all instances of x:

$$f(x) = (x-4)^2 + x$$
$$f(9) = (9-4)^2 + 9 \rightarrow 5^2 + 9 \rightarrow 25 + 9 \rightarrow 34$$

A twist on a regular function question occurs when two functions are manipulated after each is evaluated:

10. For all numbers x, let the function f be defined by $f(x) = (x-4)^2 + x$. What is the value of $f(9) - f(3)$?

(A) 6
(B) 10
(C) 30
(D) 38
(E) 68

Start by evaluating each function separately:

$$f(x) = (x-4)^2 + x$$
$$f(9) = (9-4)^2 + 9 \rightarrow 5^2 + 9 \rightarrow 25 + 9 \rightarrow 34$$

$$f(x) = (x-4)^2 + x$$
$$f(3) = (3-4)^2 + 3 \rightarrow -1^2 + 3 \rightarrow 1 + 3 \rightarrow 4$$

Now subtract $f(3)$ from $f(9)$:

$f(9) = 34$ and $f(3) = 4$
$f(9) - f(3)$
$34 - 4 = 30$ The correct answer is (C).

The College Board may also make functions more difficult by placing the keys in answer choices:

When the answer choices contain keys, evaluate each one separately.

12. For all numbers x, let the function f be defined by $f(x) = (x-4)^2 + x$. Which of the following is equal to $f(3)$?

(A) $f(-3)$
(B) $f(0)$
(C) $f(4)$
(D) $f(6)$
(E) $f(10)$

Again, begin by evaluating the function in the question first:

$f(x) = (x-4)^2 + x$
$f(3) = (3-4)^2 + 3 \rightarrow -1^2 + 3 \rightarrow 1 + 3 \rightarrow 4$

Now, which of the answer choices results in 4? BACKPLUG each answer choice until you find the one that equals 4:

Puzzle: $f(x) = (x-4)^2 + x$
Keys:
(A) $f(-3) = (-3-4)^2 + -3 \rightarrow -7^2 + -3 \rightarrow 49 - 3 \rightarrow 46$ No
(B) $f(0) = (0-4)^2 + 0 \rightarrow -4^2 \rightarrow 16$ No
(C) $f(4) = (4-4)^2 + 4 \rightarrow 0^2 + 4 \rightarrow 0 + 4 \rightarrow 4$ ✓

There is no need to evaluate (D) or (E), as answer choice (C) equals 4.

One type of function question often confuses test takers because it is missing a key. In these questions, the function is given an equivalent, and you are required to solve for a variable. Consider an example:

14. For all n, the function h is defined by $h(n) = n^2 - 2n$? When $h(n) = 15$, what is one possible value of $2n$?

(A) 5
(B) 10
(C) 15
(D) 25
(E) 30

Remember, when a problem uses the phrase "what is one possible value of...," there is more than one answer. Suspect a square root, absolute value, or quadratic equation.

This questions contains two puzzles, rather than a puzzle and a key:

Puzzle: $h(n) = n^2 - 2n$
Puzzle: $h(n) = 15$

Since both expressions equal $h(n)$, the expressions themselves are also equal:

$h(n) = h(n)$, so $n^2 - 2n = 15$

Now, solve for n by recognizing the quadratic equation:

$n^2 - 2n = 15$
$n^2 - 2n - 15 = 0$
$(n - \quad)(n + \quad) = 0$
$(n-5)(n+3) = 0$
$n = 5$ or $n = -3$

Now that you have found n, calculate $2n$:

If $n = 5$, $2n = 10$ and if $n = -3$, $2n = -6$.

Answer choice (B) is correct.

Another question type that can cause confusion is a function within a function. Consider the following student-produced response question:

14. For all numbers x, the function g is defined by $g(x) = 2x + 1$, and the function h is defined by $h(x) = g(x) - 3$. What is the value of $h(100)$?

This question has an exaggerated difficultly level:

To solve a function within a function, simply substitute the value of the internal function for the function itself.

$$g(x) = 2x + 1$$
$$h(x) = g(x) - 3 \quad \rightarrow \quad h(x) = (2x + 1) - 3 \quad \rightarrow \quad h(x) = 2x - 2$$

Now solve for $h(100)$:

$$h(x) = 2x - 2$$
$$h(100) = 2(100) - 2 \quad \rightarrow \quad 200 - 2 \quad \rightarrow \quad 198$$

The correct answer is 198. Although this question was easily solved, most students fail to correctly substitute for the function within a function. You must substitute $2x + 1$ for $g(x)$ to find $h(100)$.

The College Board will sometimes use a table to display value values for x and $f(x)$. The answer choices are functions, and you must determine which of the functions is represented by the table:

x	–2	0	2	5
$f(x)$	16	10	4	–5

9. In the table above, values are given for the linear function f for selected values of x. Which of the following defines f?

(A) $f(x) = x - 1$
(B) $f(x) = 2x + 1$
(C) $f(x) = -3x + 10$
(D) $f(x) = 4x + 10$
(E) $f(x) = -4x - 7$

Notice that the question uses the phrase, "which of the following?" This signals a BACKPLUGGING problem. Select a corresponding set of numbers from the table to BACKPLUG into each of the equations. Choose an x value from the table that will make for easy calculations, such as $x = 0$. Then BACKPLUG 0 for x in each answer choice, to find the answers that produce $f(0) = 10$.

For a table function question, it is important to plug your chosen value into all five answer choices as more than one answer choice may produce the desired result. But only one of the answer choices will work with all of the values in the table.

(A) $f(x) = x - 1 \rightarrow f(0) = 0 - 1 \rightarrow f(0) = -1$ No
(B) $f(x) = 2x + 1 \rightarrow f(0) = 2(0) + 1 \rightarrow f(0) = 1$ No
(C) $f(x) = -3x + 10 \rightarrow f(0) = -3(0) + 10 \rightarrow f(0) = 10$ ✓
(D) $f(x) = 4x + 10 \rightarrow f(0) = 4(0) + 10 \rightarrow f(0) = 10$ ✓
(E) $f(x) = -4x - 7 \rightarrow f(0) = -4(0) - 7 \rightarrow f(0) = -7$ No

Eliminate choices (A), (B), and (E), as they did not produce 10.

Because both (C) and (D) resulted in 10, choose another set of values from the table to test these two equations:

$x = 2, f(x) = 4$

(C) $f(x) = -3x + 10 \rightarrow f(2) = -3(2) + 10 \rightarrow f(2) = 4$ ✓
(D) $f(x) = 4x + 10 \rightarrow f(2) = 4(2) + 10 \rightarrow f(2) = 18$ No

Only choice (C) works when $x = 0$ and $x = 2$, so it is the function that defines the values in the table.

Confidence Quotation
"The man who has no imagination has no wings."
—Muhammad Ali, World Heavyweight Champion boxer

The final type of function in our discussion occurs in a word problem, which may provide a function and ask for a value:

$$c(n) = 100 + 5n$$

8. To host a field trip for n students, the dollar amount a museum charges is given by the function $c(n)$ above. How many dollars does the museum charge to host a field trip for 32 students?

(A) $8.13
(B) $13.60
(C) $60.00
(D) $160.00
(E) $260.00

Or provide a the situation and ask for the function:

These two questions are different versions of the same problem.

11. To host a field trip, a museum charges a base rate of $100 plus $5 per every student in attendance. Which of the following functions gives the charge, c, in dollars, of a trip if n students attend?

(A) $c(n) = 100(5n)$
(B) $c(n) = 100 + 5n$
(C) $c(n) = 100 - 5n$
(D) $c(n) = 105 + n$
(E) $c(n) = 105(n)$

To solve the first question, plug 32 in for n into the function:

$c(n) = 100 + 5n$
$c(32) = 100 + 5(32) \rightarrow c(32) = 100 + 160 \rightarrow c(32) = 260$

Answer choice (E) is correct.

To solve the second question, use TRANSLATION to create the function:

total charge is a base rate of $100 plus $5 per student
$c(n) = \$100 + \$5(n)$

The correct answer is (B).

Functions Problem Set

Solve the following multiple-choice questions by selecting the best answer from the five answer choices. For grid-in questions, write your answer in the grids and completely mark the corresponding ovals. Answers begin on page 265.

1. For all numbers x, let the function g be defined by $g(x) = (x + 5)(x - 1)$. Which of the following has a positive value?

(A) $g(1)$
(B) $g(0)$
(C) $g(-1)$
(D) $g(-4)$
(E) $g(-6)$

2. Let the function f be defined as $f(x) = \frac{x^2 + 6}{5}$. For what value of x does $f(x) = 30$?

$30 = \frac{x^2+6}{5}$
$150 = x^2+6$
$144 = x^2$
$12 = x$

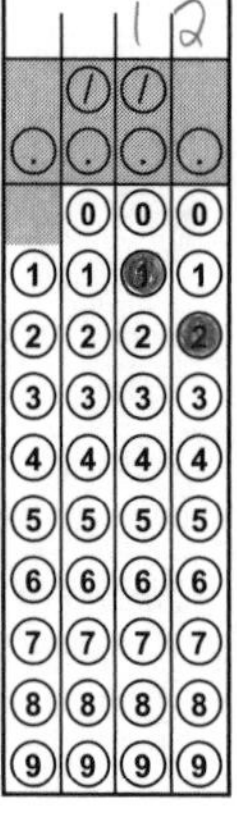

x	$k(x)$
–6	12
–4	4
2	4
10	60

3. In the table above, values are given for the linear function k for selected values of x. Which of the following defines k ?

(A) $k(x) = 2x + 15$

(B) $k(x) = x^2 - x$

(C) $k(x) = \frac{1}{2}x^2 + x$

(D) $k(x) = -x - 20$

(E) $k(x) = -\frac{x}{2}$

Functions Problem Set

Solve the following multiple-choice questions by selecting the best answer from the five answer choices. For grid-in questions, write your answer in the grids and completely mark the corresponding ovals. Answers begin on page 265.

4. Izzy was paid $40 per day when she first started working at a farm. Over time, her pay increased by 20 percent each year. Which of the following functions gives Izzy's pay, p, in dollars, after n years of working at the farm?

(A) $p(n) = 40(0.2n)$
(B) $p(n) = 40(0.2)^n$
(C) $p(n) = 40n^{0.2}$
(D) $p(n) = 40(1.2n)$
(E) $p(n) = 40(1.2)^n$

$$f(x) = zg(x) - 10$$

5. In the function above, f is defined in terms of another function g for all values of x and z is a constant. If h is the number for which $f(h) = 38$ and $g(h) = 15$, what is the value of z ?

6. For all values of x, the function f is defined as $f(x) = \frac{x}{2} + 4$. Which of the following is equal to $f(6) - f(-12)$?

(A) $f(18)$
(B) $f(10)$
(C) $f(6)$
(D) $f(0)$
(E) $f(-4)$

ALGEBRA II MASTERY ANSWER KEY

Quadratic Equations Problem Set—Page 242

1. (B) Medium

You should recognize the classic quadratic form:

$$y^2 - z^2 = (y - z)(y + z)$$

Since y and z are integers, their difference $(y - z)$ and their sum $(y + z)$ will also be integers. That means that $y - z$ must be a factor of 24. The only factor of 24 in the answer choices is 4; if $y - z = 4$, then $y + z = 6$.

$$(y - z)(y + z) = 24 \quad \rightarrow \quad (4)(6) = 24$$

2. (D) Medium

This "fake" quadratic leads many students astray. There is no need to FOIL, as it is clear that $k = 5$:

$$(k - 5)^2 = 0 \quad \rightarrow \quad \sqrt{k-5} = \sqrt{0} \quad \rightarrow \quad k - 5 = 0 \quad \rightarrow \quad k = 5$$

Now plug $k = 5$ into the required expression:

$$(k + 3)(k - 3) \quad \rightarrow \quad (5 + 3)(5 - 3) \quad \rightarrow \quad (8)(2) \quad \rightarrow \quad 16$$

Remember, answers like (E) have never been correct on any of the tests analyzed by PowerScore.

3. (C) Medium

You must FOIL two sets of binomials. Be sure to use brackets so that you remember to distribute the negative in the second set of binomials:

$$(x + 6)(x + 5) - (x - 3)(x - 2) = 0$$
$$[x^2 + 5x + 6x + 30] - [x^2 - 2x - 3x + 6] = 0$$
$$[x^2 + 11x + 30] - [x^2 - 5x + 6] = 0$$
$$x^2 + 11x + 30 - x^2 + 5x - 6 = 0$$
$$16x + 24 = 0$$
$$16x = -24$$
$$x = -1.5$$

4. (D) Hard

Notice that the two binomials use different variables (a and b). This indicates a "fake" quadratic, so spend some time studying the equation:

$$(a + 6)(b - 6) = 0$$

Either $a + 6 = 0$ or $b - 6 = 0$ in order for the entire equation to equal 0. Evaluate both possibilities:

If $a + 6 = 0$:

$$a + 6 = 0 \quad \rightarrow \quad a = -6$$

Since you are looking for the smallest possible value of $a^2 + b^2$, make b as small as possible: $b = 0$

$$a^2 + b^2 \quad \rightarrow \quad -6^2 + 0^2 \quad \rightarrow \quad 36 + 0 \quad \rightarrow \quad 36$$

Now try the other possibility to see if it creates a smaller value:

If $b - 6 = 0$:

$$b - 6 = 0 \quad \rightarrow \quad b = 6$$

Since you are looking for the smallest possible value of $a^2 + b^2$, make a as small as possible: $a = 0$

$$a^2 + b^2 \quad \rightarrow \quad 0^2 + 6^2 \quad \rightarrow \quad 0 + 36 \quad \rightarrow \quad 36$$

For both possibilities, 36 is the least value of $a^2 + b^2$.

Direct and Inverse Variation Problem Set—Page 246

1. (A) Medium

When two variables vary inversely, as one variable increases, the other decreases. For this reason, you can eliminate (B), (C), (D), and (E), as their variables both increase. Only (A) represents an inverse relationship.

2. (E) Medium

Use the formula for direct variation, where c is the constant:

$$y = cx \quad \rightarrow \quad y = c\sqrt{x}$$

Find c using the first set of value for x and y:

$$\frac{1}{3} = c\sqrt{\frac{4}{9}} \quad \rightarrow \quad \frac{1}{3} = c\frac{2}{3} \quad \rightarrow \quad \frac{3}{2} \times \frac{1}{3} = c \quad \rightarrow \quad \frac{1}{2} = c$$

Now use the value of c to find x when $y = 2$:

$y = cx \rightarrow y = c\sqrt{x}$

$2 = \frac{1}{2}\sqrt{x} \rightarrow \frac{2}{1} \times 2 = \sqrt{x} \rightarrow 4 = \sqrt{x} \rightarrow 4^2 = \left(\sqrt{x}\right)^2 \rightarrow 16 = x$

The correct answer is (E).

Symbolic Functions Problem Set—Page 252

1. 4 Easy

Set up the puzzle and key:

Puzzle: (circle with w | x over y | z) $= \frac{w - z}{y - x}$

Key: (circle with 22 | 5 over 9 | 6) $= \frac{22 - 6}{9 - 5} \rightarrow \frac{16}{4} \rightarrow 4$

2. (E) Medium

Start with the key in parentheses:

Puzzle: $x \spadesuit y = \frac{x + 5}{y - 5}$

Key: $13 \spadesuit 11 = \frac{13 + 5}{11 - 5} \rightarrow \frac{18}{6} \rightarrow 3$

Plug the new value, $13 \spadesuit 11 = 3$, into the next key, $3 \clubsuit 4$:

Puzzle: $x \clubsuit y = xy + 5$

Key: $3 \clubsuit 4 = (3)(4) + 5 \rightarrow 12 + 5 \rightarrow 17$

3. (B) Hard

This question is confusing only because the key inside the parentheses is the same as the puzzle.

Puzzle: $d\triangle = \frac{d^2}{2}$

Key: $d\triangle = \frac{d^2}{2}$

If $d\triangle = \frac{d^2}{2}$, then $d\triangle\ (\triangle) =$

Puzzle: $d\triangle = \frac{d^2}{2}$

Key: $\frac{d^2}{2}\ (\triangle) = \frac{\left(\frac{d^2}{2}\right)^2}{2} \rightarrow \frac{\frac{d^4}{4}}{2} \rightarrow \frac{d^4}{4} \div 2 \rightarrow \frac{d^4}{4} \times \frac{1}{2} \rightarrow \frac{d^4}{8}$

Functions Problem Set—Pages 260-261

1. (E) Easy

The phrase "which of the following" should signify a BACKPLUGGING problem. Plug each answer choice into the function to find the one that produces a positive result:

$g(x) = (x + 5)(x - 1)$
(A) $g(1) = (1 + 5)(1 - 1) \rightarrow g(1) = (6)(0) \rightarrow g(1) = 0$ No
(B) $g(0) = (0 + 5)(0 - 1) \rightarrow g(0) = (5)(-1) \rightarrow g(0) = -5$ No
(C) $g(-1) = (-1 + 5)(-1 - 1) \rightarrow g(-1) = (4)(-2) \rightarrow g(-1) = -6$ No
(D) $g(-4) = (-4 + 5)(-4 - 1) \rightarrow g(-4) = (1)(-5) \rightarrow g(-4) = -5$ No
(E) $g(-6) = (-6 + 5)(-6 - 1) \rightarrow g(-6) = (-1)(-7) \rightarrow g(-6) = 7$ ✓

The correct answer is (E).

2. 12 Medium

Two quantities equal $f(x)$:

$f(x) = \frac{x^2 + 6}{5}$

$f(x) = 30$

Therefore, set them equal and solve for x:

$\frac{x^2 + 6}{5} = 30 \rightarrow x^2 + 6 = 150 \rightarrow x^2 = 144 \rightarrow x = 12$

3. (C) Medium

From the table, choose a value for x that will make calculations easy. If you choose $x = 2$, then $k(x) = 4$. Plug $x = 2$ into all answer choices to find the one or ones that equal 4:

(A) $k(x) = 2x + 15 \rightarrow k(2) = 2(2) + 15 \rightarrow k(2) = 19$ No
(B) $k(x) = x^2 - x \rightarrow k(2) = 2^2 - 2 \rightarrow k(2) = 2$ No
(C) $k(x) = \frac{1}{2}x^2 + x \rightarrow k(2) = \frac{1}{2}2^2 + 2 \rightarrow k(2) = 4$ ✓
(D) $k(x) = -x - 20 \rightarrow k(2) = -2 - 20 \rightarrow k(2) = -22$ No
(E) $k(x) = -\frac{x}{2} \rightarrow k(2) = -\frac{2}{2} \rightarrow k(2) = -1$ No

The correct answer is (C).

4. (E) Medium

There are many ways to solve this question. The first is to understand yearly compound interest and its equation, where a is the final amount, p is the initial amount, r is the percentage increase, and n is the number of years:

$$a = p(1 + r)^n$$

If we plug 40 in for p and 0.2 in for r, we are left with:

$$a = 40(1.2)^n$$

Answer choice (E) is correct.

However, since this formula is not provided on the test, let us examine the BACKPLUGGING solution. First, determine what Izzy's pay would be after a certain amount of time. SUPPLY $n = 1$ year. If Izzy made \$40 on her first day, one year later she made \$40 plus 20% of \$40:

After one year: \$40 plus 20% of \$40 → $40 + .2(40)$ → $40 + 8$ → 48
After two years: \$48 plus 20% of \$48 → $48 + .2(48)$ → $48 + 9.60$ → 57.60
After three years: \$57.60 plus 20% of \$57.60 → $57.6 + .2(57.6)$ → $57.6 + 11.52$ → 69.12

As you can see it would continue to grow in this manner each year. However, you do not need to know her pay every year; just one year will help you solve this question. Do not choose $n = 1$ because 1 has special properties when multiplied or raised to a power.

So if $n = 2$, $p(n) = 57.60$. Which answer choice will give you this result?

(A) $p(n) = 40(0.2n)$ → $57.6 = 40(0.2)(2)$ → $57.6 = 16$ False
(B) $p(n) = 40(0.2)^n$ → $57.6 = 40(0.2)^2$ → $57.6 = 1.6$ False
(C) $p(n) = 40n^{0.2}$ → $57.6 = 40(2)^{0.2}$ → $57.6 = 45.94792$ False
(D) $p(n) = 40(1.2n)$ → $57.6 = 40(1.2)(2)$ → $57.6 = 96$ False
(E) $p(n) = 40(1.2)^n$ → $57.6 = 40(1.2)^2$ → $57.6 = 57.6$ ✓

Only (E) creates a true mathematical sentence.

5. 3.2 or $\frac{16}{5}$ Medium

This is a function within a function. Begin by changing $f(x)$ to $f(h)$:

$$f(x) = zg(x) - 10 \rightarrow f(h) = zg(h) - 10$$

Now sub $g(h) = 15$ into the function:

$$f(h) = zg(h) - 10 \rightarrow f(h) = z(15) - 10$$

We have two quantities that equal $f(h)$. Set them equal to solve for z:

$f(h) = z(15) - 10$ and $f(h) = 38$
$z(15) - 10 = 38 \rightarrow z(15) = 48 \rightarrow z = 3.2$

6. (B) Hard

Find $f(6)$:

$$f(x) = \frac{x}{2} + 4 \rightarrow f(6) = \frac{6}{2} + 4 \rightarrow f(6) = 3 + 4 \rightarrow f(6) = 7$$

Then find $f(-12)$:

$$f(x) = \frac{x}{2} + 4 \rightarrow f(-12) = \frac{-12}{2} + 4 \rightarrow f(-12) = -6 + 4 \rightarrow f(-12) = -2$$

Now solve $f(6) - f(-12)$:

$$7 - -2 \rightarrow 7 + 2 = 9$$

Which answer choice also results in 9?

(A) $f(18) = \frac{18}{2} + 4 \rightarrow f(18) = 9 + 4 \rightarrow f(18) = 13$ No

(B) $f(10) = \frac{10}{2} + 4 \rightarrow f(10) = 5 + 4 \rightarrow f(10) = 9$ ✓

The correct answer is (B).

CHAPTER EIGHT:
GEOMETRY MASTERY

Geometry accounts for a large portion of SAT math questions. The concepts behind these questions come from basic geometry, but their very nature makes them prime questions for testing critical reasoning skills. Classroom math textbooks emphasize the importance of basic formulas, and any student can find c in $a^2 + b^2 = c^2$. But what if you have to find b? Still a relatively simple task, but one that can be easily flubbed when you have spent the last four years solving for a different variable. And all students can find the area of a circle given the radius. But can you find the circumference when given the area? It's easy if you know to find the radius using the area, but most students are overwhelmed by this upper level Medium question.

Geometry is the branch of mathematics that deals with the properties, measurements, and relationships of lines, circles, polygons, and solid figures.

SAT geometry questions are not difficult, but they are often designed to make students think "backwards" or "outside of the box." Your key to success is to confidently understand the fundamentals of geometry and to become familiar with the patterns of SAT geometry questions. We will cover both aspects in the following chapter.

Based on PowerScore's extensive analysis of real tests, you can use the following key to predict the general frequency of each type of geometry question:

Rating	Frequency
5	Extremely High: 3 or more questions typically appear on every SAT
4	High: at least 2 questions typically appear on every SAT
3	Moderate: at least one question typically appears on every SAT
2	Low: one question typically appears on every two or three SATs
1	Extremely Low: one question appears infrequently and without a pattern

Remember that there are eight solution strategies you can employ on SAT math questions:

1. ANALYZE the Answer Choices
2. BACKPLUG the Answer Choices
3. SUPPLY Numbers
4. TRANSLATE from English to Math
5. RECORD What You Know
6. SPLIT the Question into Parts
7. DIAGRAM the Question
8. SIZE UP the Figures

An answer key follows the chapter and should be used to study the solutions to questions that you miss.

The Formula Box

Unlike other college entrance exams, the SAT provides most of the geometry formulas needed on the test. The following formula box appears at the start of every math section:

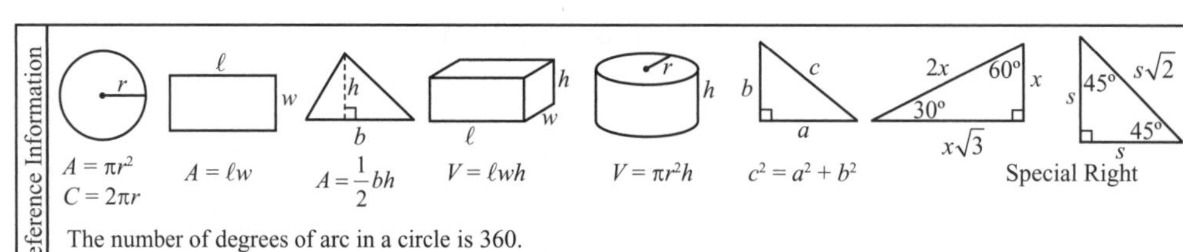

Reference Information
$A = \pi r^2$
$C = 2\pi r$
$A = \ell w$
$A = \frac{1}{2}bh$
$V = \ell wh$
$V = \pi r^2 h$
$c^2 = a^2 + b^2$
Special Right
The number of degrees of arc in a circle is 360.
The sum of the measures in degrees of the angles of a triangle is 180.

MEMORY MARKER:
PowerScore recommends that all students memorize the formulas and relationships in the Formula Box. Each of these formulas is provided in the flash cards at www.powerscore.com/satmathbible.

Most students will rely on this formula box, and return to it many times during each section. Prepared test takers, however, will know these formulas and relationships without question. By memorizing the information in this box, you are saving valuable time that would otherwise be wasted flipping pages to refer back to the formulas. Plus, a student who memorizes these formulas understands the relationships in them, and does not waste time re-orienting himself with the information presented. For optimal test results, you should memorize and understand each of the formulas and relationships represented in the formula box.

To help you become acquainted with the formula box and begin memorizing the information there, fill in each of the requested formulas. Answers are on page 336.

The area of a circle: $A = \pi r^2$

The circumference of a circle: $C = 2\pi r$

The area of a rectangle: $A = \ell w$

The area of a triangle: $A = \frac{1}{2}bh$

The volume of a rectangular solid: $V = \ell wh$

The volume of a right circular cylinder: $V = \pi r^2 h$

The Pythagorean Theorem: $c^2 = a^2 + b^2$

Complete each of the following:

In a 30:60:90 triangle, the side opposite the 30° angle is represented as x, the side opposite the 60° angle is represented as $x\sqrt{3}$, and the side opposite the 90° angle is represented as $2x$.

In a 45:45:90 triangle, the sides opposite the 45° angles are represented as x and the side opposite the 90° angle is represented as $x\sqrt{2}$.

The number of degrees of arc in a circle is 360°.

The sum of the interior angles of a triangle is 180°.

Lines and Angles

Frequency Guide: 4

Lines and angles may seem like basic concepts, but they are at the foundation of all geometry questions. Therefore, it is imperative that you fully understand the following points about lines and angles.

Required Knowledge and Skill Set

1. There are 180° of arc in a straight line. Therefore, any angles created around a straight line, called supplementary angles, must add up to 180°:

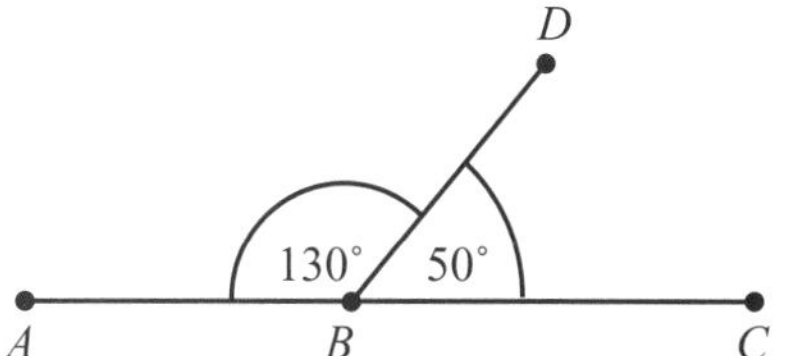

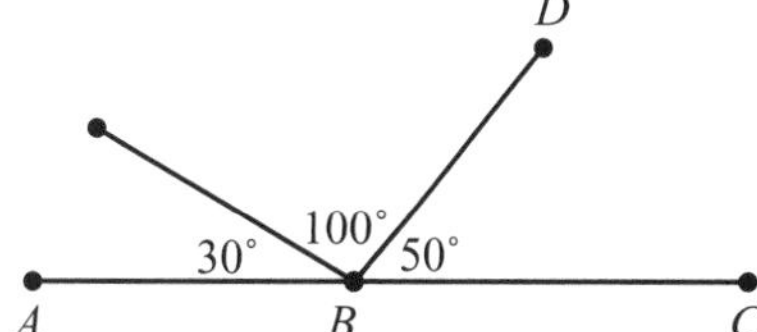

2. Vertical angles are opposite angles that share the same vertex and are created by the same pair of lines, thus sharing the same degree measurement. Two lines that intersect create supplementary and vertical angles:

You do not need to know the terms "supplementary angles" or "vertical angles," but you must understand the relationships represented by these terms.

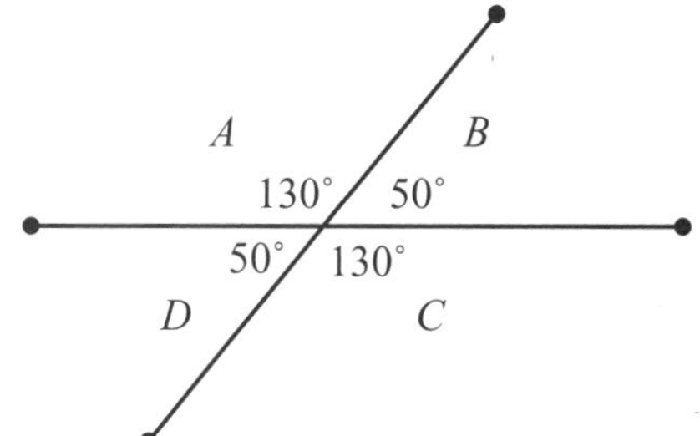

Supplementary Angles:

$\angle A$ and $\angle B$ $\quad$ $\angle B$ and $\angle C$

$\angle A$ and $\angle D$ $\quad$ $\angle C$ and $\angle D$

Vertical Angles:

$\angle A$ and $\angle C$ $\quad$ $\angle B$ and $\angle D$

3. When two parallel lines are intersected by a third line, supplementary and vertical angles are created around both parallel lines. All of the smaller angles have equal measurements, and all of the larger angles have equal measurements. This relationship, called corresponding angles, is tested on nearly every SAT.

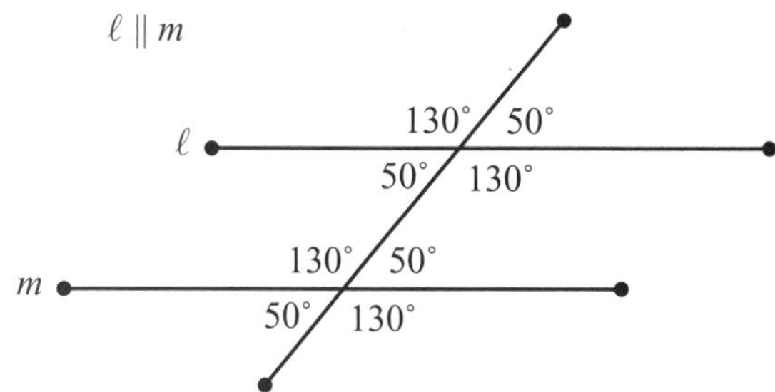

You will likely see a figure similar to this one on the SAT!

This relationship exists even if parts of the figure are shortened or removed:

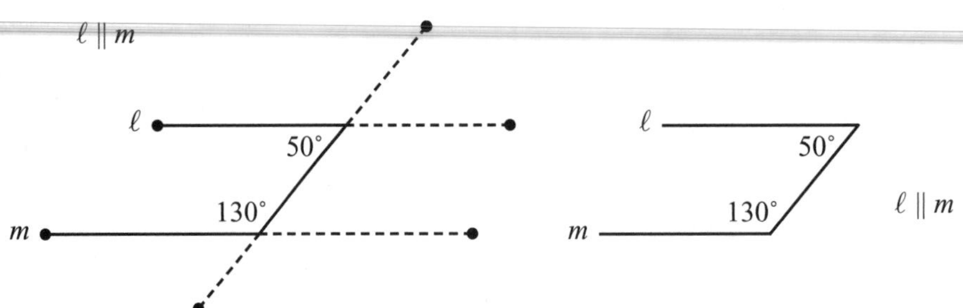

4. A right angle contains 90° and is marked by a special box-shaped symbol. Mark all 90° angles on the test:

Some right angles will be marked or labeled in the figure, while others will only be mentioned in the text of the question. If the angle is discussed in the text, make sure it is labeled on the figure.

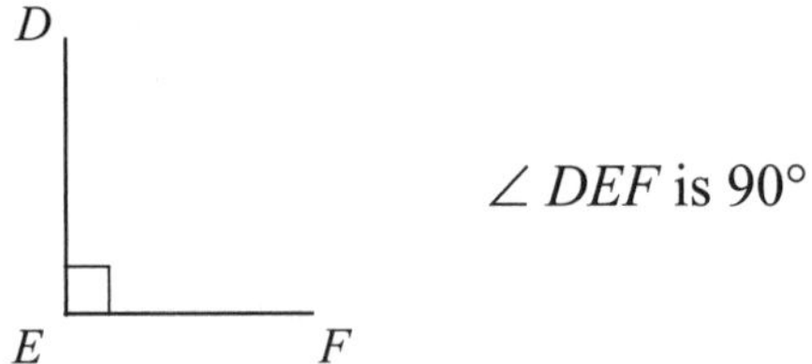

$\angle DEF$ is 90°

5. Perpendicular lines create right angles:

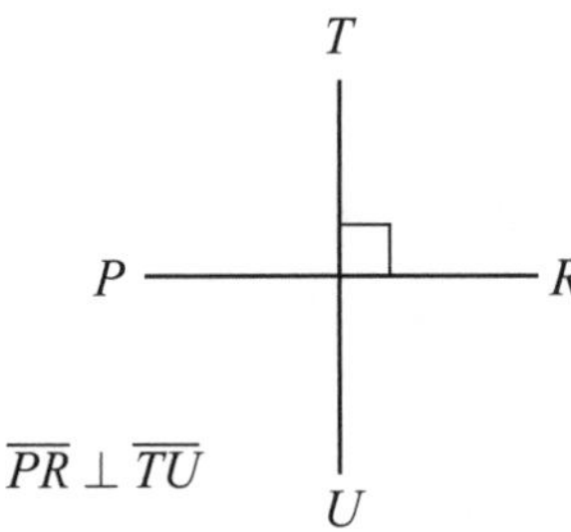

6. An angle is bisected by a line if two equal angles are created. In the following figure, line P bisects angle NMO:

The term "bisect" is used on the SAT.

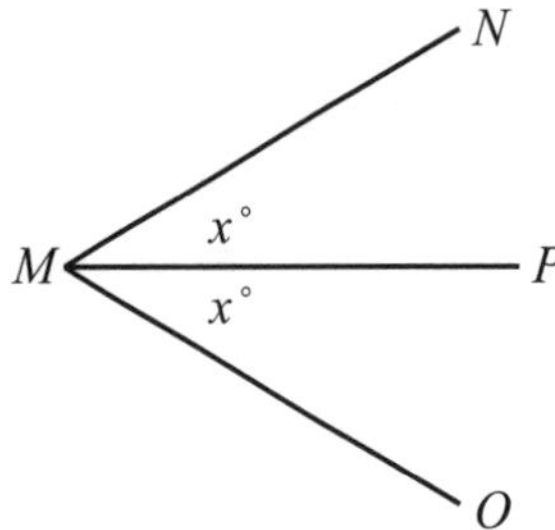

Application on the SAT

The principles of lines and angles may be used in any geometric figure. For example, a line may bisect an angle of a triangle or a line may be drawn that is perpendicular to the diameter of a circle. However, you should expect some questions solely about the relationships concerning lines and angles.

Corresponding angles are tested on nearly every single SAT. There are three basic variations of the corresponding angle figure.

The first one, which we examined in the previous section, is the most basic. Two parallel lines are intersected by a third line (called a transversal):

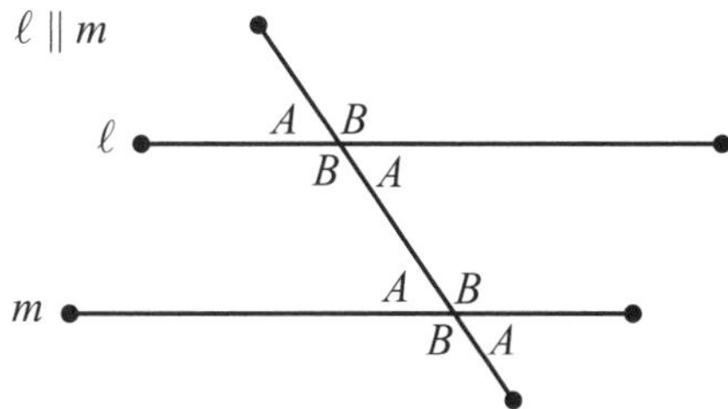

The College Board may make this diagram more difficult by adding a second transversal:

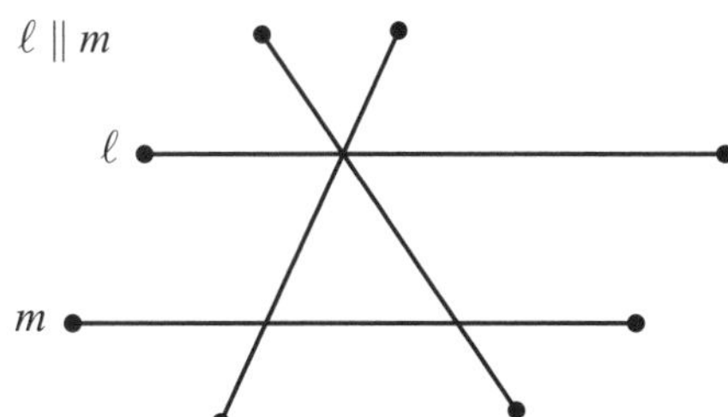

In order to correctly identify corresponding angles without making mistakes, redraw the diagram as two distinct figures and separate the transversals:

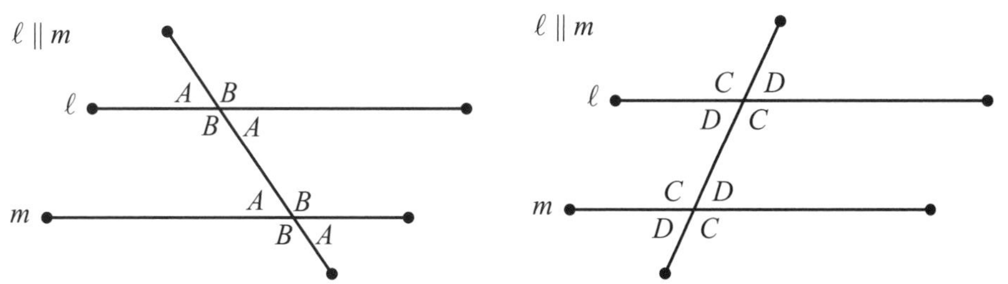

By separating the figure into two unique diagrams, you avoid the confusing intersection of three lines in the original figure.

<u>*Confidence Quotation*</u>
"The difference between the impossible and the possible lies in a man's determination."
—Tommy Lasorda, Major League Baseball manager

CAUTION: SAT TRAP!
Avoid angle confusion by separating figures of parallel lines with two transversals.

The third version of this diagram involves three parallel lines and two transversals:

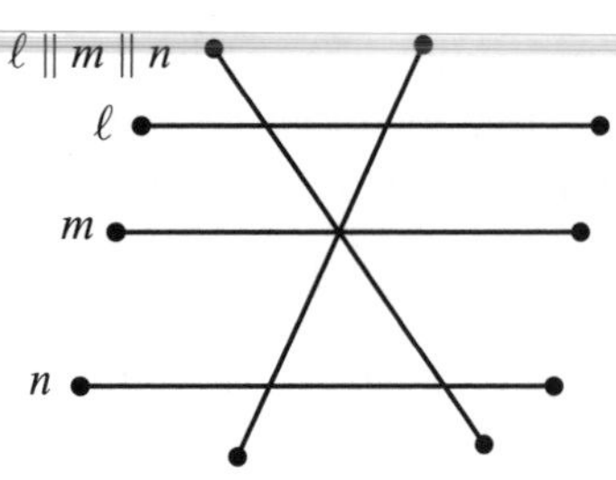

Again, you must divide this figure into two separate diagrams for each transversal:

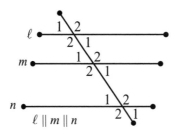

Corresponding angles are tested on nearly every SAT administration.

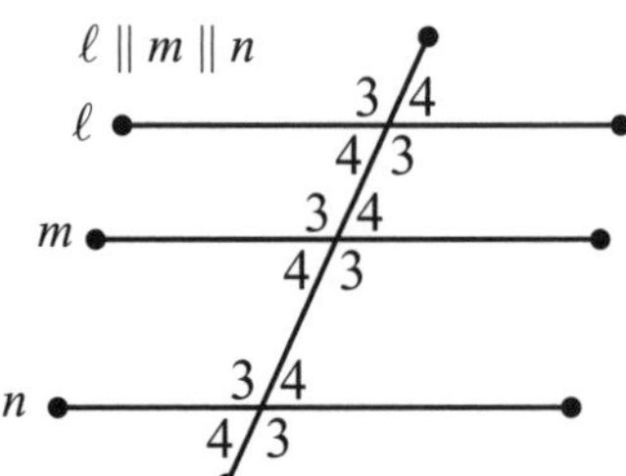

Note that corresponding angles are created with three parallel lines as well. Let's look at how these types of problems are used on the SAT:

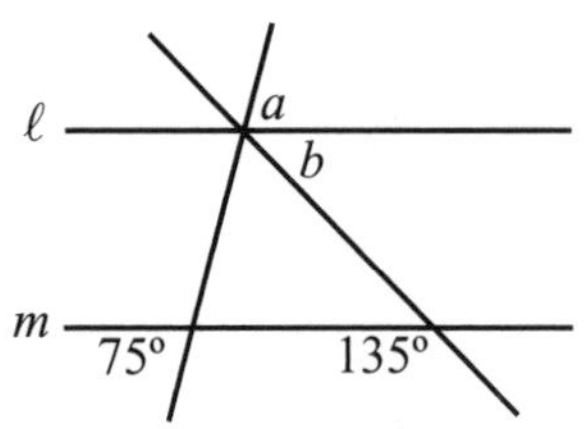

9. In the figure above, lines ℓ and m are parallel. What is the value of $a - b$?
(A) 15
(B) 30
(C) 40
(D) 60
(E) 75

To solve this question, isolate each transversal in its own diagram and find a and b:

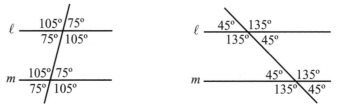

The first diagram shows that angle $a = 75$ and the second reveals that angle $b = 45$:

$a - b \rightarrow 75 - 45 = 30$ Answer choice (B) is correct.

Some line and angle questions will test your ability to calculate angle measurements given supplementary and vertical relationships:

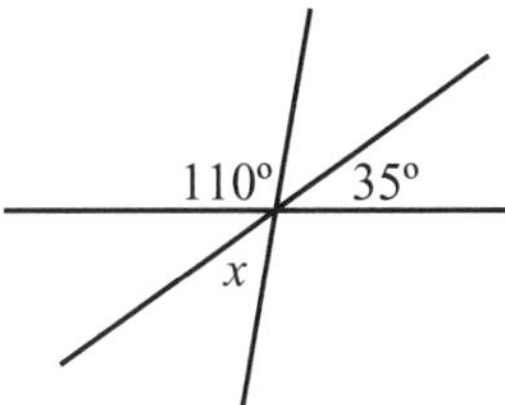

4. In the figure above, three lines intersect. What is the value of x ?

 (A) 25
 (B) 35
 (C) 75
 (D) 110
 (E) It cannot be determined from the information given.

Remember, "it cannot be determined" has never been a correct answer choice on any of the SATs analyzed by PowerScore.

This is a two part solution. For the first component, determine the measurement of the angle between the 110° and 35° angles, using your knowledge of supplementary angles. Let's call that angle a:

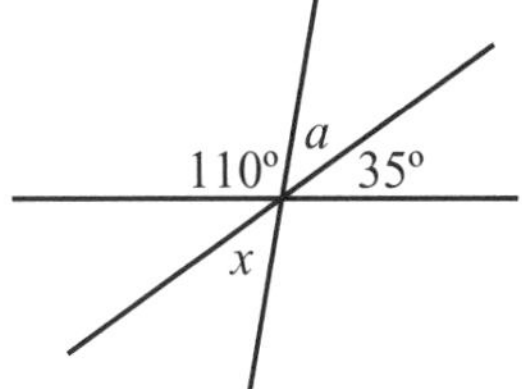

$110° + a + 35° = 180°$
$145° + a = 180°$
$a = 35°$

Now move onto the second part of the solution. If a equals 35, then x also equals 35, because a and x are vertical angles. The correct answer is (B).

As you can see, your knowledge of the principles given in the previous section may be tested in a variety of shapes and forms. Another type of question you should expect to see involves line segments, as discussed previously in the section on Number Lines and again on the next page.

Line segments compose a large portion of lines and angles questions. A common line segment question requires you to find a distance based on midpoints. Consider an example:

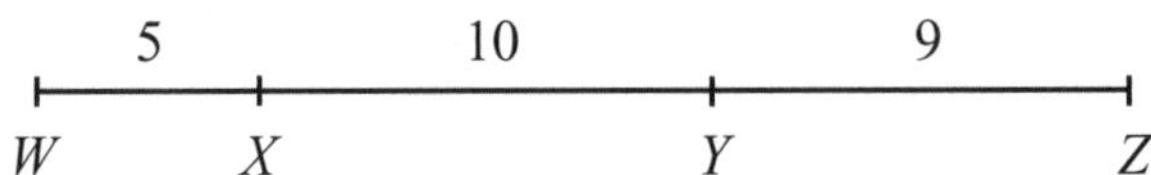

8. In the figure above, line segment WZ is divided by X and Y as shown. Point A, which is not shown, is located at the midpoint of WX and Point B, also not shown, is located at the midpoint of YZ. What is the length of line segment AB ?

    (A) 12
    (B) 13
    (C) 15
    (D) 16
    (E) 17

Be sure to DIAGRAM questions without an existing figure. It is much more difficult to solve questions about line segments without a figure.

To begin, plot points A and B on the line:

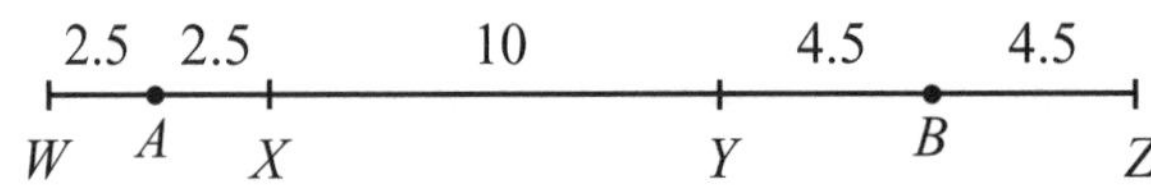

And then add the measurements of the line segments between A and B:

$$AX + XY + YB = AB$$
$$2.5 + 10 + 4.5 = 17$$

The correct answer is (E).

To solve questions about line segments, transfer information from the text to the figure. The more clearly you can label the diagram, the easier the problem is to solve. If a question does not have a figure, be sure to draw one and carefully label its parts.

Lines and Angles Problem Set

Solve the following multiple-choice questions by selecting the best answer from the five answer choices. For grid-in questions, write your answer in the grids and completely mark the corresponding ovals. Answers begin on page 336.

1. The midpoint of line segment AB is C. The midpoint of line segment AC is D and the midpoint of line segment AD is E. If the length of EC is 6, what is the length of AB ?

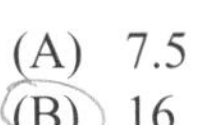

(A) 7.5
(B) 16
(C) 18
(D) 24
(E) 36

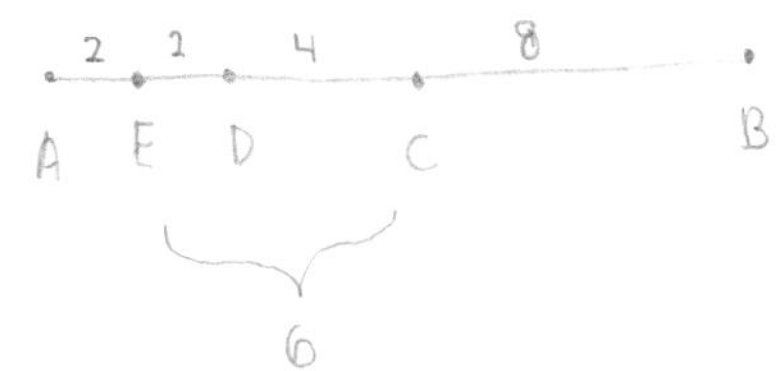

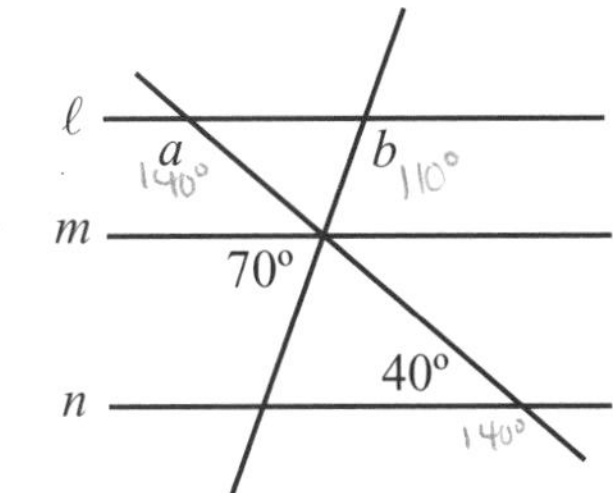

2. In the figure above, lines ℓ, m, and n are parallel. What is the value of $a + b$?

(A) 220
(B) 230
(C) 240
(D) 250
(E) 260

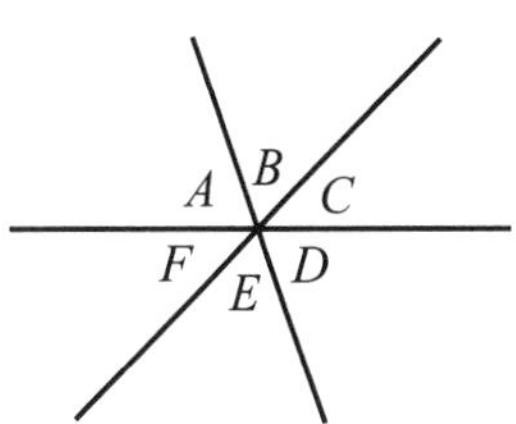

3. In the figure above, three lines intersect at a point as shown. All of the following pairs of angles are sufficient for determining all six angle measures EXCEPT

(A) A and C
(B) B and C
(C) B and E
(D) D and E
(E) D and F

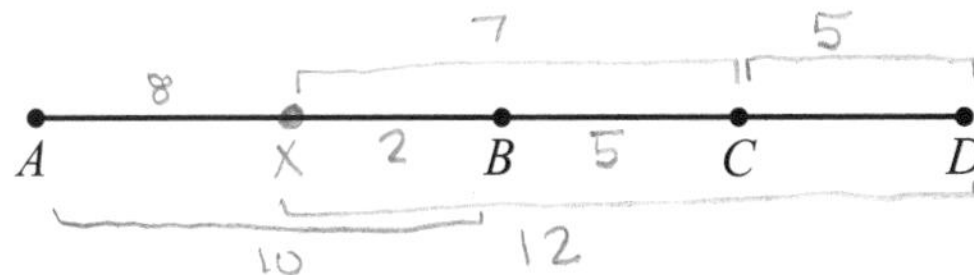

4. In the figure above, B is the midpoint of AD and C is the midpoint of BD. Point X is located on line segment AD between A and B so that $XC = 7$ and $XD = 12$. What is the length of line segment AX ?

(A) 8
(B) 9
(C) 10
(D) 11
(E) 12

Basic Triangles

Frequency Guide: 5

Triangles are by far the most represented polygon on the SAT. We are going to divide them into two groups: Basic Triangles and Special Triangles. This portion on basic triangles refers to the properties of all triangles, while the section on special triangles will look at the properties specific to equilateral, isosceles, and right triangles.

Required Knowledge and Skill Set

The relationships presented in this section apply to all triangles.

1. The interior angles of a triangle add up to 180°. If you are given the degree measure of two angles, you can find the remaining angle measurement:

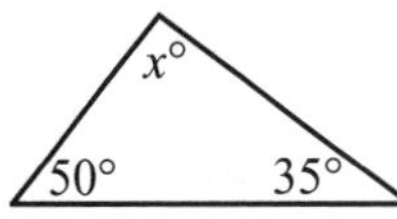

$x° + 50° + 35° = 180°$ $\qquad x = 95°$

2. Knowing the interior measures of a triangle can help you find the measurements of angles outside the triangle and vice versa:

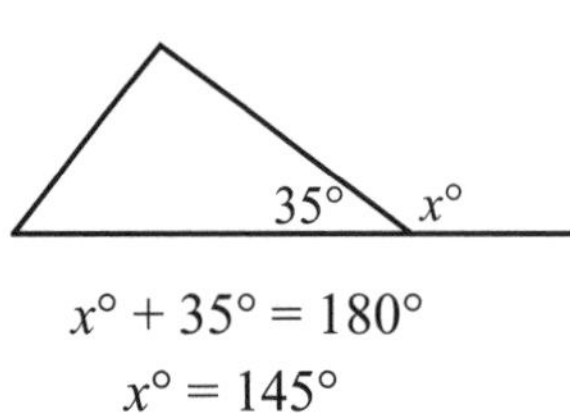

$x° + 35° = 180°$

$x° = 145°$

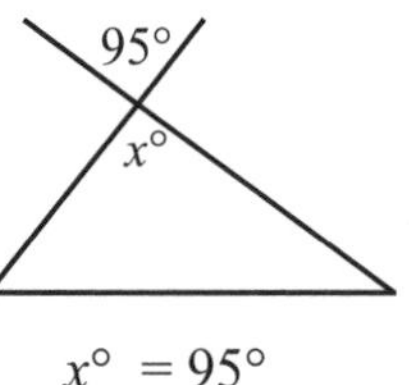

$x° = 95°$

3. The longest side of a triangle is opposite the largest angle; the shortest side of a triangle is opposite the smallest angle:

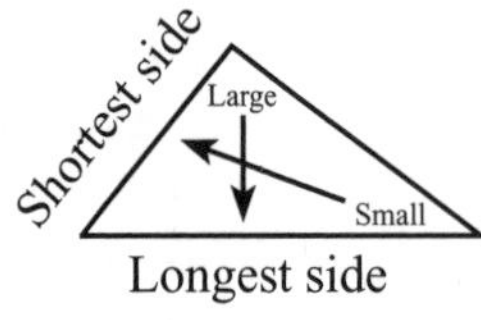

This rule can help eliminate answer choices.

ARITHMETRICK
This simple fact is the key to the solution for some Hard level questions that are actually quite easy to solve.

4. The sum of the lengths of any two sides of a triangle is always greater than the length of the remaining side:

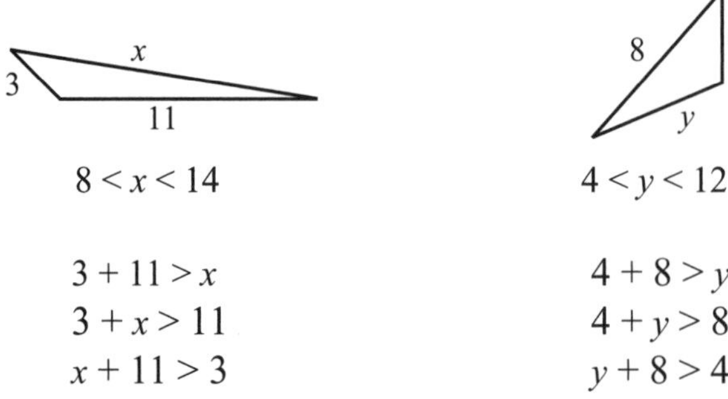

$8 < x < 14$ $\qquad 4 < y < 12$

$3 + 11 > x$
$3 + x > 11$
$x + 11 > 3$

$4 + 8 > y$
$4 + y > 8$
$y + 8 > 4$

5. The height of a triangle is any perpendicular line segment from a base to a vertex:

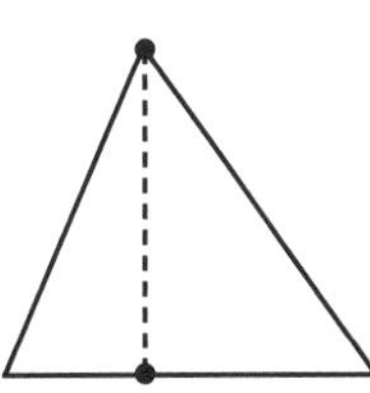
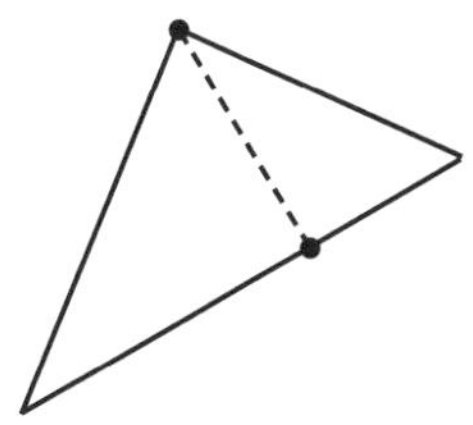
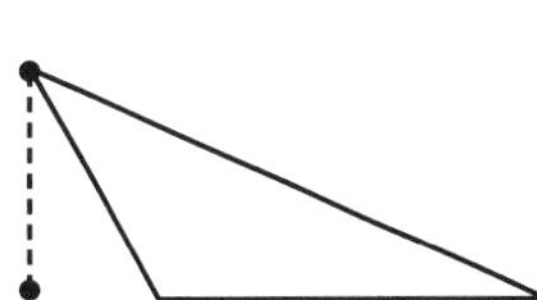

☠ CAUTION: SAT TRAP!
Many students forget that the height can be found outside of the triangle, as in the third figure. The test makers know this, and thus may ask for the area of a triangle like the triangle in the third image.

6. The formula for the area of a triangle, one-half of the base times the height ($\frac{1}{2}bh$), is provided in the Formula Box at the beginning of the math section.

   The maximum area of a triangle occurs when a triangle has equal side lengths. If the side lengths are all different, the maximum area occurs when the triangle is a right triangle.

7. The perimeter of a triangle is determined by adding all three side lengths.

   Perimeter $= s_1 + s_2 + s_3$

8. Triangles that have the exact same shape but different area are called similar triangles. The corresponding angle measurements of similar triangles are equal, and the corresponding side lengths are proportional:

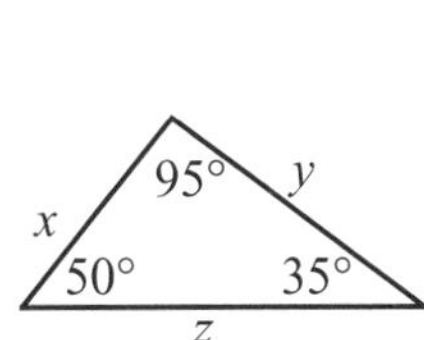

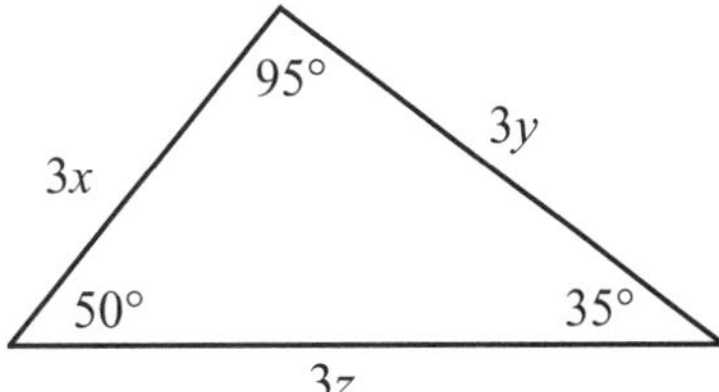

The makers of the SAT may attempt to disguise similar triangles:

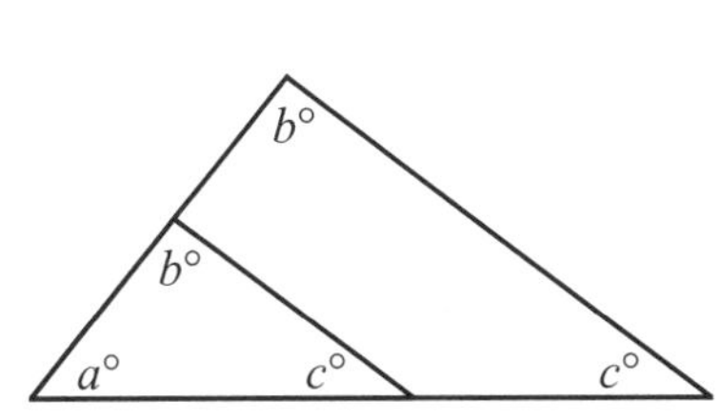

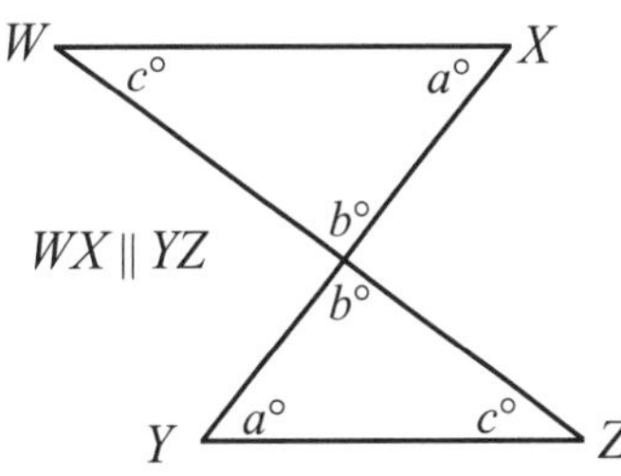

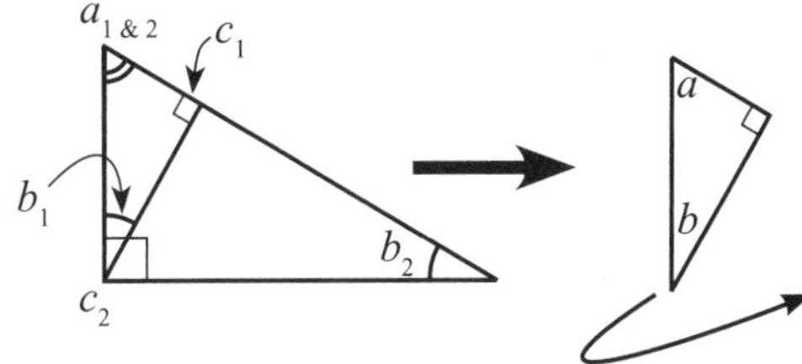

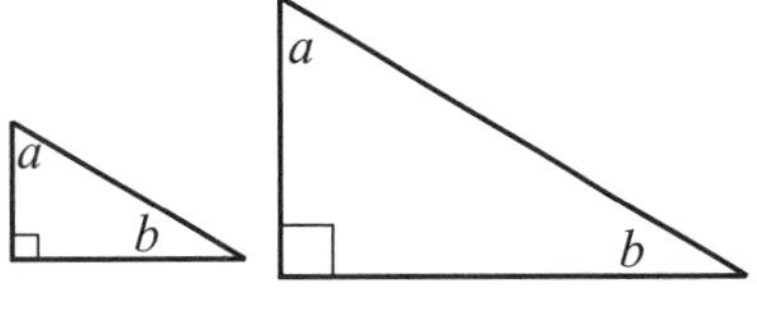

Students who can immediately recognize similar triangles in these figures will shave several seconds or even minutes off their time.

Application on the SAT

Questions about basic triangles often ask you to find an angle measurement or a side length, using the principles presented in the Required Knowledge and Skill Set. To find an angle measurement, use the rule that the sum of the interior angles is 180º. You may also have to use your knowledge of supplementary and vertical angles to find the measurement of an interior angle.

When in doubt, do not be afraid to SIZE UP the figure and make a guess. This question is drawn to scale, so SIZING UP the figure is easy. Angle x is smaller than 60º, so only (A) or (B) are viable answers. You should only SIZE UP a figure, though, when you do not know how to solve a problem.

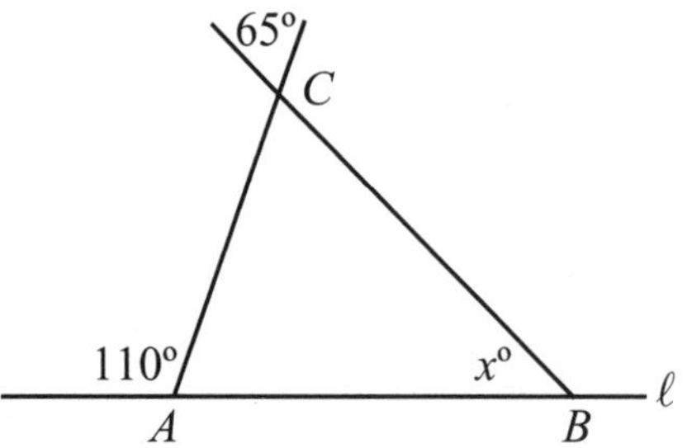

7. In the figure above, side $\overline{AB}$ of $\triangle ABC$ is on line ℓ. What is the value of x ?

(A) 45
(B) 50
(C) 65
(D) 70
(E) 135

To solve this question, you must use supplementary angles to compute $\angle CAB$ and vertical angles to find $\angle ACB$:

As you become more comfortable with the properties of triangles, you should begin to be able to solve these questions without having to create equations for every step of the process.

$110^\circ + \angle CAB = 180^\circ$
$\angle CAB = 70^\circ$

$\angle ACB$ is 65º because its vertical angle is 65º.

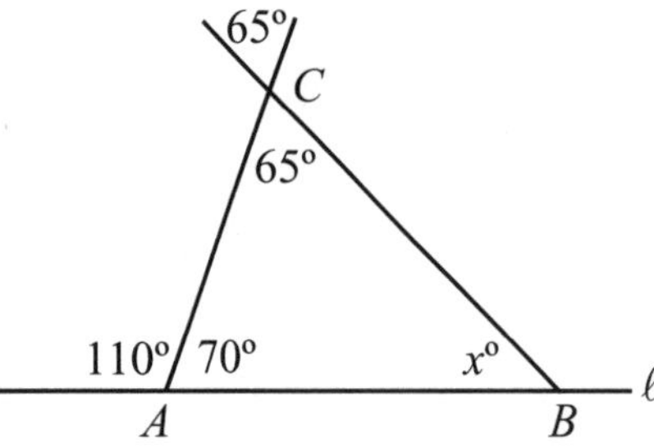

Because the sum of the interior angles of a triangle is 180º, you can find x:

$\angle CAB + \angle ABC + \angle BCA = 180^\circ$
$70^\circ + x^\circ + 65^\circ = 180^\circ$
$135^\circ + x^\circ = 180^\circ$
$x^\circ = 45^\circ$

The correct answer is (A).

The makers of the test may add difficulty by using variables as angle measurements:

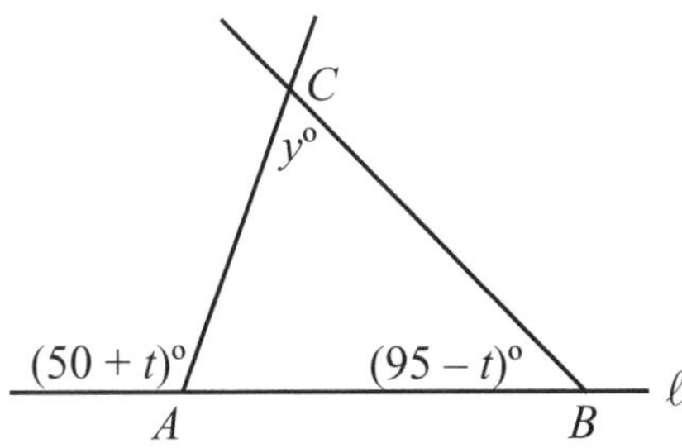

19. In the figure above, side $\overline{AB}$ of $\triangle ABC$ is on line ℓ. What is y in terms of t ?

(A) $-2t - 45$
(B) $-2t - 145$
(C) $-2t + 45$
(D) $2t - 45$
(E) $2t - 145$

The process is the same, but it now involves variables instead of numbers. Find $\angle CAB$ using supplementary angles:

$$(50 + t)^{o} + \angle CAB = 180^{o}$$
$$\angle CAB = 180^{o} - (50 + t)^{o}$$
$$\angle CAB = 180^{o} - 50^{o} - t^{o}$$
$$\angle CAB = 130^{o} - t^{o}$$

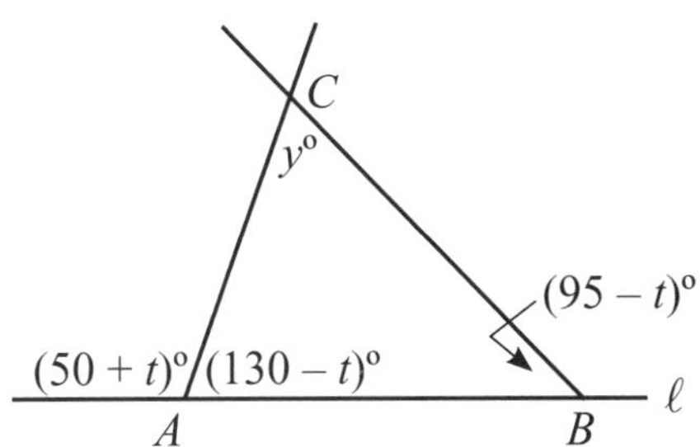

Find y, using the sum of the interior angles:

$$\angle CAB + \angle ABC + \angle BCA = 180^{o}$$
$$(130 - t)^{o} + (95 - t)^{o} + y = 180^{o}$$
$$225^{o} - 2t^{o} + y^{o} = 180^{o}$$
$$-2t^{o} + y^{o} = -45^{o}$$
$$y^{o} = 2t^{o} - 45^{o}$$

The correct answer is (D).

This question has a Hard level difficulty, but is not difficult as long as you remember that the interior angles of a triangle equal 180 degrees. Most students see the variables and forget this simple rule.

☠ CAUTION: SAT TRAP!
Avoid the panic caused by variables! Follow the same steps as you would if the angles were given as numbers.

<u>Confidence Quotation</u>
"Many of life's failures are people who did not realize how close they were to success when they gave up."
—Thomas A. Edison, inventor and scientist

Another Hard level question involves the rule stating that the sum of the lengths of any two sides of a triangle is always greater than the length of the remaining side:

17. In triangle XYZ, the length of $\overline{XY}$ is 4 and the length of $\overline{YZ}$ is 6. Which of the following could be the length of $\overline{XZ}$?

    (A) 9
    (B) 11
    (C) 12
    (D) 14
    (E) 16

Imagine the triangle. It may look like any of the following:

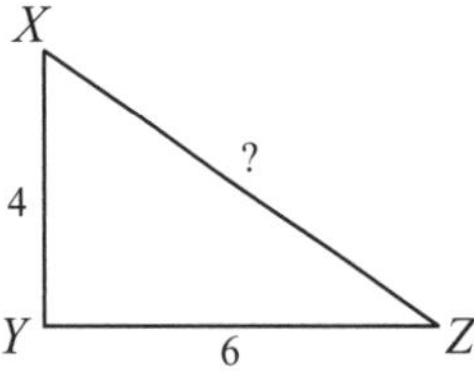

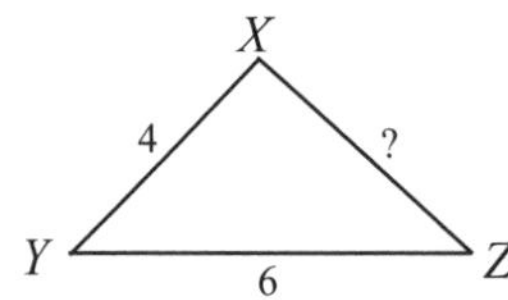

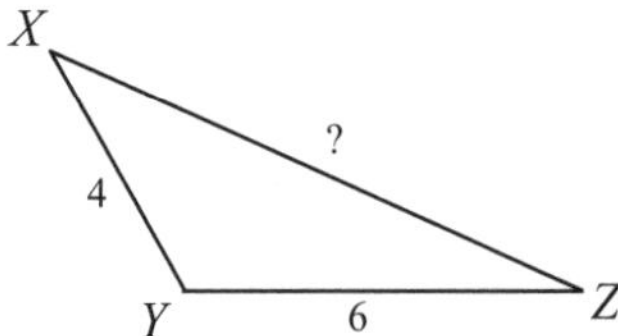

When an SAT triangle question gives the length of two sides and asks about a possible length for the third side, the question is assessing whether you know the property about the sum of two side lengths.

This relatively simple concept—that the sum of two sides of a triangle must be greater than the third side—is a bit of a "brain buster." It can make your head hurt thinking about the possible lengths of a triangle's side! Critical reasoning, the most important skill on the SAT, can often make your brain feel worn out or "overused." However, with repeated practice, critical reasoning becomes much easier and more automatic.

Since the two given side lengths add up to 10, the length of the third side must be less than 10. Answer choice (A) is the only viable option.

A similar question might state that the length of $\overline{XZ}$ is an integer and ask for the smallest possible value of $\overline{XZ}$:

17. In triangle XYZ, the length of $\overline{XY}$ is 4, the length of $\overline{YZ}$ is 6, and the length of $\overline{XZ}$ is an integer. What is the smallest possible value of $\overline{XZ}$?

    (A) 1
    (B) 2
    (C) 3
    (D) 4
    (E) 5

Again, you must use the sum of two side lengths property to find the answer:

$$\overline{XY} + \overline{XZ} > \overline{YZ}$$
$$4 + \overline{XZ} > 6$$
$$\overline{XZ} > 2$$

The correct answer is (C), as 3 is the first integer greater than 2.

In the previous problem, you determined that $\overline{XZ}$ is less than 10 and in this question you determined that $\overline{XZ}$ must be greater than 2:

$$2 < \overline{XZ} < 10$$

Questions about the area and perimeter of a triangle occur frequently with special triangles, like equilateral and right triangles. We will examine these questions more closely in the next section. But you may see an area or perimeter question involving basic triangles, too. For example, the following question gives two side lengths and asks for the maximum area of the triangle:

10. In $\triangle XYZ$, the length of $\overline{XY}$ is 9 and the length of $\overline{YZ}$ is 4. What is the greatest possible area of triangle $\triangle XYZ$?

    (A) 9
    (B) 13
    (C) 18
    (D) 36
    (E) 72

Maximum area of a triangle occurs when all of the sides are the same length. However, in this question, the two given sides have different lengths—4 and 9. Therefore, the greatest area is attained when those two lengths form a right angle, or a right triangle. Any other angle creates a smaller height, thus resulting in a smaller area:

This is an important concept that is occasionally tested. An equilateral triangle provides maximum area, but when side lengths are different, the maximum area is attained in a right triangle.

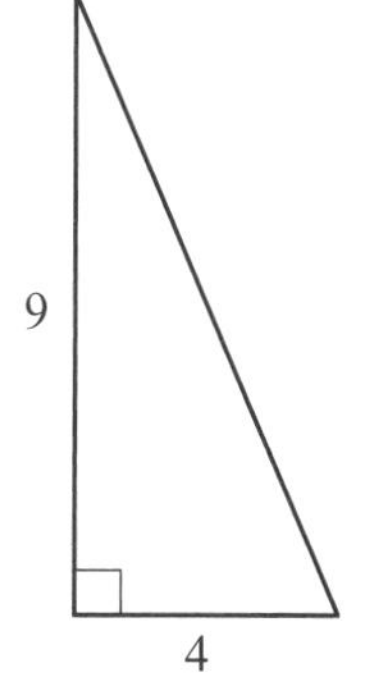

base = 4
height = 9
Maximum Area

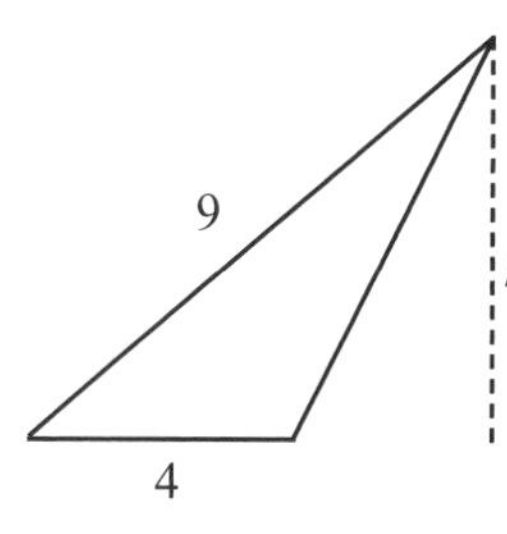

base = 4
height = less than 9

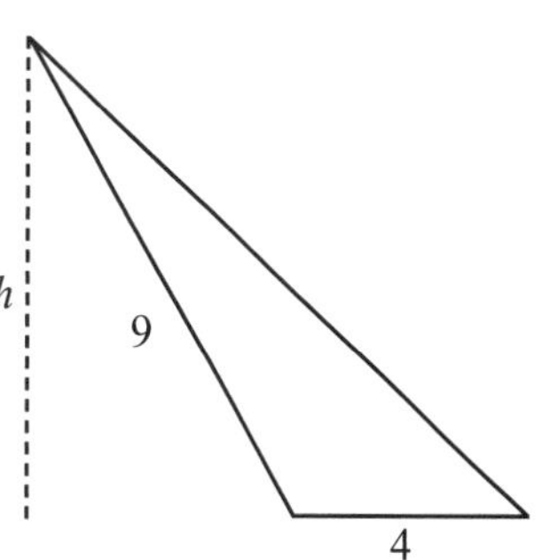

base = 4
height = less than 9

So find the area using 4 as the base and 9 as the height (or vice versa):

$$\text{Area} = \frac{1}{2}bh$$

$$\text{Area} = \frac{1}{2}(4)(9) \quad \rightarrow \quad (2)(9) \quad \rightarrow \quad 18$$

The correct answer is (C).

Similar triangles are used to assess your ability to find a missing side length. The question will give the side measurement of one of the smaller triangle sides and its corresponding side length of the larger triangle. Then it will present one more side length—on either the small or large triangle—and ask you to produce its corresponding measurement on the other triangle:

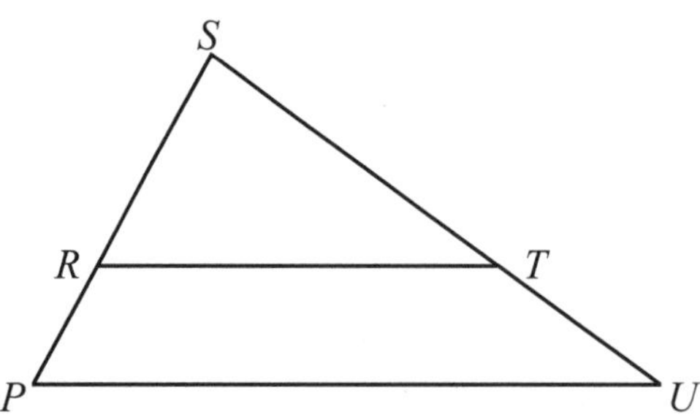

Note: The figure is not drawn to scale.

7. In the figure above, PSU is a triangle. The length of $\overline{RP}$ is k, and line $\overline{RT}$ bisects $\overline{PS}$ and $\overline{SU}$. If $\overline{PU} = 24$, what is the length of $\overline{RT}$?

   (A) 8
   (B) $8k$
   (C) 12
   (D) $12k$
   (E) $\frac{24}{k}$

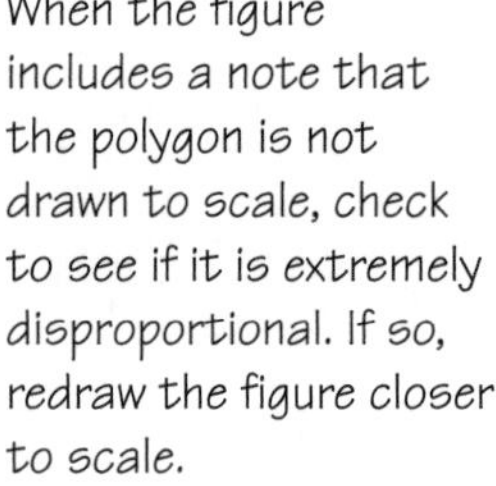
When the figure includes a note that the polygon is not drawn to scale, check to see if it is extremely disproportional. If so, redraw the figure closer to scale.

There are two reasons to redraw this diagram. For one, the figure is grossly disproportionate. According to the question, $PR = RS$ and $ST = TU$, but this diagram misrepresents these relationships. Plus, when similar triangles are hidden, redraw them as two separate triangles, adding information from the text in the question:

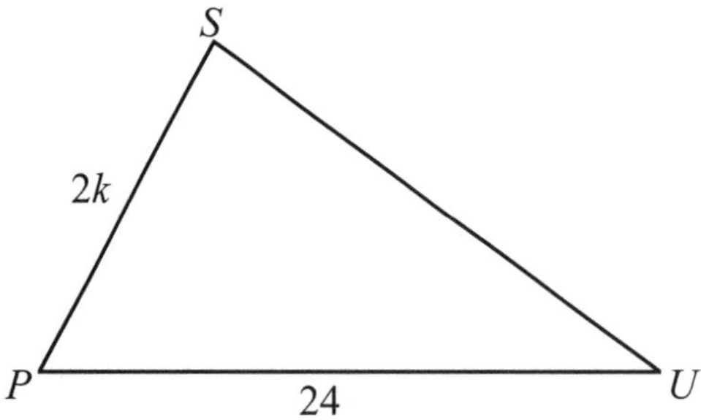

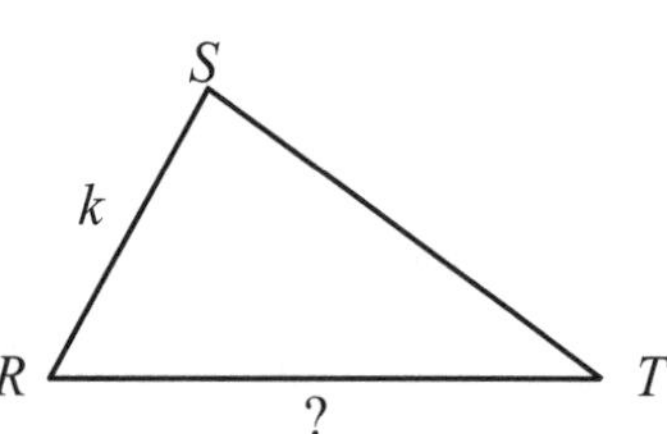

Because PS is twice the length of SR, PU is twice the length of RT. To prove this, set up a proportion:

$$\frac{2k}{24} = \frac{k}{?} \quad \rightarrow \quad (2k)(?) = (24)(k) \quad \rightarrow \quad (2k)(?) = 24k \quad \rightarrow \quad ? = 12$$

The correct answer is (C).

Basic Triangles Problem Set

Solve the following multiple-choice questions by selecting the best answer from the five answer choices. For grid-in questions, write your answer in the grids and completely mark the corresponding ovals. Answers begin on page 338.

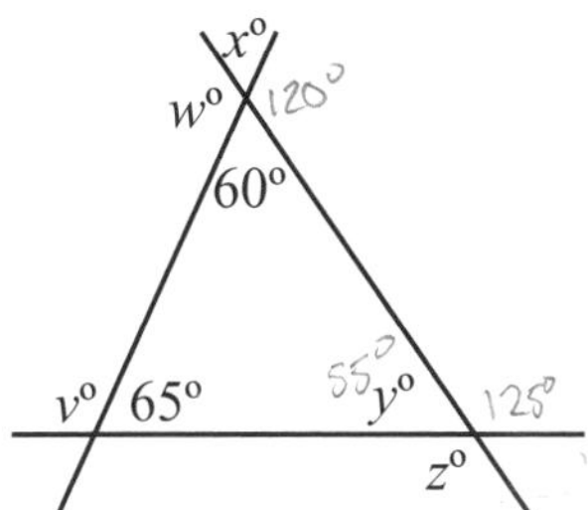

1. Which of the following angle measurements is greatest in the figure above?

(A) v
(B) w
(C) x
(D) y
(E) z

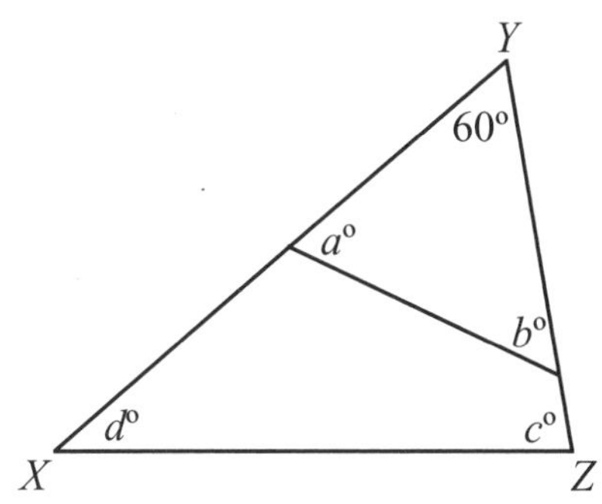

2. In the triangle XYZ above, what is the value of $\frac{a+b}{c+d}$?

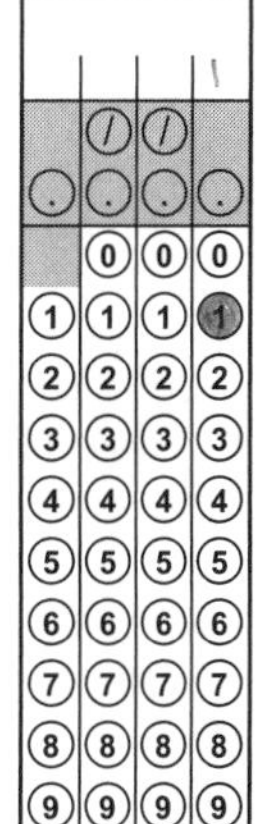

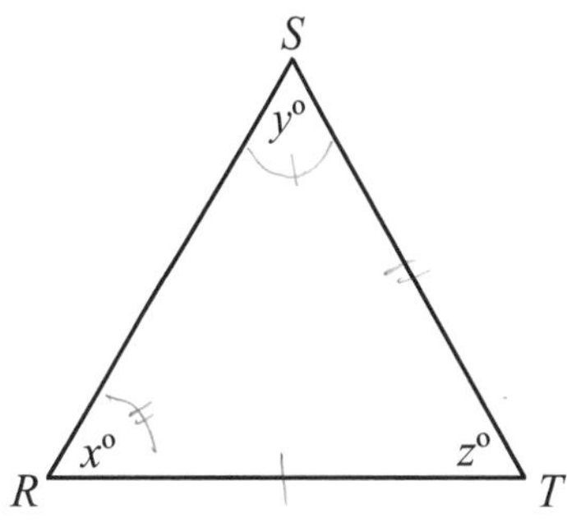

Note: Figure is not drawn to scale.

3. In the figure above, $x > y$. Which of the following statements must be true ?

(A) $z > y$
(B) $y > z$
(C) $RT > ST$
(D) $RT < ST$
(E) $RS > RT$

Basic Triangles Problem Set

Solve the following multiple-choice questions by selecting the best answer from the five answer choices. For grid-in questions, write your answer in the grids and completely mark the corresponding ovals. Answers begin on page 338.

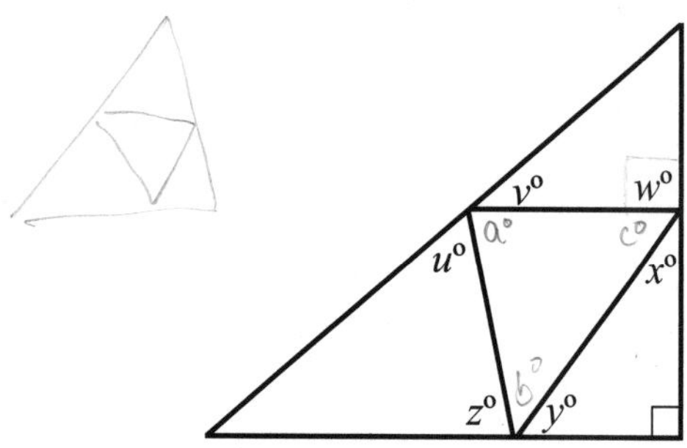

4. In the figure above, what is the sum of u, v, w, x, y, and z ?

(A) 160
(B) 180
(C) 360
(D) 420
(E) 640

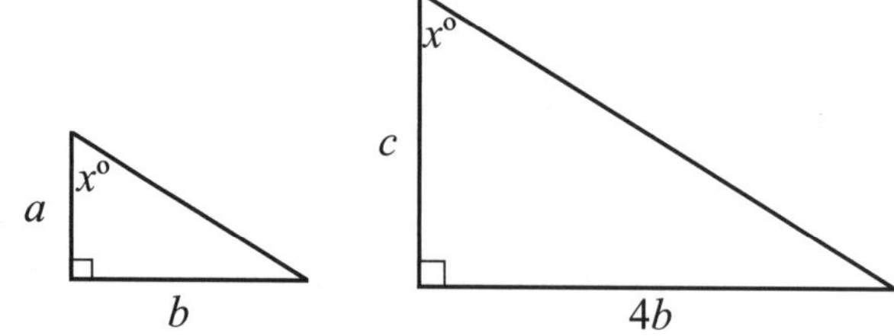

Note: Figures are not drawn to scale.

5. In the figure above, what is the value of a in terms of c ?

(A) $\frac{1}{4}c$

(B) $2\sqrt{c}$

(C) $2c$

(D) $4c$

(E) $16c$

Basic Triangles Problem Set

Solve the following multiple-choice questions by selecting the best answer from the five answer choices. For grid-in questions, write your answer in the grids and completely mark the corresponding ovals. Answers begin on page 338.

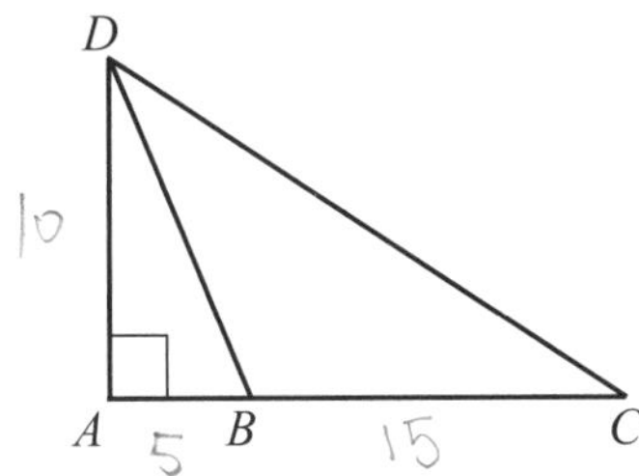

Note: Figure is not drawn to scale.

6. In the figure above, if $AB = 5$, $AC = 20$, and $AD = 10$, what is the area of $\triangle BDC$?

(A) 25
(B) 45
(C) 75
(D) 90
(E) 100

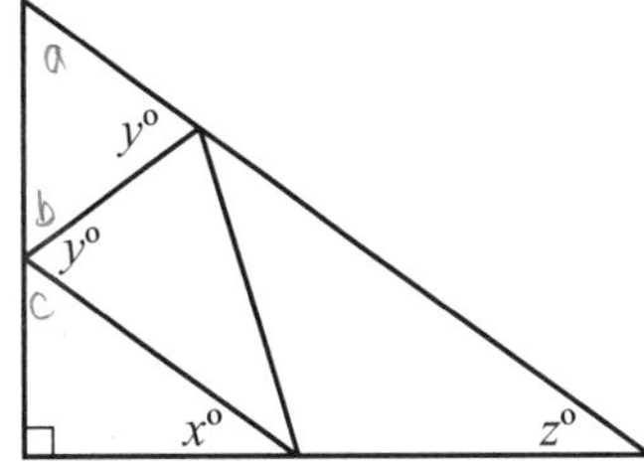

7. In the figure above, what is the value of z in terms of x and y?

(A) x
(B) $x + 180$
(C) $2x + 180$
(D) $90 - 2y$
(E) $90 - x + 2y$

Special Triangles

Frequency Guide: 3

The title "special triangles" refers to those triangles—specifically, equilateral triangles, isosceles triangles, and right triangles—that have special properties.

Required Knowledge and Skill Set

Because there are rules specific to each type of special triangle, we will examine each one separately.

Right Triangles

1. Right triangles have a 90° angle:

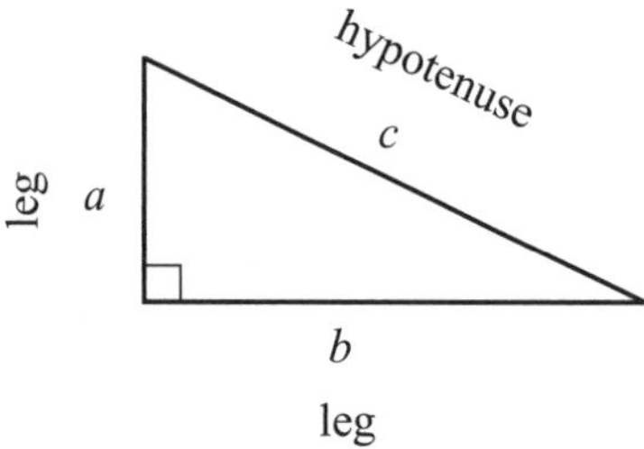

   In a right triangle, the side opposite the right angle is called the *hypotenuse*, and is usually labeled c for the convenience of formulas. The other two sides, called *legs*, are labeled a and b.

Right triangles are the most common geometric figure on the SAT. It is imperative that you recognize, understand, and demonstrate the Pythagorean Theorem.

2. If you know the lengths of two sides of a right triangle, you can find the length of the third side using the Pythagorean Theorem, a special formula that is given in the Formula Box:

   Pythagorean Theorem: $a^2 + b^2 = c^2$

   To find a missing side length, plug the values you know into the formula:

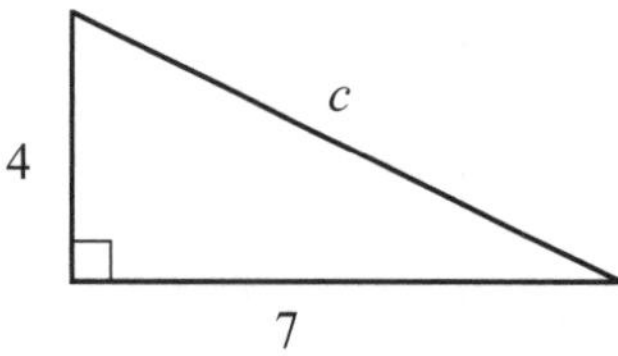

$$a^2 + b^2 = c^2$$
$$4^2 + 7^2 = c^2$$
$$16 + 49 = c^2$$
$$65 = c^2$$
$$\sqrt{65} = c$$

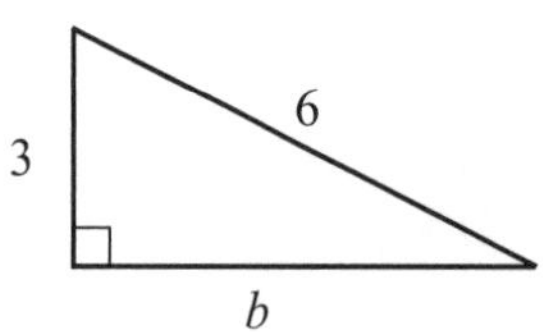

$$a^2 + b^2 = c^2$$
$$3^2 + b^2 = 6^2$$
$$9 + b^2 = 36$$
$$b^2 = 27$$
$$b = \sqrt{27} = 3\sqrt{3}$$

3. In order to make the SAT accessible to students who do not own a calculator, the test makers often rely on Pythagorean Triples. Pythagorean Triples are sets of integers that satisfy the Pythagorean Theorem. The most common triangle using Pythagorean Triples is the 3 : 4 : 5:

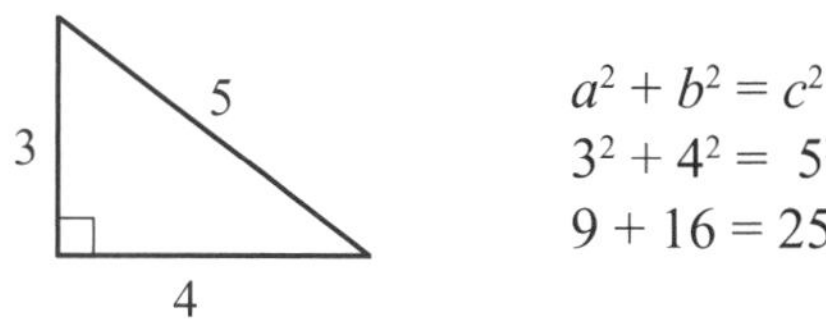

$a^2 + b^2 = c^2$
$3^2 + 4^2 = 5^2$
$9 + 16 = 25$

MEMORY MARKER:
If you can recognize these triangle shapes on figures drawn to scale, you can save time by identifying the two most common Pythagorean Triples.

Another common Pythagorean Triple used on the SAT is the 5 : 12 : 13:

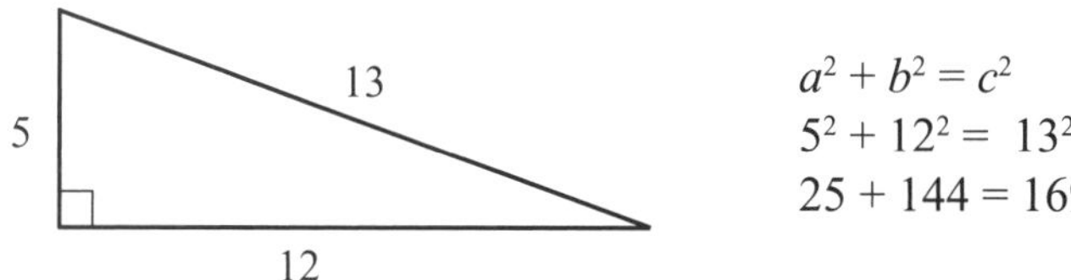

$a^2 + b^2 = c^2$
$5^2 + 12^2 = 13^2$
$25 + 144 = 169$

You should study the shape of these two triangles for easy recognition on the test.

Note that the multiples of Pythagorean Triples are also Pythagorean Triples themselves:

Original	Multiple	New Triple	Proof
3 : 4 : 5	× 2	6 : 8 : 10	$6^2 + 8^2 = 10^2$
3 : 4 : 5	× 3	9 : 12 : 15	$9^2 + 12^2 = 15^2$
3 : 4 : 5	× 4	12 : 16 : 20	$12^2 + 16^2 = 20^2$
5 : 12 : 13	× 2	10 : 24 : 26	$10^2 + 24^2 = 26^2$

PowerScore recommends memorizing the following sets of Pythagorean Triples and their multiples:

Triple:	*Multiples:*			
3 : 4 : 5	6 : 8 : 10	9 : 12 : 15	12 : 16 : 20	15 : 20 : 25
5 : 12 : 13	10 : 24 : 26			
7 : 24 : 25	14 : 48 : 50			
8 : 15 : 17	16 : 30 : 34			
9 : 40 : 41				
12 : 35 : 37				
20 : 21 : 29				

MEMORY MARKER:
By memorizing Pythagorean Triples, you may avoid performing the Pythagorean Theorem, thus saving time and preventing calculation errors.

The smaller the value of the integers, the more likely that Triple is to appear on the SAT. Whenever you find that the Pythagorean Theorem is required on an SAT math question, check the triangle's two given side lengths to see if they fit one of the Pythagorean Triple combinations.

4. There are two special right triangles featured in the formula box: the 30º:60º:90º triangle and the 45º:45º:90º triangle.

A 30º:60º:90º triangle has angle measurements of 30 degrees, 60 degrees, and 90 degrees. It is a right triangle, and some students might be tempted to use the Pythagorean Theorem to find a missing side length. However, this triangle has a special ratio that allows you to find two side lengths when given the length of just a single side.

The length of the sides have a ratio of $x : 2x : x\sqrt{3}$. The shortest side of the triangle, opposite the 30º angle, is x. The hypotenuse is $2x$, and the remaining side, opposite the 60º angle, is $x\sqrt{3}$. The size and position of the triangle do not matter as long as the interior angles are 30º, 60º, and 90º:

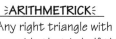

ARITHMETRICK
Any right triangle with one side that is half the value of the hypotenuse is a 30º:60º:90º.

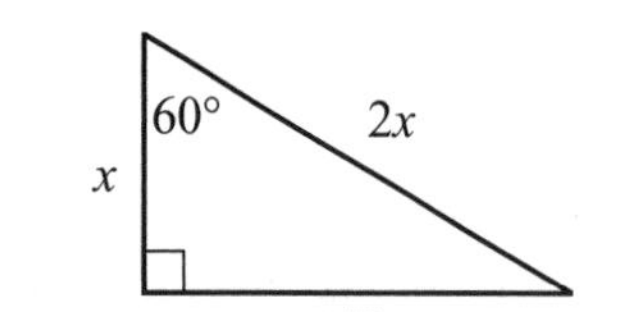

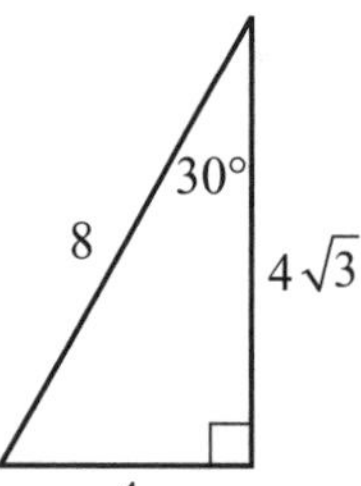

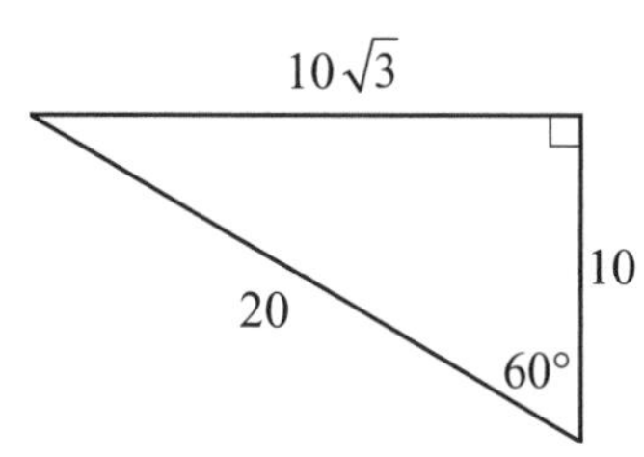

Given one side length, you can determine the other two side lengths:

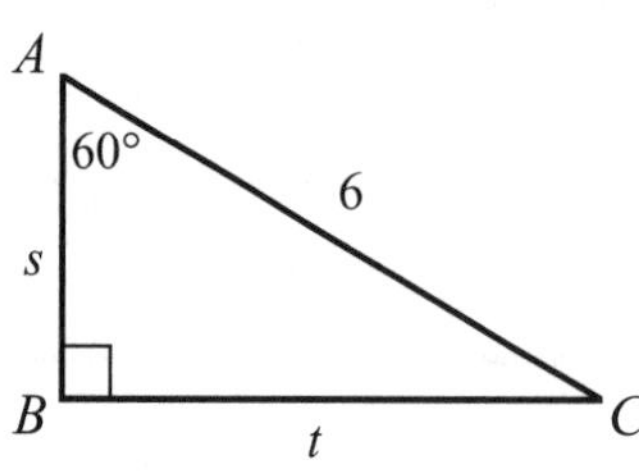

$\angle A + \angle B + \angle C = 180°$
$60° + 90° + \angle C = 180°$
$150° + \angle C = 180°$
$\angle C = 30°$

$s = 3$
$t = 3\sqrt{3}$

5. Every equilateral triangle contains two hidden 30º:60º:90º triangles. You can use this information to find the height or side length of the triangle, which can help you find the area of a triangle:

ARITHMETRICK
The height of an equilateral triangle is one-half the side length times the square root of 3.

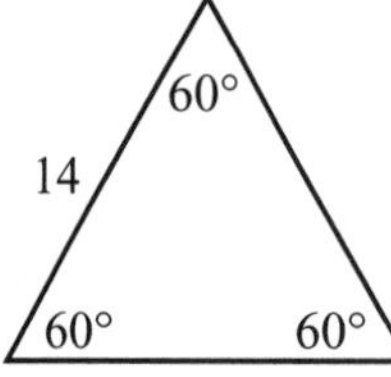

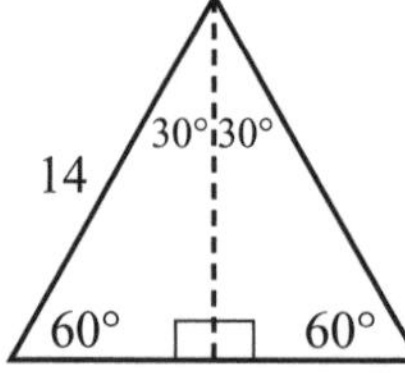

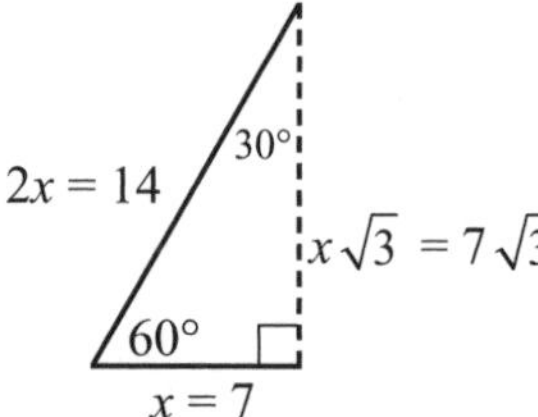

6. If answer choices contain $\sqrt{3}$, the question most likely involves a 30º:60º:90º triangle, even if the question appears to be about circles, rectangles, or other polygons. Look for a hidden triangle.

7. A 45°:45°:90° triangle has angle measurements of 45 degrees, 45 degrees, and 90 degrees. It also has a special side ratio, $s : s : s\sqrt{2}$, that helps you determine side length when only one side is given. The two shortest sides of the triangle, opposite the two 45° angles, are equal to each other and are both represented by s. The hypotenuse is $s\sqrt{2}$. The size and position of the triangle do not matter as long as the interior angles are 45°, 45°, and 90°. Notice that the 45° angles do not have to be labeled if the two equal sides are revealed, as in the second figure.

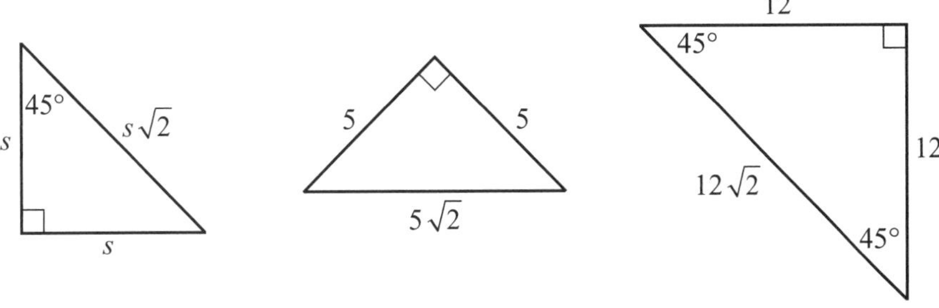

ARITHMETRICK

In a 45°:45°:90° triangle, the length of the hypotenuse is the length of a side times the square root of 2.

Given one side length, you can determine the other two side lengths:

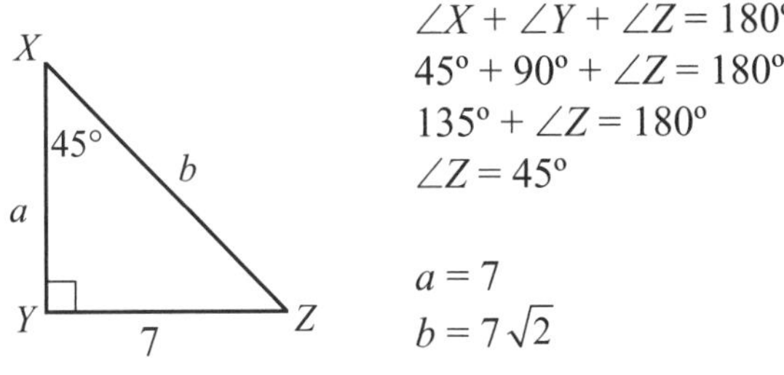

$\angle X + \angle Y + \angle Z = 180°$
$45° + 90° + \angle Z = 180°$
$135° + \angle Z = 180°$
$\angle Z = 45°$

$a = 7$
$b = 7\sqrt{2}$

8. Every square conceals two 45°:45°:90° triangles. You can use this information to find the length of the diagonal of the square:

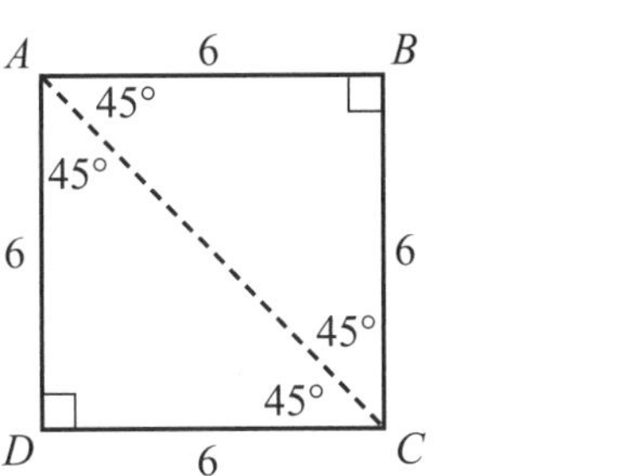

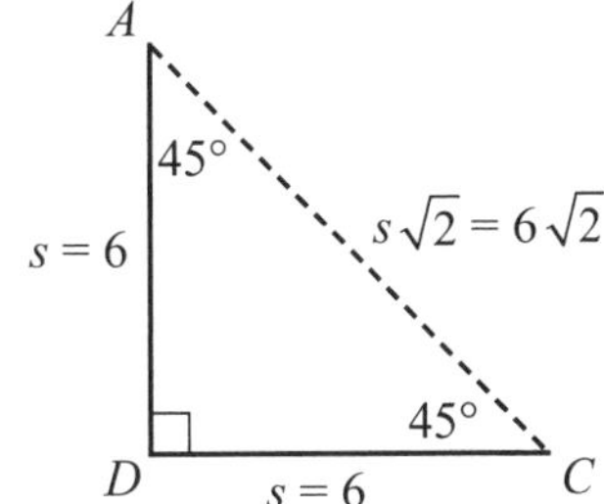

9. If answer choices contain $\sqrt{2}$, the question most likely involves a 45°:45°:90° triangle, even if the question appears to be about circles, squares, or other polygons. Look for a hidden triangle. Some questions may have answers with $\sqrt{3}$ or $\sqrt{2}$. If this is the case, determine whether the triangle in the question is a 30°:60°:90° or 45°:45°:90°.

Always ANALYZE the answer choices before starting the solution, as the answers to geometry questions often point you in the direction of the solution.

Equilateral Triangles

1. Equilateral triangles have equal side lengths and equal angle measurements. Since the interior angles of a triangle add up to 180˚, the three angles of an equilateral triangle must each equal 60°. If you encounter an equilateral triangle on the SAT, immediately label each of the angles as 60°:

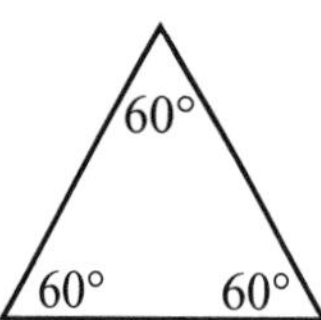

Perimeter of an equilateral triangle = side × 3

If you are given the length of one side of an equilateral triangle, be sure to label the all three sides of the triangle with this value.

2. Remember, equilateral triangles contain two 30º:60º:90º triangles, and the height of the equilateral triangle can be determined by using the side ratios of those 30º:60º:90º triangles.

Students who miss questions involving equilateral triangles or isosceles triangles usually fail to fully DIAGRAM the question. Equilateral triangles have three 60º angles and three equal sides, and isosceles triangles share two side lengths and two angle measurements.

Isosceles Triangles

1. An isosceles triangle has two sides of equal length and two angles of equal size. The two equal angles are opposite the two equal-length sides:

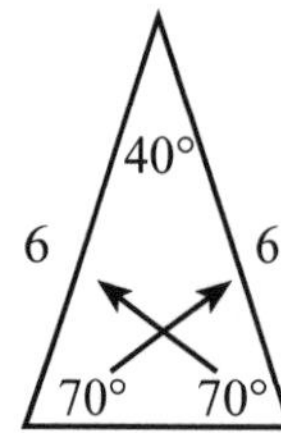

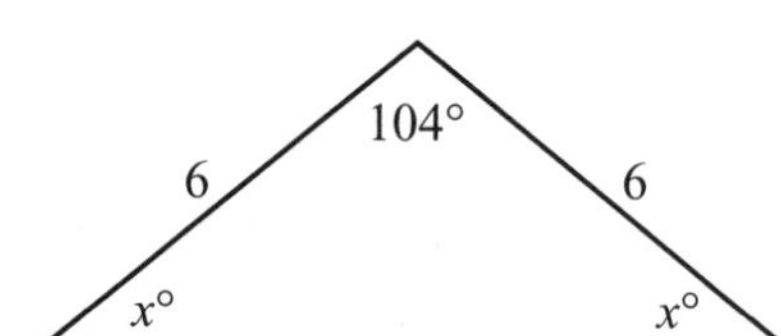

If the question tells you that a triangle is isosceles or that two angles or side lengths are equal, complete the figure with as much information as possible.

Application on the SAT

Expect to use the Pythagorean Theorem on the SAT. Often, finding the length of a leg of a right triangle may be used for computing another calculation:

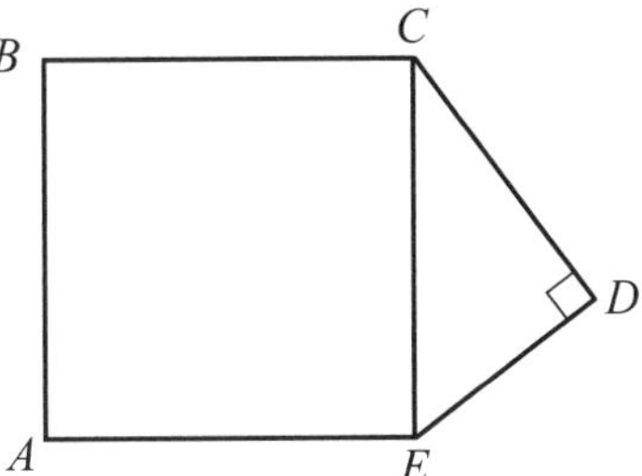

Note: Figure is not drawn to scale.

15. In the figure above, $ABCE$ is a square. If $CD = 4$ and $DE = 2$, what is the area of $ABCE$?

    (A) $\sqrt{18}$
    (B) $\sqrt{20}$
    (C) 8
    (D) 20
    (E) 24

Did you ANALYZE the answer choices before solving the problem? If so, the two answers in square root form should have indicated that the Pythagorean Theorem was a possibility.

The area of a square is simply the length of one side squared. There is no point in trying to find the length of AB, BC, or AE, as there is nothing to indicate that you can find those lengths. But CE is a part of a right triangle, so turn your attention to this side of the square. Use the Pythagorean Theorem to find the length of CE:

$a^2 + b^2 = c^2$
$DE^2 + CD^2 = CE^2$
$2^2 + 4^2 = CE^2$
$4 + 20 = CE^2$
$20 = CE^2$
$\sqrt{20} = CE$

Then find the area of the square:

Area $= s^2$
Area $= (\sqrt{20})^2 = 20$

The formulas for squares, circles, and other figures discussed in this section will be examined more closely later in this chapter.

The correct answer is (D).

In this question, the test makers assessed your ability to perform the calculations required for the Pythagorean Theorem. But they may also use a Pythagorean Triple to assess your ability to save time and avoid calculations.

Consider the same question with different measurements:

Before looking at the solution, try solving this question without using pencil and paper.

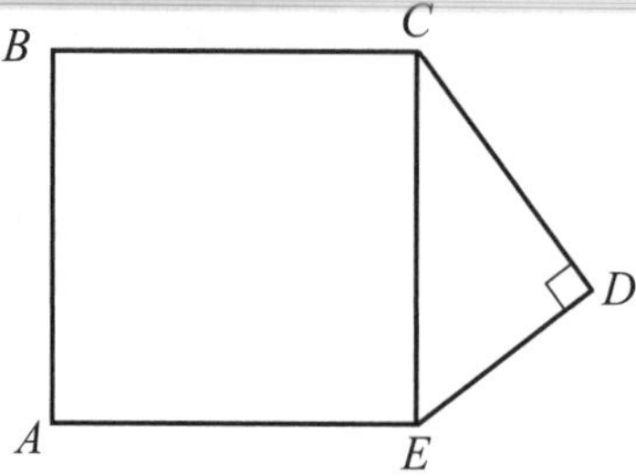

Note: Figure is not drawn to scale.

15. In the figure above, $ABCE$ is a square. If $CD = 8$ and $DE = 6$, what is the area of $ABCE$?

(A) 10
(B) 14
(C) 81
(D) 100
(E) 196

Savvy students will recognize the two legs of the Pythagorean Triple 6 : 8 : 10. If the length of CE is 10, then the area of $ABCE$ is 10^2 or 100. The correct answer, (D), was discovered without a single calculation on paper. If you can solve this problem as quickly, you will earn valuable time to use on the remaining questions.

Expect to see at least one question with a 30º:60º:90º triangle, a 45º:45º:90º triangle, or both triangles. Because of the special side ratios, these questions are also used to assess your ability to save time and avoid calculations. A common special triangle question asks for the area of an equilateral triangle:

When ANALYZING the answer choices, what does the presence of the square root of 3 indicate?

13. Triangle XYZ is an equilateral triangle. If side $XY = 8$, what is the area of $\triangle XYZ$?

(A) $8\sqrt{3}$
(B) 24
(C) $16\sqrt{3}$
(D) 32
(E) $32\sqrt{3}$

The height of an equilateral triangle is also the leg of a hidden 30º:60º:90º triangle. DIAGRAM the question:

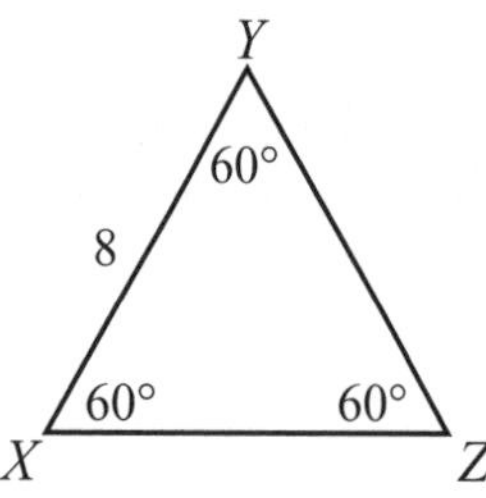

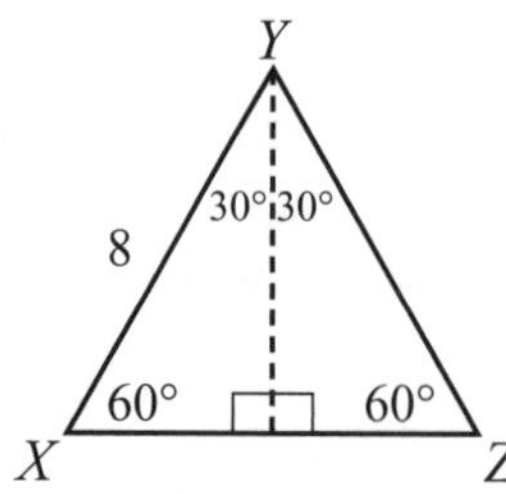

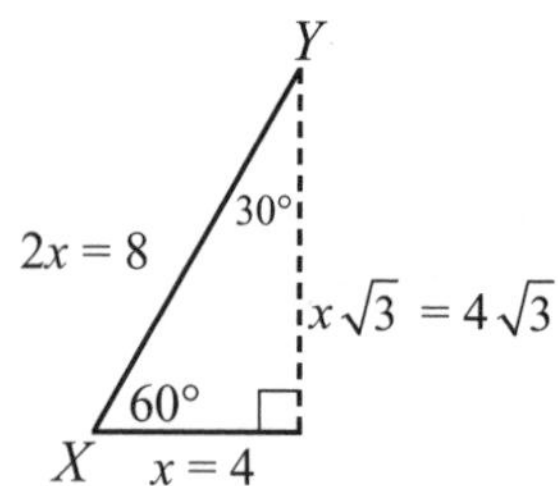

As you can see, the height of the triangle is $4\sqrt{3}$. Now, use the formula for the area of a triangle:

$$\text{Area} = \frac{1}{2}bh$$

$$\text{Area} = \frac{1}{2}(8)(4\sqrt{3}) \rightarrow (4)(4\sqrt{3}) \rightarrow 16\sqrt{3}$$

Hidden triangles are the key to many geometry questions. A similar question, involving a hidden 45°:45°:90° triangle, requires the area of a circle given an inscribed square:

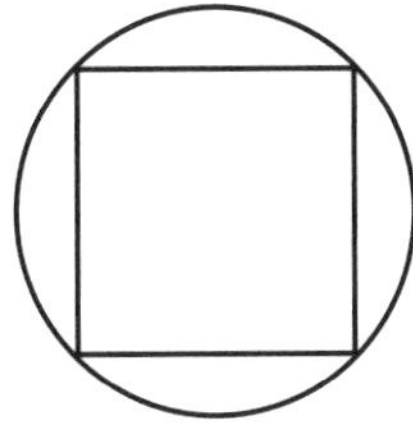

20. In the figure above, a square with sides of length 10 is inscribed in a circle. What is the area of the circle?

(A) 10π
(B) 25π
(C) 50π
(D) 75π
(E) 100π

The formula for the area of a circle is πr^2. In order to find the radius of the circle, we must rely on the diagonal of the square, which is the same as the diameter of the circle:

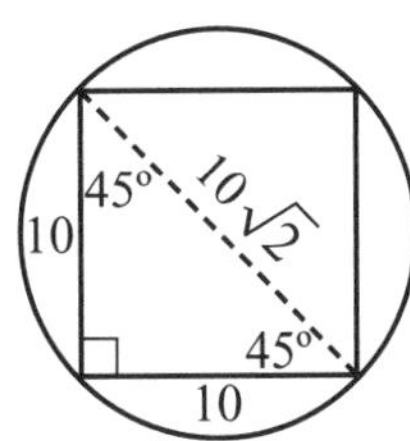

By drawing the diagonal of the square, you have created two 45°:45°:90° triangles. Using the special ratios of a 45°:45°:90° triangle, you can determine that the diagonal is $10\sqrt{2}$.

If the diagonal is $10\sqrt{2}$, than the diameter is also $10\sqrt{2}$. The radius is half of the diameter, or $5\sqrt{2}$. Now find the area of the circle:

$$\text{Area} = \pi r^2$$
$$\text{Area} = \pi(5\sqrt{2})^2 \rightarrow \pi(25 \times 2) \rightarrow 50\pi$$

Most 30°:60°:90° and 45°:45°:90° triangles will not be so secretive and hidden. But it is important to be on the lookout for them in any geometry problem.

Questions involving 30°:60°:90° and 45°:45°:90° triangles often require you to manipulate square roots. If you are not confident in your ability to add, subtract, multiply, and divide square roots, review the Operation Mastery chapter.

Because so many of the special triangle questions deal with finding the lengths of a triangle side, expect to see some perimeter questions. This is especially true for isosceles triangles.

5. A triangle has a side of length 5 and a perimeter of 19. The lengths of the two remaining sides are equal. What is the length of each of these remaining sides?

(A) 5
(B) 6
(C) 7
(D) 8
(E) 9

Confidence Quotation
"The thing always happens that you really believe in; and the belief in a thing makes it happen."
—Frank Lloyd Wright, American architect

DIAGRAM the question:

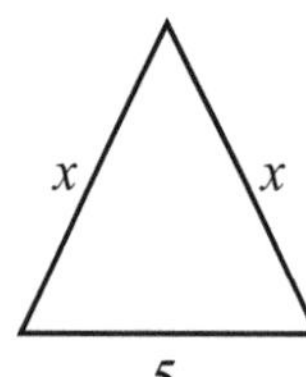

Perimeter = $s_1 + s_2 + s_3$
Perimeter = 19

And then solve:

Perimeter = $s_1 + s_2 + s_3$
$19 = x + x + 5 \quad \rightarrow \quad 14 = 2x \quad \rightarrow \quad 7 = x$

Another isosceles asks you to solve for the missing angle:

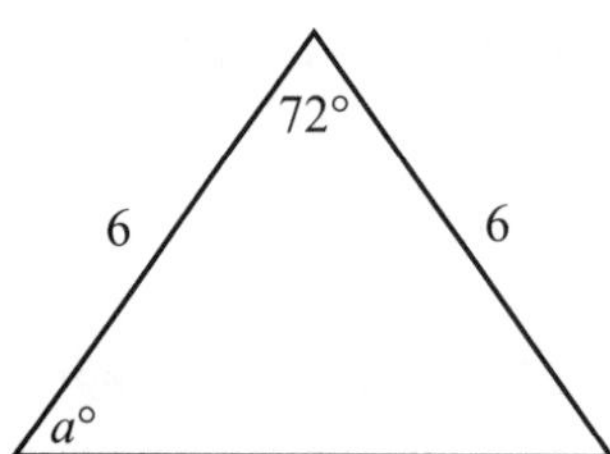

5. In the figure above, what is the value of a ?

(A) 54
(B) 60
(C) 72
(D) 108
(E) It cannot be determined from the information given.

Notice that neither question on this page refers to the triangles as "isosceles." You must realize this vital fact on your own, given the information provided.

The angles opposite the sides of length 6 must be the same measurement. So the unlabeled third angle is also a°:

$a° + a° + 72° = 180°$
$2a° = 108°$
$a° = 54$ The correct answer is (A).

Although we have separated "Special Triangles" from "Basic Triangles," the properties of triangles are not exclusive and you should be prepared to use all of this knowledge in a single problem. Let's look at one final example:

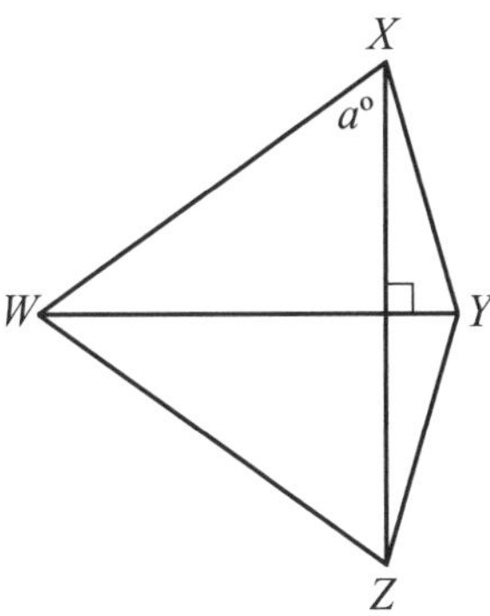

Note: Figure is not drawn to scale.

20. In the figure above, $WZ = WY = WX$ and $XY = YZ$. The measure of angle WYZ is 75°. What is the measure of angle a ?

    (A) 40
    (B) 50
    (C) 60
    (D) 70
    (E) 80

DIAGRAM the question. Start with $WZ = WY = WX$ and $XY = YZ$. This creates two isosceles triangles, and isosceles triangles have equal angles opposite the two equal sides (as in the first figure):

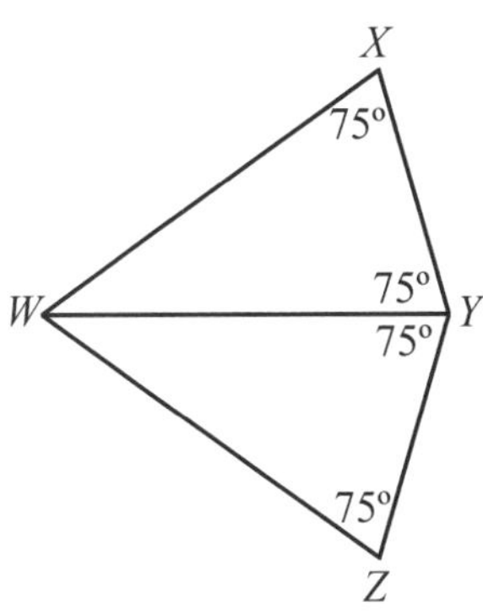

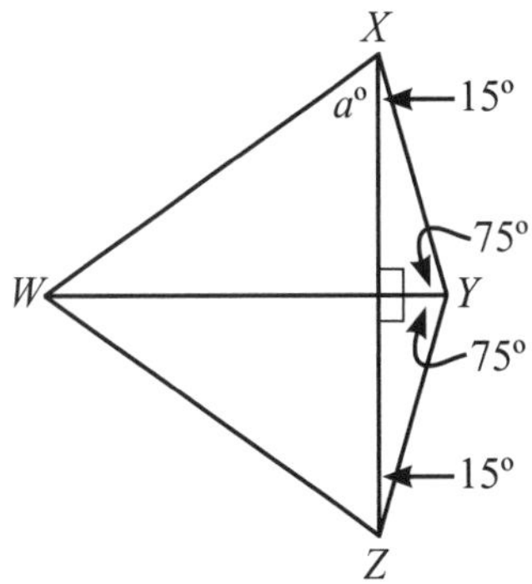

ARITHMETRICK

When you are stuck on a difficult geometry question with a complicated figure, try disassembling the figure to find a more basic shape.

Then use the right angles in the original figure to find the measurement of ZXY and XZY ($15° + 75° + 90° = 180°$).

Now find a:

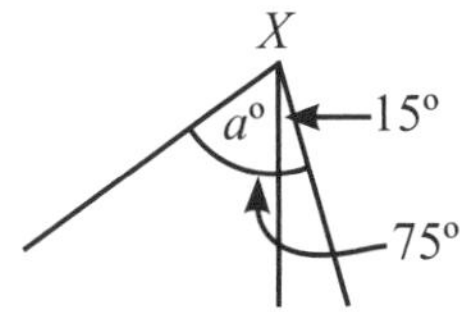

$75° - 15° = a°$

$60° = a°$

The correct answer is (C).

Special Triangles Problem Set

Solve the following multiple-choice questions by selecting the best answer from the five answer choices. For grid-in questions, write your answer in the grids and completely mark the corresponding ovals. Answers begin on page 341.

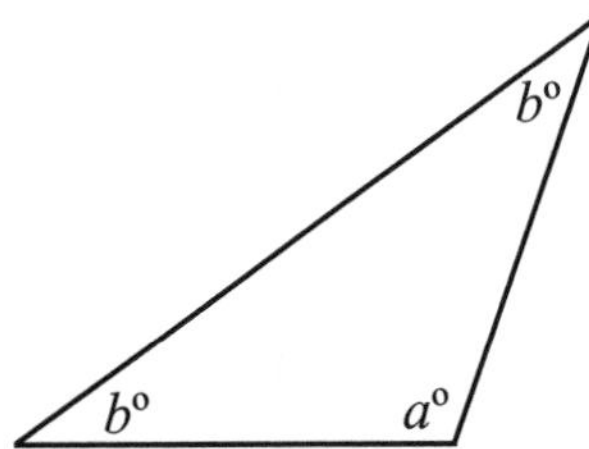

1. In the figure above, if a and b are integers and $32 < b < 36$, what is one possible value of a ?

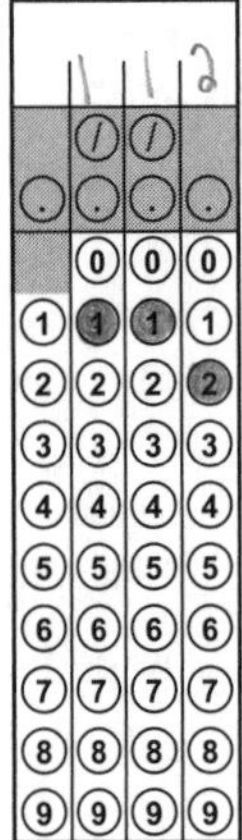

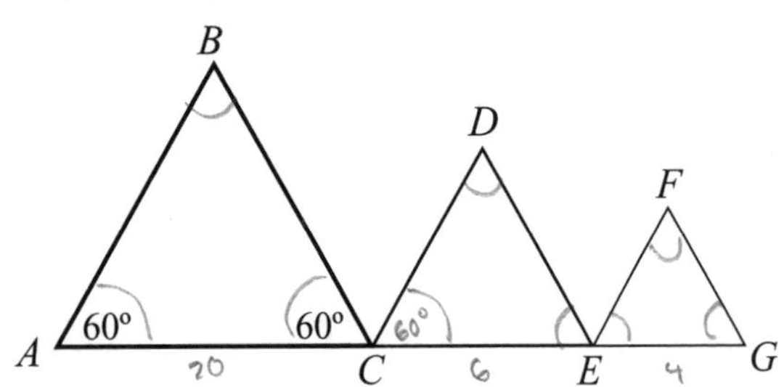

2. In the figure above, triangles CDE and EFG are similar to triangle ABC. If the length of line segment AG is 30, what is the sum of the perimeters of all three triangles?

(A) 90
(B) 100
(C) 110
(D) 120
(E) 130

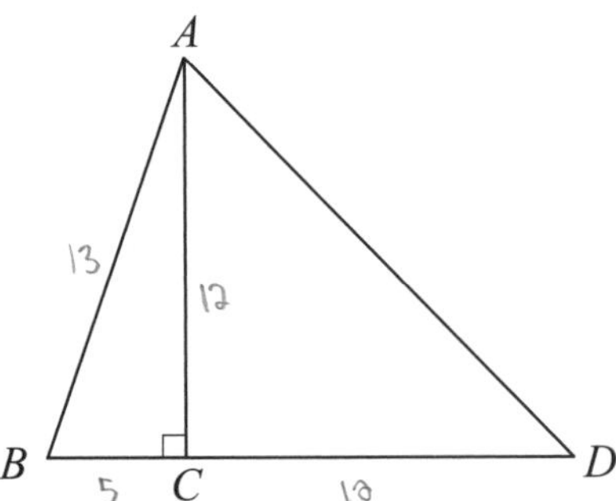

3. In the figure above, $AB = 13$, $BC = 5$, and $AC = CD$. What is the length of AD ?

(A) 12
(B) $12\sqrt{2}$
(C) $12\sqrt{3}$
(D) $18\sqrt{2}$
(E) $18\sqrt{3}$

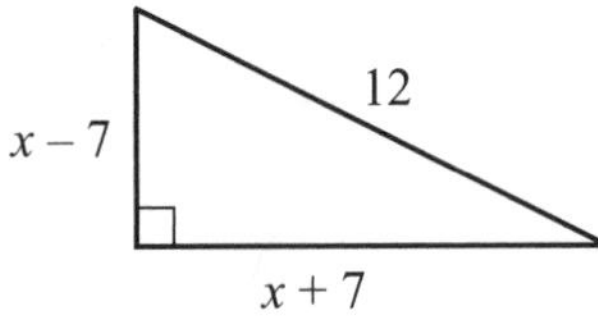

Note: Figure is not drawn to scale.

4. In the right triangle above, what is the value of $x^2 + 49$?

(A) 25
(B) 72
(C) 84
(D) 96
(E) 144

Special Triangles Problem Set

Solve the following multiple-choice questions by selecting the best answer from the five answer choices. For grid-in questions, write your answer in the grids and completely mark the corresponding ovals. Answers begin on page 341.

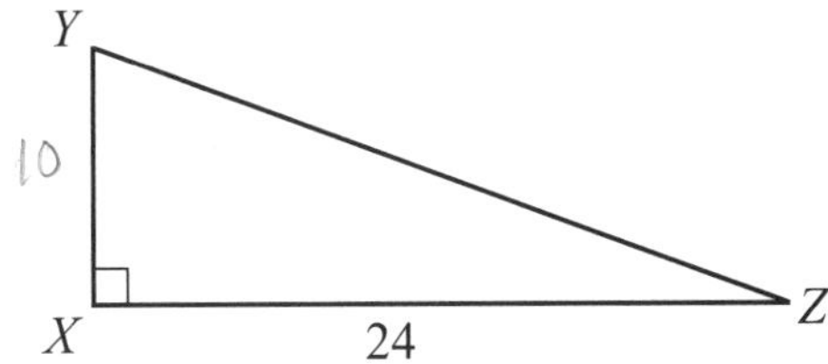

Note: Figure is not drawn to scale.

5. In the figure above, the area of triangle XYZ is 120. What is the length of YZ?

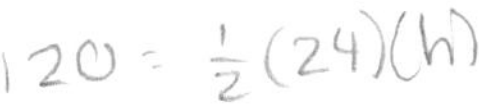

(A) 26
(B) 28
(C) 30
(D) 32
(E) 34

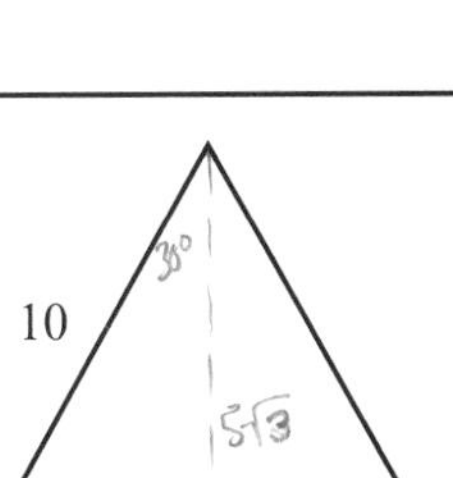

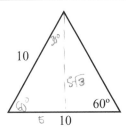

6. In the figure above, what is the area of the triangle?

(A) $5\sqrt{3}$
(B) $25\sqrt{2}$
(C) $25\sqrt{3}$
(D) $50\sqrt{2}$
(E) $50\sqrt{3}$

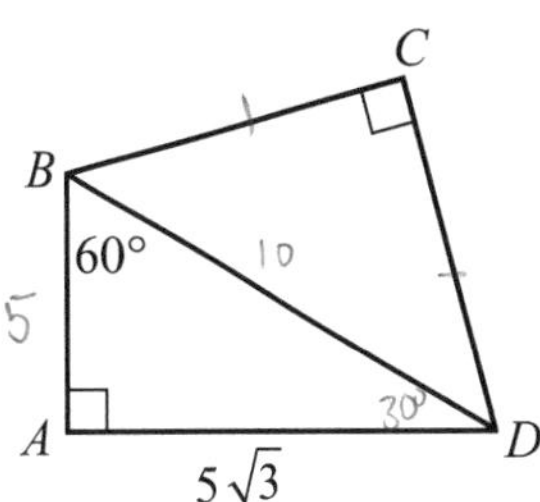

7. In the figure above, $BC = CD$. What is the value of CD?

(A) 5
(B) 7
(C) $5\sqrt{2}$
(D) 10
(E) $10\sqrt{2}$

Quadrilaterals

Frequency Guide: 3

A quadrilateral is a four-sided figure. There are three special quadrilaterals that appear on the SAT—rectangles, squares, and parallelograms.

Required Knowledge and Skill Set

1. A rectangle is a four-sided figure with four right angles. The formula for the area (Area = length × width) is provided in the formula box. The formula for the perimeter is not included in the formula box, but is merely the sum of each side.

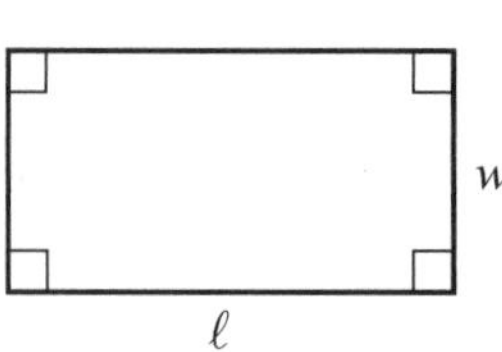

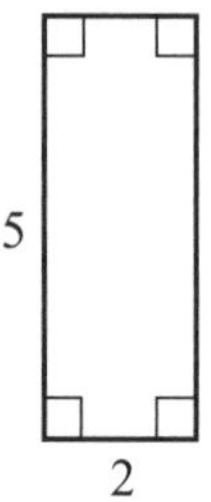

Area = $\ell \times w$
Perimeter = $\ell + w + \ell + w = 2\ell + 2w$

Area = $2 \times 5 = 10$
Perimeter = $2(5) \times 2(2) = 14$

MEMORY MARKER:
The formula for the perimeter of a rectangle is included in the free flash cards at www.powerscore.com/satmathbible.

2. By drawing the diagonal of a rectangle, two congruent triangles are created. You can determine the length of the diagonal by using the Pythagorean Theorem:

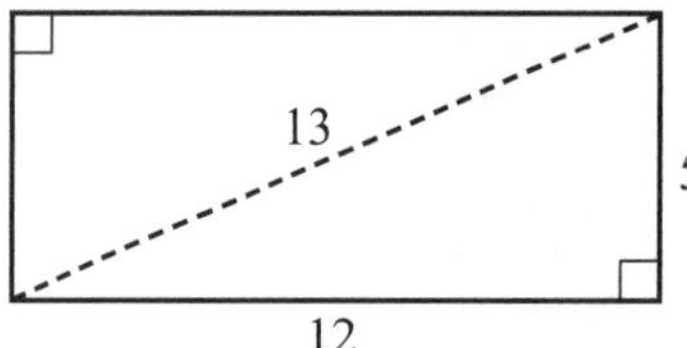

$a^2 + b^2 = c^2$
$5^2 + 12^2 = c^2$
$25 + 144 = c^2$
$169 = c^2$
$13 = c$

3. A square has four right angles and four equal sides. The area of a square can also be found by multiplying the length by the width. However, since the length and width are equal, you can simply square the length of one side to find the area (Area = side^2). The perimeter is the sum of the four sides or 4 times one side.

Because a square is a rectangle, the formulas for a rectangle will work for a square. However, the formulas for a square will NOT work for a rectangle, because a rectangle is not a square.

A square IS a rectangle; a rectangle IS NOT a square.

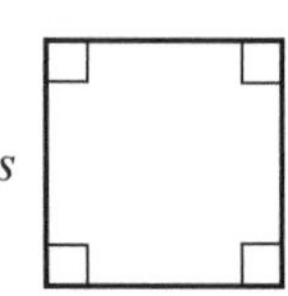

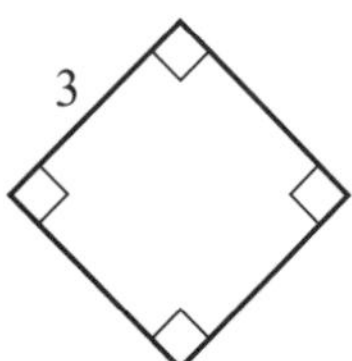

Area = $\ell \times w = s^2$
Perimeter = $s + s + s + s = 4s$

Area = $3^2 = 9$
Perimeter = $4(3) = 12$

4. As discussed in the section on special triangles, drawing one diagonal of a square creates two 45°:45°:90° triangles. Drawing both diagonals creates four 45°:45°:90° triangles, as the two diagonals create 90° angles at their intersection:

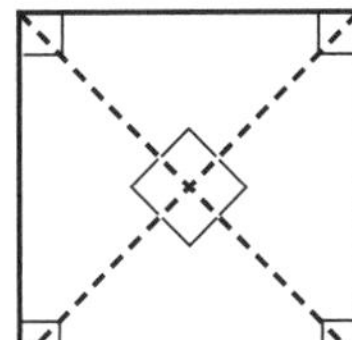

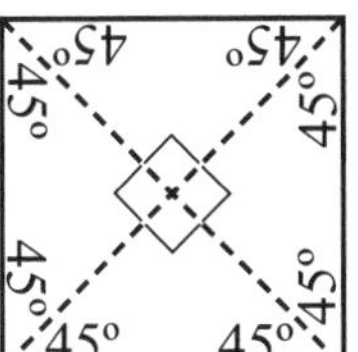

5. In a parallelogram, each of the opposite sides is parallel and equal length. The opposite angles are equal. Although squares and rectangles are parallelograms, most students think of parallelograms as "leaning rectangles."

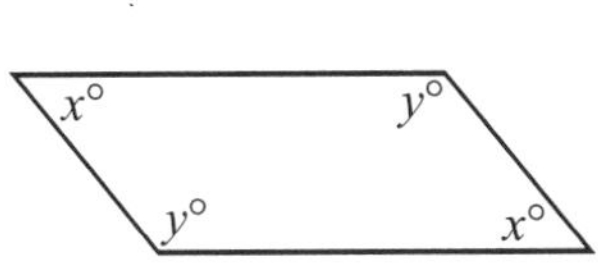

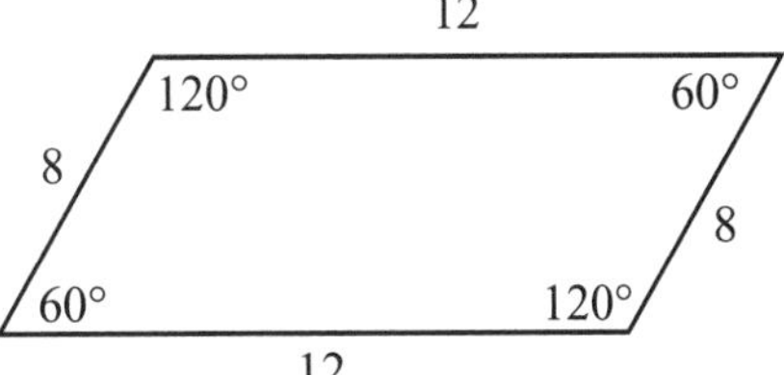

6. The formula for the area of a parallelogram (Area = length × height) is not given in the formula box (but is rarely tested on the SAT). To understand how the area of a parallelogram compares to the area of a rectangle, think of a parallelogram as a rearranged rectangle:

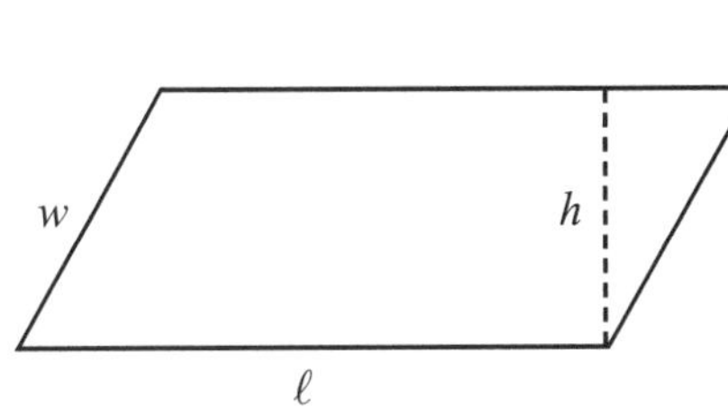

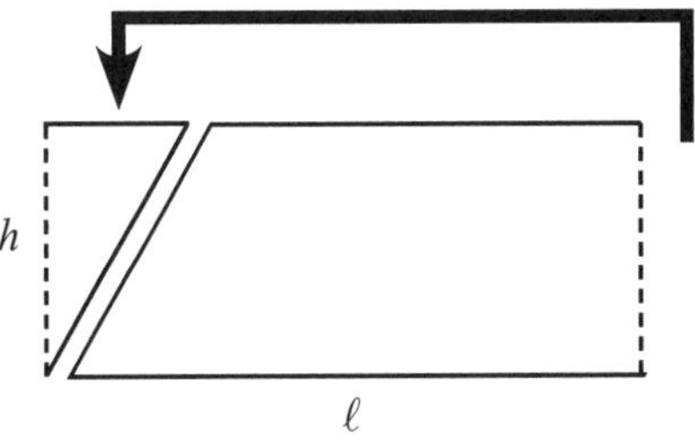

$\text{Area} = \ell \times h$

$\text{Perimeter} = \ell + w + \ell + w = 2\ell + 2w$

MEMORY MARKER: The formula for the area of a parallelogram is not provided in the formula box.

7. Two parallelograms with the same height and base share the same area, even if they have different shapes:

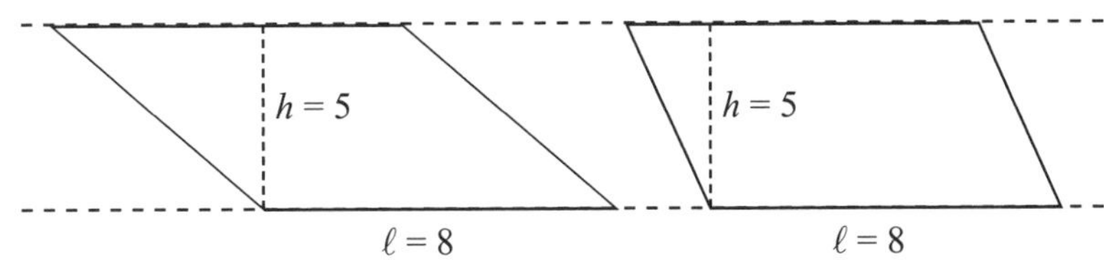

$\text{Area} = 8 \times 5 = 40$ $\qquad$ $\text{Area} = 8 \times 5 = 40$

8. The diagonals of a parallelogram bisect each other. In other words, they divide each other into equal parts:

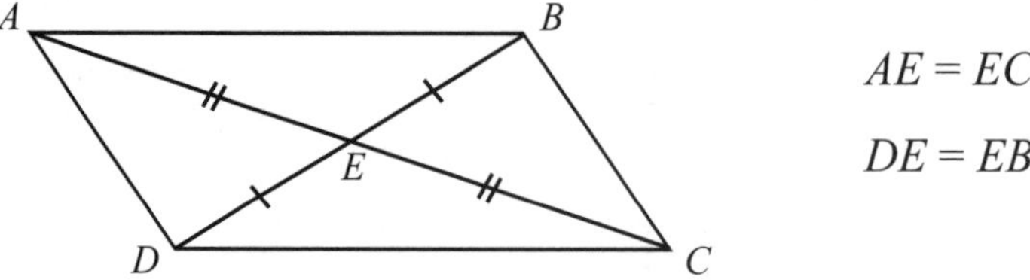

Parallelograms can be divided into two congruent triangles and one rectangle.

9. Many problems involving rectangles, squares, and parallelograms are actually triangle questions in disguise. Be on the lookout for hidden triangles in questions involving quadrilaterals.

10. The properties of other quadrilaterals, such as a trapezoid or rhombus, are not tested on the SAT. If you encounter an unknown four-sided figure, divide the figure into rectangles and triangles to determine area or perimeter:

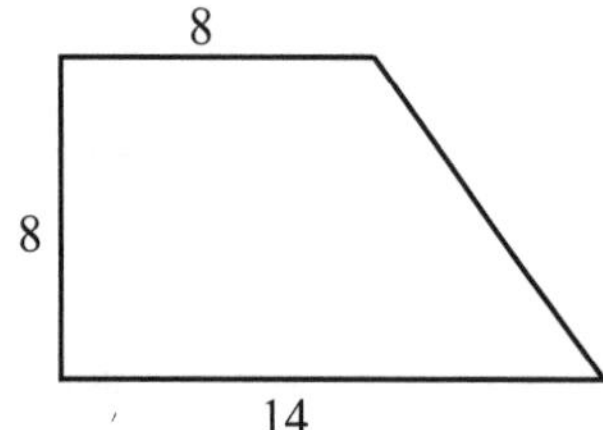

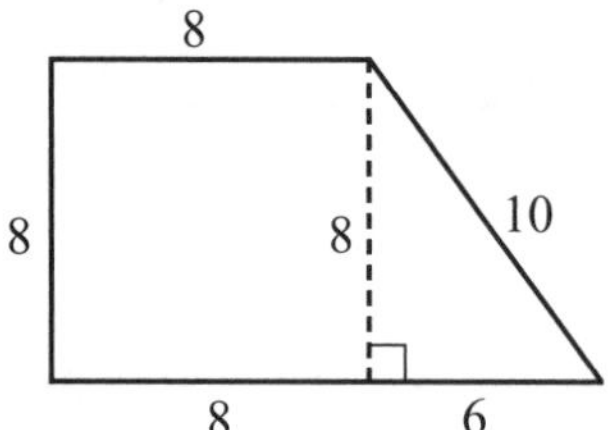

Application on the SAT

Nearly all quadrilateral questions deal with finding the area or the perimeter of the figure. There are many ways these questions can be presented though. In the chapter "Introduction to SAT Math," we looked at an area and perimeter question as a typical example of how the makers of the SAT intentionally make test questions more difficult than classroom math questions. This type of question gives the area and asks for the perimeter or vice versa. It bears examining again:

CAUTION: SAT TRAP! In school, students are trained to find the area and perimeter of a rectangle given the length and width. The SAT will require you to "work backwards," extracting the length from the area in order to find the perimeter or vice versa.

9. The width of a rectangle is 2 inches and the area is 16 square inches. What is the perimeter of the rectangle in inches?

(A) 14
(B) 18
(C) 20
(D) 32
(E) 36

To find the perimeter of a rectangle, you must have the length and width. Since you only have the width, use the area to find the length:

$$\text{Area} = \ell w \rightarrow 16 = \ell(2) \rightarrow 8 = \ell$$

Now find the perimeter:

$$\text{Perimeter} = 2\ell + 2w \rightarrow 2(8) + 2(2) \rightarrow 16 + 4 \rightarrow 20 \text{ inches}$$

Many area and perimeter questions are word problems. It may help to DIAGRAM these story problems, as they may not have an accompanying figure. You can also use TRANSLATION as needed. The most difficult word problems involve an odd-shaped tile and a rectangular floor:

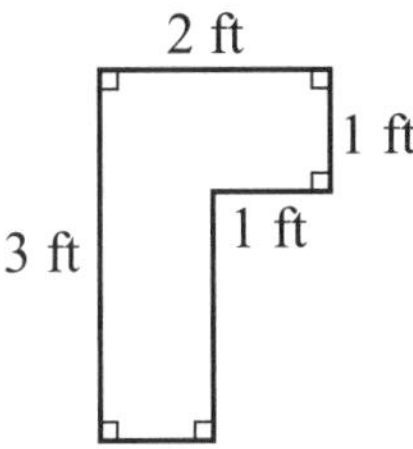

16. A rectangular floor has a length of 40 feet and a width of 20 feet. If a flooring company is using the tiles above, how many tiles are needed to completely cover the floor?

    (A) 150
    (B) 200
    (C) 250
    (D) 300
    (E) 360

To solve these questions, begin by figuring the area of the floor:

$$\text{Area} = \ell w \quad \rightarrow \quad (40)(20) \quad \rightarrow \quad 800 \text{ square feet}$$

Then, assemble tiles to make a rectangle. It may take two, three, four, or more tiles. For this problem, it takes two tiles to create a rectangle:

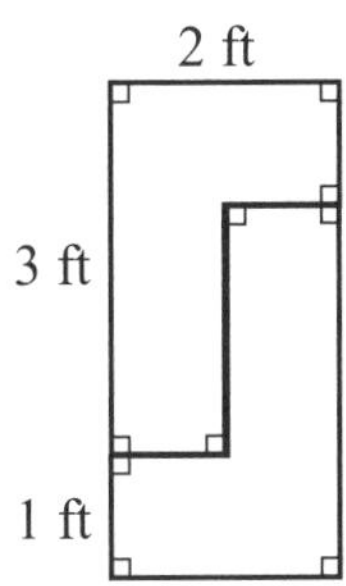

When the two tiles combine, the length of the rectangle is 4 feet and the width is 2 feet. Therefore, these two tiles cover an area of 8 square feet ($4 \times 2 = 8$).

DIAGRAMMING the question is an essential strategy for nearly all geometry questions. If a figure is provided, fill in as much information as possible. If no figure is present, draw one.

Now, figure how many rectangles at an area of 8 square feet are required to cover 800 square feet:

800 sq. ft. ÷ 8 sq. ft. = 100 rectangles

How many tiles are in each rectangle? Two:

100 rectangles × 2 tiles = 200 tiles

It takes 200 tiles to cover the floor. Answer choice (B) is correct.

If the area of the rectangle is 60, can you SIZE UP the area of the shaded region? It's less than half of the rectangle, so it must be less than 30.

Another type of area and perimeter question involves a shaded region of a quadrilateral. Consider the following grid-in question:

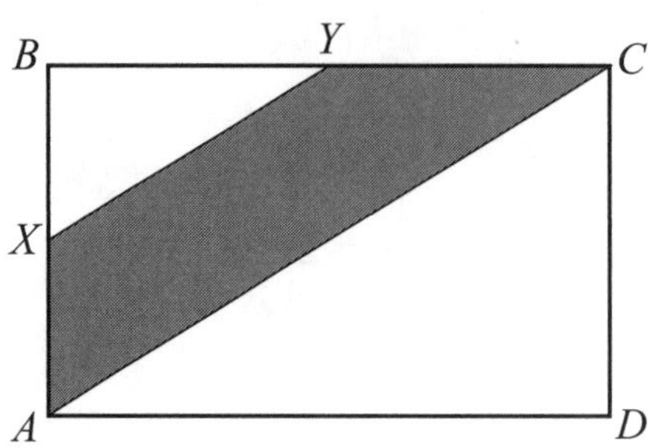

17. In the figure above, $AD = 10$, $CD = 6$, $BX = XA$ and $BY = YC$. What is the area of the shaded region?

There are several ways to solve this question. The fastest solution method involves dividing the figure into fractions, but it does not always work for every figure. However, since it provides such a quick solution, it is worth an attempt. Draw the lines that bisect the height and width.

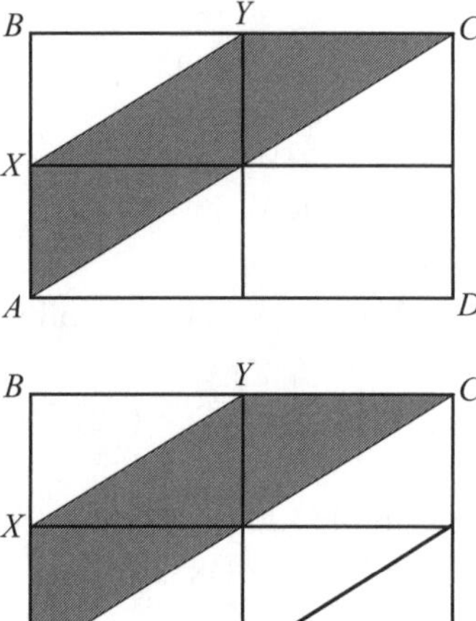

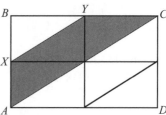

Then, draw the diagonal of the bottom right rectangle so there are 8 equal triangles in the large rectangle.

Of the 8 small triangles, 3 of them are shaded. So $\frac{3}{8}$ of the total area is shaded. Find the total area:

$$\text{Area} = \ell w \quad \rightarrow \quad (10)(6) \quad \rightarrow \quad 60$$

Now find three-eighths of the total area:

$$\frac{3}{8} \text{ of } 60 = \frac{3}{8} \times 60 = 22.5$$

That was easy! But not all shaded figures can be divided into fractional parts. When you cannot divide the figure into equal parts, then you must solve using the areas of each part of the figure.

For the ease of calculations, DIAGRAM the question:

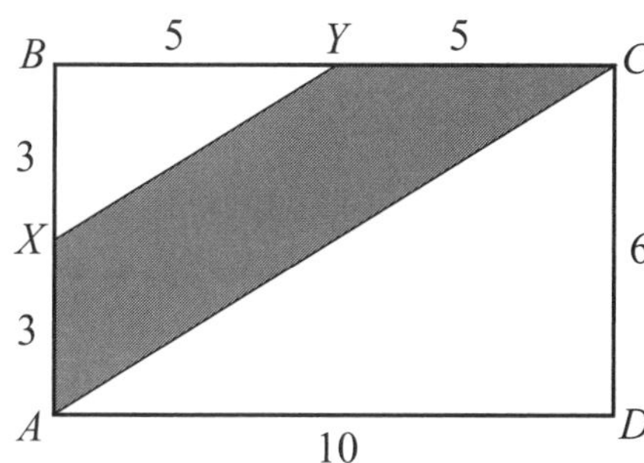

Then write a word equation that calculates the sum of the individual areas:

Area of XBY + shaded area + area of ACD = area of the rectangle

Find the area of the two triangles:

Area of $XBY = \frac{1}{2}bh \rightarrow \frac{1}{2}(5)(3) \rightarrow 7.5$

Area of $ACD = \frac{1}{2}bh \rightarrow \frac{1}{2}(10)(6) \rightarrow 30$

Find the area of the rectangle:

Area $= \ell w \rightarrow (10)(6) \rightarrow 60$

Now plug all of the values back into the word equation:

Area of XBY + shaded area + area of ACD = area of the rectangle
7.5 + shaded area + 30 = 60
37.5 + shaded area = 60
shaded area = 22.5

Questions about shaded areas are usually higher difficulty levels, but with two different ways to solve them, you should be able to quickly find an answer.

Let's return to the same figure one more time. What if you had to find the perimeter of the shaded region?

Perimeter = $XA + AC + YC + XY$

You can use the right triangles XBY and ACD to find the lengths of their hypotenuses ($a^2 + b^2 = c^2$), which are also the missing lengths of the perimeter.

In the content sections on basic and special triangles, we discussed the triangle's ability to hide in other types of geometry questions. This question demonstrates how triangles can factor into questions about shaded portions of quadrilaterals. They can also play an important part in questions involving "broken quadrilaterals," as we will examine on the next page.

On the test, you do not have to write every word of your equation. Abbreviate or use symbols to save time: Sm△ + Sh. + Lg△ = □

Confidence Quotation
"The soul never thinks without a picture."
—Aristotle,
Greek philosopher

"Broken quadrilateral" questions ask test takers to find the area or perimeter of a figure in which part of a quadrilateral is removed. Consider an example:

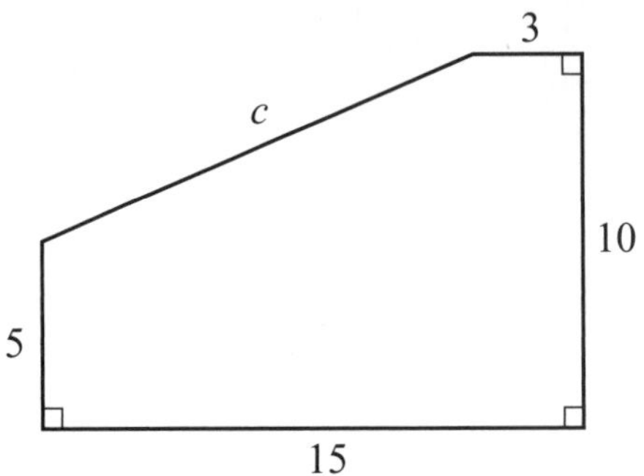

Remember, when the College Board uses italics, underlining, or bold lettering in a question, they are kindly warning you about a trap into which many students fall. In this question, some students mistakenly find the area of the figure. Two of the answer choices—(D) and (E)—are designed to trick students who mistakenly find the area.

9. What is the *perimeter* of the five-sided figure above?

(A) 13
(B) 46
(C) 50
(D) 120
(E) 150

Even though the figure is a five-sided figure, the key to the solution is the broken quadrilateral. Complete the rectangle and find the lengths of its missing perimeter:

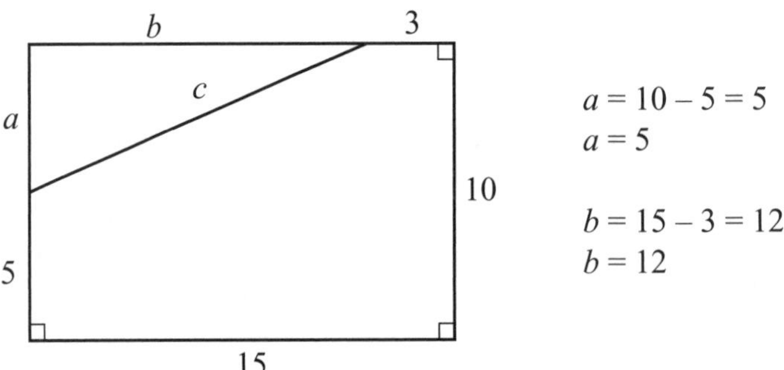

$a = 10 - 5 = 5$
$a = 5$

$b = 15 - 3 = 12$
$b = 12$

By drawing the quadrilateral, you created a right triangle with sides a, b, and c. Students who studied the Pythagorean Triples will immediately know that $c = 13$. But you can also find the length of c using the Pythagorean Theorem to find c:

$a^2 + b^2 = c^2$
$5^2 + 12^2 = c^2$
$25 + 144 = c^2$
$169 = c^2$
$13 = c$

Now find the perimeter of the original figure:

Perimeter $= 5 + 15 + 10 + 3 + c \quad \rightarrow \quad 5 + 15 + 10 + 3 + 13 \quad \rightarrow \quad 46$

Answer (B) is correct. When faced with a odd figure requiring area or perimeter, turn the figure into a square or rectangle.

Quadrilaterals Problem Set

Solve the following multiple-choice questions by selecting the best answer from the five answer choices. For grid-in questions, write your answer in the grids and completely mark the corresponding ovals. Answers begin on page 343.

1. A rectangle has an area of 30. If the length and width of the rectangle are integers, what is one possible value for its perimeter?

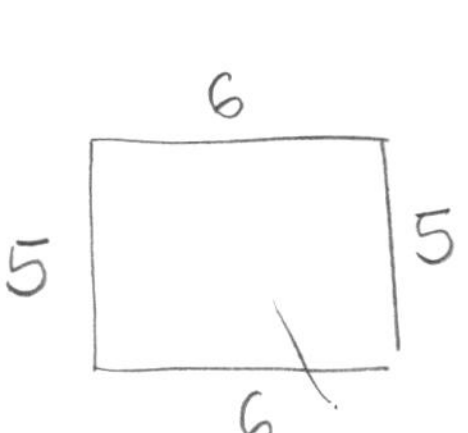

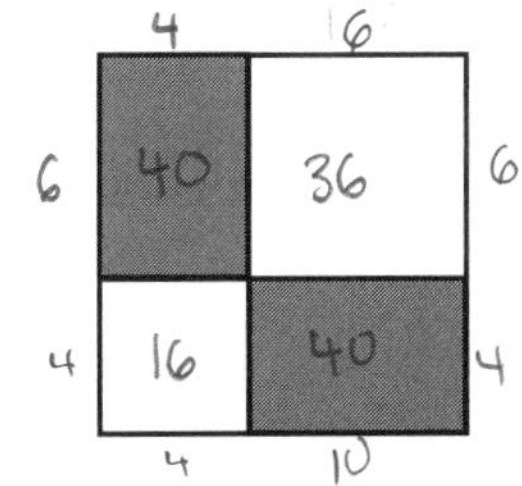

2. In the figure above, a square is divided into two shaded rectangles and two non-shaded squares. The perimeters of the non-shaded squares are 16 and 24. Which of the following has the greatest value?

    (A) The sum of the perimeters of the shaded rectangles
    (B) The sum of the perimeters of the non-shaded squares
    (C) The sum of the area of the shaded rectangles
    (D) The sum of the area of the non-shaded squares
    (E) The perimeter of the entire figure

3. If the length of a rectangle is one-sixteenth of the perimeter, then the width of the rectangle is how many times the length?

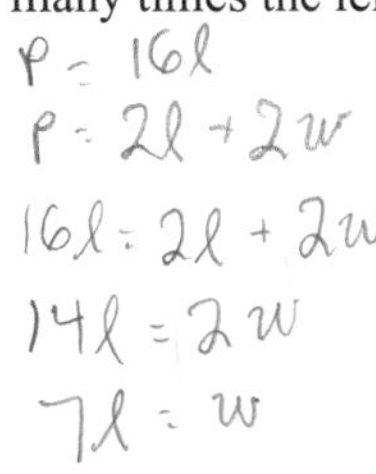

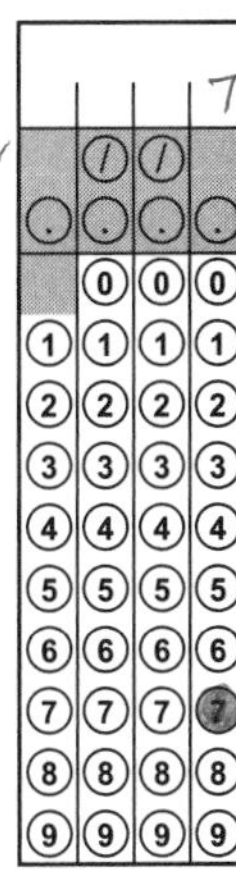

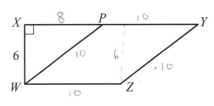

4. In the figure above, each side of quadrilateral $WPYZ$ has a length of 10. What is the area of $WXYZ$?

    (A) 84
    (B) 96
    (C) 108
    (D) 124
    (E) 148

Polygons

Frequency Guide: 2

A polygon is any closed figure with three or more sides. A triangle is a three-sided polygon; a square is a four-sided polygon. But since triangles, squares, and rectangles have their own unique properties, we use the term polygon to describe figures with five or more sides.

Required Knowledge and Skill Set

1. Some polygons have irregular shapes and angles:

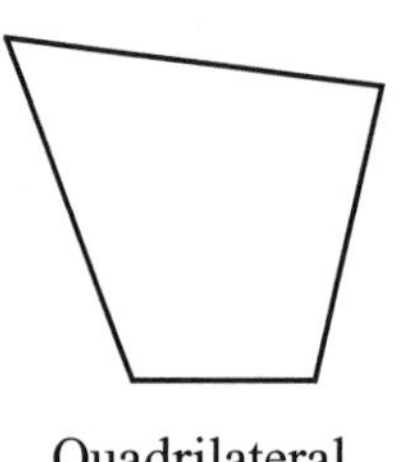

Quadrilateral

Pentagon

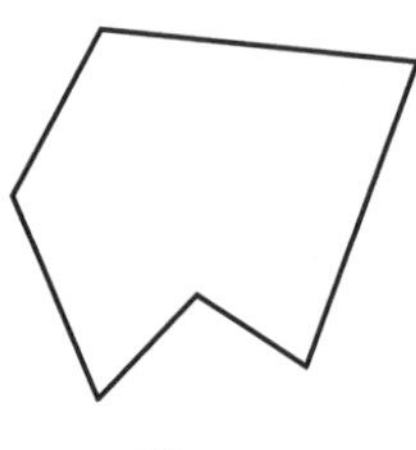

Hexagon

It is unlikely that you will encounter polygons with irregular shapes. If you do, divide the polygon into rectangles and right triangles as discussed in the section on quadrilaterals.

The phrase "regular polygon" is used on the SAT, so you must understand its meaning.

2. Polygons that have equal side lengths and equal angle measurements are called regular polygons:

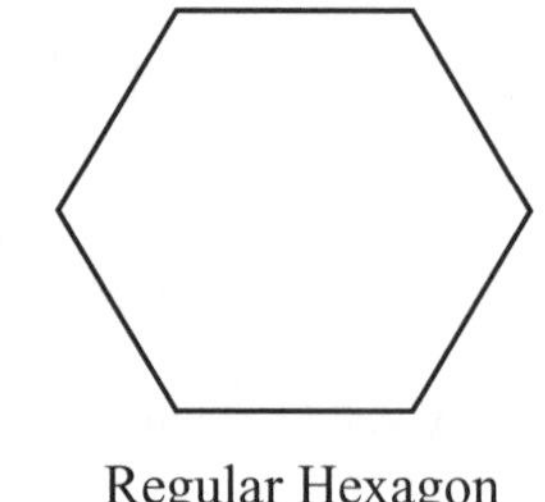

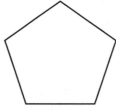

Square Regular Pentagon Regular Hexagon

You can expect to see regular polygons on the SAT because of their special properties.

The most commonly occurring polygons (not counting triangles and quadrilaterals) are pentagons and hexagons.

3. Polygons present many opportunities for using the properties of hidden triangles:

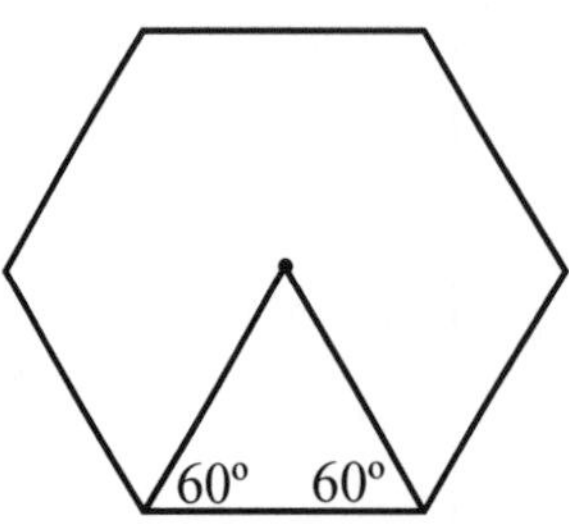

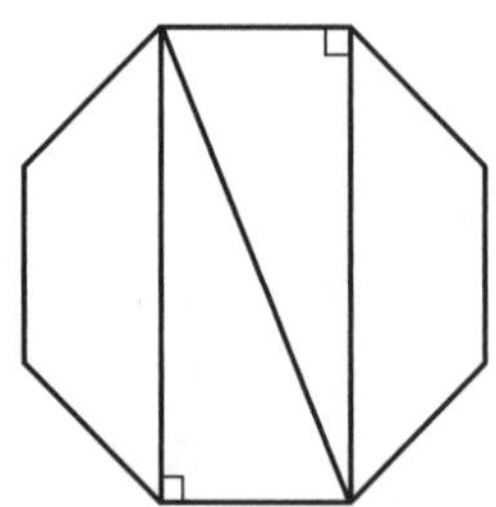

4. You must know the sum of the interior angle measurements of regular polygons. The easiest way to find these sums is to memorize the following table:

Regular Figure	No. of Sides	Sum of the Interior Angles	Measure of Each Interior Angle
Triangle	3	180º	180º ÷ 3 = 60º
Quadrilateral	4	360º	360º ÷ 4 = 90º
Pentagon	5	540º	540º ÷ 5 = 108º
Hexagon	6	720º	720º ÷ 6 = 120º
Octagon	8	1080º	1080º ÷ 8 = 135º

MEMORY MARKER: Be sure to memorize the sum of the interior angles and the measure of each angle for all of the polygons in the table.

Do you notice a pattern in the table? For every side that is added, another 180º is added to the sum of the interior angles! If you are unable to memorize the sums of the interior angles, you can still calculate them. Divide a regular polygon into triangles with a common vertex. Since the interior angles of a triangle equal 180°, multiply the number of triangles by 180°:

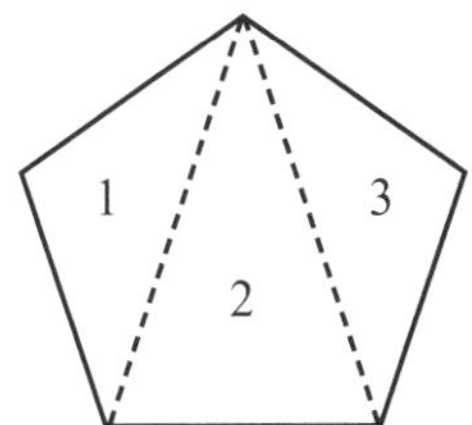

Sum of the interior angles:
3 triangles × 180° = 540°

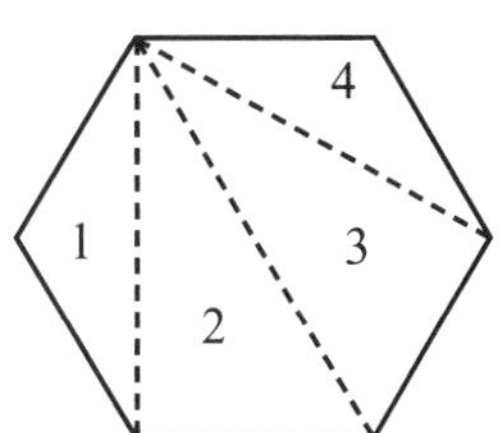

Sum of the interior angles:
4 triangles × 180° = 720°

5. Once you have the sum of the interior angles of a regular polygon, you can determine the measurement of each angle. Divide the sum by the number of angles:

You can also memorize these measurements as presented in the table above.

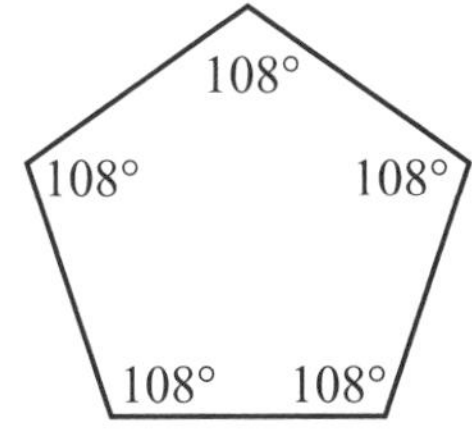

Sum of the interior angles = 540°
540° ÷ 5 angles = 108°

6. A regular polygon with an even number of sides (square, hexagon, octagon, etc.) has diagonals that bisect each other:

Note that a line drawn from the center of a regular polygon to a vertex is a bisector of the interior angle.

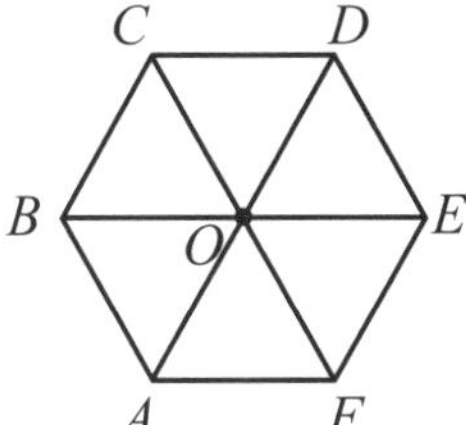

$OA = OD$
$OB = OE$
$OC = OF$

$OA = OD = OB = OE = OC = OF$

Application on the SAT

There are three basic types of question about polygons. The first tests your knowledge of the properties of polygons, and is usually the most difficult. It may include a picture to DIAGRAM or it may be a Roman numeral question like the following:

15. If all six interior angles of a hexagon have the same measure, which of the following statements must be true?

    I. All sides of the hexagon have the same length.
    II. The diagonals of the hexagon bisect each other.
    III. The measure of each interior angle of the hexagon is 60º.

    (A) I only
    (B) I and II
    (C) I and III
    (D) II and III
    (E) I, II, and III

A hexagon with equal interior angles is a regular hexagon. Therefore, the six sides must also be the same length. So (I) must always be true. This eliminates (D).

Be sure the polygon has an even number of sides (4, 6, 8, etc.) before applying this property.

The second Roman numeral deals with the diagonals of the hexagon. As you learned in the Required Knowledge and Skill Set section, a regular polygon with an even number of sides has diagonals that bisect each other. Since a hexagon has six sides, (II) must be true. This eliminates (A) and (C).

To prove or disprove the third Roman numeral, rely on your memorization of the table on the previous page to know that the interior angles of a regular hexagon are each 120º. If you do not remember the measurement from the table, you can find the measure of each interior angle of a hexagon. Divide a regular hexagon into triangles using a single vertex:

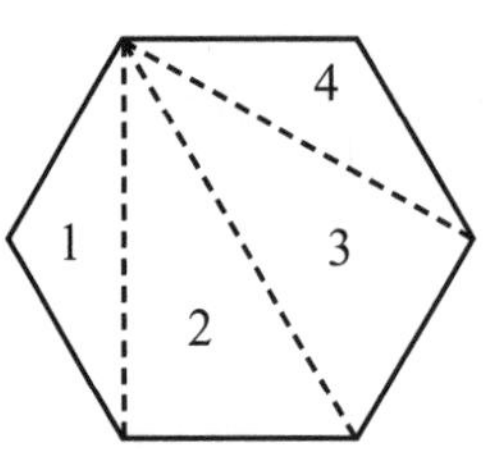

There are 4 triangles. Each triangle has 180º:

$$4 \times 180° = 720°$$

The sum of the interior angles is 720º. There are 6 angles:

$$720° \div 6 = 120°$$

Each interior angle is 120º. So Roman numeral III is not true. This eliminates (E) and reveals the correct answer to be (B).

The second type of polygon question also deals with the measurement of the interior angles of a regular polygon. The easiest ones will simply ask for the measurement of one of the angles, similar to Roman numeral III above. As an added level of difficulty, other questions may ask for the degree measurement of a bisected angle.

Consider an example:

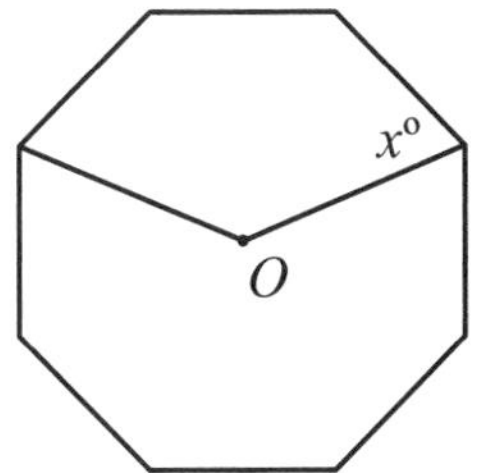

11. The figure above is a regular octagon. What is the value of x ?

(A) 60
(B) 67.5
(C) 70
(D) 78.5
(E) 135

Each interior angle is 135. You know this because you have memorized the sum of the interior angles, right? If not, you can find this information by dividing the octagon into triangles with a shared vertex.

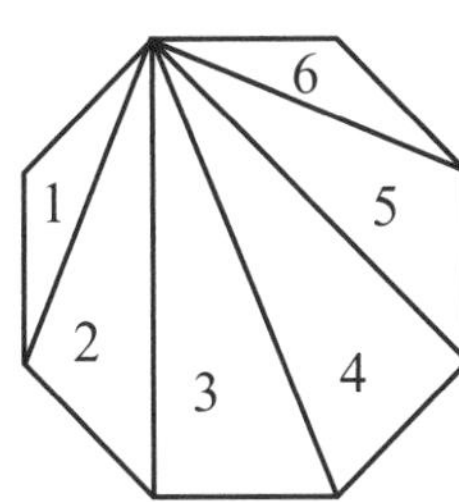

There are 6 triangles. Each triangle has 180°:

$$6 \times 180^\circ = 1080^\circ$$

The sum of the interior angles is 1080°. There are 8 angles:

$$1080^\circ \div 8 = 135^\circ$$

Remember that any line drawn from the center of the regular polygon to a vertex is a bisector of the interior angle. So x is one-half of 135°, or 67.5°

The third and final type of polygon question asks you to find the measurement of an angle with the center of the polygon as the vertex:

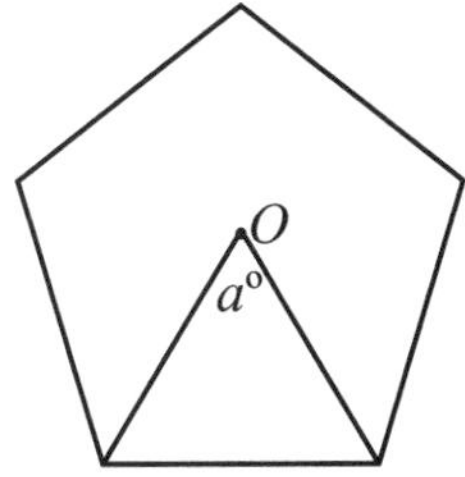

13. The figure above is a regular pentagon. What is the value of a ?

(A) 40
(B) 56
(C) 60
(D) 68
(E) 72

Confidence Quotation
"I was always looking outside myself for strength and confidence, but it comes from within. It is there all the time."
—Anna Freud, psychoanalyst

This is true for all polygons—even and odd numbered sides alike!

There are several ways to solve this question, but the easiest method involves thinking of the center of the polygon as a circle:

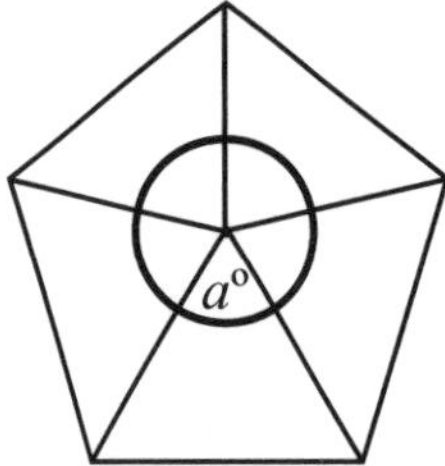

There are 360º of arc in a circle. This circle is divided by 5:

$360^{\circ} \div 5 = 72^{\circ}$ $a = 72^{\circ}$

This solution method would also work if the question was looking for the sum of two or more central angles:

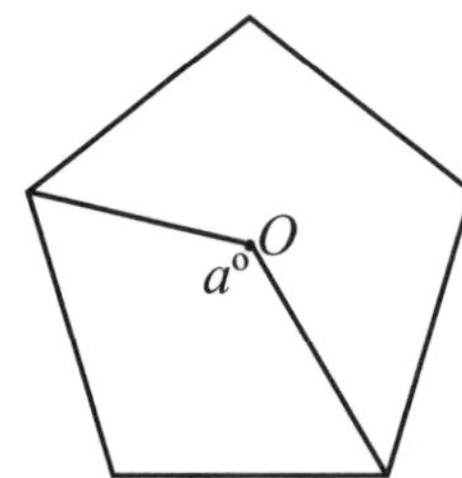

14. The figure above is a regular pentagon. What is the value of a ?

(A) 72
(B) 100
(C) 128
(D) 144
(E) 160

The concept is the same:

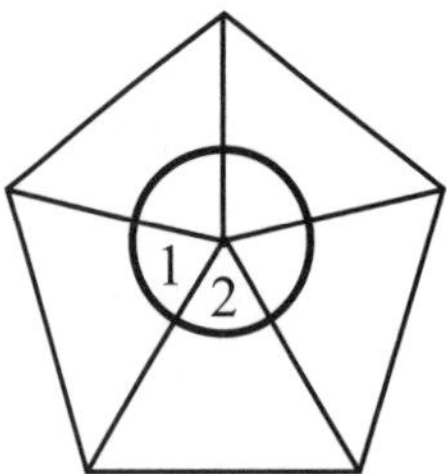

There are 360º of arc in a circle. This circle is divided by 5:

$360^{\circ} \div 5 = 72^{\circ}$
Angle 1 = 72º
Angle 2 = 72º
Angle 1 + Angle 2 = 72º + 72º = 144º

This question can also be solved using the properties of quadrilaterals and pentagons. The four interior angles of the quadrilateral must add up to 360º.

We have presented the most efficient way for solving questions about polygons, but there are many others. If you forget a solution method, you can often rely on your knowledge of other figures, such as triangles, quadrilaterals, and circles to solve a polygon question.

Polygons Problem Set

Solve the following multiple-choice questions by selecting the best answer from the five answer choices. For grid-in questions, write your answer in the grids and completely mark the corresponding ovals. Answers begin on page 345.

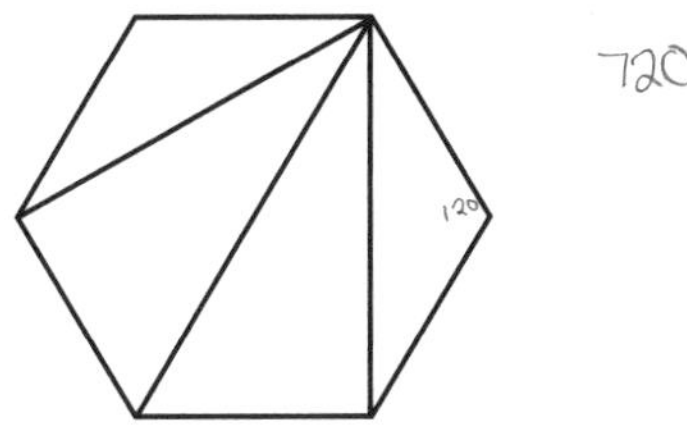

1. The figure above is a regular hexagon that is divided into four nonoverlapping triangles. Which of the following is true about all four triangles?

   (A) They are isosceles.
   (B) They have equal perimeters.
   (C) They are similar.
   (D) They each have at least one 30° angle.
   (E) They have equal areas.

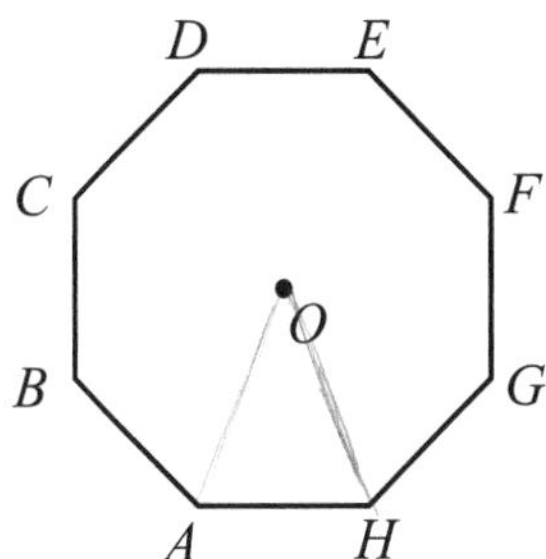

2. The figure above is a regular octagon. What is the degree measure of $\angle AOH$ (not shown) ?

   (A) 40
   (B) 45
   (C) 50
   (D) 55
   (E) 60

Circles

Frequency Guide: 4

Circles are the second most commonly tested geometric figure on the SAT.

Required Knowledge and Skill Set

1. The radius of a circle is the length from the center of the circle to any point on the circle. The diameter is any line between two points on the circle that crosses through the center of the circle. The diameter is twice the radius.

CAUTION: SAT TRAP! The test makers may provide the diameter rather than the radius. Be sure to divide the diameter by 2 to find the radius before using any circle formulas.

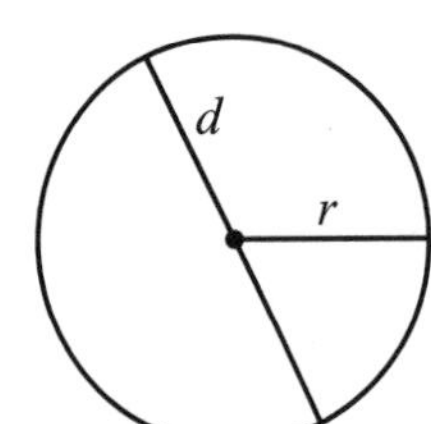

If the test question provides the diameter, be sure to use the radius for the formulas in the Formula Box.

2. The formulas for a circle's circumference and area are provided in the Formula Box. The circumference is the perimeter of the circle. The circumference is also the distance of one revolution of the circle.

Just because a formula is in the Formula box does not mean you do not have to memorize it! Memorization will build confidence and save time.

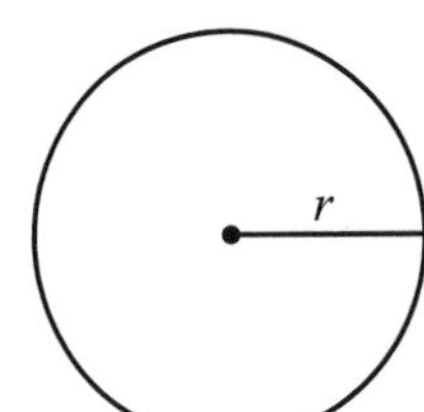

Circumference = $2\pi r$
Area = πr^2

If $r = 5$
Circumference = $2\pi r = 2\pi(5) = 10\pi$
Area = $\pi r^2 = \pi(5)^2 = 25\pi$

3. If *pi* is in your answer, do not convert it to 3.14; the SAT leaves the *pi* symbol (π) in the answer choices. Furthermore, if you encounter a question with the *pi* symbol in any of the answer choices, you should know to look for the area or circumference of a circle.

4. A tangent is a line that touches a circle at only one point. A radius or diameter drawn to that point is perpendicular to the tangent.

Tangent lines create 90° angles.

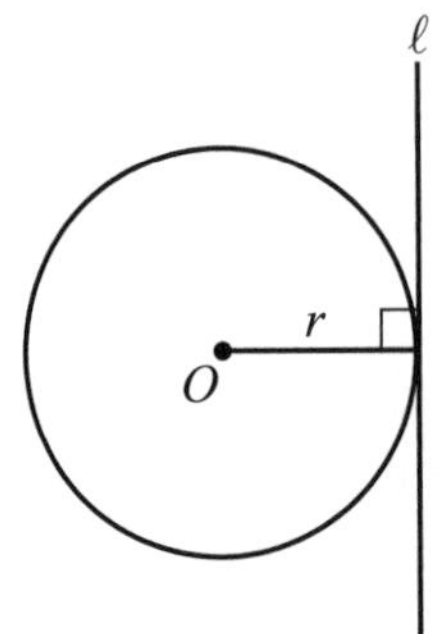

Line ℓ is tangent to Circle O.

5. An arc is a segment of a circle between any two points on the circle. There are 360° of arc in a circle, which is noted in the Formula Box. The circle below has two arcs: the small arc *AB* which corresponds to $x°$, and the large arc *BA*, which corresponds to $360° - x$.

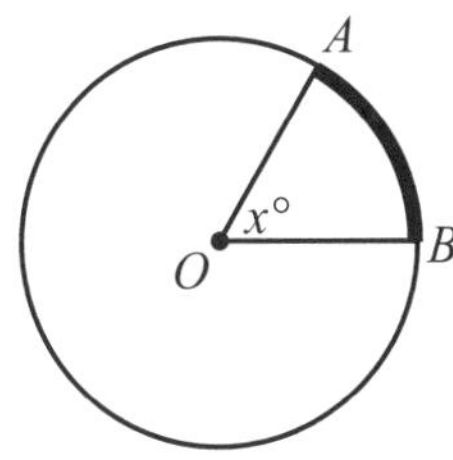

If $x = 60$, then x represents $\frac{60}{360}$ or $\frac{1}{6}$ of the interior degrees of the circle, so the length of arc *AB* is $\frac{60}{360}$ or $\frac{1}{6}$ of the total circumference.

The formula for finding the length of an arc is the total circumference of the circle multiplied by the fraction of the circle in question:

$$\text{The length of an arc} = \frac{x^{\circ}}{360^{\circ}}(2\pi r)$$

An arc is a fraction of the circumference; a sector is a fraction of the area.

MEMORY MARKER: The formula for the length of an arc is the circumference of the circle multiplied by the fraction of the circle in the arc.

6. A sector of a circle is a "wedge" formed by a central angle. The circle below has the shaded sector *AOB*, formed by central angle x.

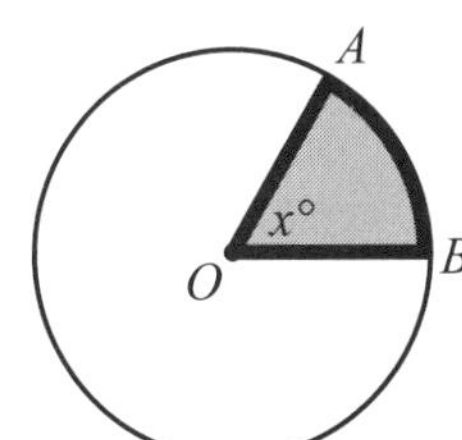

If $x = 60$, then x represents $\frac{60}{360}$ or $\frac{1}{6}$ of the interior degrees of the circle, so the area of the sector is $\frac{60}{360}$ or $\frac{1}{6}$ of the total area.

The formula for finding the area of a sector is the total area of the circle multiplied by the fraction of the circle in question:

$$\text{The area of a sector} = \frac{x^{\circ}}{360^{\circ}}(\pi r^2)$$

MEMORY MARKER: The formula for the area of a sector is the area of the circle multiplied by the fraction of the circle in the sector.

7. Triangles that have two vertices on the circle and one at the center of the circle are always isosceles triangles. Triangles are often hidden in circle questions so be prepared to use the properties of triangles.

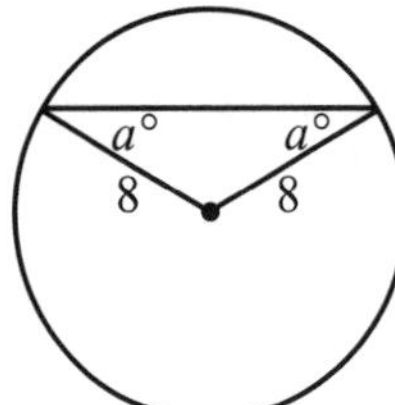

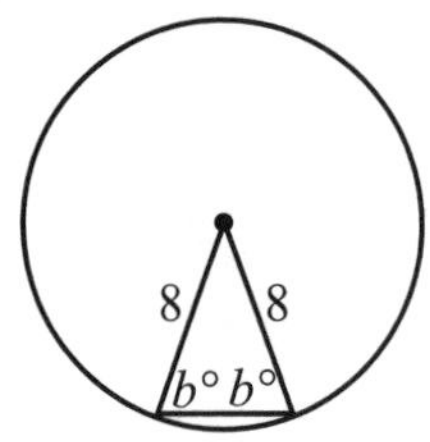

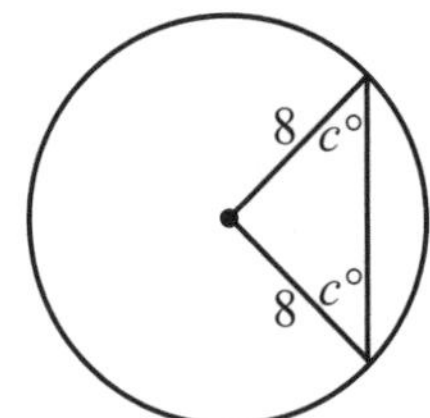

8. If a square is inscribed in a circle, you can find the diameter and radius of the circle by determining the diagonals of the square.

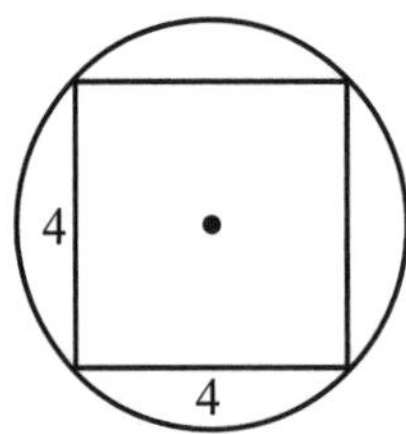

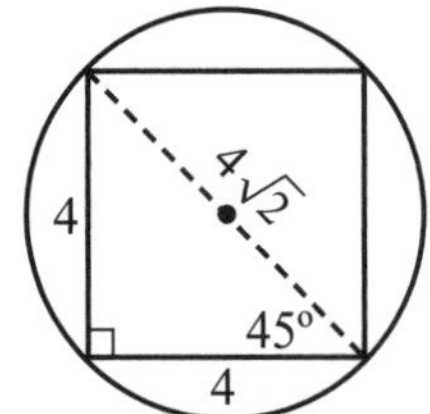

diameter = $4\sqrt{2}$

radius = $\frac{1}{2}$ diameter = $\frac{1}{2}\left(4\sqrt{2}\right) = 2\sqrt{2}$

Expect to see circles used with triangles, squares, rectangles, and other polygons, where the properties of both figures are required to find a solution.

ARITHMETRICK

When a square is inscribe in a circle, the radius of the circle is one half of the square's side length times the square root of 2.

Application on the SAT

The most common circle questions involve area and circumference. One such question type is a variation of the quadrilateral questions we examined earlier. The test makers will give you the area and ask for the circumference, or provide the circumference and require the area:

9. If the area of a circle is 16π, what is the circumference of the circle?

   (A) 2π
   (B) 4π
   (C) 6π
   (D) 8π
   (E) 32π

RECORD what you know:

Area = πr^2
Area = 16π

Solve for the radius by setting the values equal:

$\pi r^2 = 16\pi \rightarrow r^2 = 16 \rightarrow r = 4$

Knowing that the radius is 4, find the circumference:

Circumference = $2\pi r \rightarrow 2\pi(4) \rightarrow 8\pi$ The correct answer is (D).

One way that the test makers may try to trick you is by using *pi* as the length of the radius or diameter. Do not allow *pi* to throw you off; solve the question the same way you would if the radius was 4 or 9 or 12:

13. If the radius of a circle is π, what is the area of the circle?

   (A) π^2
   (B) π^3
   (C) 2π
   (D) $2\pi^2$
   (E) $2\pi^3$

RECORD what you know:

Area = πr^2
Radius = π

Now solve:

Area = = $\pi r^2 \rightarrow (\pi)(\pi^2) \rightarrow \pi^3$

Choice (B) is the correct answer. This is an easy problem that catches many students because of the unconventional radius.

☠ CAUTION: SAT TRAP!
You are trained to find the circumference or area given the radius. The test makers know this, and thus will force you to "work backwards" to find the radius given the circumference or area, and then apply that radius to the other formula.

☠ CAUTION: SAT TRAP!
The use of pi as the radius, circumference, or area is a known trick on the SAT. Students may become confused when pi is used in a way they are not used to. Treat it the same way you would treat any number.

Another way that the test makers increase the difficulty of area and circumference questions is by using ratios. In these questions, you will be asked to find the ratio of the measurements between two or more circles:

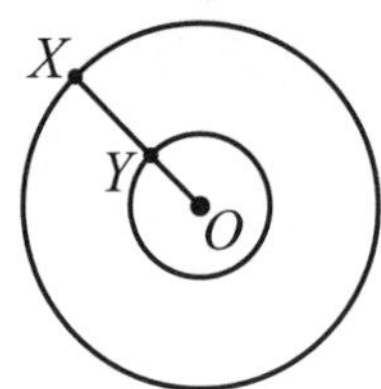

9. If the figure above, two circles have a center at O. Point Y lies on XO. If $XO = 10$ and $XY = 6$, what is the ratio of the area of the larger circle to the area of the smaller circle?

   (A) 15:2
   (B) 25:4
   (C) 5:2
   (D) 9:4
   (E) 6:4

Confidence Quotation
"To follow, without halt, one aim; there is the secret of success."
—Anna Pavlova, Russian ballerina

DIAGRAM the question:

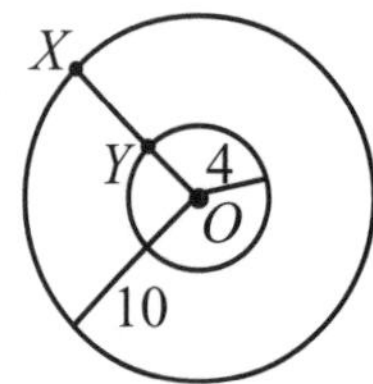

Radius of large circle = 10

Radius of small circle = 4

The ratio is comparing the area of the large circle to the area of the small circle. Begin by finding the values being compared:

Area = πr^2
Area of large circle = $\pi(10^2)$ → 100π
Area of small circle = $\pi(4^2)$ → 16π

The ratio of the area of the large circle to the area of the small circle is $100\pi{:}16\pi$. This ratio can be reduced:

Writing a ratio as a fraction may make it easier to reduce.

$$\frac{100\pi}{16\pi} \rightarrow \frac{50}{8} \rightarrow \frac{25}{4}$$

The correct answer, 25:4, is choice (B).

The SAT may also test your ability to find the shaded area of a figure involving a circle. For these questions, the circle will be combined with a polygon, creating a shaded region that is not a regular figure.

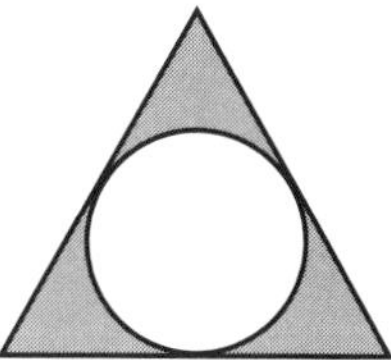

Note: Figure is not drawn to scale

16. If the figure above, a circle is inscribed in an equilateral triangle. The radius of the circle is 4 and the length of one side of the triangle is 14. What is the area of the shaded region?

(A) $14\sqrt{3} - 4\pi$
(B) $14\sqrt{3} - 16\pi$
(C) $49\sqrt{3} - 4\pi$
(D) $49\sqrt{3} - 16\pi$
(E) $98\sqrt{3} - 4\pi$

To solve these questions, begin by writing a statement the isolates the area of the shaded region. Consider how the areas of the figures can be added or subtracted to find the shaded area:

Area of the triangle – area of the circle = area of the shaded region

The word equation can also be written as a sum:

○ + shaded = △

Remember to use your own shorthand to save time.

Now find the area of the triangle using the two hidden 30°:60°:90° triangles:

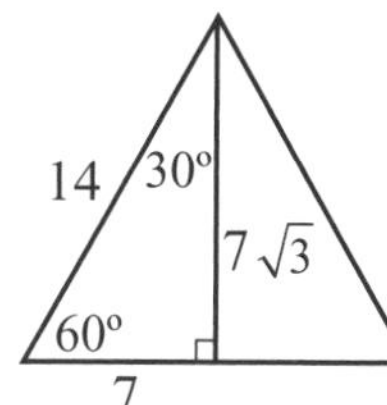

Area = $\frac{1}{2}bh$ base = 14 height = $7\sqrt{3}$

Area = $\frac{1}{2}(14)(7\sqrt{3}) \rightarrow (7)(7\sqrt{3}) \rightarrow 49\sqrt{3}$

The answer is likely to be (C) or (D), as these both have $49\sqrt{3}$ as part of the answer. Now find the area of the circle:

Area = πr^2
Radius = 4

Area = $\pi 4^2 \rightarrow 16\pi$

Return to your word equation and supply values:

Area of the triangle – area of the circle = area of the shaded region

$49\sqrt{3} - 16\pi$ = area of the shaded region

Answer choice (D) is correct.

Questions about arc length and sector area tend to be among the most difficult circle questions, but only because students forget the formulas needed to solve these questions. Consider an example:

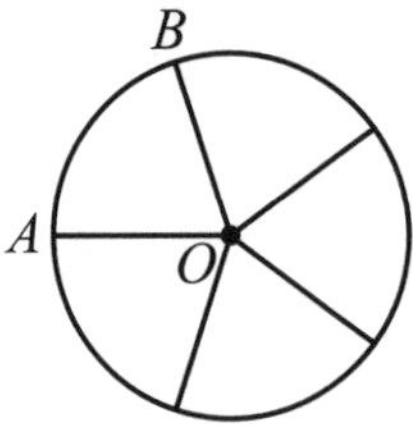

"Congruent" means "equal."

19. In the figure above, the circle has an area of 25π and is divided into 5 congruent sectors. What is the perimeter of AOB ?

(A) $10 + \dfrac{\pi}{5}$

(B) $10 + 2\pi$

(C) $10 + 5\pi$

(D) $10 + 10\pi$

(E) $10 + 20\pi$

The perimeter has two sides and an arc:

$$\text{Perimeter of } AOB = AO + BO + AB$$

The length of AO and BO is also the length of the radius. Use the circle's area to find the radius:

$$\text{Area} = \pi r^2$$
$$\text{Area} = 25\pi$$
$$\pi r^2 = 25\pi \quad \rightarrow \quad r^2 = 25 \quad \rightarrow \quad r = 5$$

$$\text{Perimeter of } AOB = 5 + 5 + AB \quad \rightarrow \quad 10 + AB$$

To find the length of arc AB, you must figure out the measure of $\angle AOB$:

$$360° \text{ of arc in a circle} \div 5 \text{ equal parts} = 72°$$

Or you must know that the $\angle AOB$ is one-fifth of the entire circle.

The length of the arc is just a fraction of the circumference of the circle. You can find this fraction by determining the number of equal parts of the circle, or by dividing the degree measure of the sector by 360°.

Now use the arc length formula:

$$\text{The length of an arc} = \frac{x°}{360°}(2\pi r) \quad \rightarrow \quad \frac{72°}{360°}(2\pi 5) \quad \rightarrow \quad \frac{1}{5}(10\pi) \quad \rightarrow \quad 2\pi$$

$$\text{Perimeter of } AOB = 10 + 2\pi$$

Choice (B) is the correct answer.

The final type of circle question we will investigate involves revolutions of a disk or a wheel. The circumference of the wheel equals the distance traveled in a single revolution of the wheel. If you could "unroll" a circle, you would see that its circumference is the distance traveled as the circle unrolled:

6

Circumference = $2\pi r$ → $2\pi 6$ → 12π

3π 6π 9π 12π

To see how this principle is tested, consider the following question:

20. A wheel on a truck has a diameter of 4 feet. If the wheel spins at a rate of 330 revolutions per minute, how many miles will a point on the wheel travel in 2 hours? (Note: 5280 feet = 1 mile)

    (A) 15π
    (B) 18π
    (C) 20π
    (D) 24π
    (E) 30π

DIAGRAM the question and find the circumference of the wheel:

Circumference = $2\pi r$ → $2\pi 2$ → 4π feet

With one revolution, the wheel travels 4π feet. How many feet does it travel after 330 revolutions?

$$\frac{330 \text{ revolutions}}{1 \text{ minute}} \times \frac{4\pi \text{ feet}}{1 \text{ revolution}} = \frac{1320\pi \text{ feet}}{1 \text{ minute}}$$

Notice how each calculation is clearly labeled. This is extremely important for revolution questions in order to keep the information separate and manageable.

How far would the wheel go in 2 hours?

$$\frac{1320\pi \text{ feet}}{1 \text{ minute}} \times \frac{120 \text{ minutes}}{2 \text{ hours}} = \frac{158,400\pi \text{ feet}}{2 \text{ hours}}$$

Finally, calculate the number of miles in $158,400\pi$ feet:

$$\frac{158,400\pi \text{ feet}}{2 \text{ hours}} \times \frac{1 \text{ mile}}{5280 \text{ feet}} = \frac{30\pi \text{ miles}}{2 \text{ hours}}$$

The wheel travels 30π miles in 2 hours. Answer (E) is correct.

Circles Problem Set

Solve the following multiple-choice questions by selecting the best answer from the five answer choices. For grid-in questions, write your answer in the grids and completely mark the corresponding ovals. Answers begin on page 346.

1. If the circumference of a circle is 36π, what is the area of the circle?

(A) 12π
(B) 18π
(C) 64π
(D) 128π
(E) 324π

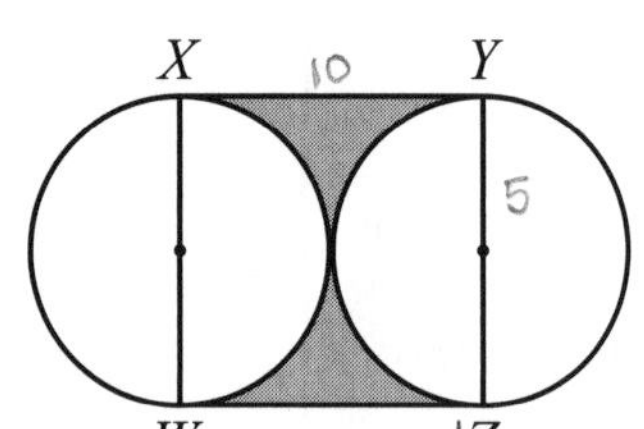

2. In the figure above, *WXYZ* is a square with sides of length 10. If *XW* and *YZ* pass through the centers of the two circles, what is the are of the shaded region?

(A) 50
(B) 100
(C) 75π
(D) $100 - 25\pi$
(E) $100 - 100\pi$

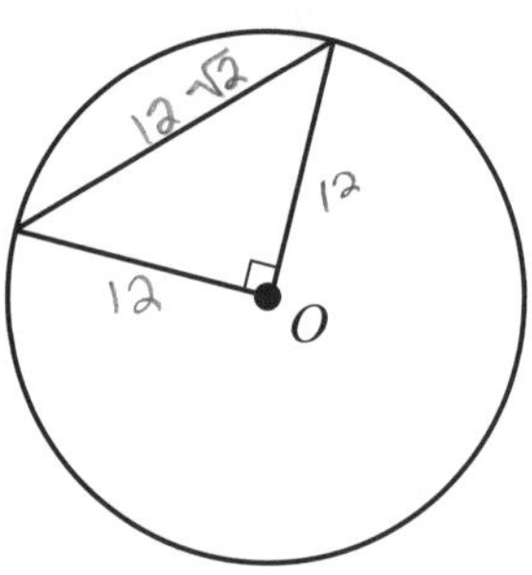

3. In the figure above, the center of the circle is O. If the hypotenuse of the right triangle is $12\sqrt{2}$, what is the circumference of the circle?

(A) 6π
(B) 12π
(C) 24π
(D) 48π
(E) 144π

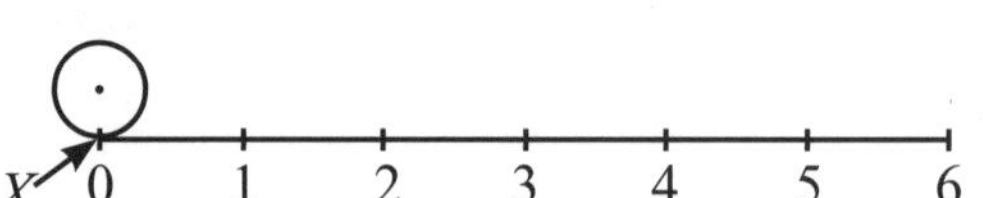

4. In the figure above, point *X* of a circle sits on the line at point 0. The circle has a radius of $\frac{1}{\pi}$. If the circle rolls so that point *X* travels from point 0 to point 6, how many revolutions will the circle make?

(A) 2
(B) 3
(C) 4
(D) 5
(E) 6

Circles Problem Set

Solve the following multiple-choice questions by selecting the best answer from the five answer choices. For grid-in questions, write your answer in the grids and completely mark the corresponding ovals. Answers begin on page 346.

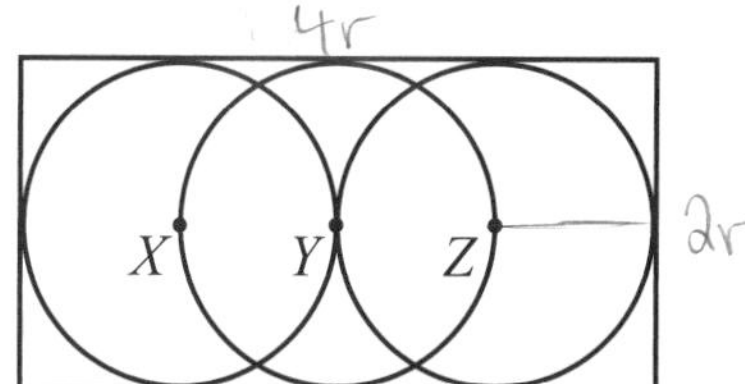

5. In the figure above, the three circles have centers at X, Y, and Z. If the circumference of each circle is $\frac{\pi}{3}$, what is the area of the rectangle ?

(A) $\frac{2}{9}$

(B) $\frac{8}{9}$

(C) $\frac{4}{3}$

(D) $\frac{8}{3}$

(E) $\frac{32}{3}$

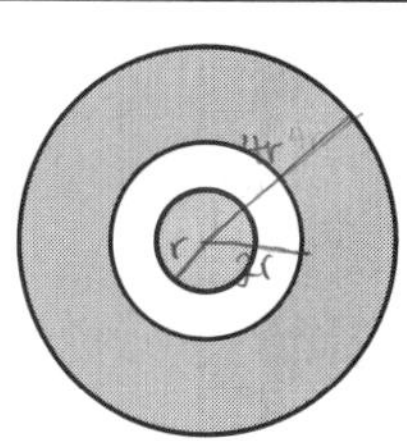

6. In the figure above, three circles share the same center. The radius of the middle circle is twice the radius of the smallest circle and the radius of the largest circle is twice the radius of the middle circle. What is the ratio of the area of the small shaded circle to the area of the large shaded ring?

(A) 1:4
(B) 1:12
(C) 3:5
(D) 3:8
(E) 5:12

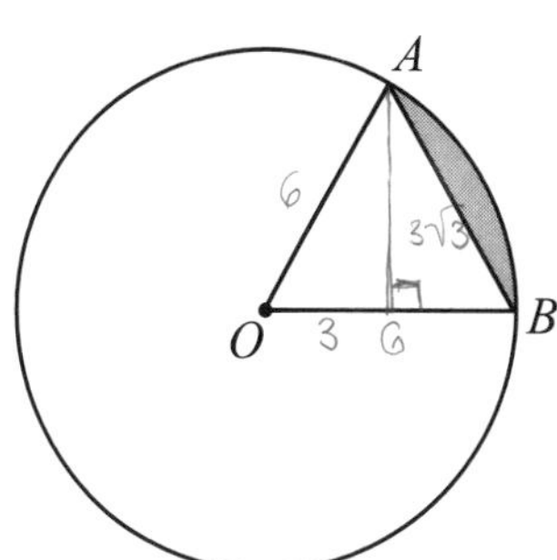

7. In the figure above, ABO is an equilateral triangle. If the straight line distance between AB is 6, what is the area of the shaded region?

(A) $6\pi + 3\sqrt{2}$
(B) $6\pi + 3\sqrt{3}$
(C) $6\pi - 9\sqrt{3}$
(D) $9\pi - 3\sqrt{2}$
(E) $9\pi - 9\sqrt{3}$

Geometric Solids

Frequency Guide: 3

A geometric solid is a three dimensional shape. The most commonly tested geometric solids are cubes, rectangular solids, and right circular cylinders. Other solids may appear, but their properties are not tested; they are usually just a foil for the properties of other shapes and solids.

Required Knowledge and Skill Set

1. Cubes and rectangular solids are six-sided "boxes." The formula for the volume of a rectangular solid is provided in the formula box.

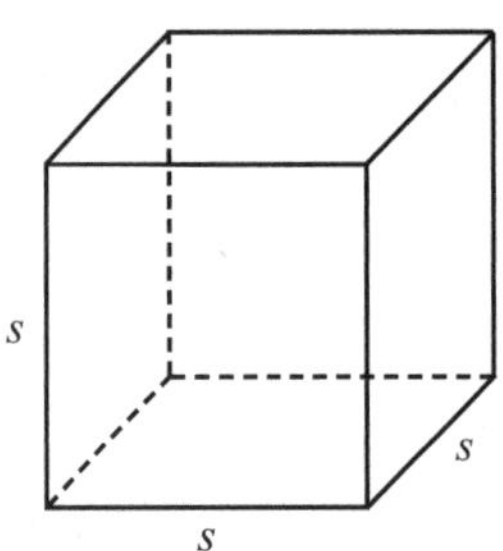

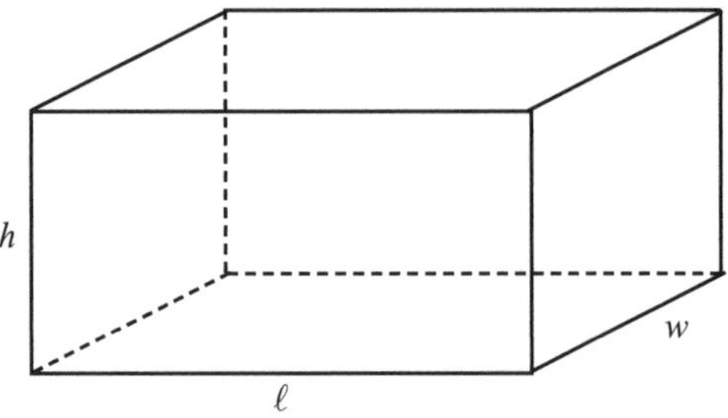

MEMORY MARKER: The formula for the volume of a cube and the formulas for surface area are not provided in the formula box.

Volume = s^3
Surface area = $6s^2$
(s = the length of a side)

Volume = ℓwh
Surface area = sum of areas of six faces
Surface area = $2\ell w + 2hw + 2h\ell$

2. The surface area of a cube or a rectangular solid is the sum of the areas of the six faces. This formula is not provided in the formula box.

3. When rectangular solids or cubes are stacked on top of each other, surface area pertains only to exposed portions of the solids. Unexposed portions of the figure must be subtracted from the total surface area of the individual components.

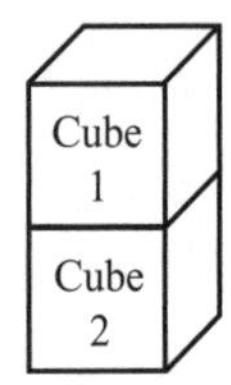

Unexposed Faces:
The bottom of Cube 1
and the top of Cube 2

If the side length of each cube is 3, the surface area of a freestanding cube would be 54:

Total surface area of one cube = $6s^2 \rightarrow 6(3^2) \rightarrow 6(9) \rightarrow 54$

However, the area of one side must be discounted:

Surface area of Cube 1 = $5s^2 \rightarrow 5(3^2) \rightarrow 5(9) \rightarrow 45$
Surface area of Cube 2 = $5s^2 \rightarrow 5(3^2) \rightarrow 5(9) \rightarrow 45$

The total surface area of the stacked figure is the sum of the surface areas of Cube 1 and Cube 2: $45 + 45 = 90$.

4. There are 8 vertices in a cube or rectangular solid. The farthest distance from a single vertex is its diagonal vertex:

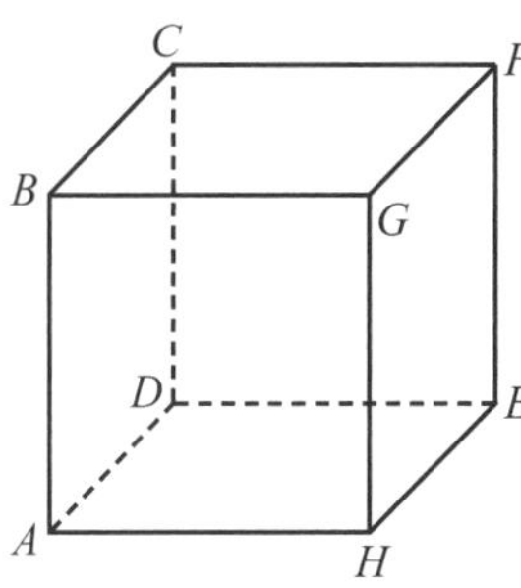

$AB = AD = AH$
$AC = AE = AG$
$AF > AB$
$AF > AC$

5. To find the length of a diagonal in a cube or rectangular solid, use the following formula:

Length of a diagonal in a rectangular solid = $\sqrt{l^2 + w^2 + h^2}$

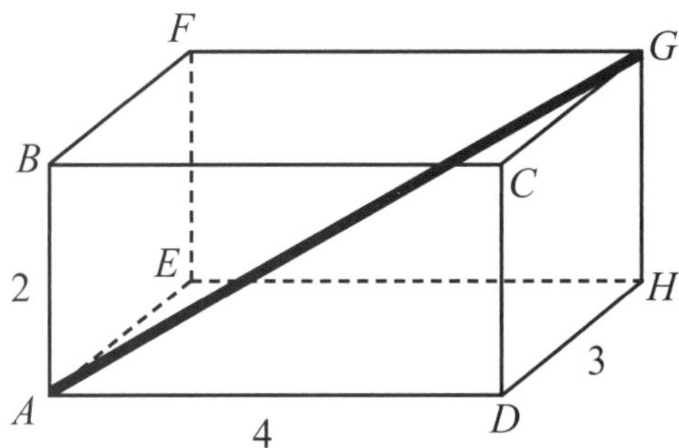

$$\begin{aligned} \text{Diagonal} &= \sqrt{l^2 + w^2 + h^2} \\ &= \sqrt{4^2 + 3^2 + 2^2} \\ &= \sqrt{16 + 9 + 4} \\ &= \sqrt{29} \end{aligned}$$

If you forget the formula, use hidden right triangles to determine the length of the diagonal:

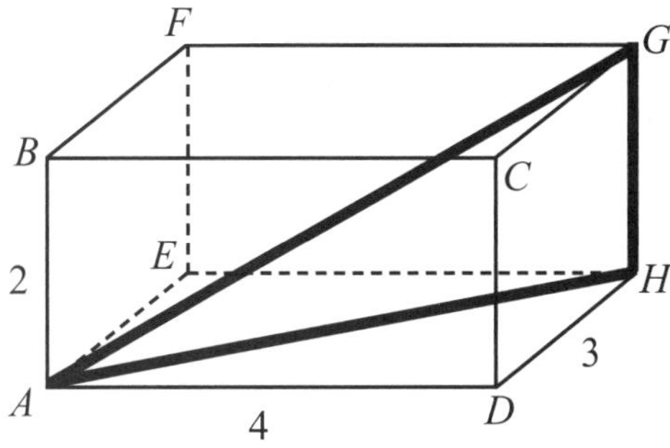

Length of AH:
$a^2 + b^2 = c^2$
$AD^2 + DH^2 = AH^2$
$3^2 + 4^2 = AH^2$
$9 + 16 = AH^2$
$25 = AH^2$
$5 = AH$

Length of AG:
$a^2 + b^2 = c^2$
$AH^2 + GH^2 = AG^2$
$5^2 + 2^2 = AG^2$
$25 + 4 = AG^2$
$29 = AG^2$
$\sqrt{29} = AG$

Memorizing the formula is more efficient, but hidden triangles can bail you out if you fail to recall the formula.

MEMORY MARKER:
The formula for the length of a diagonal in a rectangular solid should be memorized. Finding this distance using triangles can be a lengthy process.

Can you draw a rectangular solid and right circular cylinder? Questions about these two solids often are not accompanied by a figure, thus requiring you to DIAGRAM the figure. Practice drawing each one in the space below.

6. The SAT frequently tests right circular cylinders, which are geometric solids with equal circular bases joined by curved surfaces that form right angles at the circles. The formula for a volume of a right circular cylinder is provided in the Formula Box.

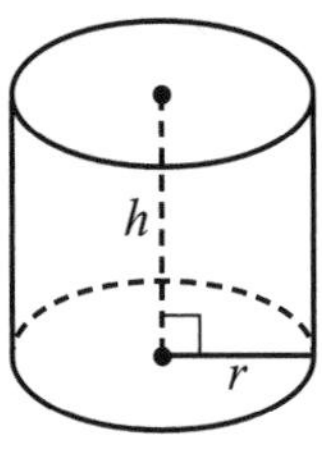

Volume = $\pi r^2 h$

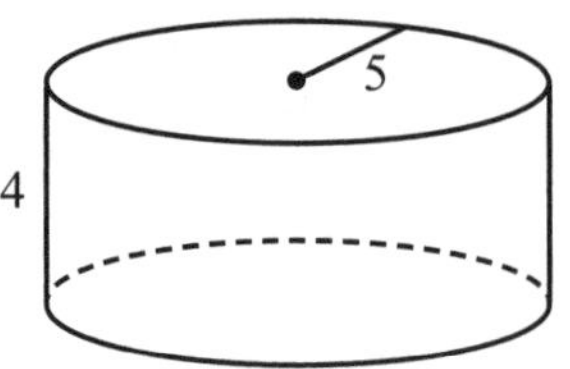

Volume = $\pi r^2 h$
Volume = $\pi(5^2)(4)$
Volume = 100π

7. Right circular cylinders are often used to hide right triangles::

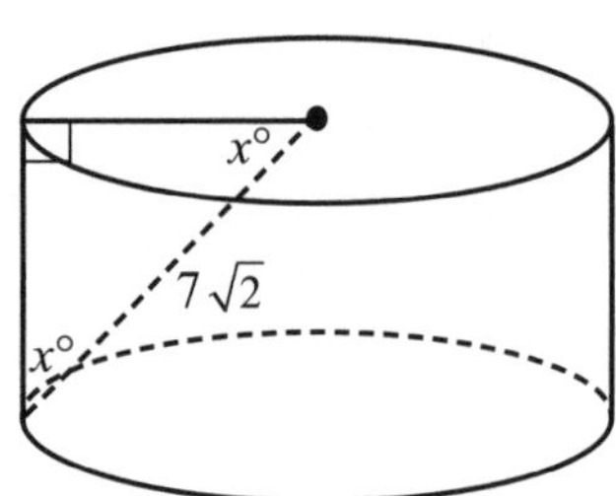

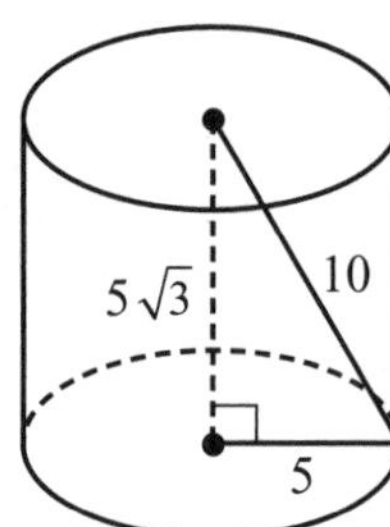

8. You may be asked to work with spheres, cones, or pyramids. You do not need to know any properties of these solids, but as always, you should be on the lookout for hidden triangles or other polygons.

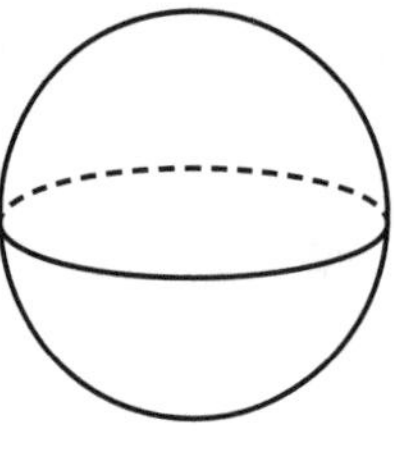

Sphere

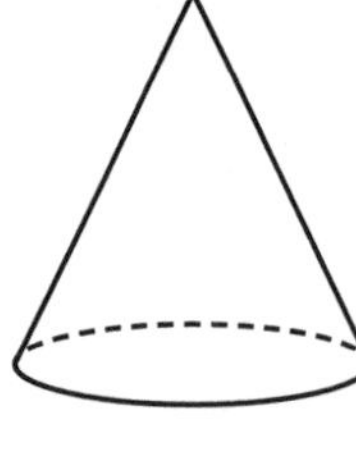

Cone

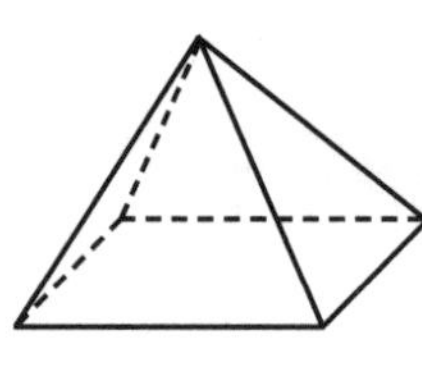

Pyramid

Note that the center of the sphere is a circle, the base of the cone is a circle, and the base of the pyramid is either a square or a rectangle. If these figures are used on the SAT, you will likely be asked to evaluate some part of the base using the formulas for these previously covered polygons.

Application on the SAT

The majority of geometric solid questions involve the volume of the figure. The easiest questions will simply ask for the volume of a cube, rectangular solid, or right circular cylinder:

6. A right circular cylinder has a height of 4 and a base with a radius of 3. What is the volume of the figure?

   (A) 12π
   (B) 24π
   (C) 36π
   (D) 48π
   (E) 144π

Remember to ANALYZE the answer choices! The only formulas that involve π are formulas for circles, sectors or arcs, and cylinders.

To solve, plug the values into the formula:

$$\text{Volume} = \pi r^2 h \quad \rightarrow \quad \pi(3^2)(4) \quad \rightarrow \quad \pi(9)(4) \quad \rightarrow \quad 36\pi$$

Volume questions involving cubes and rectangular solids are often worked "backwards," much like the circle questions that give the area and require the circumference. The volume of the solid is given, but the area of a single face is requested. Consider the following question:

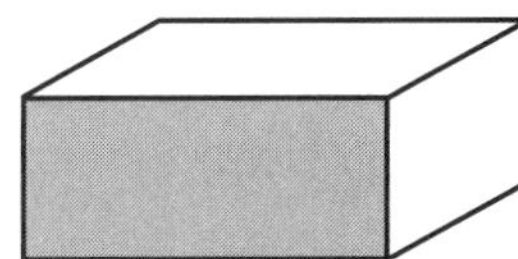

9. In the figure above, the rectangular solid has a depth of 5 centimeters and a volume of 150 cubic centimeters. What is the area, in square centimeters, of the shaded face?

   (A) 10
   (B) 15
   (C) 20
   (D) 30
   (E) 50

Confidence Quotation
"Skill and confidence are an unconquered army."
—George Herbert, Welsh poet and advisor to King James I

DIAGRAM the question:

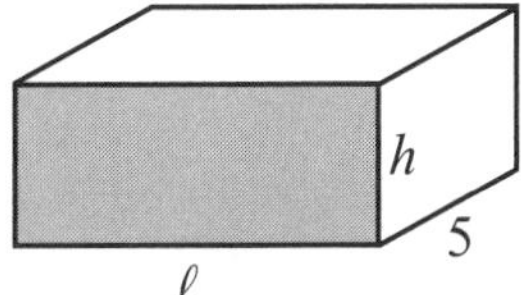

Volume = ℓwh
$150 = \ell(5)h$
$30 = \ell h$

The area of a rectangle is the length times the width. Because this rectangle is part of a rectangular solid, the area of the shaded face is the length times the height. Therefore, the area is 30. Answer choice (D) is correct.

Knowing that there are only two types of questions about right circular cylinders makes these questions easier to solve. If the text of the question does not use the word "volume," look for hidden triangles.

Questions about right circular cylinders fall into two categories: (1) questions involving volume and (2) questions involving triangles. The volume questions may be like the one we discussed on the previous page, where the volume must be determined, or they may give the volume and ask for the height of the cylinder or radius of the base. Right circular cylinder questions involving triangles usually have a Hard difficulty level, but they are easy if you remember to use the Pythagorean Theorem:

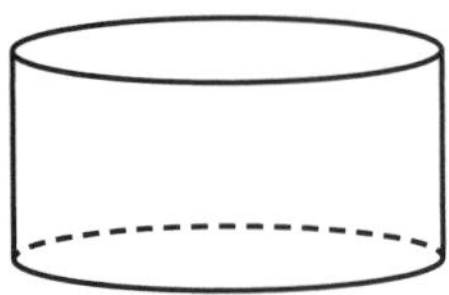

20. In the figure above, a cup used to hold drinking straws is a right circular cylinder with a height of 5 inches. The base of the cup has a radius of 6 inches. Of the following straw lengths, which is the longest one that can fit entirely in the can?

    (A) 11.2 inches
    (B) 12.8 inches
    (C) 13.4 inches
    (D) 14.6 inches
    (E) 15.0 inches

The longest straw that would fit in the cup would stretch from opposite points on the two bases. DIAGRAM the question to discover the triangle:

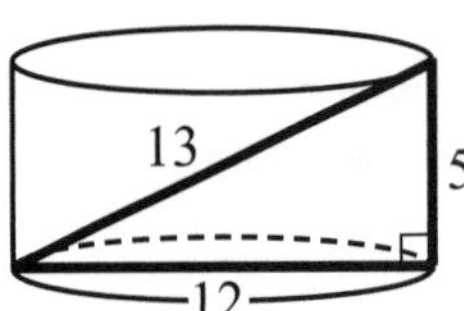

$a^2 + b^2 = c^2$
$5^2 + 12^2 = c^2$
$25 + 144 = c^2$
$169 = c^2$
$13 = c$

You should have immediately recognized the Pythagorean Triple, 5:12:13. If you did not, you can use the Pythagorean Theorem to find the distance between opposite points on the bases.

The longest straw that could fit in the cup is 13 inches. Therefore, the longest in the list of answer choices that would fit is answer choice (B), 12.8 inches. The straw in choice (C), 13.4 inches, is 0.4 inches too long.

If you encounter a question with a right circular cylinder, first assess whether it involves volume (these are usually early Medium questions). If the question does not use volume, find the hidden triangle (these are usually Hard questions).

Similarly, Easy and Medium questions about cubes and rectangular solids tend to deal with volume. The more difficult questions address surface area. Some of these questions ask you to find the surface area of a stacked figure, as demonstrated in the Required Knowledge and Skill Set. Others are word problems that involve surface area and volume:

20. In math class, students are given small plastic cubes. Each small plastic cube has a surface area of 24 square inches. If the teacher gives the students a large cube with a surface area of 864 square inches, how many small cubes can fit into the large cube?

(A) 18
(B) 36
(C) 72
(D) 144
(E) 216

Begin with the information you are given: surface area. Find the length of the sides of the cubes using the surface area formula.

Surface area of small cube = $6s^2$
$24 = 6s^2 \quad \rightarrow \quad 4 = s^2 \quad \rightarrow \quad 2 = s$

Surface area of large cube = $6s^2$
$864 = 6s^2 \quad \rightarrow \quad 144 = s^2 \quad \rightarrow \quad 12 = s$

Think of surface area as the amount of wrapping paper needed to cover a box.

Think of volume as the amount of water needed to fill an aquarium.

Now use the side length to find the volume of each sized cube:

Volume of small cube = $s^3 \quad \rightarrow \quad 2^3 \quad \rightarrow \quad 8$

Volume of large cube = $s^3 \quad \rightarrow \quad 12^3 \quad \rightarrow \quad 1728$

If the volume of the large cube is 1728, how many cubes of volume 8 can it hold? Imagine the large cube filled with small cubes:

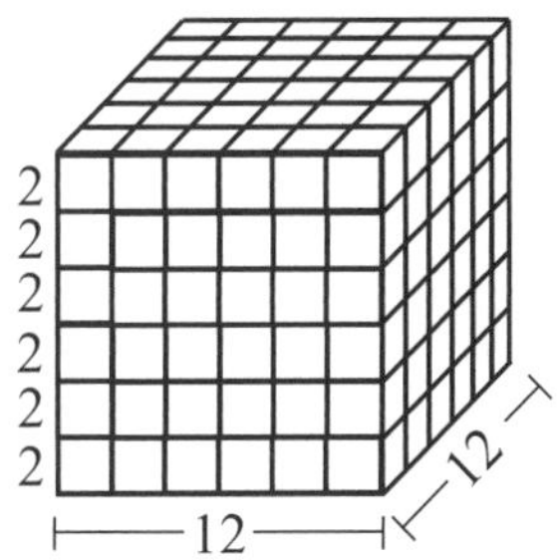

$1728 \div 8 = 216$

Choice (E) is the correct answer choice.

Geometric Solids Problem Set

Solve the following multiple-choice questions by selecting the best answer from the five answer choices. For grid-in questions, write your answer in the grids and completely mark the corresponding ovals. Answers begin on page 350.

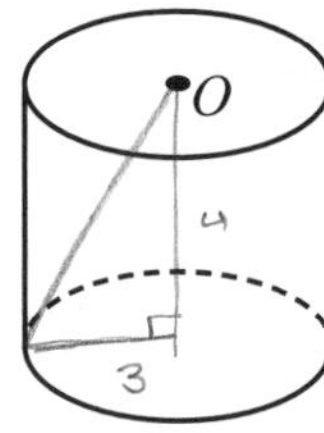

1. In the figure above, a cylinder has a height of 4 and a diameter of 6. If point O is the center of the top circular base, what is the distance from O to any point on the edge of the bottom base?

   (A) 2
   (B) $\sqrt{5}$
   (C) 5
   (D) $2\sqrt{13}$
   (E) 10

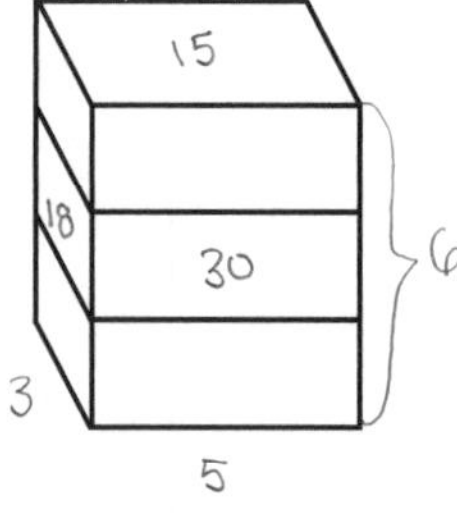

2. In the figure above, each of the three individual boxes has a depth of 3 inches, length of 5 inches, and height of 2 inches. When the three boxes are stacked together as shown, what is the surface area, in inches, of the resulting figure?

   (A) 60
   (B) 90
   (C) 96
   (D) 126
   (E) 186

3. The volume of a right circular cylinder is 200π. If the base of the cylinder has an circumference of 10π, what is the height of the cylinder?

   (A) 2
   (B) 4
   (C) 8
   (D) 12
   (E) 20

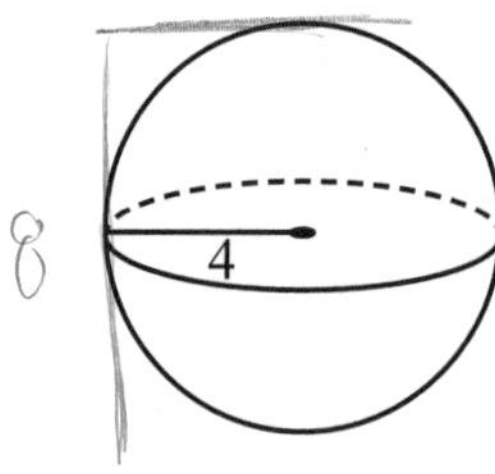

4. A sports equipment manufacturer created a spherical wooden ball that has a radius of 4 centimeters. The manufacturer wants to place the ball into the smallest box possible. If the box is a cube, what is the volume, in cubic centimeters, of the smallest box that will contain the entire ball?

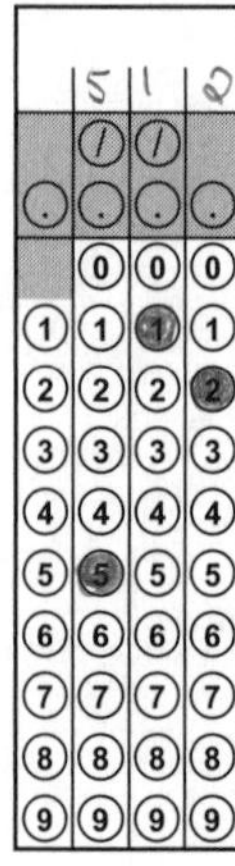

Geometric Perception

Frequency Guide: 2

Questions that require you to visualize a geometric situation or pattern based on provided information are categorized as geometric perception questions. Consider an example:

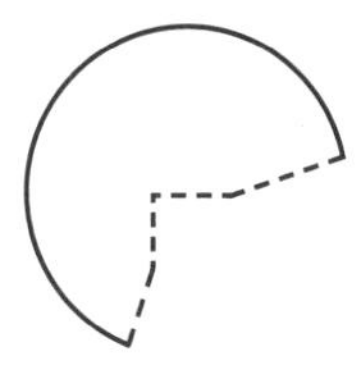

A circular disc is broken into two pieces, one of which is shown on the left. Which of the following could be the second piece of the disc?

(A)

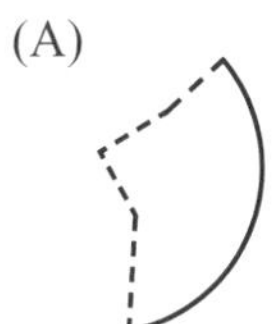

(B)

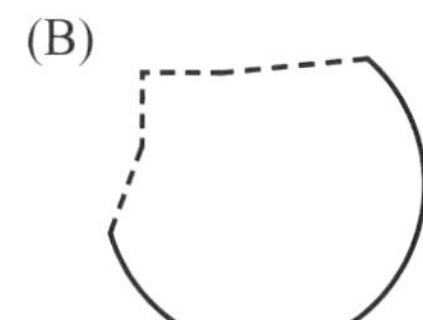

(C)

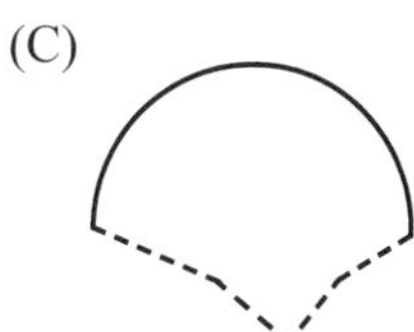

(D)

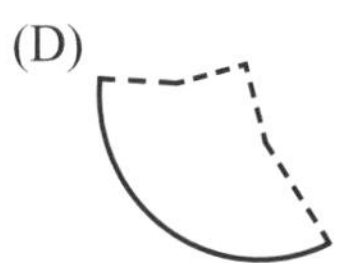

(E)

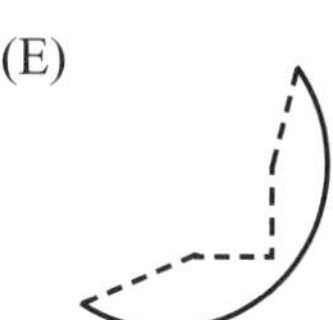

Do you know the answer? It's choice (D).

Required Knowledge and Skill Set

1. There are no properties or formulas to help you solve geometric perception questions.

2. If possible, DIAGRAM the question or draw a picture of the resulting figure.

3. Use items in the testing room as references. If the question involves a cylinder, use a pencil cup or trash can as a visual reference. Need a rectangular solid? Look for a tissue box or file cabinet. Using items as visual references can help you more easily judge distances and relationships.

4. If the question involves a pattern, draw or plot all stages of the pattern.

Confidence Quotation
"Some things have to be believed to be seen."
—Ralph Hodgson, English poet

Application on the SAT

There are many forms of geometric visualization problems, but two particular types appear frequently: "completed figures" and "rotated figures."

Consider an example of a completed figures question:

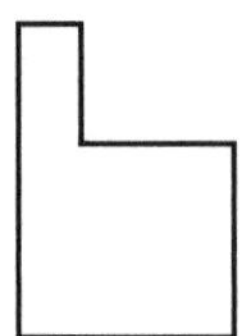

7. A square was cut into two pieces, one of which is shown above. Which of the following could be the other piece of the square?

(A) 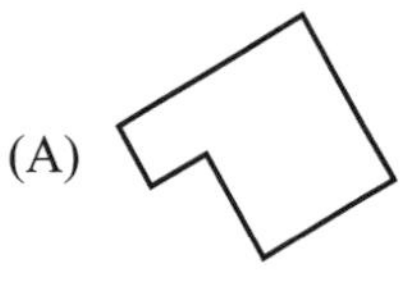

(D) 

(B) 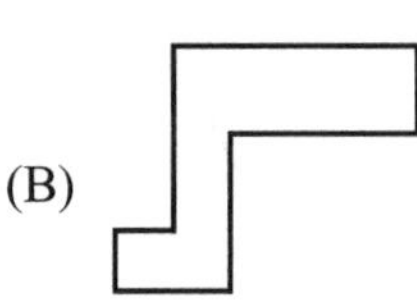

(E) 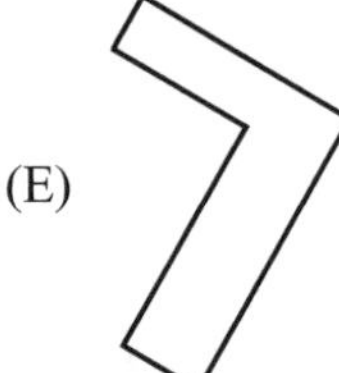

(C) 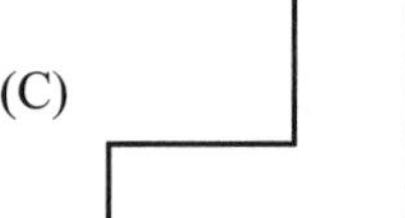

Some students can visualize these questions without DIAGRAMMING. If you can quickly solve these without drawing a completed picture, by all means do! But if you have a hard time visualizing the completed figure, spend a few seconds creating a sketch.

Imagine putting each of the answer choices with the figure at the top of the question:

(A) 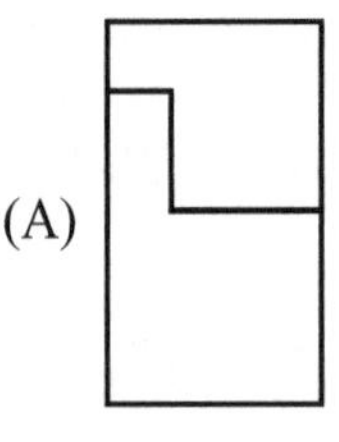

(D) 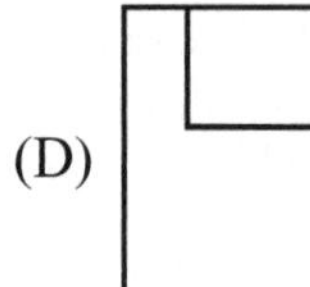

(B) 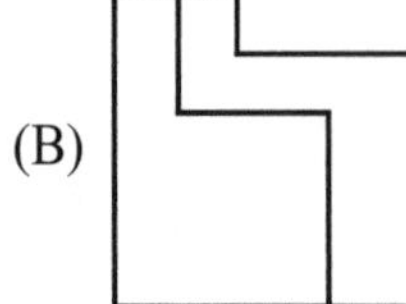

(E) 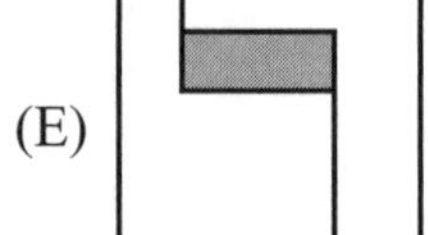

(C) 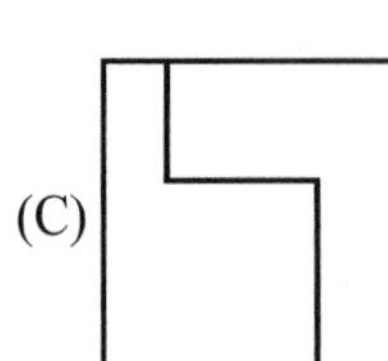

Only choice (C) makes a true square. Choice (E) may look like a square, but there is an area in the center of the figure that is left open.

The other type of geometric visualization question deals with the rotation of a figure:

9. The figure above is a regular hexagon divided into equal triangles. If the hexagon is rotated 240 degrees in a clockwise direction, which of the following is the result?

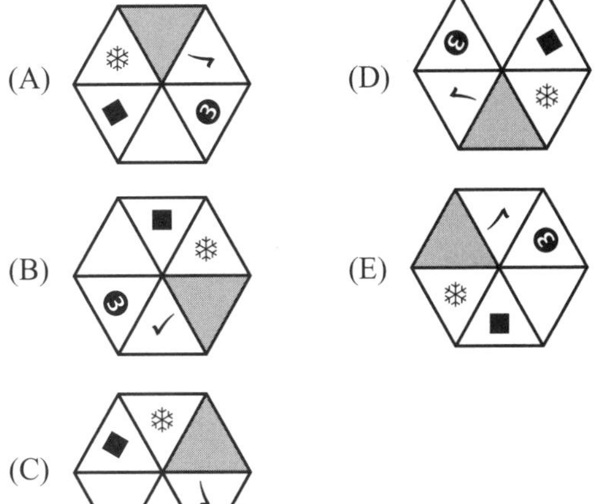

To redraw the entire hexagon would take a lot of time! Instead, pick one section of the hexagon, such as the shaded section. Then study each answer choice to see how many degrees the shaded portion has rotated.

Each time the hexagon rotates one side to the right, it moves 60°:

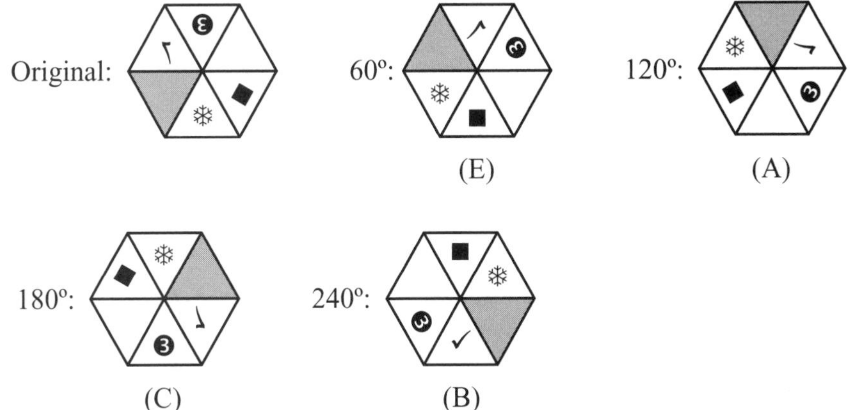

Choice (B) rotates the hexagon 240 degrees.

When dealing with patterns or rotation, plot each step to avoid confusion.

Geometric Perception Problem Set

Solve the following multiple-choice questions by selecting the best answer from the five answer choices. For grid-in questions, write your answer in the grids and completely mark the corresponding ovals. Answers are on page 352.

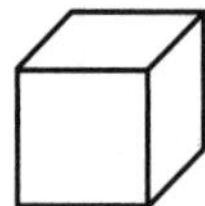

1. The figure above is a cube-shaped box. Which of the following patterns, when folded, could create the cube-shaped box?

(A) 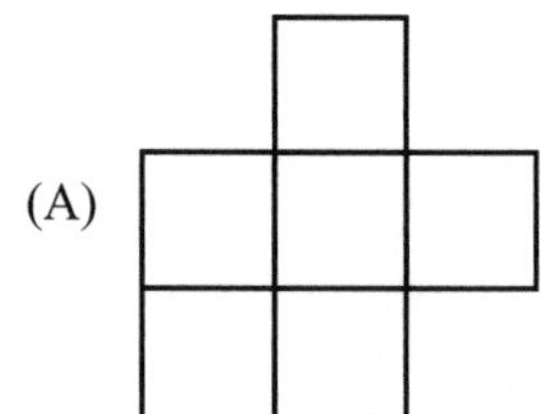

(B) 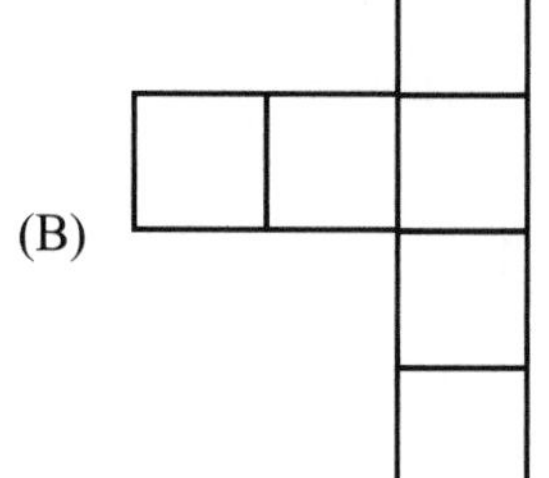

(C) 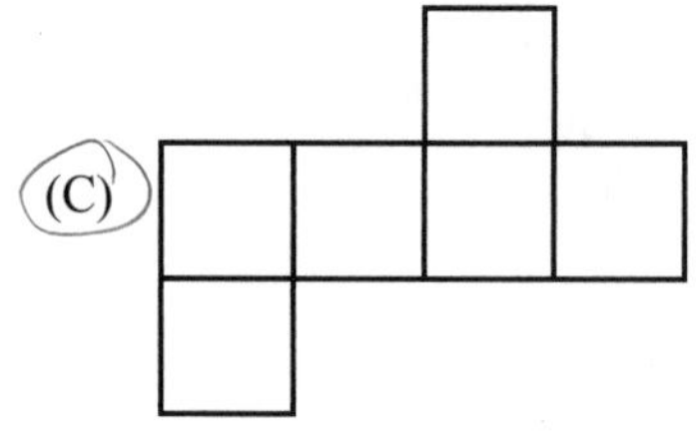

(D) 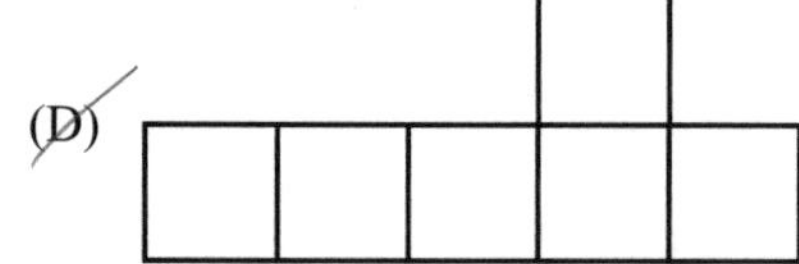

(E) 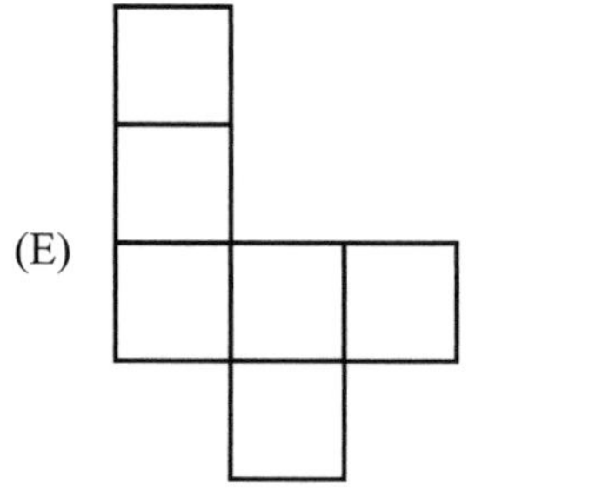

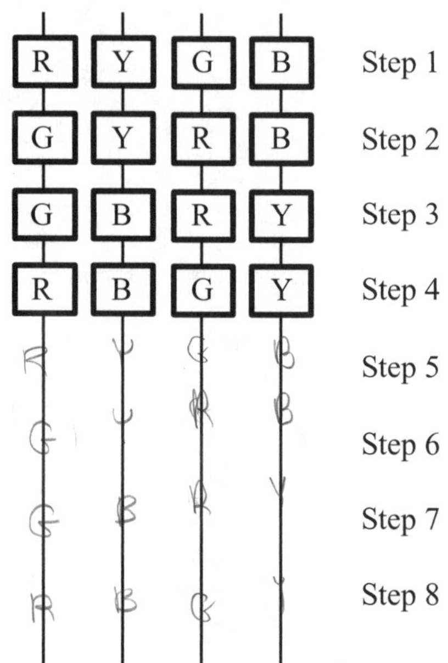

2. Maria is making a bracelet using four strands of beads and four different bead colors: red (R), yellow (Y), green (G), and blue (B). The bracelet has a particular pattern and the first four steps are show above. For example, Step 3 requires the colors green (G), blue (B), red (R), and yellow (Y) in that order. If Maria follows the pattern correctly, what will be the order of the beads in Step 8?

(A) 

(B) 

(C) 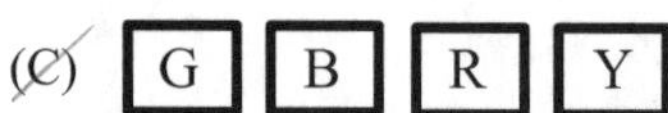

(D) 

(E) B G Y R

Notes

GEOMETRY MASTERY ANSWER KEY

Formula Box Exercise—Page 270

The area of a circle: πr^2

The circumference of a circle: $2\pi r$

The area of a rectangle: ℓw

The area of a triangle: $\frac{1}{2}bh$

The volume of a rectangular solid: ℓwh

The volume of a right circular cylinder: $\pi r^2 h$

The Pythagorean Theorem: $a^2 + b^2 = c^2$

In a 30:60:90 triangle, the side opposite the 30º angle is represented as x, the side opposite the 60º angle is represented as $x\sqrt{3}$, and the side opposite the 90º angle is represented as $2x$.

In a 45:45:90 triangle, the sides opposite the 45º angles are represented as s and the side opposite the 90º angle is represented as $s\sqrt{2}$.

The number of degrees of arc in a circle is 360º.

The sum of the angles of a triangle is 180º.

Lines and Angles Problem Set—Page 277

1. (B) Easy

DIAGRAM the question:

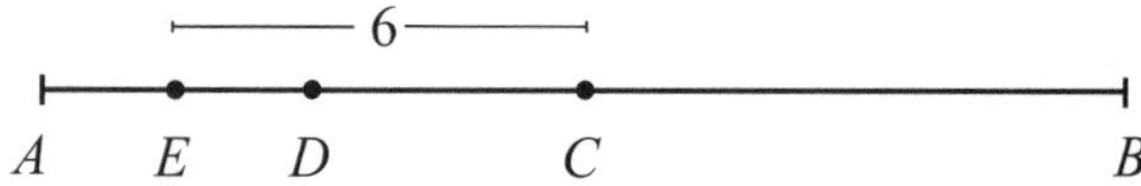

Using the information in the figure, write an equation to find the length of AD and DC:

$$ED = \frac{1}{2}DC$$

$$\frac{1}{2}DC + DC = 6 \rightarrow 1.5DC = 6 \rightarrow DC = 4$$

If $DC = 4$, then $AD = 4$ and $CB = 8$.

$$AD + DC + CB = AB \rightarrow 4 + 4 + 8 = 16$$

Answer choice (B) is correct.

2. (D) Medium

Separate the figure into two distinct diagrams:

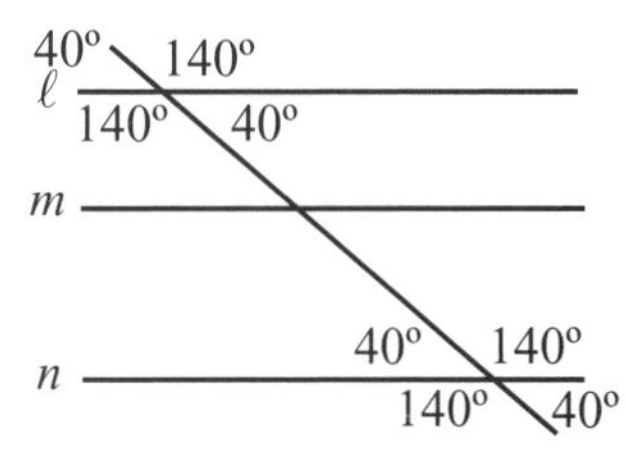

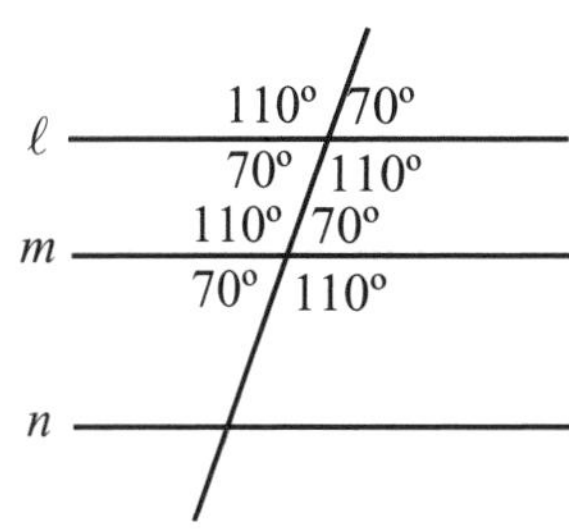

Using corresponding angles, you see that $a = 140$ and $b = 110$.

$a + b = 140 + 110 = 250$ Choice (D) is correct.

3. (C) Medium

SUPPLY numbers for each pair of angles in each answer choice, and then work through each answer choice to find the one with a set of angles that does not reveal all 6 angles:

(A) If $A = 80$ and $C = 40$ (Supplied numbers)
$B = 180 - 80 - 40 = 60$ (Supplementary Angles)
$D = A = 80$ (Vertical angles)
$E = B = 60$ (Vertical angles)
$F = C = 40$ (Supplementary and Vertical Angles)

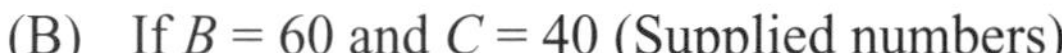

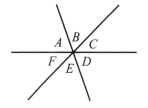

(B) If $B = 60$ and $C = 40$ (Supplied numbers)
$A = 180 - 60 - 40 = 80$ (Supplementary Angles)
$D = A = 80$ (Vertical angles)
$E = B = 60$ (Vertical angles)
$F = C = 40$ (Supplementary and Vertical Angles)

(C) If $B = 60$ and $E = 60$
It is impossible to determine A, C, D, or F

4. (A) Hard

DIAGRAM the question:

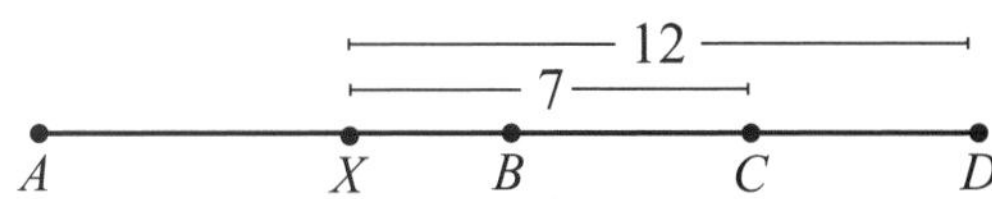

From this original diagram, you can see that $CD = 5$, because $7 + 5 = 12$, or the distance from X to D.

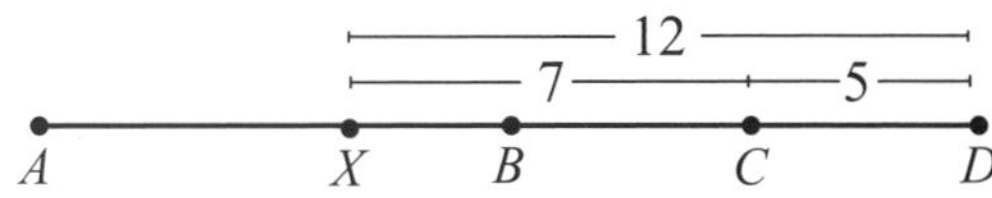

Now fill in the rest of the diagram:

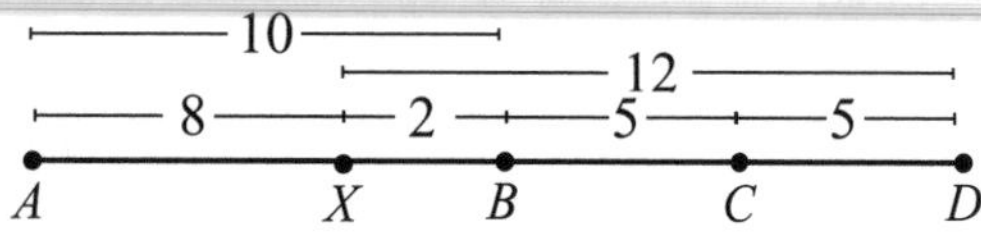

$BC = CD$, so $BC = 5$. Therefore, $BD = 10$. Since $AB = BD$, $AB = 10$. $BX = 2$, so $AX = 8$

The correct answer choice is (A).

Basic Triangles Problem Set—Pages 285-287

1. (E) Easy

Find each angle using supplementary anges, vertical angles, or the sum of the interior angles:

$v + 65 = 180$
$v = 115$

$w + 60 = 180$
$w = 120$

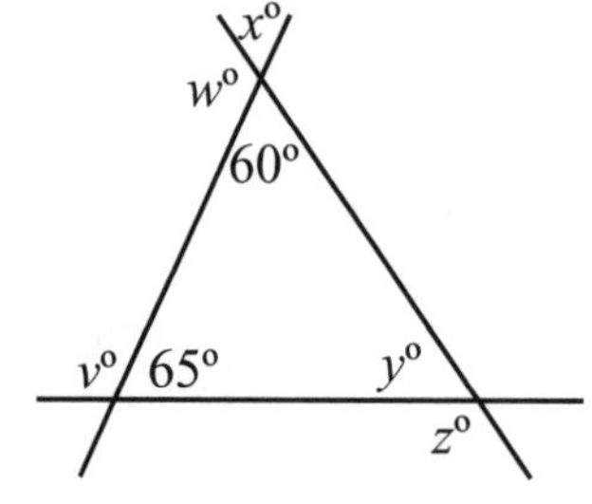

x is a vertical angle with 60, so $x = 60$

$y + 65 + 60 = 180$
$y + 125 = 180$
$y = 55$

$z + y = 180$
$z + 55 = 180$
$z = 125$

The largest angle is z. Answer choice (E) is correct.

2. 1 Medium

It is impossible to find the exact value of a, b, c, or d. However, you can find the sum of $a + b$ and $c + d$ because the sum of the interior angles of a triangle is 180:

$a + b + 60 = 180$
$a + b = 120$

$c + d + 60 = 180$
$c + d = 120$

$$\frac{a+b}{c+d} \rightarrow \frac{120}{120} = 1$$

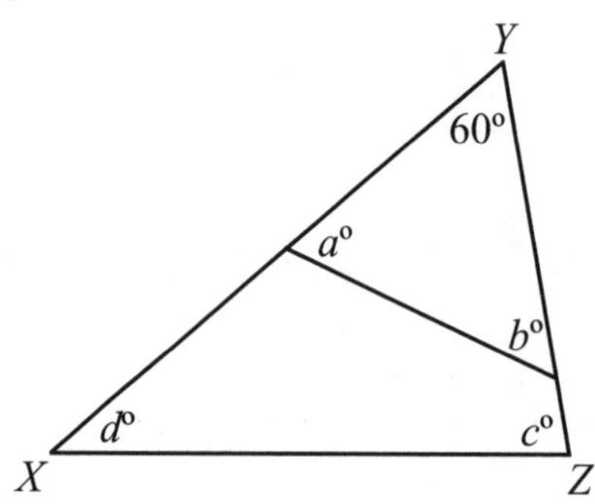

3. (D) Medium

If you know that one angle in a triangle is larger than another, you also know that the side opposite the larger angle is longer than the side opposite the shorter angle. However, you do not know anything about the third angle or the third side length.

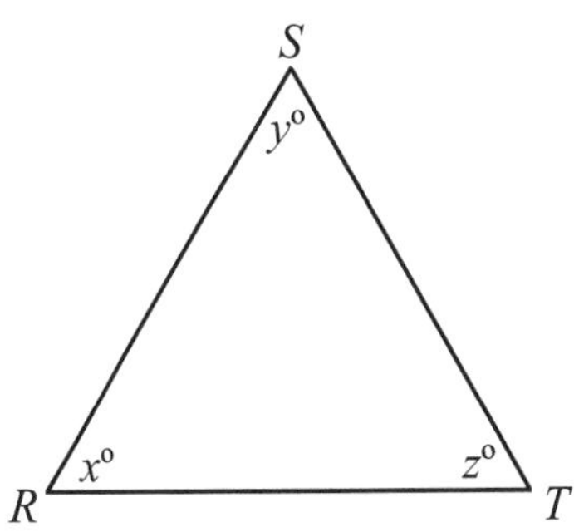

4. (C) Medium

Assign variables a, b, and c to the angles of the internal triangle for the convenience of discussion:

You know that $a + b + c = 180$, so SUPPLY numbers to these values. It does not matter if you assign the correct value; just be sure that they add up to 180:

$a = 80$
$b = 40$
$c = 60$

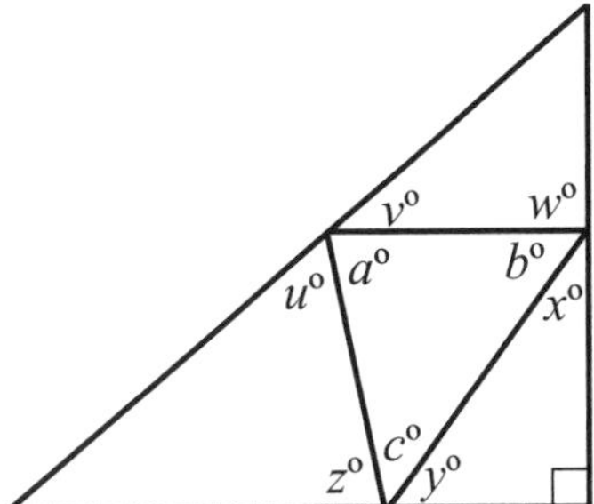

Therefore:

$u + a + v = 180$
$u + 80 + v = 180$
$u + v = 100$

$w + b + x = 180$
$w + 40 + x = 180$
$w + x = 140$

$y + c + z = 180$
$y + 60 + z = 180$
$y + z = 120$

Now find the sum of u, v, w, x, y, and z:

$$u + v + w + x + y + z \quad \rightarrow \quad (u + v) + (w + x) + (y + z) \quad \rightarrow \quad 100 + 140 + 120 \quad \rightarrow \quad 360$$

To prove that this works with any three numbers with a sum of 180, try it on your own when $a = 1$, $b = 19$, and $c = 160$.

5. (A) Medium

The triangles are similar triangles. We know this because the triangles have the same angle measurements. According to the base of the triangles (b and $4b$), the larger triangle side lengths are 4 times the size of the smaller triangle side lengths. TRANSLATE this sentence and solve for a:

$$c = 4a \quad \rightarrow \quad \frac{c}{4} = a \quad \rightarrow \quad \frac{1}{4}c = a$$

6. (C) Hard

This question is only Hard because students forget that the height of a triangle can be found outside the triangle. The base, BC, is 15. The height, DA, is 10.

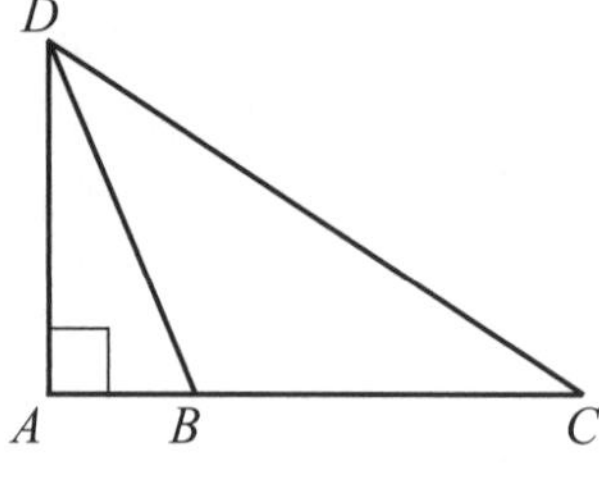

$$\text{Area} = \frac{1}{2}bh$$

$$\text{Area} = \frac{1}{2}(15)(10) \rightarrow (15)(5) \rightarrow 75$$

7. (A) Hard

Use the rules of supplementary angles and interior angles to solve this question. For the convenience of discussion, we are labeling other important angles with a, b, and c:

Start with a:

$90 + x + a = 180$ (the small bottom triangle)
$x + a = 90$
$a = 90 - x$

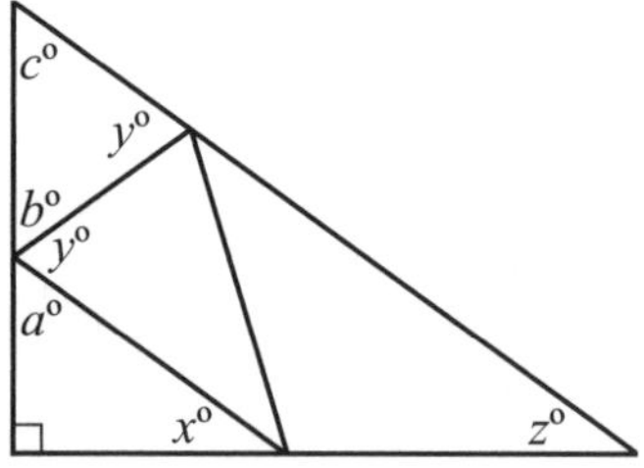

Now, use the expression you found for a to find b:

$a + y + b = 180$ (the vertical line or height)
$(90 - x) + y + b = 180$
$90 - x + y + b = 180$
$-x + y + b = 90$
$b = 90 + x - y$

Use the expression for b to find c:

$b + c + y = 180$ (the top small triangle)
$(90 + x - y) + c + y = 180$
$90 + x - y + c + y = 180$
$x + c = 90$
$c = 90 - x$

Now use the expression for c to find z:

$c + z + 90 = 180$ (the largest triangle)
$(90 - x) + z + 90 = 180$
$90 - x + z + 90 = 180$
$180 - x + z = 180$
$-x + z = 0$
$z = x$

Special Triangles Problem Set—Pages 298-299

1. 110, 112, or 114 Medium

SUPPLY a number for b to satisfy $32 < b < 36$.

If $b = 34$, then $34 + 34 + a = 180$

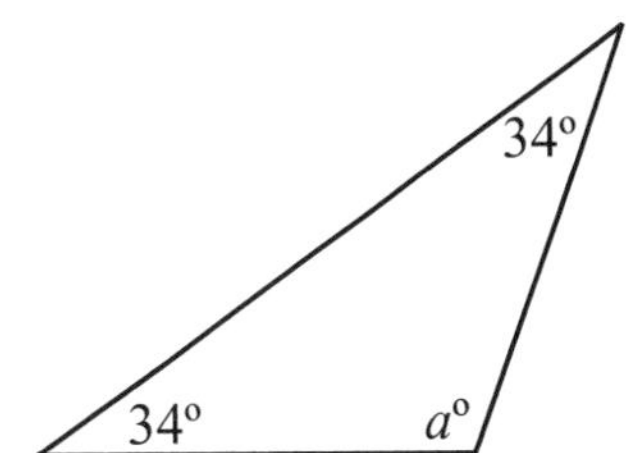

$34 + 34 + a = 180$
$68 + a = 180$
$a = 112$

If $b = 33$, $a = 114$
If $b = 34$, $a = 112$
If $b = 35$, $a = 110$

2. (A) Medium

Some students will recognize that the perimeter of all three triangles is simply 30×3. However, if you do not understand the reasoning behind this calculation, you can SUPPLY numbers:

The first triangle is an equilateral triangle (because it has three 60 degree angles). Since all thee triangles are similar, they are all equilateral triangles.

SUPPLY a base length for each triangle so that the bases add up to 30:

The perimeter of the largest triangle is $15 + 15 + 15 = 45$
The perimeter of the middle triangle is $10 + 10 + 10 = 30$
The perimeter of the smallest triangle is $5 + 5 + 5 = 15$

The total perimeter is $45 + 30 + 15 = 90$, or answer choice (A).

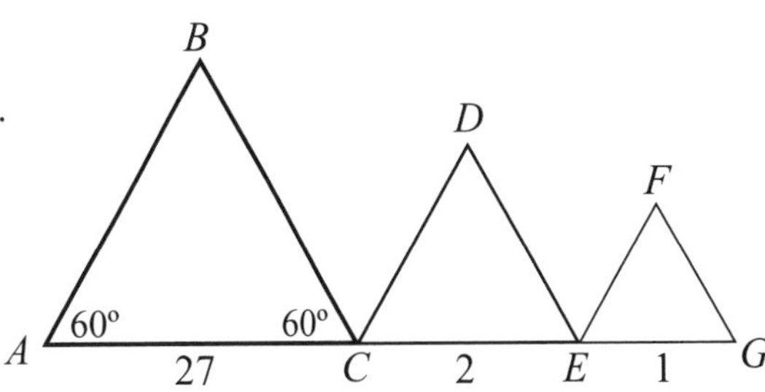

It does not matter what values you use to supply the base lengths, as long as their sum is 30:

The perimeter of the largest triangle is $27 + 27 + 27 = 81$
The perimeter of the middle triangle is $2 + 2 + 2 = 6$
The perimeter of the smallest triangle is $1 + 1 + 1 = 3$

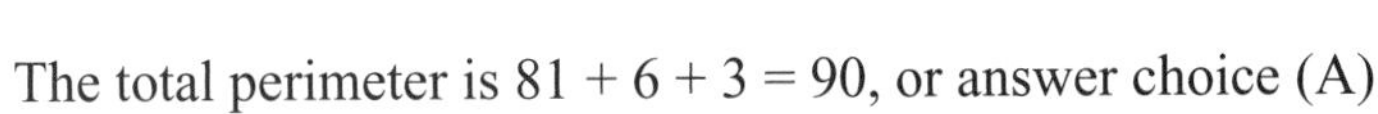

The total perimeter is $81 + 6 + 3 = 90$, or answer choice (A).

3. (B) Medium

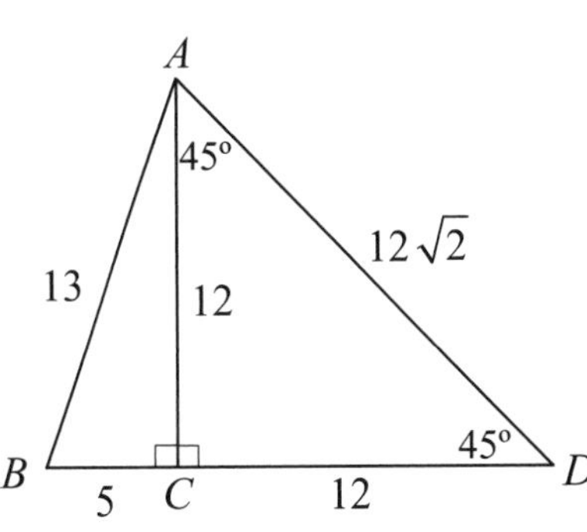

The length of AC (12) can be found by recognizing the Pythagorean Triple (5:12:13) or performing the Pythagorean Theorem.

Triangle ACD is a 45:45:90 because it is a right triangle in which the legs are the same length. The hypotenuse of a 45:45:90 triangle is the length of a leg multiplied by the square root of 2.

4. (B) Medium

Can you guess the most common wrong answer? It's (E). Students see $x^2 + 49$ and just assume that you square 12 to get 144. But this is wrong. The only way to solve this question is by performing the Pythagorean Theorem.

$$a^2 + b^2 = c^2$$
$$(x + 7)^2 + (x - 7)^2 = 12^2$$
$$(x + 7)(x + 7) + (x - 7)(x - 7) = 144$$
$$(x^2 + 7x + 7x + 49) + (x^2 - 7x - 7x + 49) = 144$$
$$x^2 + 14x + 49 + x^2 - 14x + 49 = 144$$
$$2x^2 + 98 = 144$$
$$2(x^2 + 49) = 2(72)$$
$$x^2 + 49 = 72$$

Be sure to stop at this step. Many students go on to find x, only to plug the value back into $x^2 + 49$. But as you can see, $x^2 + 49 = 72$.

5. (A) Medium

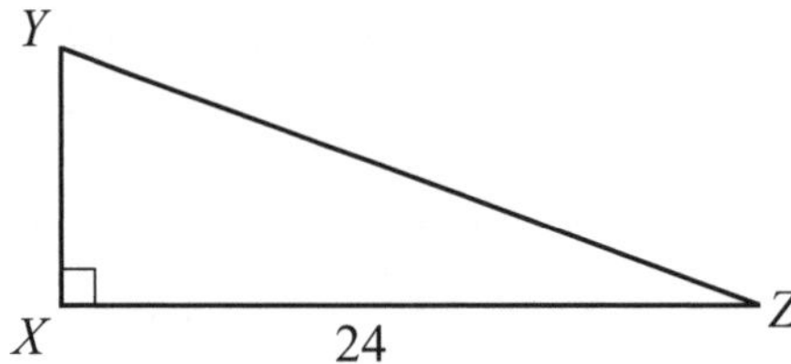

Use the area to find the height:

$$\text{Area} = 120$$
$$\text{Area} = \frac{1}{2}bh$$

$$120 = \frac{1}{2}24h \quad \rightarrow \quad 120 = (12)(h) \quad \rightarrow \quad 10 = h$$

The height is 10 and the base is 24. Do you recognize a Pythagorean Triple (10:24:26)? If not, use the Pythagorean Theorem to find YZ:

$$a^2 + b^2 = c^2$$
$$10^2 + 24^2 = c^2$$
$$100 + 576 = c^2$$
$$676 = c^2$$
$$26 = c$$

6. (C) Medium

The triangle is an equilateral triangle. Divide it into two 30:60:90 triangles to find the height.

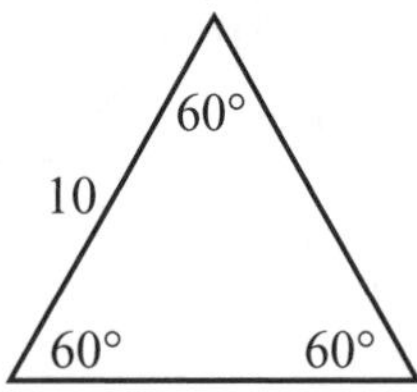

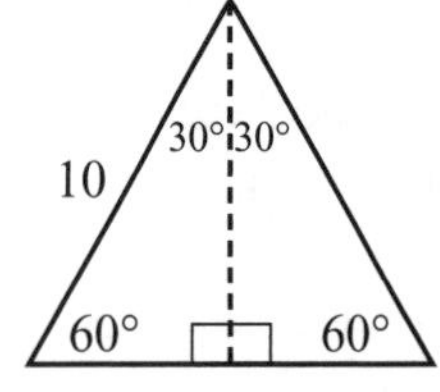

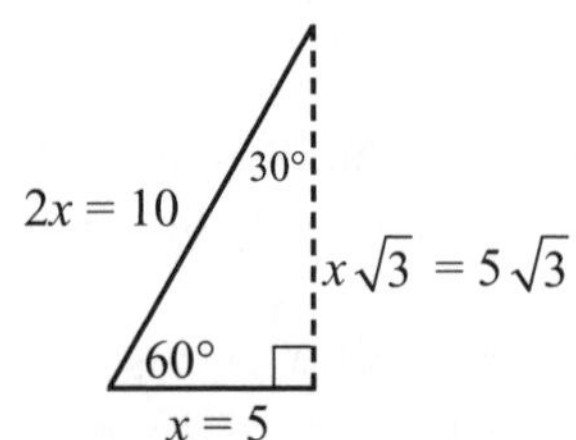

Now find the area:

$$\text{Area} = \frac{1}{2}bh$$

$$\text{Area} = \frac{1}{2}(10)(5\sqrt{3}) \rightarrow (5)(5\sqrt{3}) \rightarrow 25\sqrt{3}$$

7. (C) Medium

Begin with the bottom triangle, which is a 30:60:90 triangle. The side opposite the 30º angle is 5, so the hypotenuse is 10.

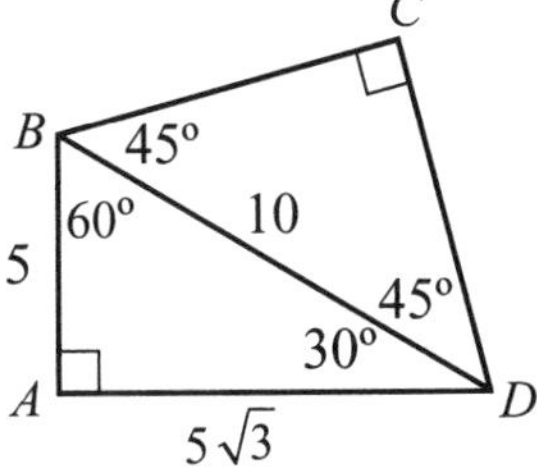

Now look at the top triangle. Since $BC = CD$, the triangle is a 45:45:90 triangle. This is where the question's difficulty plays in. Most students are used to the sides being an integer and the hypotenuse being that integer times the square root of two. However, the hypotenuse in this question is an integer. You must use your knowledge of radical equations to solve for the length of the side, s:

$$s\sqrt{2} = 10$$

$s = \dfrac{10}{\sqrt{2}}$ Since this is not an answer choice, remove the square root from the denominator:

$$(s)^2 = \left(\frac{10}{\sqrt{2}}\right)^2$$

$$s^2 = \frac{100}{2} = 50$$

$$s = \sqrt{50} = \sqrt{25} \times \sqrt{2} = 5\sqrt{2}$$

When the hypotenuse is an integer, the side is one half of that integer multiplied by the square root of 2.

Quadrilaterals Problem Set—Page 307

1. 22, 26, 34, 62 Medium

The area of a rectangle is the width times the length. So there are 4 possibilities given an area of 30:

$1 \times 30 = 30$
$2 \times 15 = 30$
$3 \times 10 = 30$
$5 \times 6 = 30$

The perimeter of a rectangle is 2 times the length plus 2 times the width. There are 4 possibilities:

$2(1) + 2(30) \rightarrow 2 + 60 = 62$
$2(2) + 2(15) \rightarrow 4 + 30 = 34$
$2(3) + 2(10) \rightarrow 6 + 20 = 26$
$2(5) + 2(6) \rightarrow 10 + 12 = 22$

2. (D) Medium

The sides of the squares must be 4 and 6:

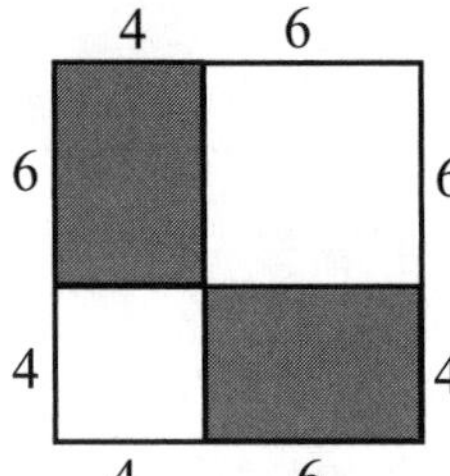

Perimeter of a square = $4s$
$16 = 4s$
$4 = s$

$24 = 4s$
$6 = s$

Now run through each answer choice to find the greatest value:

(A) The sum of the perimeters of the shaded rectangles:
$\text{Perimeter} = 2\ell + 2w \rightarrow 2(6) + 2(4) \rightarrow 12 + 8 \rightarrow 20$
Sum of two = 20 + 20 = 40

(B) The sum of the perimeters of the non-shaded squares:
$\text{Perimeter} = 4s \rightarrow 4(4) \rightarrow 16$
$\text{Perimeter} = 4s \rightarrow 4(6) \rightarrow 24$
Sum = 16 + 24 = 40

(C) The sum of the area of the shaded rectangles:
$\text{Area} = \ell w \rightarrow (6)(4) \rightarrow 24$
Sum of two = 24 + 24 = 48

(D) The sum of the area of the non-shaded squares:
$\text{Area} = s^2 \rightarrow 4^2 \rightarrow 16$
$\text{Area} = s^2 \rightarrow 6^2 \rightarrow 36$
Sum = 16 + 36 = 52

(E) The perimeter of the entire figure
$\text{Perimeter} = 4s \rightarrow 4(10) \rightarrow 40$

The greatest value is 52, so answer choice (D) is correct.

3. 7 Medium

TRANSLATE:

length is one-sixteenth of the perimeter

$$\ell = \frac{1}{16}\text{P} \rightarrow 16\ell = \text{P}$$

Now there are two things equal to perimeter. Set them equal to each other:

$\text{Perimeter} = 16\ell$
$\text{Perimeter} = 2\ell + 2w$
$16\ell = 2\ell + 2w$
$14\ell = 2w$
$7\ell = w$

4. (A) Hard

Diagram the question, and complete the rectangle:

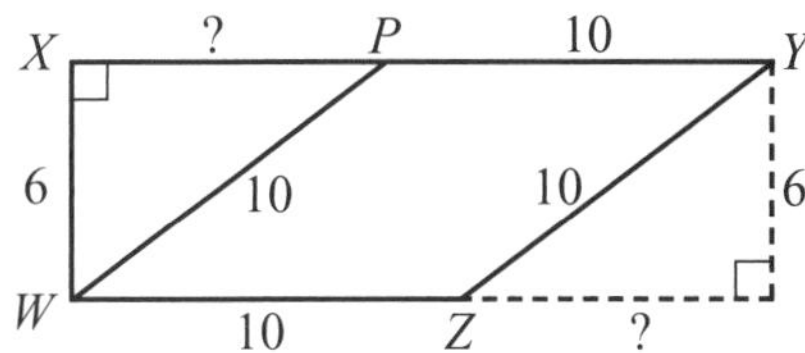

Use the Pythagorean Theorem to find the unknown leg of the right triangle (or recognize the Pythagorean Triple 6:8:10).

$a^2 + b^2 = c^2$
$6^2 + ?^2 = 10^2$
$36 + ?^2 = 100$
$?^2 = 64$
$? = 8$

The area of the original figure is the area of the rectangle minus the area of the added right triangle:

Area of the rectangle = $\ell w \rightarrow (18)(6) \rightarrow 108$
Area of triangle = $\frac{1}{2}bh \rightarrow \frac{1}{2}(8)(6) \rightarrow (4)(6) \rightarrow 24$

Area of original figure = $108 - 24 = 84$

Polygons Problem Set—Page 313

1. (D) Medium

Review each answer choice to confirm or eliminate:

(A) They are isosceles. False
While the two outer triangles are isosceles, the two inner triangles are not.
(B) The have equal perimeters. False
The two outer triangles have equal perimeters and the two inner triangles have equal perimeters, but the inner triangles have a larger perimeter than the outer triangles.
(C) They are similar. Again, the two outer triangles are similar and the two inner triangles are similar, but the inner triangles are not similar to the outer triangles.
(D) They each have at least one 30° angle.
To prove or disprove this answer choice, you must DIAGRAM the question. Because that might take some time, check out (E) first. If you can quickly eliminate it, then (D) is the answer.
(E) They have equal areas. False
The base of the outer triangles is the same as the height of the inner triangles. However, the height of the outer triangles is shorter than the base of the inner triangles, so this is not true.

The correct answer, then, is (D). To prove this, examine the diagram:

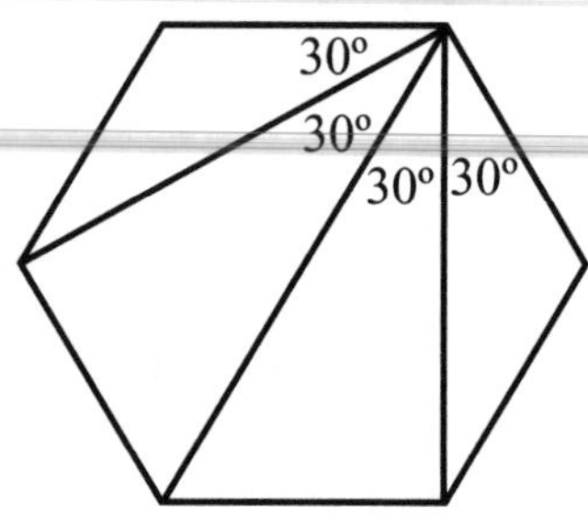

The sum of the interior angles is 720°. There are 6 angles:

$720° \div 6 = 120°$

As you can see, one interior angle is divided into 4 equal parts:

$120° \div 4 = 30°$

Therefore, all four triangles have a 30° angle.

2. (B)

DIAGRAM the question:

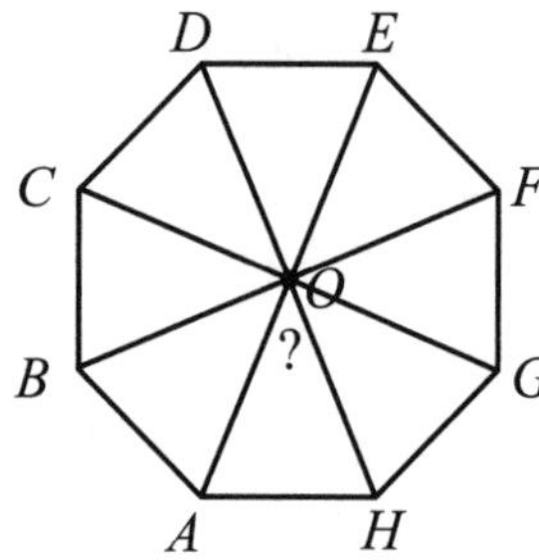

There are 360° in a circle surrounding the center of the polygon. There are 8 equal angles:

$360° \div 8 = 45°$

Circles Problem Set—Pages 322-323

1. (E) Medium

RECORD what you know:

Circumference = $2\pi r$
Circumference = 36π

Solve for the radius by setting the values equal:

$2\pi r = 36\pi \quad \rightarrow \quad 2r = 36 \quad \rightarrow \quad r = 18$

Knowing that the radius is 18, find the area:

Area = $\pi r^2 \quad \rightarrow \quad \pi(18)^2 \quad \rightarrow \quad 324\pi$

The correct answer is (E).

2. (D) Medium

DIAGRAM the question:

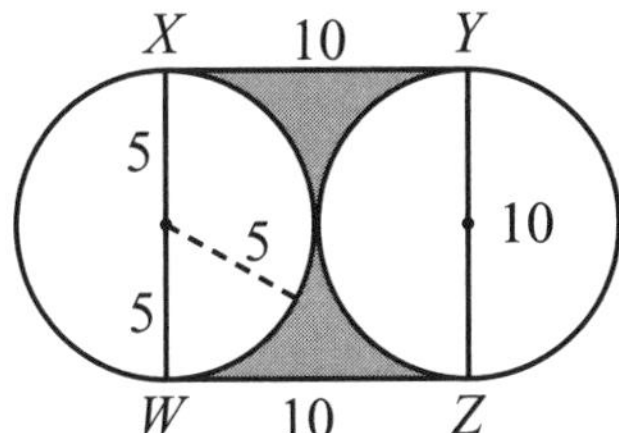

Area of a square = s^2
Area = 10^2
Area = 100

Area of a circle = πr^2
Area = $\pi 5^2$
Area = 25π

Write a statement that solves for the shaded area:

Area of the square – area of half of a circle – area of half a circle = shaded area

Two halves of a circle equal one full circle:

Area of the square – area of one circle = shaded area

Now substitute the values in place of the words:

Area of the square – area of one circle = shaded area
$100 - 25\pi$ = shaded area

3. (C) Medium

DIAGRAM the question:

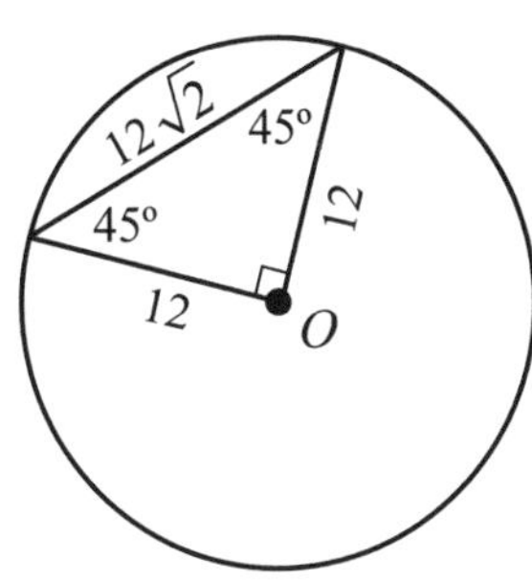

Because the triangle has one vertex on the center and the other two vertices on the circle, it is in isosceles triangle. If the angle at the center is 90º, then the other two angles must add up to 90º. Since they must have the same angle measurement, they are each 45°. The triangle is not only isosceles, but a 45º:45º:90º triangle. If the hypotenuse is 12 $\sqrt{2}$, then the legs are 12.

The legs of the triangle are the same length of the radius. The radius is 12:

Circumference = $2\pi r$ → $2\pi(12)$ → 24π

The correct answer is (C).

4. (B) Hard

Remember, the distance of one revolution is the circumference. Find the circumference:

$$\text{Circumference} = 2\pi r \quad \rightarrow \quad 2\pi(\frac{1}{\pi}) \quad \rightarrow \quad 2$$

When the wheel makes one revolution, it has rolled 2 units. When it makes two revolutions, it has rolled 4 units. When it makes three revolutions, it has rolled 6 units. Choice (B) is correct.

5. (A) Hard

The length of the rectangle is 4 radii and the width is 2 radii:

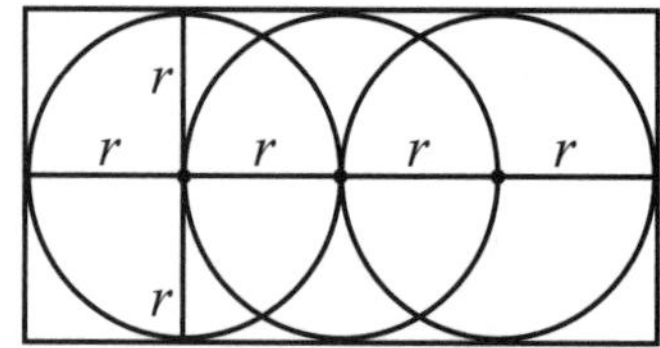

You must find the radius of a circle to find the length and width. Since the question gives the circumference formula, use it to find the radius:

$$\text{Circumference} = 2\pi r$$
$$\text{Circumference} = \frac{\pi}{3}$$

$$2\pi r = \frac{\pi}{3} \quad \rightarrow \quad (3)2\pi r = \pi \quad \rightarrow \quad 6r = 1 \quad \rightarrow \quad r = \frac{1}{6}$$

Now find the length and width:

$$\text{Length} = 4r \quad \rightarrow \quad (4)\frac{1}{6} \quad \rightarrow \quad \frac{4}{6} \quad \rightarrow \quad \frac{2}{3}$$

$$\text{Width} = 2r \quad \rightarrow \quad (2)\frac{1}{6} \quad \rightarrow \quad \frac{2}{6} \quad \rightarrow \quad \frac{1}{3}$$

Now find the area of the rectangle:

$$\text{Area} = \ell w \quad \rightarrow \quad \frac{2}{3} \times \frac{1}{3} \quad \rightarrow \quad \frac{2}{9}$$

The correct answer is (A).

6. (B) Hard

SUPPLY numbers for the radii and DIAGRAM the question:

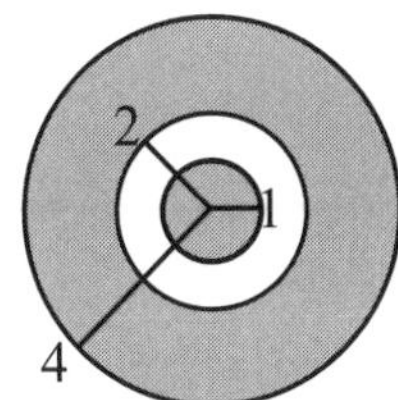

You can use any values for the radii as long as you follow the rules: the radius of the middle circle is twice as large as the radius of the smallest circle and the radius of the largest circle is twice as large as the radius of the middle circle.

Now, find the area of the small shaded circle:

Area = πr^2 → $\pi(1)^2$ → 1π

To find the area of the large shaded ring, write a word equation that solves for the shaded area:

Area of the largest circle – area of the middle circle = area of shaded ring

Find the area of the largest circle and the area of the middle circle:

Area of largest circle = πr^2 → $\pi(4)^2$ → 16π
Area of the middle circle = πr^2 → $\pi(2)^2$ → 4π

Now substitute the values for words:

Area of the largest circle – area of the middle circle = area of shaded ring
$16\pi - 4\pi = 12\pi$

Compare the area of the small shaded circle to the area of the large shaded ring:

Small: 1π
Large: 12π

$1\pi : 12\pi$ → 1:12

The correct answer is (B).

7. (C) Hard

Begin by DIAGRAMMING the equilateral triangle:

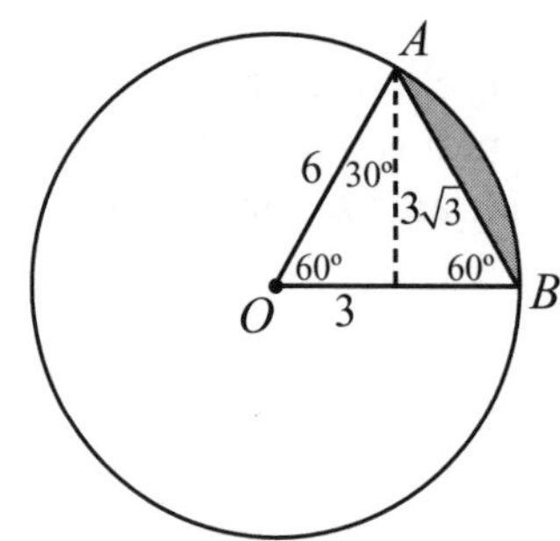

Every equilateral triangle has two hidden 30°:60°:90° triangles. You can use one of these 30°:60°:90° triangles to find the height of the equilateral triangle.

Write a word equation to solve for the shaded portion:

Area of sector *ABO* – area of the equilateral triangle = area of the shaded region

Find the area of the sector:

The area of a sector $= \frac{x^\circ}{360^\circ}(\pi r^2) \rightarrow \frac{60^\circ}{360^\circ}(\pi 6^2) \rightarrow \frac{1}{6}(36\pi) \rightarrow 6\pi$

Find the area of the equilateral triangle:

Area $= \frac{1}{2}bh \rightarrow \frac{1}{2}(6)(3\sqrt{3}) \rightarrow (3)(3\sqrt{3}) \rightarrow 9\sqrt{3}$

Now change the word equation:

Area of sector *ABO* – area of the equilateral triangle = area of the shaded region
$6\pi - 9\sqrt{3}$ = area of the shaded region

The correct answer is (C).

Geometric Solids Problem Set—Page 330

1. (C) Hard

DIAGRAM the question to see the right triangle:

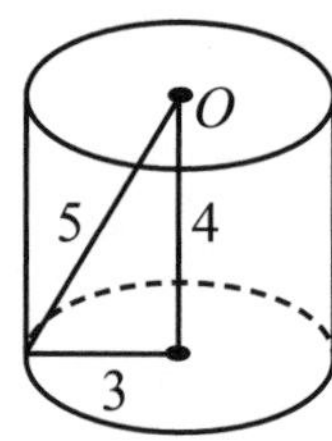

You should immediately recognize the Pythagorean Triple, 3:4:5. If not, use the Pythagorean Theorem:

$a^2 + b^2 = c^2$
$3^2 + 4^2 = c^2$
$9 + 16 = c^2$
$25 = c^2$
$5 = c$

2. (D) Hard

Find the surface area of the three individual boxes as if they were floating in space and not stacked together:

Surface area = $2\ell w + 2\ell h + 2wh$
Surface area = $2(5)(3) + 2(5)(2) + 2(3)(2)$
Surface area = $30 + 20 + 12 = 62$

Surface area of 3 boxes = $62 \times 3 = 186$

Now identify the unexposed faces:

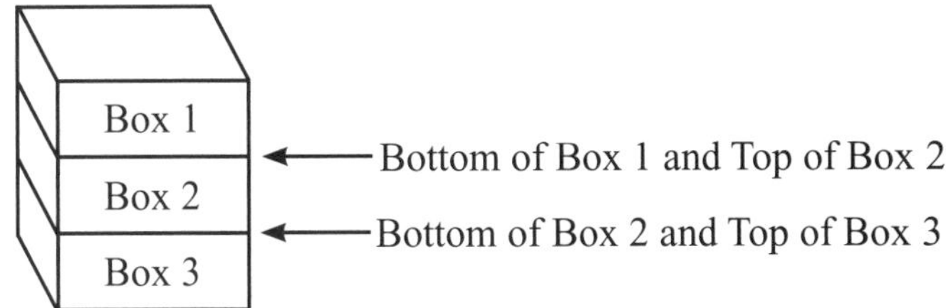

There are 4 unexposed faces, and they are all have the same area (length × width):

Area of unexposed face = $\ell w = (5)(3) = 15$

Area of 4 unexposed faces = $4(15) = 60$

Now subtract the area of the unexposed faces from the surface area of three floating boxes:

$186 - 60 = 126$ The correct answer is (D).

3. (C) Hard

This is a classic "work backwards" SAT question. Use the circumference to find the radius:

Circumference = $2\pi r$
Circumference = 10π

$2\pi r = 10\pi \rightarrow 2r = 10 \rightarrow r = 5$

The only formula that uses the height of a cylinder is the volume formula. Use the radius and the volume to find the height:

Volume = $\pi r^2 h$
Volume = 200π Radius = 5

$\pi(5^2)h = 200\pi \rightarrow 25h = 200 \rightarrow h = 8$

The correct answer is (C).

4. 512 Hard

DIAGRAM the question:

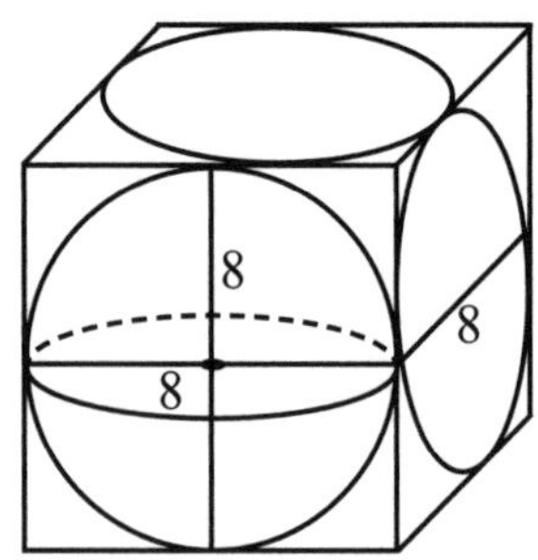

The length, width, and height of the cube are all 8 centimeters.

Volume = s^3 → 8^3 → 512 cubic centimeters

Geometric Perception Problem Set—Page 334

1. (C) Medium

The correct answer is (C). The four connect squares create the four sides of the box, and the two "flaps" create the top and bottom of the box.

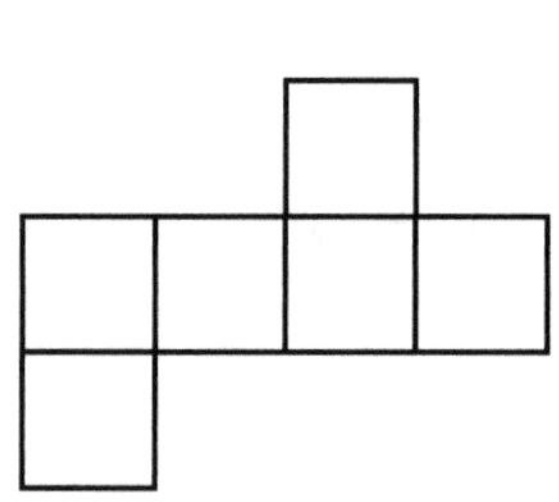

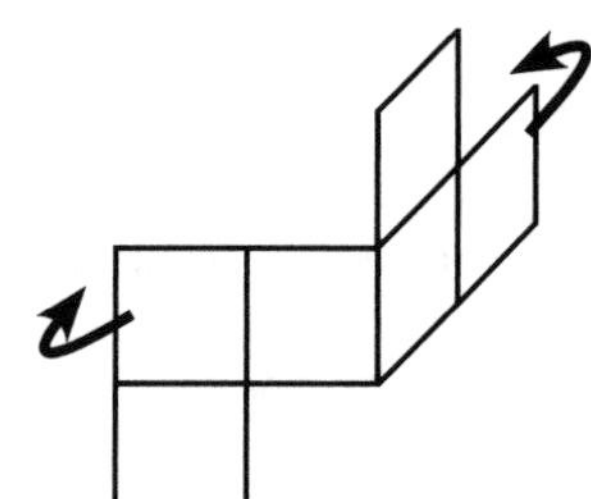

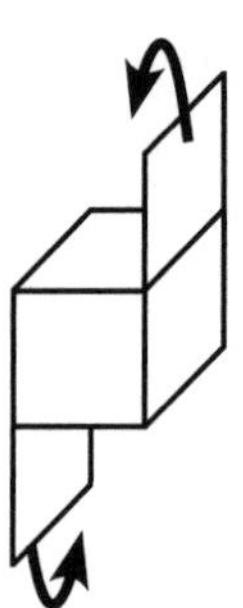

2. (A) Medium

Complete each step of the pattern:

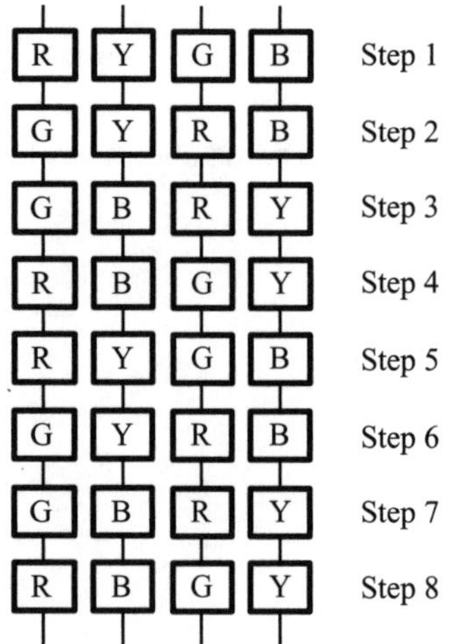

Notes

CHAPTER NINE:
COORDINATE GEOMETRY MASTERY

Coordinate geometry questions comprise anywhere from 5% to 10% of the total number of math questions on the SAT. These questions involve the coordinate plane, a two-dimensional plane with two perpendicular axes. You must understand how to plot and measure lines, as well as integrate regular geometry and Algebra II, onto the coordinate plane.

Coordinate geometry is a branch of mathematics in which algebra is used to study geometry on a coordinate plane.

At the beginning of each content area, in the notes column near the edge of the page, is a frequency guide where the content is assigned a number. This number indicates the likelihood that this content will be tested on your SAT. Based on PowerScore's extensive analysis of real tests, you can use the following key to predict the general frequency of each question type:

Rating	Frequency
5	Extremely High: 3 or more questions typically appear on every SAT
4	High: at least 2 questions typically appear on every SAT
3	Moderate: at least one question typically appears on every SAT
2	Low: one question typically appears on every two or three SATs
1	Extremely Low: one question appears infrequently and without a pattern

Remember that there are eight solution strategies you can employ on SAT math questions:

1. ANALYZE the Answer Choices
2. BACKPLUG the Answer Choices
3. SUPPLY Numbers
4. TRANSLATE from English to Math
5. RECORD What You Know
6. SPLIT the Question into Parts
7. DIAGRAM the Question
8. SIZE UP the Figures

It is important to review any of the questions that you miss after each content section. Answers and full explanations appear that the end of the chapter, and should be used to compare your work. The PowerScore solution may be more efficient than the method you have chosen, and should be studied in the event you are faced with similar questions in the future.

Coordinate Geometry Review

The SAT requires a basic knowledge of coordinate geometry:

Confidence Quotation

"Perpetual optimism is a force multiplier."
—Colin Powell, four-star army general and Chairman of the Joint Chiefs of Staff

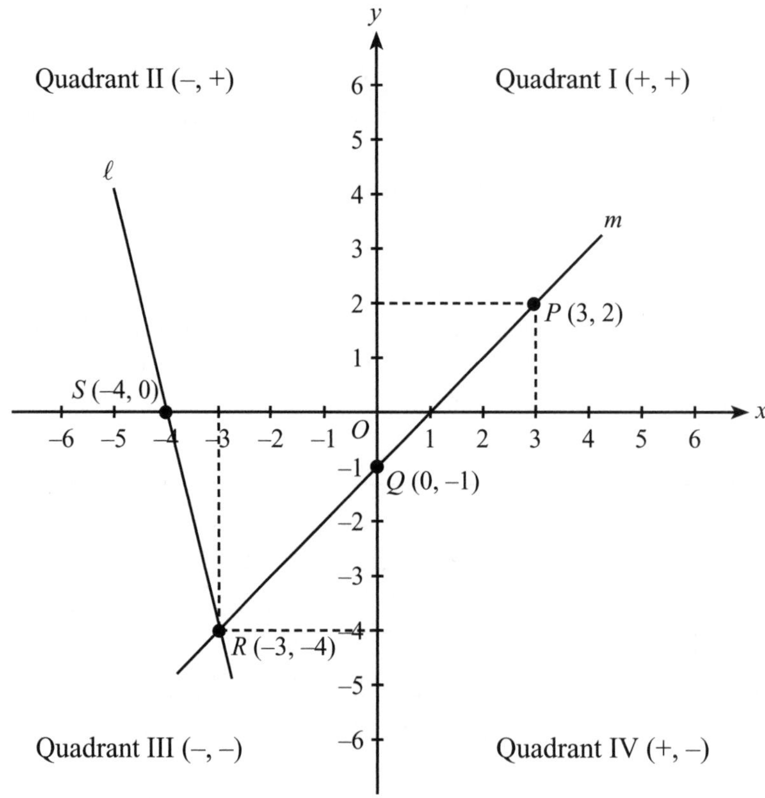

These basics of coordinate geometry are the foundation of all SAT questions involving the coordinate plane.

1. The x-axis is the horizontal line on the graph. The y-axis is the vertical line on the graph.

2. The meeting point of the axes is called the origin, marked as O on the diagram.

3. The tick marks on the x- and y-axes indicate distance from the origin.

4. The x- and y-axes divide the plane into four quadrants. Quadrant 1 contains both positive x- and y-coordinates. Quadrant 2 contains negative x-coordinates and positive y-coordinates. Quadrant 3 has negative coordinates for both x and y. And Quadrant 4 has positive x-coordinates and negative y-coordinates.

5. The figure contains two lines, line ℓ and line m.

6. Point P on line m is located at x-coordinate 3 and y-coordinate 2. This location is given using the ordered pair (3, 2). There are three other points shown on the figure: Point Q at (0, –1), Point R at (–3, –4), and Point S at (–4, 0).

7. Point R occurs at the intersection of line ℓ and line m.

8. Line ℓ crosses the x-axis at Point S (–4, 0). This point is called the x-intercept.

9. Line m crosses the y-axis at Point Q (0, –1). This point is called the y-intercept.

Lines in Coordinate Geometry

Frequency Guide: 4

Questions involving straight lines make up the bulk of coordinate geometry questions.

Required Knowledge and Skill Set

1. On the SAT, there are two ways to find the length of a line in a coordinate plane.

   The first method uses the Pythagorean Theorem. The line in question is the hypotenuse of a right triangle:

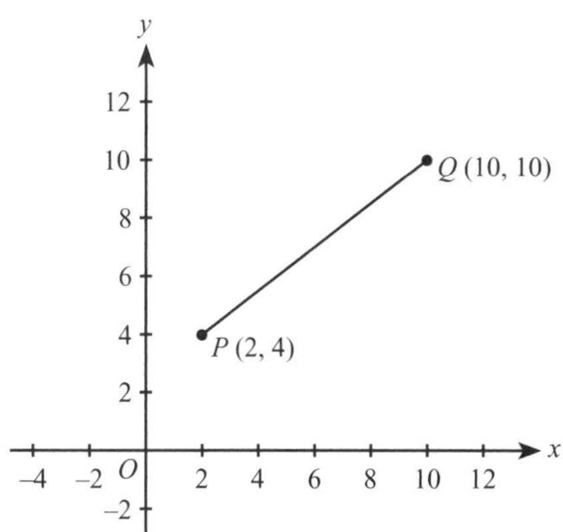

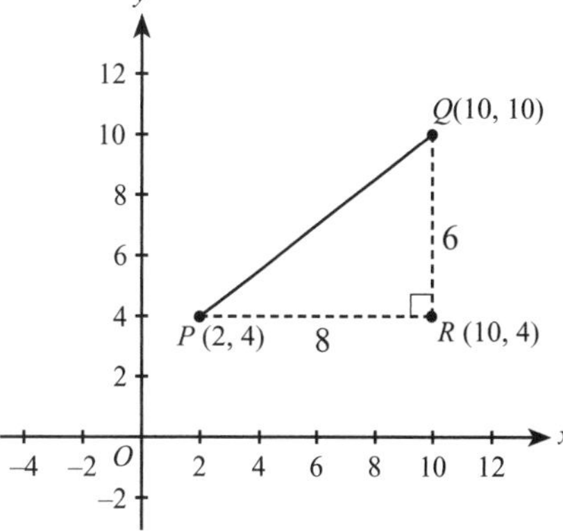

   Because the test makers prefer to use integers, the length of the line is likely part of a Pythagorean Triple, so you will not have to perform any calculations. In the diagram above, the right triangle is a 6:8:10, so the length of *PQ* is 10. Of course, if you forget the Pythagorean Triples, you can still perform the Pythagorean Theorem:

$$a^2 + b^2 = c^2$$
$$6^2 + 8^2 = c^2$$
$$36 + 64 = c^2$$
$$100 = c^2$$
$$10 = c$$

   The formulaic method for finding line length is the Distance Formula. To find the distance between two points $[(x_1, y_1)$ and $(x_2, y_2)]$, use the following formula:

MEMORY MARKER:
The Distance Formula is not provided in the formula box.

$$\text{Distance} = \sqrt{(x_2 - x_1)^2 + (y_2 - y_1)^2}$$

   Now find the distance of *PQ* given the two points, (2, 4) and (10, 10):

$$\text{Distance} = \sqrt{(10-2)^2 + (10-4)^2} \rightarrow \sqrt{8^2 + 6^2} \rightarrow \sqrt{64+36} \rightarrow \sqrt{100}$$

$$\text{Distance} = 10$$

   By tracking the movement along the x-axis and the movement along the y-axis, the Distance Formula basically creates a right triangle and applies the Pythagorean Theorem. The hypotenuse of this newly created triangle measures the distance between the two points.

MEMORY MARKER:
The Midpoint Formula may be helpful on the SAT.

2. The Midpoint Formula is used to find the exact center of a line. In this formula, the averages of the x- and y-coordinates at the start and end of a line provide the midpoint:

$$\text{Midpoint} = \left(\frac{x_1 + x_2}{2}, \frac{y_1 + y_2}{2}\right)$$

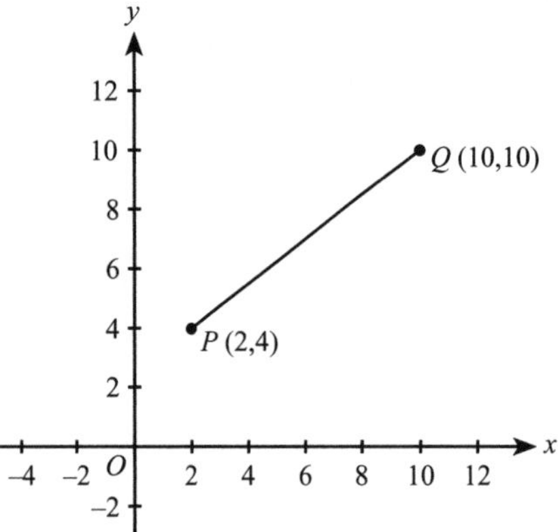

$$\text{Midpoint} = \left(\frac{2+10}{2}, \frac{4+10}{2}\right)$$

$$\text{Midpoint} = \left(\frac{12}{2}, \frac{14}{2}\right) \rightarrow (6, 7)$$

Note that it is easy to SIZE UP the midpoint on the SAT. You can estimate where the midpoint occurs on a line, and since the SAT mostly deals in integers, you will often be correct. However, this strategy should only be used when you forget the midpoint formula.

3. The slope of a line is the most tested coordinate geometry concept. The slope is determined by calculating a line's "rise" over "run." The rise refers to the number of vertical tick marks (on the y-axis) between two points on the line; the run refers to the number of horizontal tick marks (on the x-axis) between the same two points on the line.

To find the slope of a line, pick two points on the line and count the rise over the run:

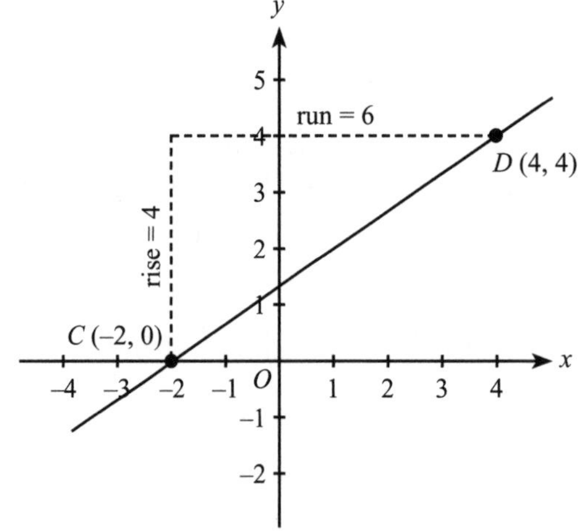

$$\text{Slope} = \frac{\text{rise}}{\text{run}} = \frac{4}{6} = \frac{2}{3}$$

MEMORY MARKER:
The Slope Formula does not appear in the formula box and must be memorized.

Or you can use the Slope Formula:

$$\text{Slope} = \frac{y_2 - y_1}{x_2 - x_1} \qquad \text{Slope} = \frac{4-0}{4--2} \rightarrow \frac{4}{6} \rightarrow \frac{2}{3}$$

Notice that this formula works regardless of the point from which you choose to start:

$$\text{Slope} = \frac{y_2 - y_1}{x_2 - x_1} \qquad \text{Slope} = \frac{0-4}{-2-4} \rightarrow \frac{-4}{-6} \rightarrow \frac{4}{6} \rightarrow \frac{2}{3}$$

4. Lines that tilt up from left to right have a positive slope:

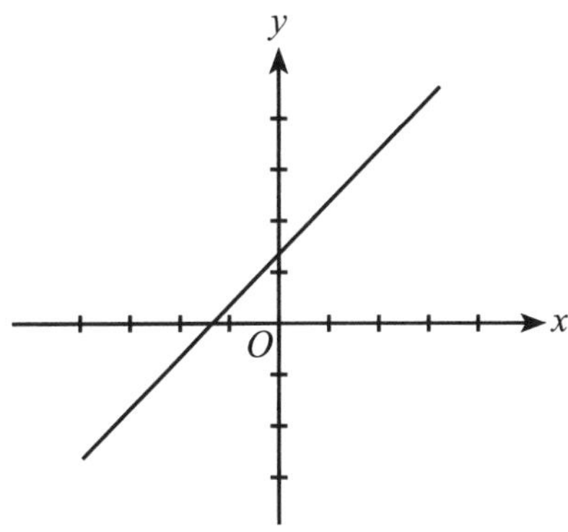

Positive Slope

Lines that tilt down from left to right have a negative slope:

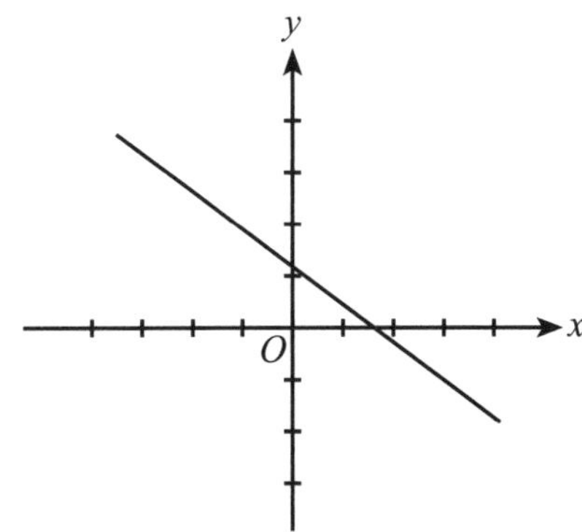

Negative Slope

MEMORY MARKER:
You may need to identify a line based solely on its positive or negative slope.

And lines that are parallel to the x-axis have a slope of 0:

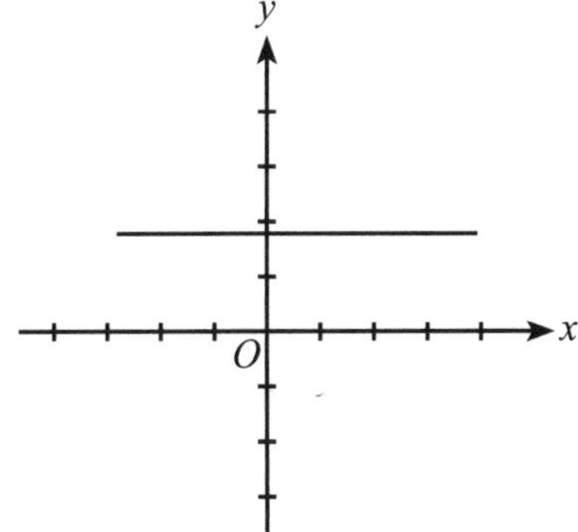

Zero Slope

5. The slopes of parallel and perpendicular lines are often tested on the SAT. Parallel lines have the same slope:

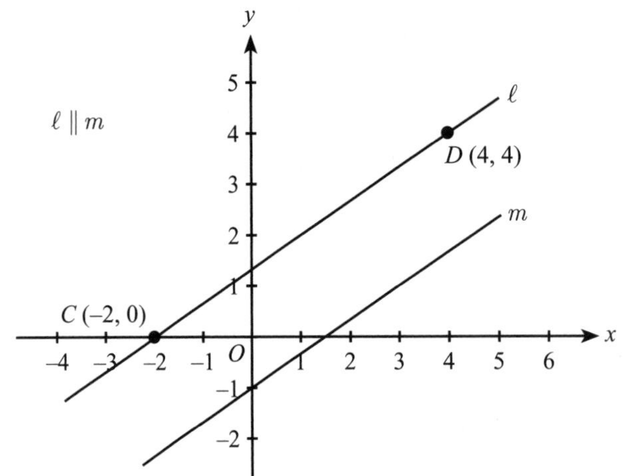

Slope of line $\ell = \frac{2}{3}$

Slope of line $m = \frac{2}{3}$

Perpendicular lines have slopes that are negative reciprocals of each other:

MEMORY MARKER:
The makers of the SAT count on you to forget these two slope properties.

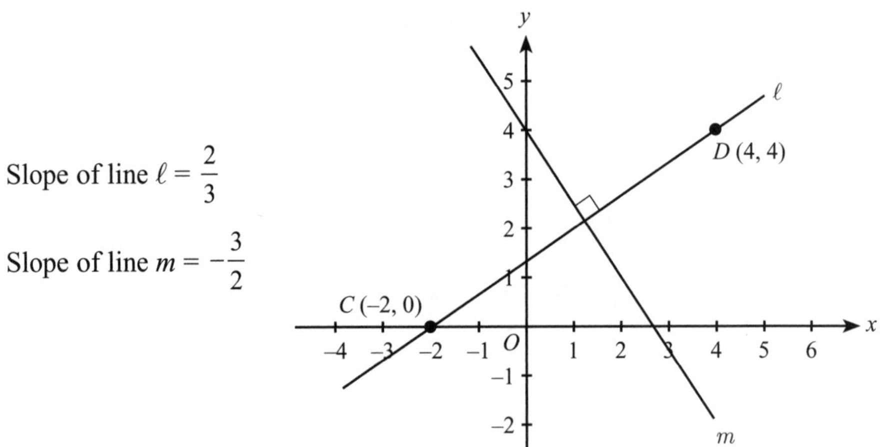

Slope of line $\ell = \frac{2}{3}$

Slope of line $m = -\frac{3}{2}$

6. The equation of a line, called a linear equation, is given by the following formula:

MEMORY MARKER:
The equation of a line is the most important concept in coordinate geometry questions.

Equation of a line: $y = mx + b$

Where:
m = slope
b = y-intercept
x and y = the x- and y-coordinate (x, y) of any point on the line

By plugging in any three of the variables, you can find the fourth missing variable.

Application on the SAT

When faced with a problem concerning the length of a line, try to avoid using the Distance Formula at first. The majority of these questions can be solved without it:

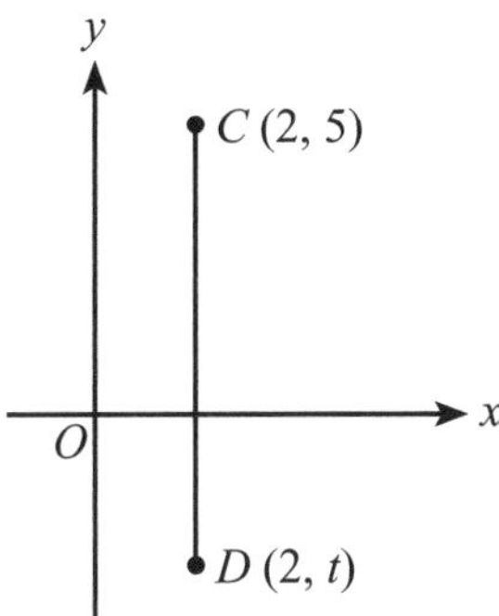

Note: Figure is not drawn to scale.

8. In the coordinate plane above, the length of CD is 8 and it is parallel to the y-axis. What is the value of t ?

   (A) –6
   (B) –3
   (C) 2
   (D) 5
   (E) 13

Students who immediately pull out the Distance Formula for this question will be wasting time and risking errors. This simple question has a simple solution. The value of the y-coordinate in D is 8 less than the y-coordinate in C:

$5 - 8 = -3$ The correct answer is (B).

Imagine if you had used the Distance Formula on this problem:

$\text{Distance} = \sqrt{(x_2 - x_1)^2 + (y_2 - y_1)^2}$

$8 = \sqrt{(2-2)^2 + (5-t)^2}$

$8 = \sqrt{0^2 + (5-t)(5-t)}$

$64 = 25 - 10t + t$

$0 = t - 10t - 39$

$0 = (t - 13)(t + 3)$

$t = 13, t = -3$ Since point D is in Quadrant IV, t must be negative (–3).

The Distance Formula is hard to remember and often difficult to compute, thanks to the inclusion of binomials and radicals. Many students find the Pythagorean Theorem an easier and safer method for finding distance.

This solution is time consuming and risky, given all of the intricate calculations.

If you encounter a problem that seems to require the Distance Formula, there is another way to solve it *without* the Distance Formula. Spend a few seconds determining whether the Distance Formula presents the most efficient solution.

Many coordinate geometry questions involve the equation of a line, even if an equation is not given in the problem. These questions give you enough information to find three of the variables in the equation of a line, and ask you to solve for the fourth variable:

Note that this figure is drawn to scale. Can you SIZE UP the figure and estimate the value of b?

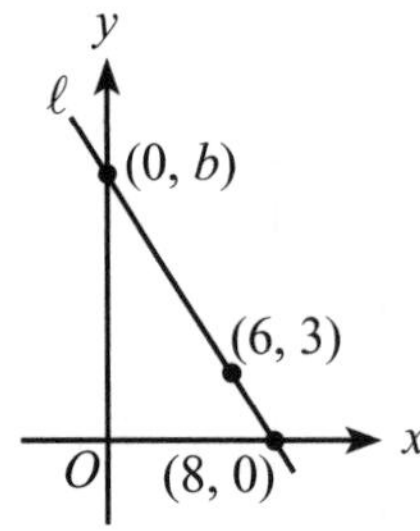

15. In the figure above, there are three points on line ℓ. What is the value of b ?

    (A) 8
    (B) 12
    (C) 14
    (D) 15
    (E) 16

Many students miss this question because they fail to realize that the equation of a line is the key to the solution. The value b is not only a y-coordinate, but also the y-intercept. It is convenient that it is labeled b in the figure, as it is also referred to as b in the equation of a line:

$$y = mx + b$$

So if we are solving for b, then we must find y, m, and x. To find (x, y), we simply use one of the points on the line. Either (6, 3) or (8, 0) will work. Choose one:

$$0 = m(8) + b$$

Now you must find m, the slope. Use the two points to find the slope of line ℓ:

$$\text{Slope} = \frac{y_2 - y_1}{x_2 - x_1} \quad \rightarrow \quad \frac{3-0}{6-8} \quad \rightarrow \quad \frac{3}{-2}$$

Now use the slope in the equation of the line to solve for b:

You can also solve this problem using just the slope formula. However, since the equation of a line is more widely used than the Slope Formula on the SAT, we chose to use the equation.

$$0 = -\frac{3}{2}(8) + b \quad \rightarrow \quad 0 = -12 + b \quad \rightarrow \quad 12 = b$$

When you are given a pair of coordinates on the SAT, you should always be on alert that you will likely need to use the equation of a line. You will only need to find one other variable; either the slope (in which case you need two pairs of coordinates) or the y-intercept.

You may also need to identify a line using its equation, or identify an equation using its line:

6. Which of the following could be the graph of $y = 3x - 4$?

(A)

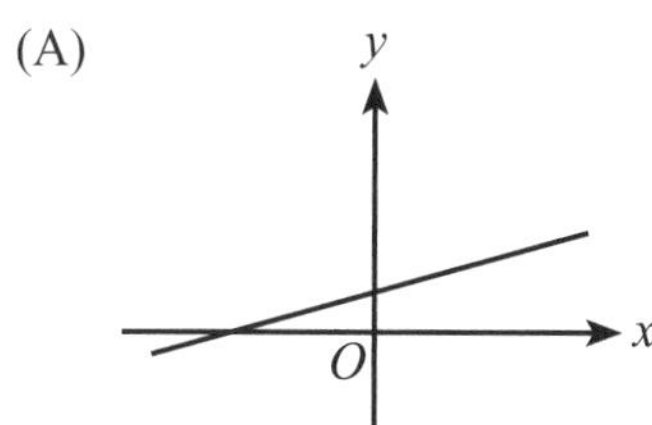

(D)

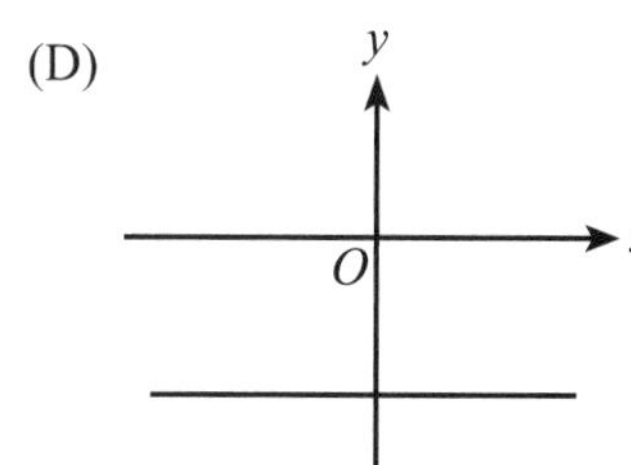

(B)

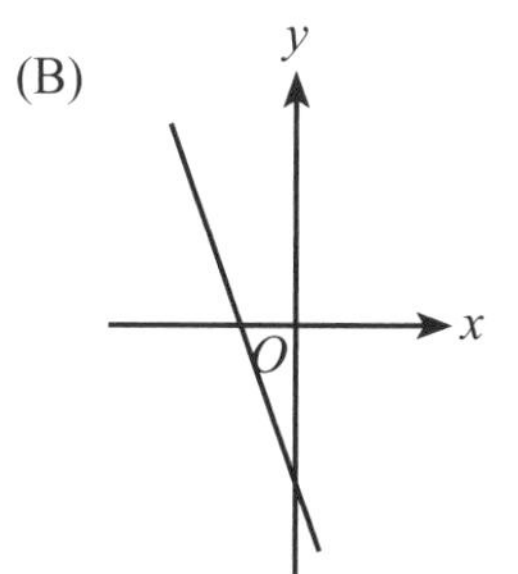

(E)

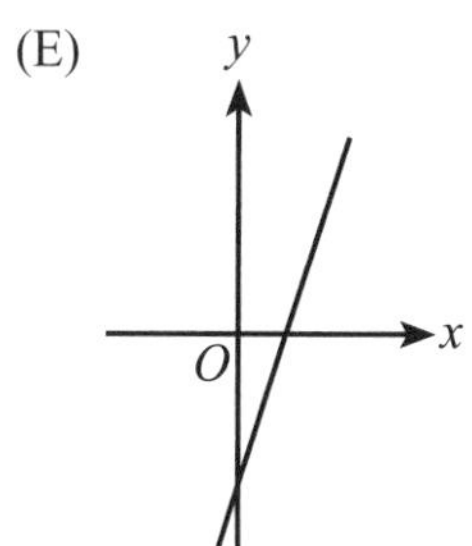

(C)

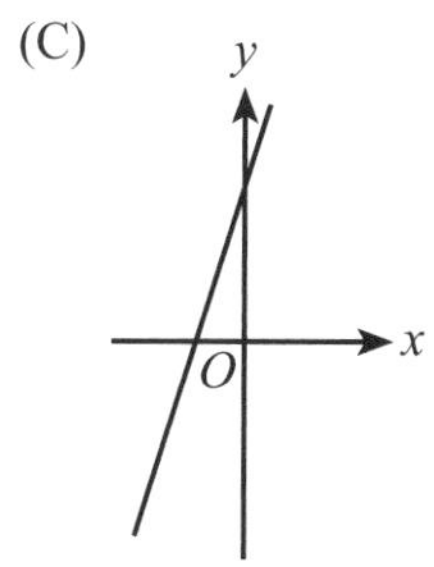

Questions like this one are best solved by eliminating wrong answer choices based on their slopes and y-intercepts.

Begin by dissecting the equation of the line:

$$y = 3x - 4$$

The slope, 3, is positive. This eliminates (D)—zero slope—and (B)—negative slope.

The y-intercept is –4. This eliminates (A) and (C), as they both have a positive y-intercept.

Therefore, the only possible answer is (E).

You may also be given a diagram of a line through the coordinate plane and asked to pick the equation of the line from five answer choices. In this case, figure out whether the graph has a positive, negative, or zero slope, and then estimate the value of its y-intercept. Eliminate the answer choices that do not match your findings.

Confidence Quotation
"Energy and persistence conquer all things."
—Benjamin Franklin, Founding Father of the United States

The test makers may try to confuse you by using variables instead of numbers for coordinates. Consider a question with variables:

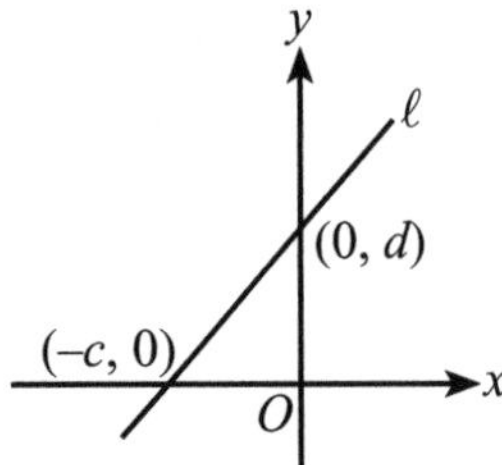

13. What is the slope of line ℓ in the figure above?

(A) $-\frac{c}{d}$

(B) $\frac{c}{d}$

(C) $-\frac{d}{c}$

(D) $\frac{d}{c}$

(E) $\frac{1}{d}$

Do not let variables throw you. Apply them to the proper formula just as you would numbers. In this case, we are looking for the slope of line ℓ:

$$\text{Slope} = \frac{y_2 - y_1}{x_2 - x_1} \rightarrow \frac{d-0}{0-(-c)} \rightarrow \frac{d}{0+c} \rightarrow \frac{d}{c}$$

The correct answer is (D). This relatively easy question has an exaggerated difficulty level because many students panic when variables are used as coordinates. Remember to use your formulas and plug in whatever information is available, whether that information be numbers or variables.

V ery
A ggravating
R uses
I ncrease
A nxiety &
B lur
L egitimately
E asy
S olutions

Can you find the midpoint of line ℓ using variables?

$$\text{Midpoint} = \left(\frac{x_1 + x_2}{2}, \frac{y_1 + y_2}{2}\right) \rightarrow \left(\frac{-c+0}{2}, \frac{0+d}{2}\right) \rightarrow \left(\frac{-c}{2}, \frac{d}{2}\right)$$

Do not allow variables to break your confidence. Use them just as you would use integers or fractions.

Some questions will test your knowledge of parallel and perpendicular slopes. If you see a coordinate geometry question that uses parallel or perpendicular lines, the slopes are the key to the solution:

Can you DIAGRAM the question and draw line m?

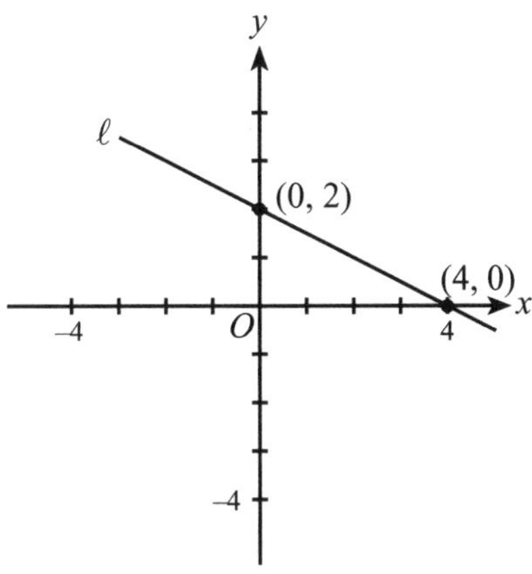

16. In the figure above, line m, which is not shown, is drawn through the origin and is perpendicular to line ℓ. Which of the following is the equation of line m ?

(A) $y = 2x$
(B) $y = -2x$
(C) $y = 2x + 2$
(D) $y = -2x + 2$
(E) $y = 2x - 2$

In order to find the equation of line m, we need the slope and the y-intercept. The slope is the negative reciprocal of the slope of line ℓ. Find the slope of line ℓ to learn the slope of line m:

$$\text{Slope of line } \ell = \frac{y_2 - y_1}{x_2 - x_1} \rightarrow \frac{2-0}{0-4} \rightarrow \frac{2}{-4} \rightarrow -\frac{1}{2}$$

$$\text{Slope of line } m = \text{reciprocal of} -\frac{1}{2} = 2$$

$$y = mx + b \rightarrow y = 2x + b$$

You can eliminate (B) and (D) because they have a negative slope.

When a line passes through the origin, its y-intercept is 0. You may want to draw line m in the figure to see this y-intercept.

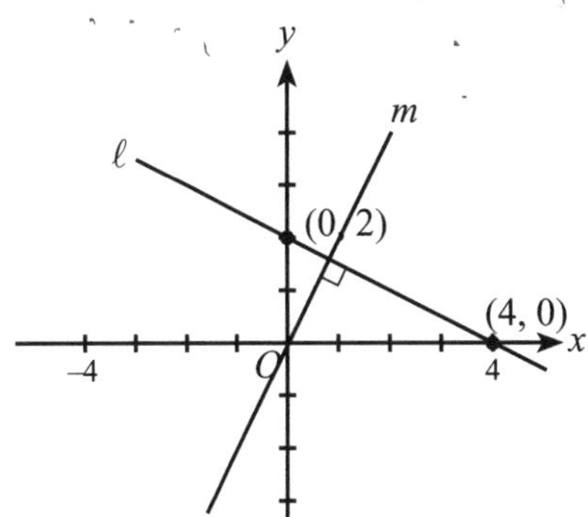

So if the y-intercept is 0, then $b = 0$:

$$y = mx + b$$
$$y = 2x + b$$
$$y = 2x + 0$$
$$y = 2x$$

The correct answer is (A).

Lines in Coordinate Geometry Problem Set

Solve the following multiple-choice questions by selecting the best answer from the five answer choices. For grid-in questions, write your answer in the grids and completely mark the corresponding ovals. Answers begin on page 388.

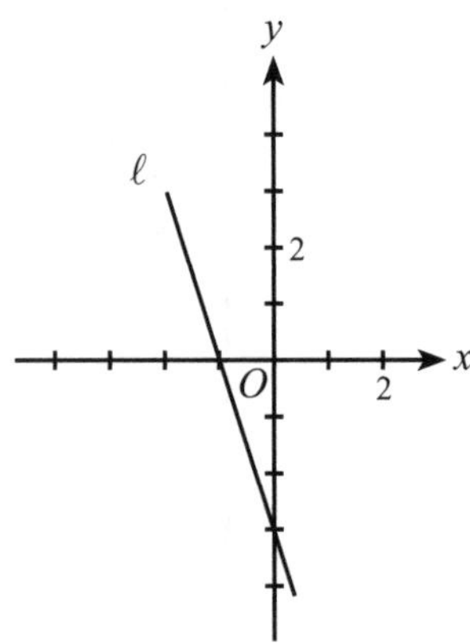

1. Which of the following is the equation of the line in the figure above?

(A) $y = -\frac{1}{3}x - 3$
(B) $y = \frac{1}{3}x - 3$
(C) $y = -3x + 3$
(D) $y = -3x - 3$
(E) $y = 3x - 3$

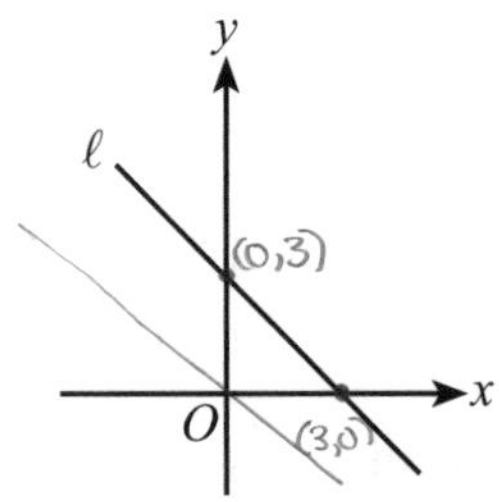

2. In the figure above, line ℓ intersects the x-axis at $x = 3$ and the y-axis at $y = 3$. Line m, which is not shown, is perpendicular to line ℓ and passes through the origin. What is the equation of line m ?

(A) $y = -x$
(B) $y = x$
(C) $y = -x + 3$
(D) $y = -3x + 3$
(E) $y = 3x - 3$

3. Line ℓ is parallel to the x-axis and is four units below the x-axis. What is the equation of line ℓ ?

(A) $y = -4$
(B) $y = 0$
(C) $y = 4$
(D) $x = -4$
(E) $x = 4$

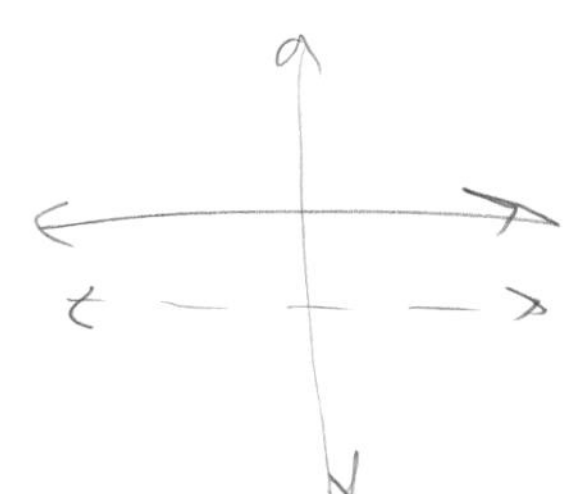

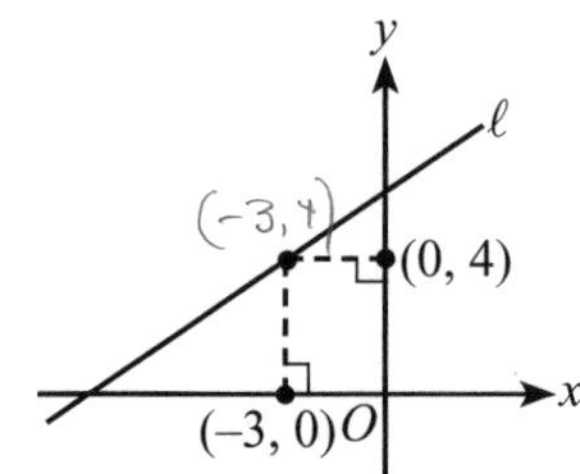

Note: Figure is not drawn to scale.

4. In the figure above, line ℓ has a slope of $\frac{2}{3}$. What is the y-intercept of line ℓ ?

(A) 6
(B) 7
(C) 8
(D) 9
(E) 10

Lines in Coordinate Geometry Problem Set

Solve the following multiple-choice questions by selecting the best answer from the five answer choices. For grid-in questions, write your answer in the grids and completely mark the corresponding ovals. Answers begin on page 388.

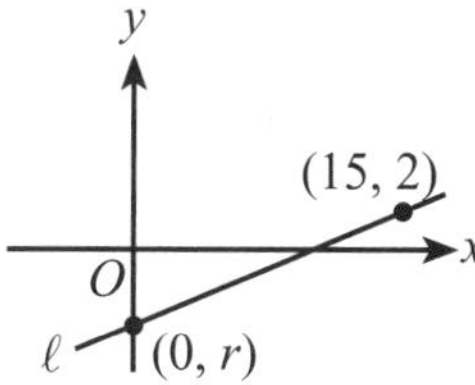

5. If the slope of line ℓ in the figure above is $\frac{2}{5}$, what is the value of r ?

(A) –3
(B) –4
(C) –5
(D) –6
(E) –7

6. In the coordinate plane, line ℓ passes through point (–4, 0) and then passes BETWEEN the points (3, 6) and (4, 6) without passing through them. What is one possible value for the slope of line ℓ ?

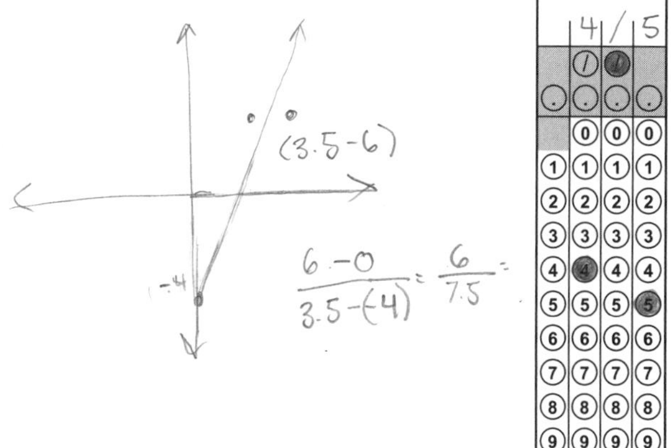

7. In the coordinate plane, the length of line RS is 17. If point R is located at (10, 9) and point S is located at $(x, 1)$, what is one possible value of x ?

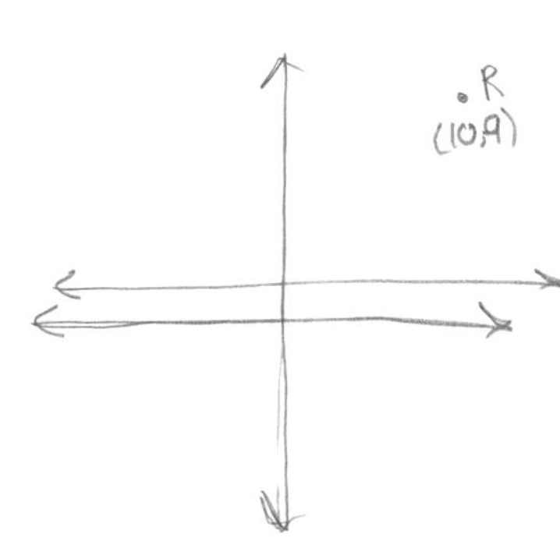

Figures in Coordinate Geometry

Frequency Guide: 3

Some SAT questions will require knowledge of both geometric figures and the coordinate plane.

Required Knowledge and Skill Set

1. All of the properties of polygons and circles may be applied in coordinate geometry. Expect to find the area and perimeter of triangles, quadrilaterals, pentagons, hexagons, octagons, and circles using the formulas in the formula box.

2. All of the properties discussed involving lines in coordinate geometry, including slope and intercepts, might be combined with questions about geometric figures, especially triangles and quadrilaterals.

Application on the SAT

One type of question involving figures in coordinate geometry asks you to find the coordinates of a point using the properties of a geometric figure:

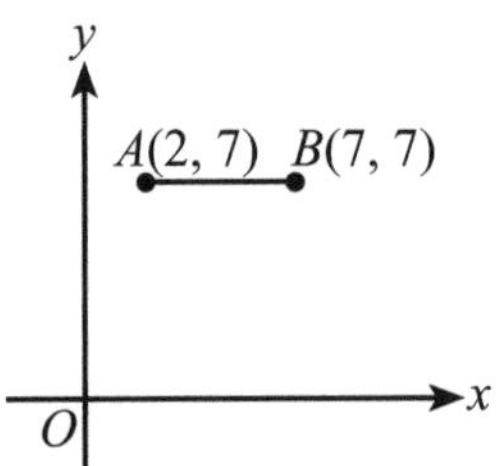

7. In the figure above, AB is one side of square $ABCD$. Which of the following could be the coordinates of point C?

(A) (2, 5)
(B) (5, 2)
(C) (5, 5)
(D) (7, 2)
(E) (7, 5)

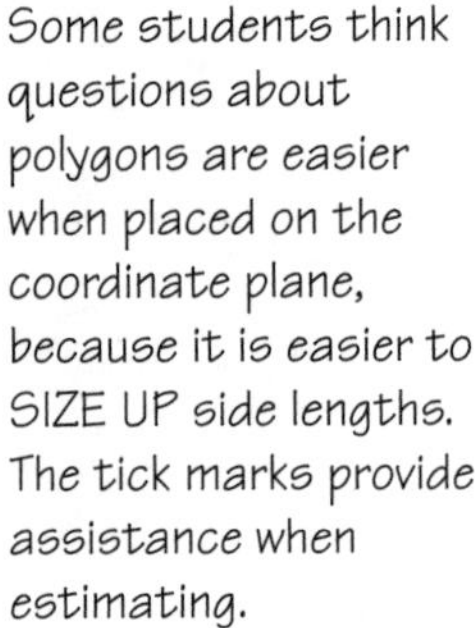

Some students think questions about polygons are easier when placed on the coordinate plane, because it is easier to SIZE UP side lengths. The tick marks provide assistance when estimating.

One property of a square states that all four sides are equal. By subtracting one x-coordinate from another, you learn that the length of AB is 5 ($7 - 2 = 5$). Imagine the possible squares in the plane:

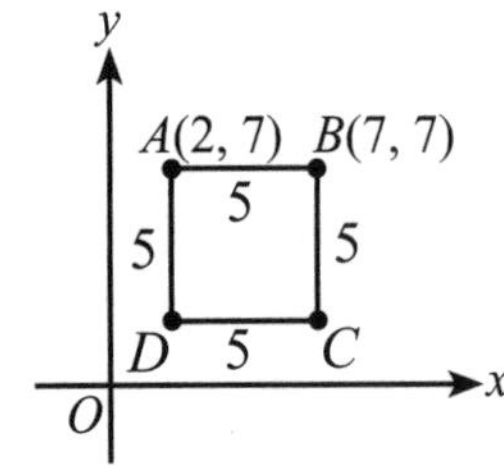

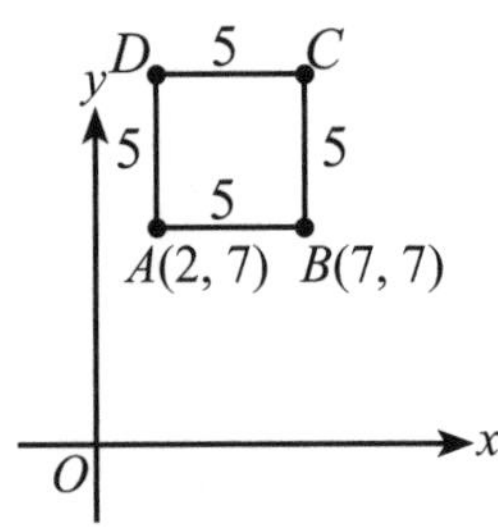

Points B and C have the same x-coordinate. Their difference is the change in the y-coordinate. Because the side of the square is 5, add or subtract 5 to the y-coordinate of B to find the y-coordinate of C:

$B\ (7, 7)$
$C\ (7, 7 + 5)$ or $(7, 7 - 5) \rightarrow C\ (7, 12)$ or $(7, 2)$

Choice (D) is correct.

These same principles are behind the area and perimeter questions. You must use your knowledge of the coordinate plane to determine the length of a part of a polygon.

11. In the coordinate plane, points $A(-4, 2)$, $B(8, 2)$, and $C(8, -3)$ are vertices of a triangle. What is the area of ABC?

    (A) 13
    (B) 15
    (C) 24
    (D) 30
    (E) 60

When a coordinate geometry question does not have a figure, consider DIAGRAMMING the question. A rough sketch of a line or figure will help you see slope, length, and intercepts.

DIAGRAM the question:

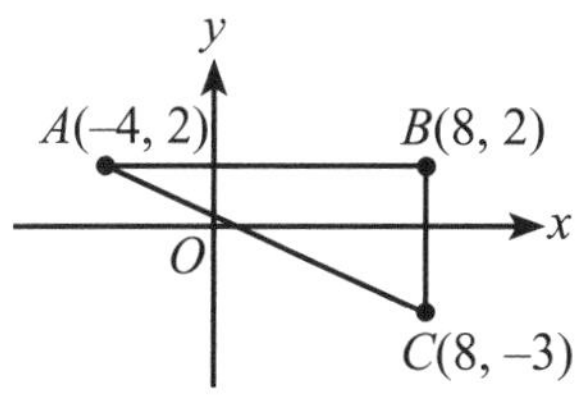

The length of AB is the change in x-coordinates: $8 - (-4) = 12$

The length of BC is the change in y-coordinates: $2 - (-3) = 5$

Now use the area of a triangle to solve the question:

$$A = \frac{1}{2}bh \rightarrow \frac{1}{2}(12)(5) \rightarrow (6)(5) \rightarrow 30$$

Can you find the perimeter of the triangle in the diagram? You have already found the lengths of AB and BC. How do you find the length of AC?

Recognize a Pythagorean Triple or use the Pythagorean Theorem:

$a^2 + b^2 = c^2$
$AB^2 + BC^2 = AC^2$
$12^2 + 5^2 = AC^2$
$144 + 25 = AC^2$
$169 = AC^2$
$13 = AC$

The perimeter is the sum of the three sides: $12 + 5 + 13 = 30$. Both the area and the perimeter are 30.

Be prepared to use the properties of a line in the coordinate plane with the properties of triangles and squares. Take the following grid-in question:

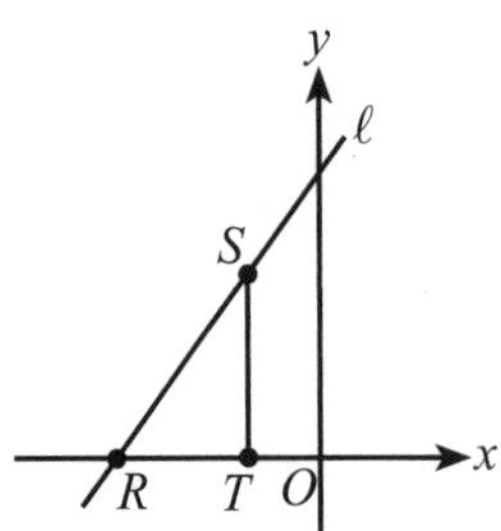

18. In the figure above, ST is parallel to the y-axis and point T is located at (–3, 0). If $RS = 10$ and $RT = 6$, what is the value of the y-intercept of line ℓ ?

Begin by DIAGRAMMING the question:

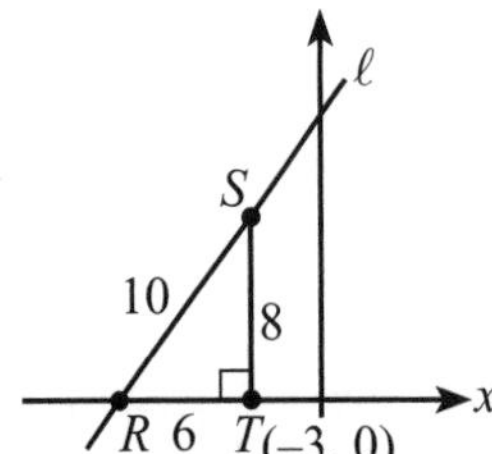

RST is a right triangle that satisfies the Pythagorean Triple 6:8:10. Therefore, $ST = 8$.

Point R (–9, 0)
Point S (–3, 8)

Have you memorized the Pythagorean Triples yet? By now you should realize how much time they can save.

Now you have a point on the line (–9, 0) or (–3, 8) and can begin using the equation of line ℓ to find the y-intercept:

$$y = mx + b$$
$$0 = m(-9) + b$$

To find the slope, calculate the rise over the run or use the Slope Formula:

$$\text{Slope} = \frac{y_2 - y_1}{x_2 - x_1} \rightarrow \frac{0-8}{-9-(-3)} \rightarrow \frac{-8}{-6} \rightarrow \frac{-4}{-3} \rightarrow \frac{4}{3}$$

Now solve for b, the y-intercept:

$$y = mx + b$$
$$0 = \frac{4}{3}(-9) + b$$
$$0 = \frac{-36}{3} + b$$
$$0 = -12 + b$$
$$12 = b$$

You can also solve this question using the properties of similar triangles.

The correct answer is 12. This difficult question combined right triangles and the equation of a line.

Finally, expect some hidden triangles in coordinate geometry questions. These triangles are the key to the solution of some difficult problems. Can you find the hidden triangle in the following question?

There are many hidden triangles in this figure, but which one will help you determine the length of one side of the pentagon?

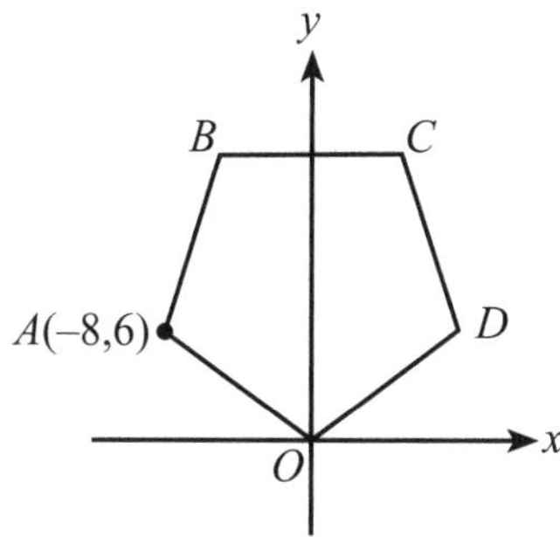

18. In the figure above, the regular pentagon $OABCD$ is located on the coordinate plane. What is the perimeter of the pentagon?

    (A) 10
    (B) 30
    (C) 40
    (D) 50
    (E) 70

To find the perimeter of a regular pentagon, you only need to find the length of one side of the pentagon and multiply it by 5. In this figure, use the side that has a coordinate:

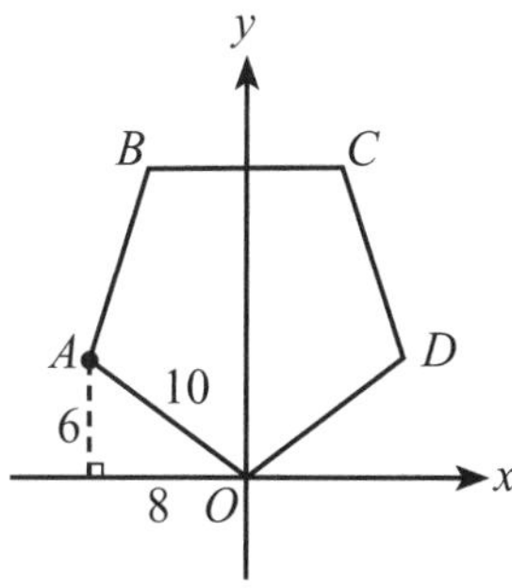

By now you should recognize the Pythagorean Triple, 6:8:10. If you still have not memorized it, use the Pythagorean Theorem.

$a^2 + b^2 = c^2$
$6^2 + 8^2 = c^2$
$36 + 64 = c^2$
$100 = c^2$
$10 = c$

Do not get tricked into choosing answer choice (A). The length of AO is 10, but the perimeter of the pentagon is 5 times the length of AO:

Perimeter $= 5s = 5(10) = 50$

The correct answer is (D).

Confidence Quotation
"In the province of the mind, what one believes to be true either is true or becomes true."
—John Lilly, psychoanalyst and philosopher

Figures in Coordinate Geometry Problem Set

Solve the following multiple-choice questions by selecting the best answer from the five answer choices. For grid-in questions, write your answer in the grids and completely mark the corresponding ovals. Answers begin on page 391.

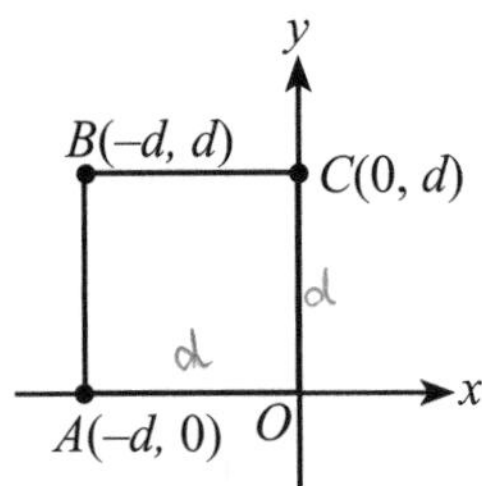

1. In the figure above, $OABC$ is a square. What is the area of the square?

   (A) d
   (B) $2d$
   (C) $4d$
   (D) d^2
   (E) $2d^2$

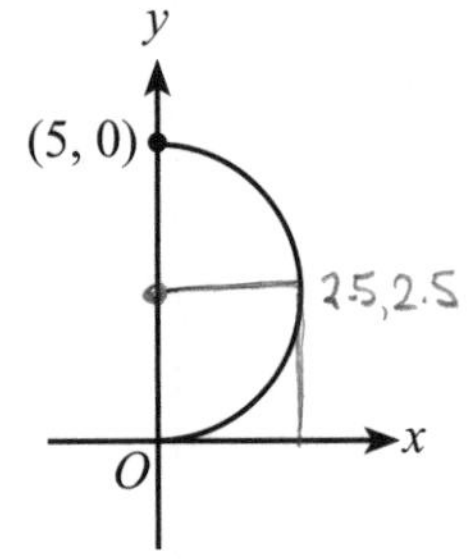

2. The figure above is a semicircle in the coordinate plane. What is the value of the x-coordinate on the semicircle that is located the farthest from the y-axis?

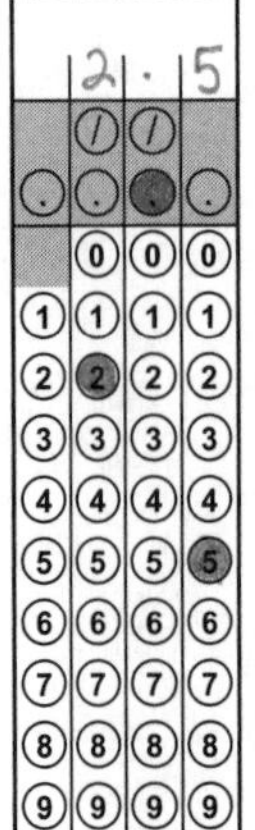

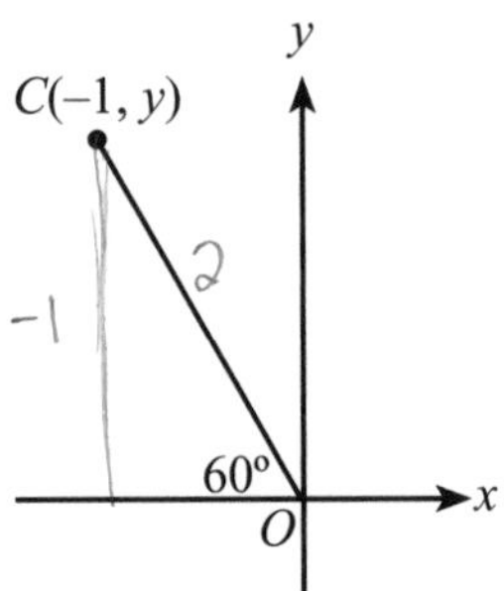

3. What is the length of CO in the figure above?

   (A) 1
   (B) $\sqrt{2}$
   (C) $\sqrt{3}$
   (D) 2
   (E) 3

Parabolas in Coordinate Geometry

Parabolas are not tested frequently on the SAT, but they do appear once every four or five SAT administrations.

Frequency Guide: 1

Required Knowledge and Skill Set

1. When a linear equation ($y = mx + b$) is graphed on the coordinate plane, the result is a straight line. When a quadratic equation is graphed on the coordinate plane, the result is a U-shaped parabola:

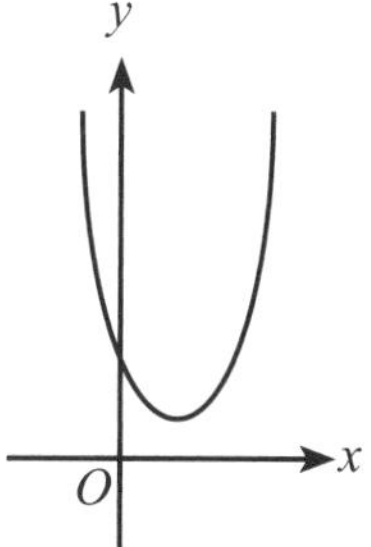

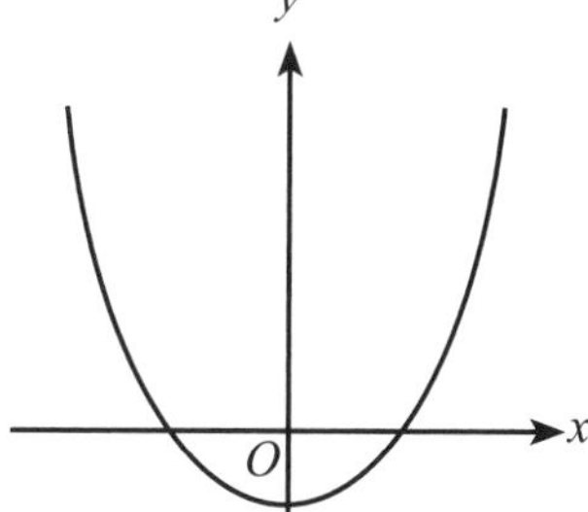

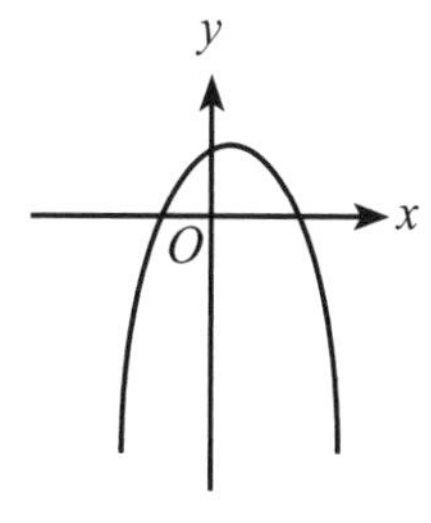

2. You must know the standard equation of a parabola and understand some of its components:

   Standard equation of a parabola: $y = ax^2 + bx + c$

   - a, b, and c are constants
   - x and y = the x- and y-coordinate (x, y) of any point on the parabola
   - $(0, c)$ is the y-intercept

   - When a is positive, the parabola opens upward
   - When a is negative, the parabola opens downward
   - When $b = 0$, the parabola is centered on the y-axis
   - When $b > 0$, the parabola moves to the left of the y-axis
   - When $b < 0$, the parabola moves to the right of the y-axis

MEMORY MARKER: You would be wise to learn and understand both equations of a parabola.

3. An equation of a parabola can also be expressed in a vertex form:

   Vertex equation of a parabola: $y = a(x - h)^2 + k$

   - (h, k) is the vertex of the parabola
   - x and y = the x- and y-coordinate (x, y) of any point on the parabola

   - When a is positive, the parabola opens upward
   - When a is negative, the parabola opens downward

   You should be prepared to work with either form of equation on the SAT.

Application on the SAT

Parabola questions on the SAT are very basic. You will either need to identify a parabola given an equation or information about the constants, or identify an equation given a parabola:

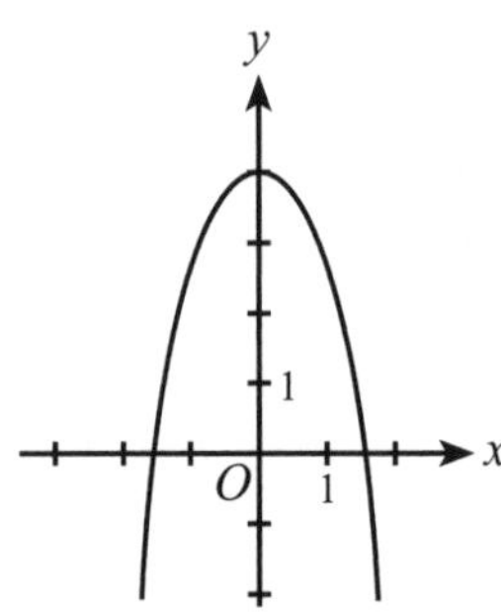

8. In the figure above, a parabola is shown in the coordinate plane. Which of the following could be the equation of the parabola?

    (A) $y = -x^2 + 4$
    (B) $y = x^2 + 4$
    (C) $y = -x^2 - 4$
    (D) $y = x^2 - 4$
    (E) $y = -4x^2$

Often on the SAT, you do not need to analyze the value of b. Notice that the value of b in all of these answer choices is 0.

You can use either the standard form or vertex form of the equation of a parabola to solve this question. Let's examine both methods, starting with the standard form:

Standard equation of a parabola: $y = ax^2 + bx + c$

The parabola opens downward, so $a < 0$. In all of the answer choices, the value of a is either 1 or –1. You can eliminate (B) and (D) as they have a positive value of a. Next, determine that the y-intercept $(0, c)$ is $(0, 4)$. This eliminates (C), where $c = -4$, and (E), where $c = 0$. The correct answer is (A).

You can also solve this question using the vertex form of the equation of a parabola:

Vertex equation of a parabola: $y = a(x - h)^2 + k$

You know that $a < 0$ because the parabola opens downward. The vertex of the parabola, (h, k), is at $(0, 4)$. Plug this information into the equation:

$$y = -a(x - 0)^2 + 4$$
$$y = -ax^2 + 4$$

How would you find the value of a? You need another point on the parabola to plug in for x and y.

Even without knowing the value of a, you can see that choice (A) is the only answer that fits the equation in vertex form.

Parabolas in Coordinate Geometry Problem Set

Solve the following multiple-choice questions by selecting the best answer from the five answer choices. For grid-in questions, write your answer in the grids and completely mark the corresponding ovals. Answers are on page 392.

1. In the quadratic equation $y = ax^2 + bx + c$, $a > 0$ and $c < 0$. Which of the following could be the graph of this equation?

(A)

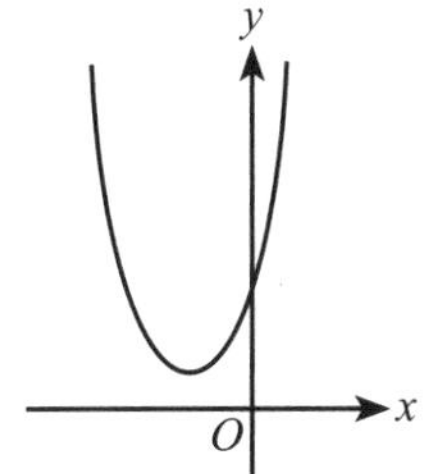

(B)

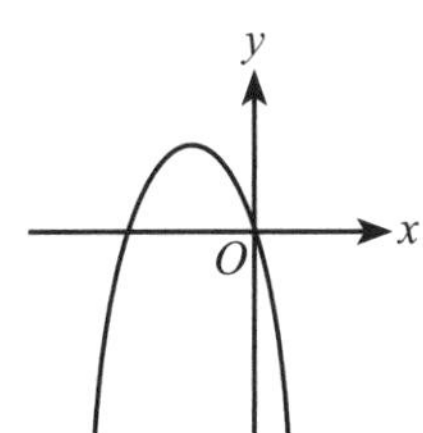

(C)

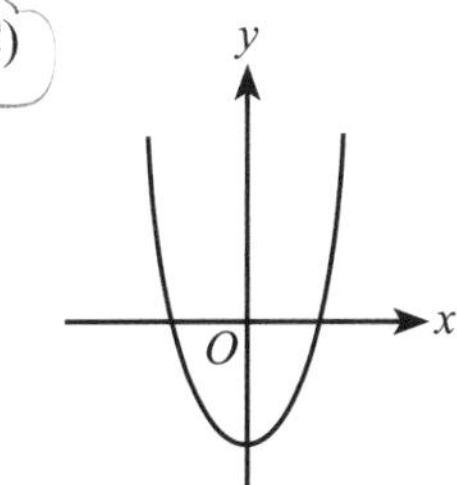

(D)

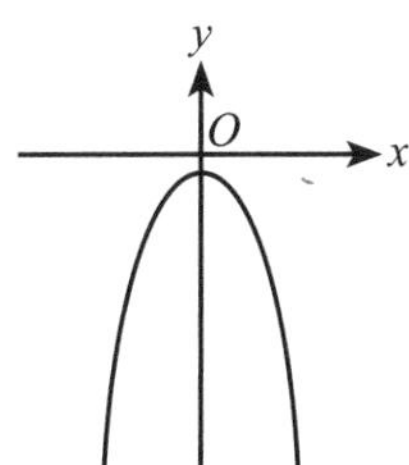

(E)

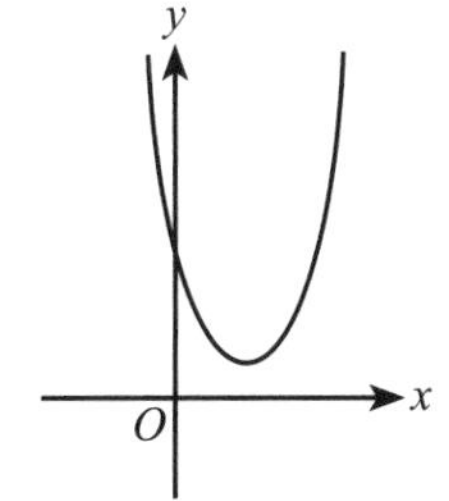

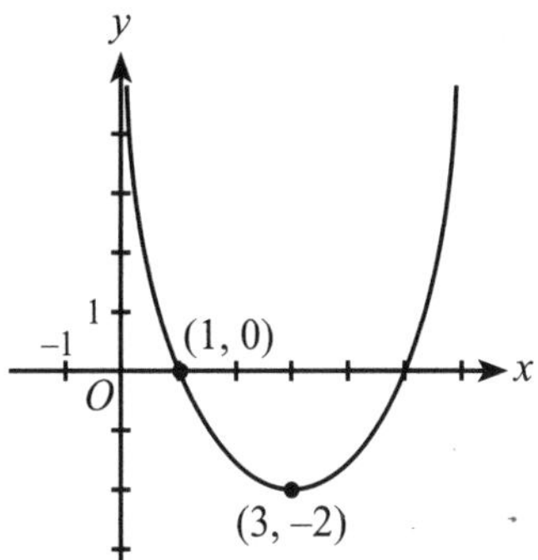

2. Which of the following is the equation of the graphed parabola in the figure above?

(A) $y = \frac{1}{2}(x^2 + 3)^2 - 2$

(B) $y = \frac{1}{2}(x^2 - 3)^2 + 2$

(C) $y = \frac{1}{2}(x^2 - 3)^2 - 2$

(D) $y = 2(x^2 + 3)^2 + 2$

(E) $y = 2(x^2 - 3)^2 - 2$

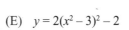

Functions in Coordinate Geometry

Frequency Guide: 2

Like linear and quadratic equations, functions can be graphed in the coordinate plane.

Required Knowledge and Skill Set

MEMORY MARKER:
If you know the equation of a line, the equation of a linear function should be a no-brainer.

1. Linear functions are graphed just like linear equations. The only difference is that the y-coordinate value is represented as $f(x)$:

 Equation of a line: $y = mx + b$
 Equation of a linear function: $f(x) = mx + b$

 Where:
 m = slope
 b = y-intercept
 x and $f(x)$ = the x- and y-coordinate (x, y) of any point on the line

 Therefore, the graph of $f(x) = 2x + 1$ is the same as the graph of $y = 2x + 1$ where the slope is 2 and the y-intercept is 1:

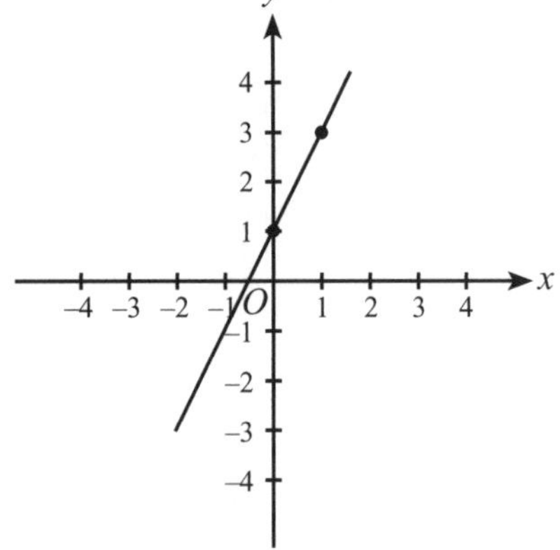

 The key to solving linear functions on the SAT is remembering that $f(x) = y$.

2. Quadratic functions are graphed just like quadratic equations, both of which result in parabolas. Again, the only difference in the equations is that the $f(x)$ replaces y, although both symbols represent the y-coordinate.

 Standard equation of a parabola: $y = ax^2 + bx + c$
 Standard equation of a quadratic function: $f(x) = ax^2 + bx + c$

 Vertex equation of a parabola: $y = a(x - h)^2 + k$
 Vertex equation of a quadratic function: $f(x) = a(x - h)^2 + k$

Application on the SAT

Some of the coordinate geometry function questions will be just like the questions we covered in "Lines in Coordinate Geometry" and "Parabolas in Coordinate Geometry." The only difference will be the use of $f(x)$ instead of y. Remember this question?

6. Which of the following could be the graph of $f(x) = 3x - 4$?

(A)

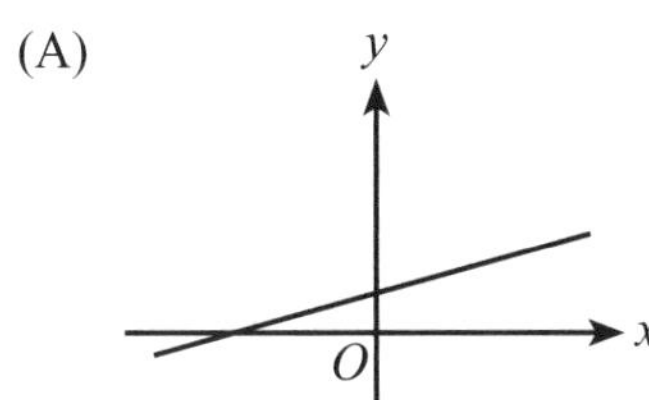

(D)

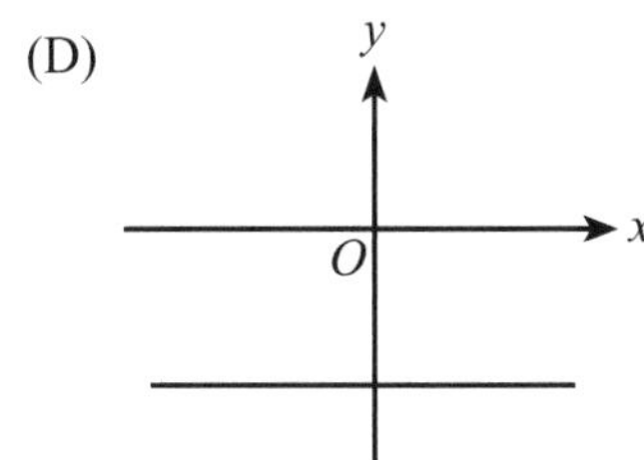

(B)

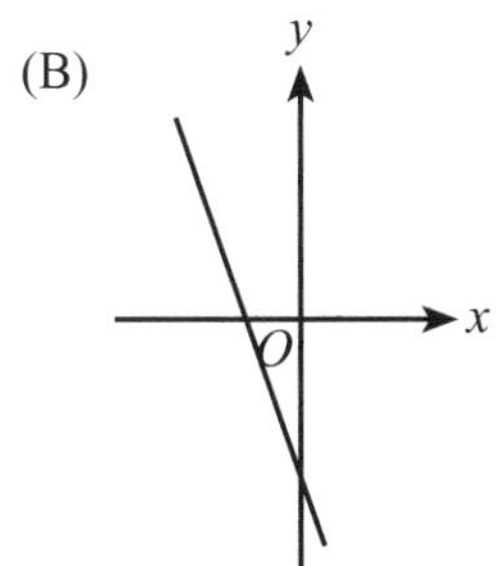

(E)

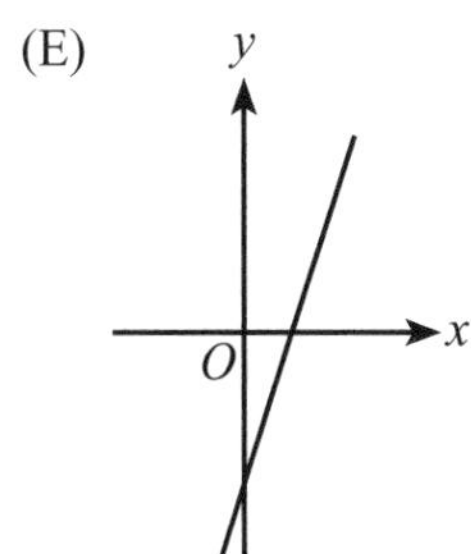

(C)

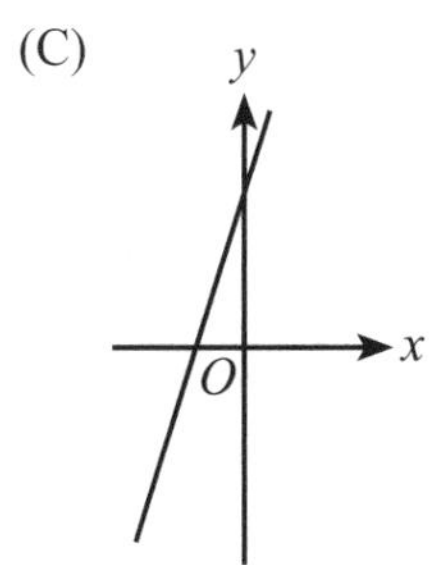

This is the same question we examined in "Lines in Coordinate Geometry." However, the equation has been changed from $y = 3x - 4$ to $f(x) = 3x - 4$. But the solution is exactly the same.

Begin by dissecting the equation of the line:

$$f(x) = 3x - 4$$

The slope, 3, is positive. This eliminates (D)—zero slope—and (B)—negative slope.

The y-intercept is –4. This eliminates (A) and (C), as they both have a positive y-intercept.

Therefore, the only possible answer is (E).

f(x) = y ...
f(x) = y ...
f(x) = y ...

Repeat this until you know it and understand it! It may be easier to rewrite functions using y instead of f(x).

Confidence Quotation
"We succeed only as we identify in life, or in war, or in anything else, a single overriding objective, and make all other considerations bend to that one objective."
—Dwight D. Eisenhower, 34th President of the United States

Functions involving parabolas are also solved the same way as quadratic equations in coordinate geometry. Consider a new question:

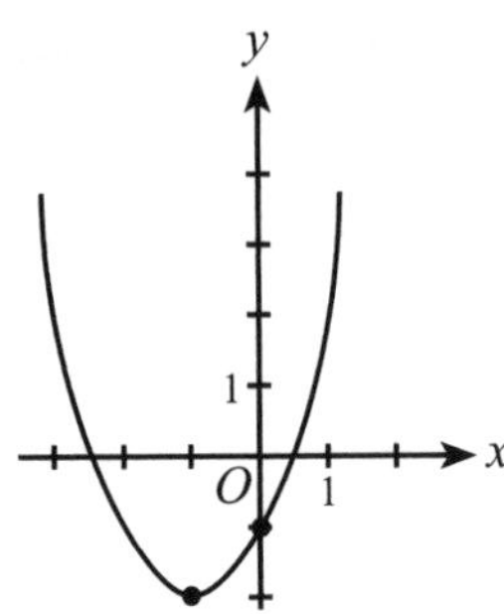

11. Which of the following equations could be represented by the graph of function f above?

    (A) $f(x) = x^2 + 2x - 1$
    (B) $f(x) = x^2 + x$
    (C) $f(x) = x^2 - x - 2$
    (D) $f(x) = x^2 - 2x + 1$
    (E) $f(x) = x^2 - 3x + 2$

Since the answer choices are in the standard form, consider the standard equations:

Standard equation of a parabola: $y = ax^2 + bx + c$
Standard equation of a quadratic function: $f(x) = ax^2 + bx + c$

Because the value of a is positive in every answer choice, we can not use this information to eliminate any answers. However, look at the y-intercept, c. It occurs at $y = -1$. The only answer choice where $c = -1$ is (A), so the correct answer is (A). The value of b ($b = 2$) supports this, as a positive b indicates that the parabola is to the left of the y-axis.

The most common function question in coordinate geometry requires you to use a graph to find a value for $f(x)$. Study the following graph of $f(x)$:

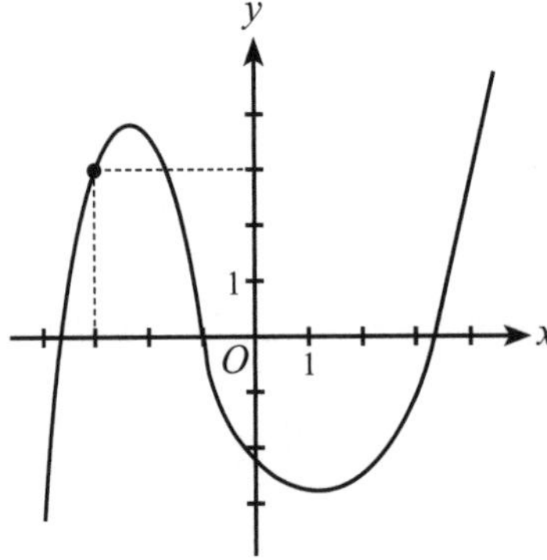

What is the value of $f(-3)$?

Remember, $f(x) = y$. This question is asking "what is the value of y when $x = -3$?"

Use the graph. When $x = -3$, the y-coordinate is 3. The coordinate is $(-3, 3)$. So $f(-3) = 3$.

Can you find $f(-1)$ and $f(3)$? When $x = -1$, $y = 0$, so $f(-1) = 0$. And when $x = 3$, $y = -1$, so $f(3) = -1$.

Let's examine how this technique is used on an SAT question:

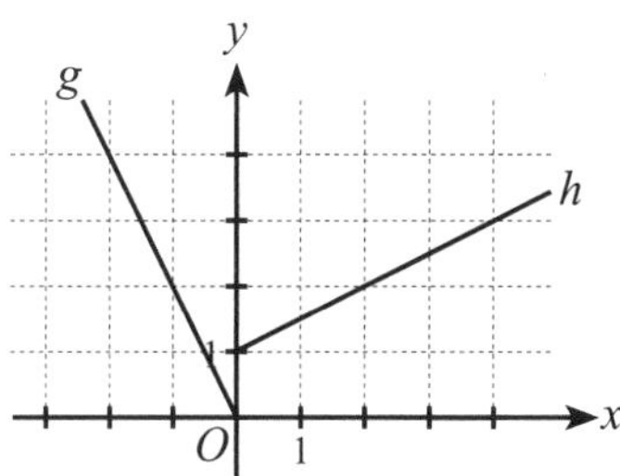

14. In the figure above, g and h are lines. What is the value of $g(-2) - h(4)$?

(A) 0
(B) 1
(C) 2
(D) 3
(E) 4

Start with each separate function.

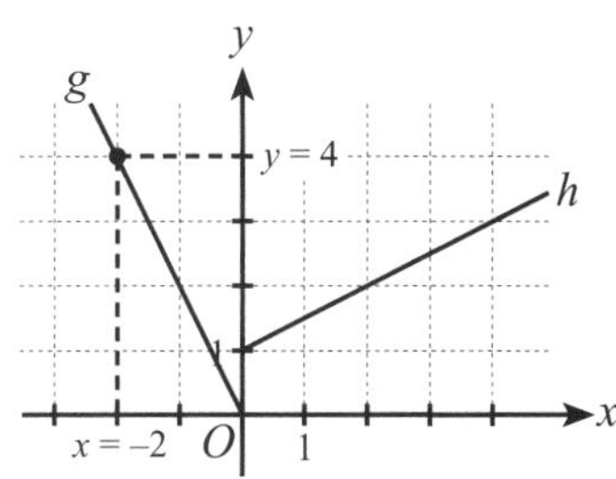

The first is $g(-2)$. This function simply asks, "What is the value of y when $x = -2$?"

Use the graph to find $x = -2$ on line g. What is the y-coordinate at that point?

$g(-2) = 4$

Because students become flustered by the use of f(x) instead of y, these questions have an exaggerated difficulty level. They should be easy points for you, though, because you know that f(x) = y.

Now examine the second function:

The function $h(4)$ asks, "What is the value of y when $x = 4$?"

Use the graph to find $x = 4$ on line h. What is the y-coordinate at that point?

$h(4) = 3$

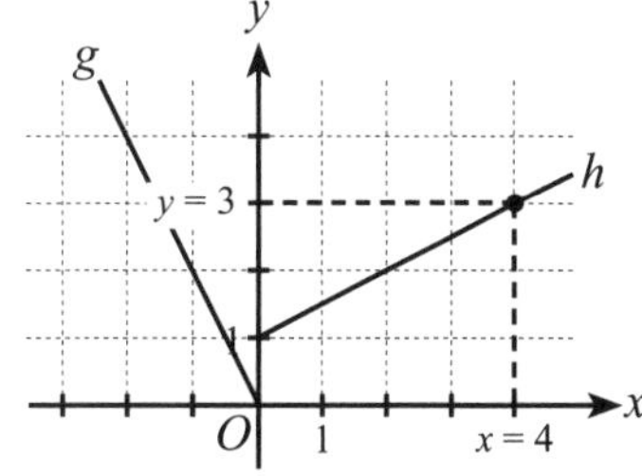

Finally, use the two values to perform the subtraction problem:

$g(-2) - h(4) \quad \rightarrow \quad 4 - 3 = 1$

The correct answer is (B).

Do not let functions in the coordinate plane cause you any anxiety. As long as you remember that $f(x) = y$, these questions become ordinary coordinate geometry questions.

Functions in Coordinate Geometry Problem Set

Solve the following multiple-choice questions by selecting the best answer from the five answer choices. For grid-in questions, write your answer in the grids and completely mark the corresponding ovals. Answers are on page 393.

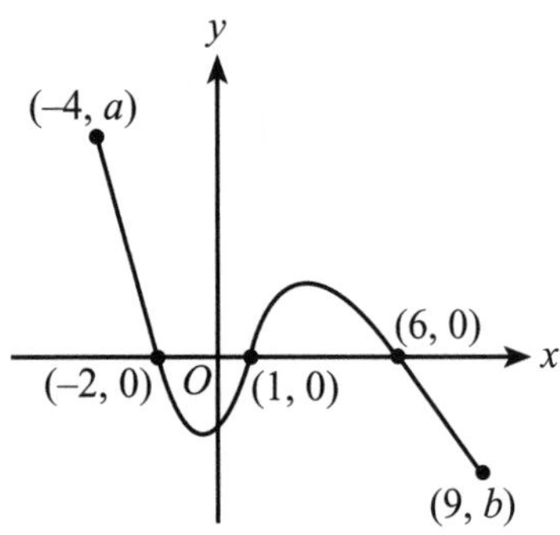

1. The function f is graphed above. For what values of x is $f(x)$ positive?

   (A) $-2 < x < 6$
   (B) $-4 \le x \le 9$
   (C) $-2 < x < 1$ or $6 < x \le 9$
   (D) $-4 \le x < -2$ or $6 < x \le 9$
   (E) $-4 \le x < -2$ or $1 < x < 6$

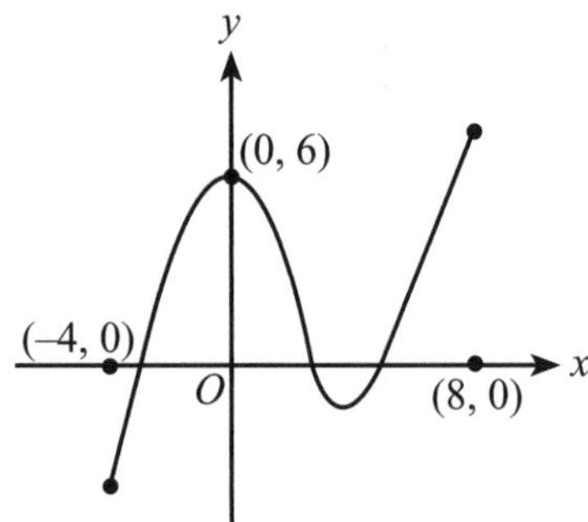

2. The function $y = f(x)$ is graphed above, where $-4 \le x \le 8$. How many values of x result in $f(x) = 3$?

   (A) One
   (B) Two
   (C) Three
   (D) Four
   (E) More than four

Transformations

A transformation is any change in a function that alters the graph of the function. The most common transformations on the SAT are translations and reflections.

Frequency Guide: 1

Required Knowledge and Skill Set

1. A translation is a shift of the graph, up or down, or left or right.

Study the original graph of $y = f(x)$ on the right.

To shift the parabola up or down, manipulate the y-coordinates by adding to or subtracting from $f(x)$. Examine the effects of these changes in the figures below.

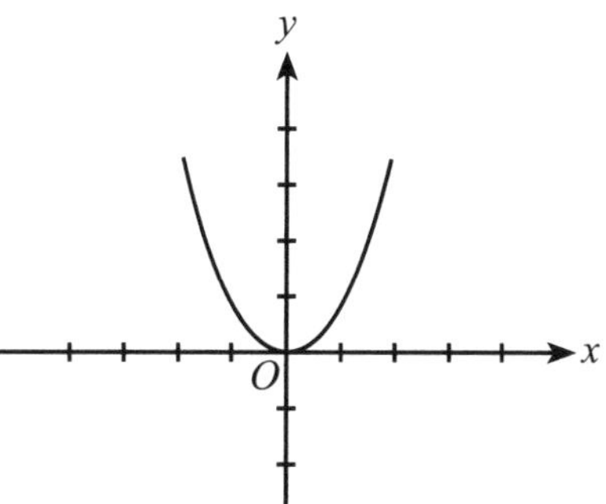

$y = f(x) + 1$
Shifts up 1 unit

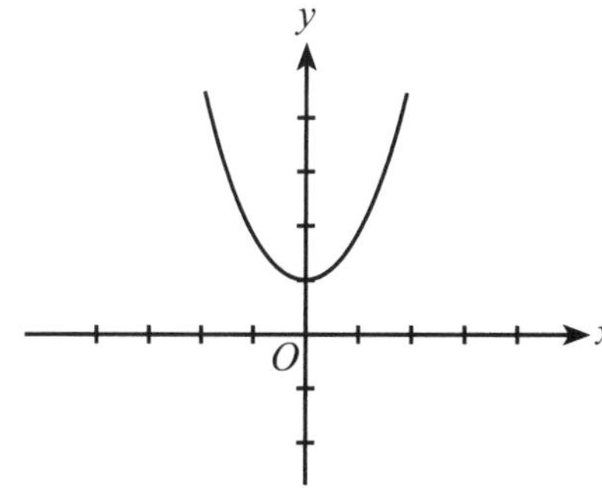

$y = f(x) - 1$
Shifts down 1 unit

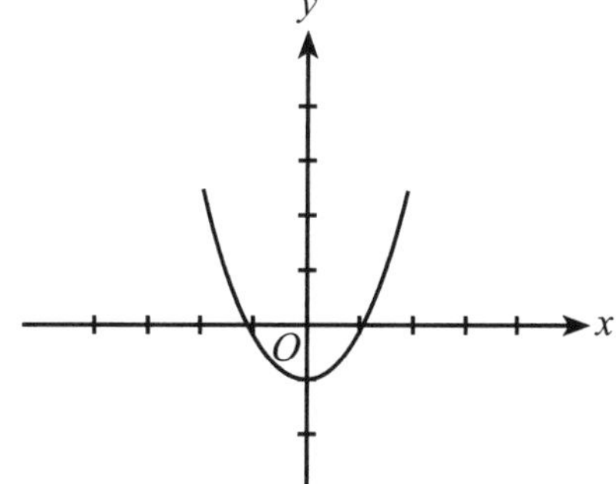

MEMORY MARKER: Be sure to memorize how graphs are affected when the x-and y-coordinates are manipulated.

To shift the parabola left and right, manipulate the x-coordinates by adding to or subtracting from x (inside the parentheses):

$y = f(x + 1)$
Shifts left 1 unit

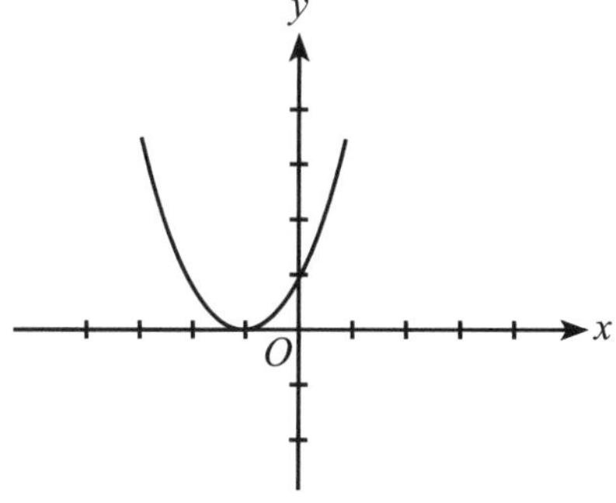

$y = f(x - 1)$
Shifts right 1 unit

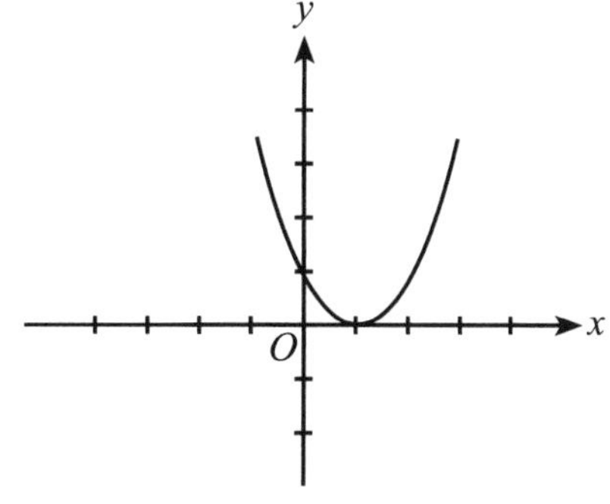

Some students think of x-coordinate manipulation as "backwards" because the plus sign moves the graph "back." Use any memory techniques that will help you with these transformations.

2. You should be able to recognize when a function is stretched or shrunk. Manipulating x by multiplication or division stretches or compresses the function along the x-axis:

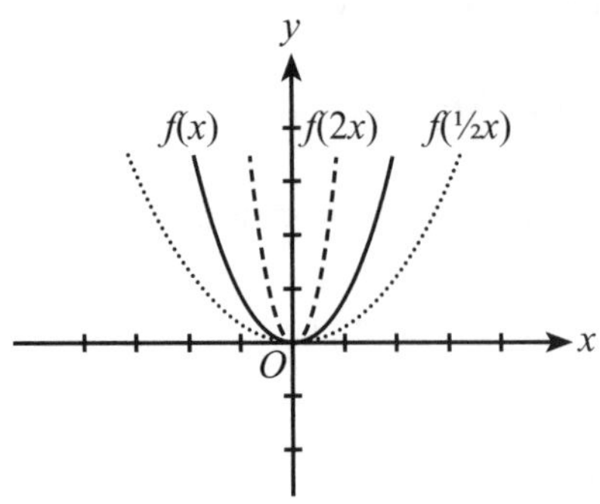

When x is greater than 1, the result is a "skinnier" parabola, but when x is less than 1, a "fatter" parabola is created.

Manipulating $f(x)$ by multiplication or division stretches or compresses the parabola along the y-axis:

Small numbers result in short and fat parabolas while larger numbers create tall and skinny parabolas.

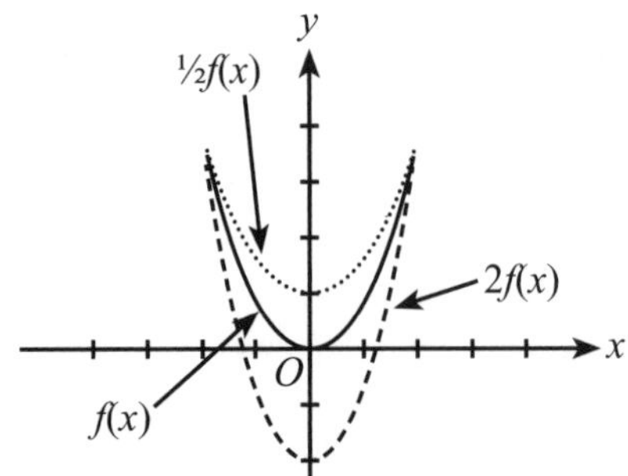

When x is greater than 1, the result is a "longer" parabola, but when x is less than 1, a "shorter" parabola is created.

3. A reflection occurs when an mirror image is created across a line of reflection. The SAT often uses the x- and y-axes as the line of reflection. Imagine holding a mirror on the line of reflection and glancing at the image in the mirror.

   If the x-axis is the line of reflection, place the mirror on the x-axis:

   $y = f(x)$ $\qquad\qquad$ $y = -f(x)$

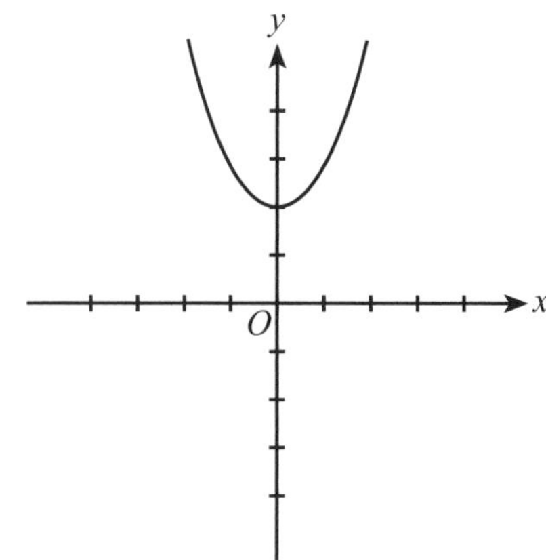

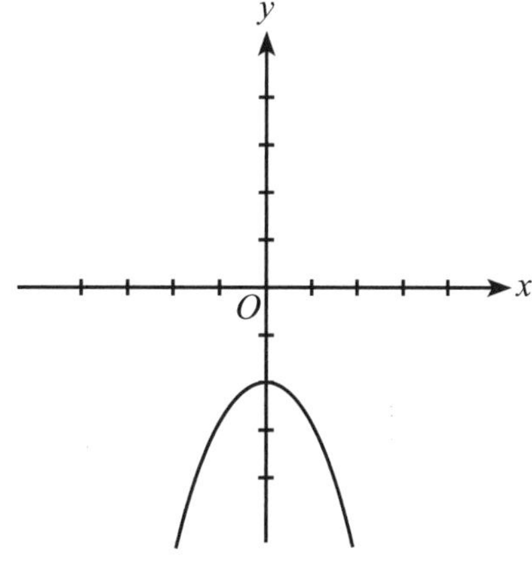

   If the y-axis is the line of reflection, place the mirror on the y-axis:

   $y = f(x)$ $\qquad\qquad$ $y = f(-x)$

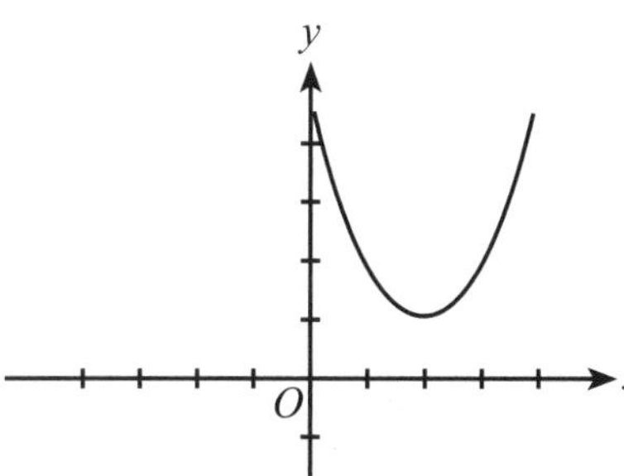

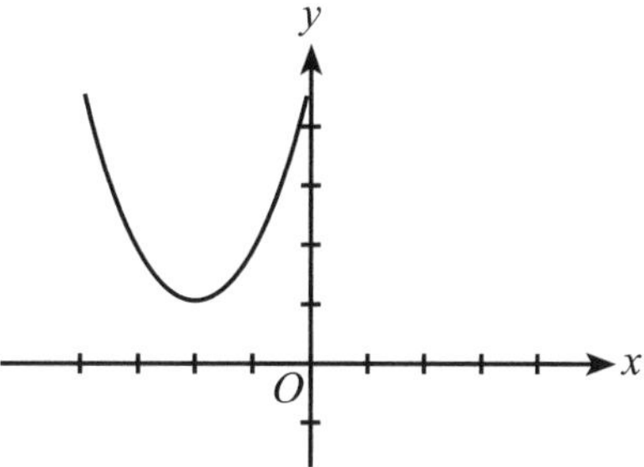

   The other two common lines of reflection are $y = x$ and $y = -x$. Consider the following, in which the figure is reflected over $y = x$:

To help you see reflections, imagine stamping an image with wet ink on the graph and then folding the paper along the line of reflection. The wet ink will make a mirror image on the folded paper. Where would the image appear when you opened the paper back up?

MEMORY MARKER:
These equivalents are included in the free flash cards at www.powerscore.com/satmathbible.

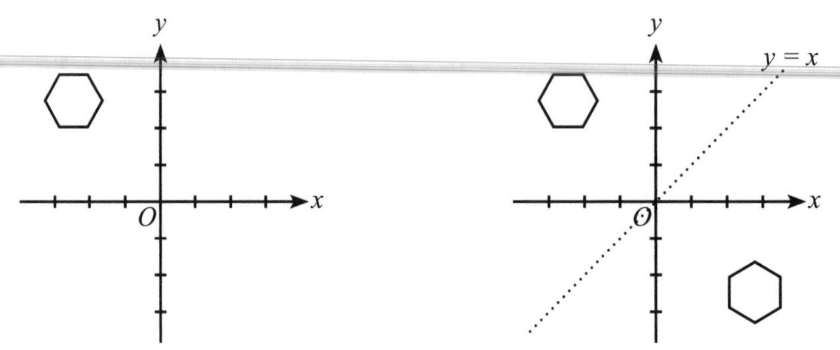

4. You should memorize the following information:

Summary of Transformations

Translations:
$y = f(x) + c$ shifts $y = f(x)$ up by c units
$y = f(x) - c$ shifts $y = f(x)$ down by c units
$y = f(x - c)$ shifts $y = f(x)$ to the right by c units
$y = f(x + c)$ shifts $y = f(x)$ to the left by c units

Shrinks and Stretches:
$y = f(cx)$ shrinks $y = f(x)$ along the x-axis
$y = f(x/c)$ stretches $y = f(x)$ along the x-axis
$y = cf(x)$ stretches $y = f(x)$ along the y-axis
$y = (1/c)f(x)$ shrinks $y = f(x)$ along the y-axis

Reflections:
$y = -f(x)$ reflects $y = f(x)$ across the x-axis
$y = f(-x)$ reflects $y = f(x)$ across the y-axis

Application on the SAT

Some transformation questions can be solved without memorizing the summary of transformations. You just need to understand that a shift replots points at a given distance from the original graph:

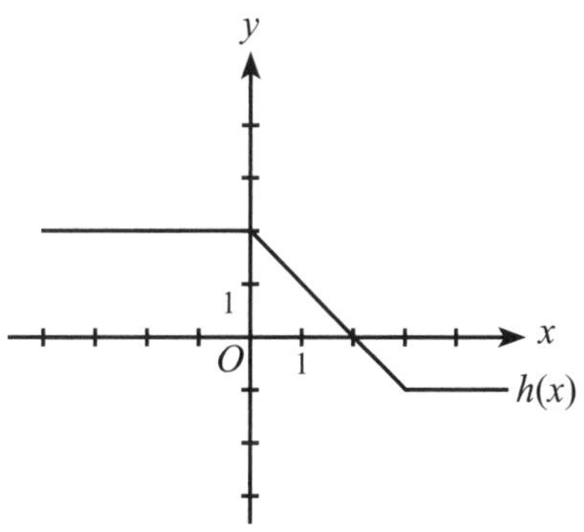

15. The figure above shows the graph of the function $h(x)$. The graph of $h(x)$ is simply the graph of $g(x)$, which is not shown, when $g(x)$ is shifted 2 units to the left. What is the y-intercept on the graph of $g(x)$?

    (A) –1
    (B) 0
    (C) 1
    (D) 2
    (E) 4

To solve this question, sketch $g(x)$:

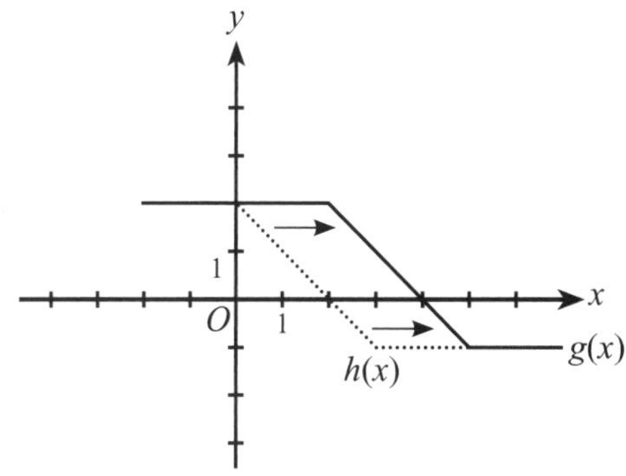

Because $h(x)$ is $g(x)$ shifted 2 units to the left, all of the points on $g(x)$ are moved 2 units to the right.

The sketch reveals that the y-intercept of $g(x)$ is 2.

The College Board may also test your knowledge of translations given information about the equation of the function. Consider the same graph with a new question:

The only reason translations have high difficulty levels is because most students forget the rules and become flustered. When the majority of students miss a question, it receives a "Hard" difficulty level.

Confidence Quotation
"Self-confidence is the first requisite to great undertakings."
—Samuel Johnson, English poet and essayist

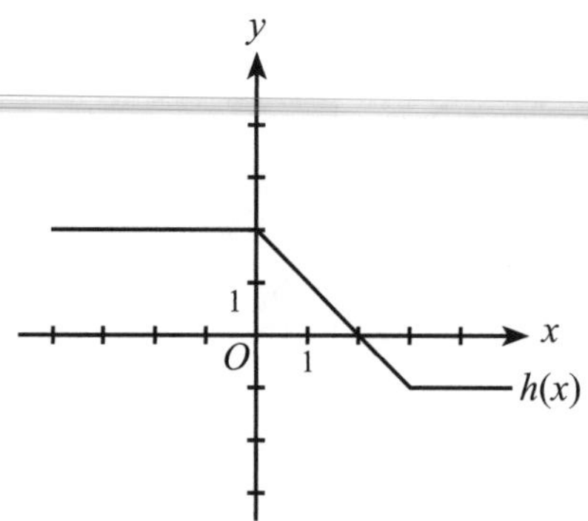

19. The figure above shows the graph of the function $y = h(x)$. What is the y-intercept of the graph of $y = h(x) - 2$?

    (A) -1
    (B) 0
    (C) 1
    (D) 2
    (E) 4

By memorizing the summary of translations, you know the following:

$y = f(x) - c$ shifts $y = f(x)$ down by c units

Therefore, $y = h(x) - 2$ shifts $y = h(x)$ down by 2 units. Sketch the graph.

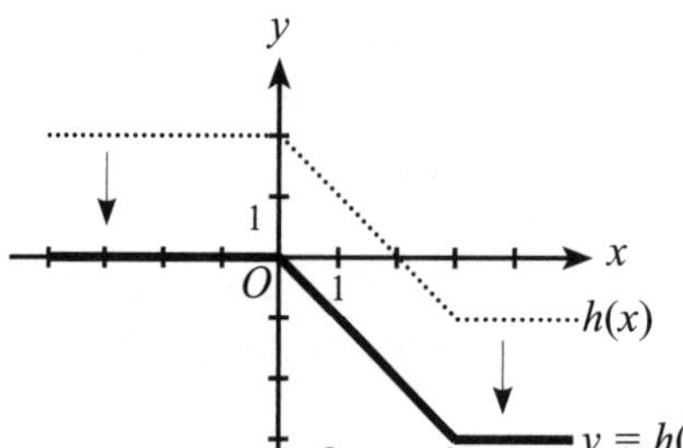

The y-intercept is 0. The correct answer is (B).

The most common transformation questions involve reflections. You may need to write the equation of a line reflected over an axis, or identify a point on a figure reflected over a line. Consider the following:

Reflections are "wet ink" problems. What happens when the paper is folded along the line of reflection?

Transformations Problem Set

Solve the following multiple-choice questions by selecting the best answer from the five answer choices. For grid-in questions, write your answer in the grids and completely mark the corresponding ovals. Answers begin on page 393.

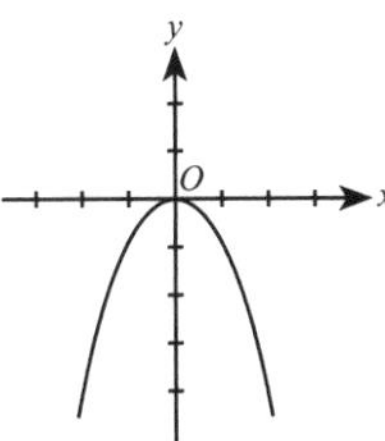

1. The function $f(x) = -x^2$ is graphed in the figure above. Which of the following could be the graph of $f(x) = -(x + 2)^2$?

(A)
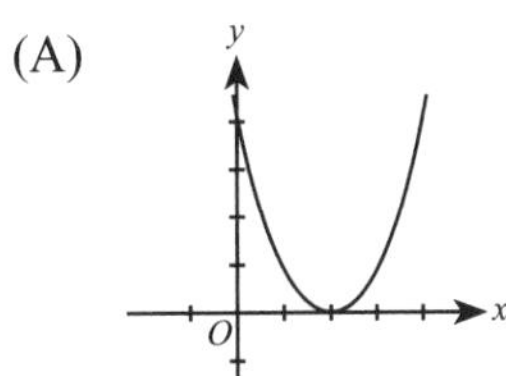

(B)
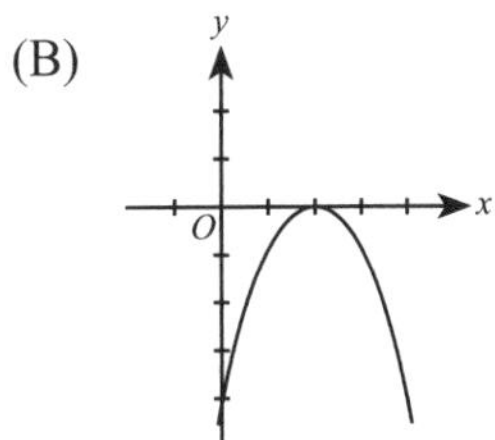

(C)
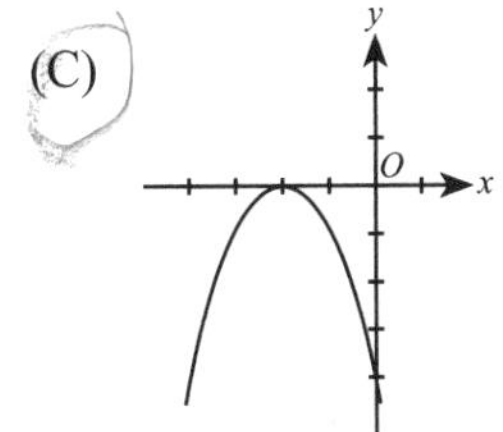

(D)
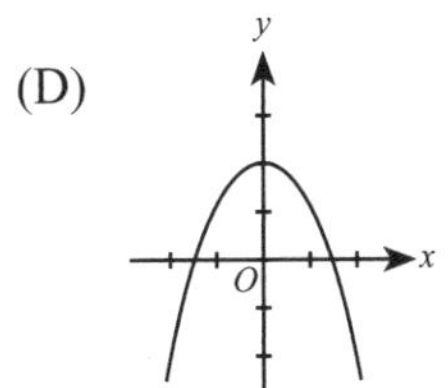

(E)
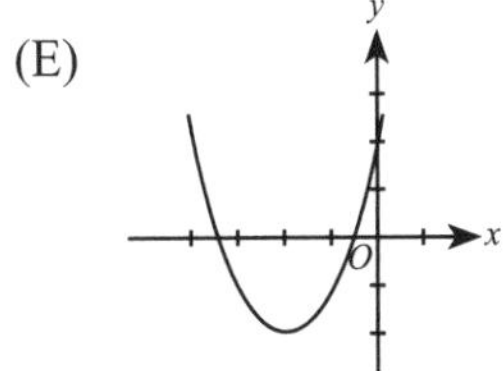

2. The equation of line ℓ, which is $y = -2x + 1$, is graphed in the coordinate plane. Line m is a reflection of line ℓ across the x-axis. Which of the following is the equation of line m ?

(A) $y = -2x - 1$

(B) $y = -\frac{x}{2} - 1$

(C) $y = \frac{x}{2} - 1$

(D) $y = 2x - 1$

(E) $y = 2x + 1$

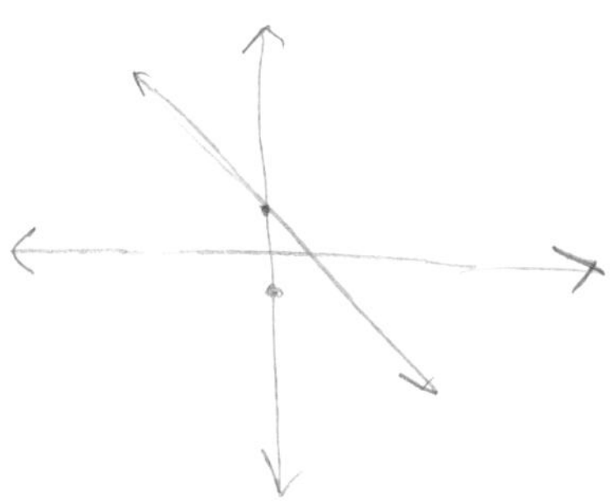

COORDINATE GEOMETRY MASTERY ANSWER KEY

Lines in Coordinate Geometry Problem Set—Pages 366-367

1. (D) Medium

Because the line is going down from left to right, it has a negative slope. This eliminates (B) and (E) because these equations have positive slopes. Line ℓ crosses the y-axis at –3, so the y-intercept is –3. This eliminates (C). That leaves (A) and (D). Is the slope one-third or negative three? You can use the slope formula and the points (–1, 0) and (0, –3), or you can use the graph to count the rise over run. The rise is –3 and the run is 1. Therefore, the slope is –3. Answer choice (D) is correct.

2. (B) Medium

DIAGRAM the question:

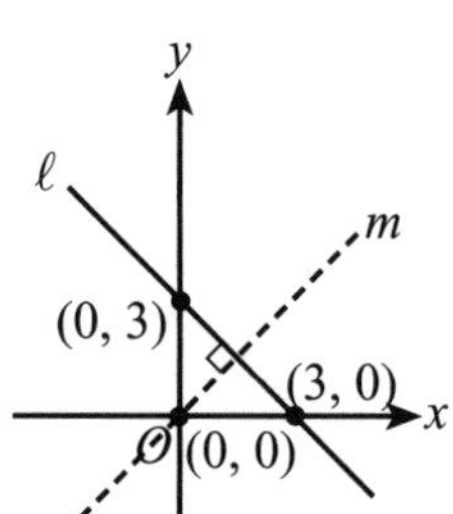

Anytime the College Board uses perpendicular lines, the reciprocal slope is the key. To find the slope of m, find the slope of ℓ. You can use the Slope Formula, or count the rise and run using (0, 3) and (3, 0):

$$\text{Slope of line } \ell = \frac{y_2 - y_1}{x_2 - x_1} \rightarrow \frac{3-0}{0-3} \rightarrow -1$$

Slope of line m = negative reciprocal of –1 = 1

Plug the slope into the equation of line m:

$y = mx + b \rightarrow y = (1)x + b$

Because line m passes through the origin, its y-intercept is 0:

$y = mx + b \rightarrow y = (1)x + 0 \rightarrow y = x$ The correct answer is (B).

3. (A) Medium

DIAGRAM the question:

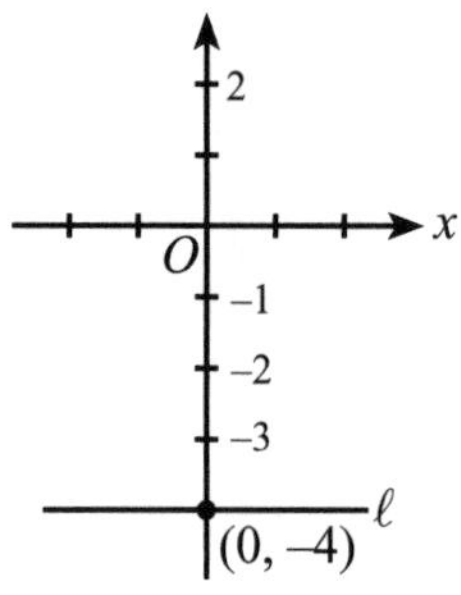

Because line ℓ is parallel to the x-axis, its slope is 0. The y-intercept is –4.

$y = mx + b \rightarrow y = (0)x - 4 \rightarrow y = -4$

The correct answer is (A).

4. (A) Hard

Because the question is asking for the y-intercept, use the equation of a line:

$y = mx + b$

The slope is provided, so you need a point on the line (x, y). Use the diagram to find that point:

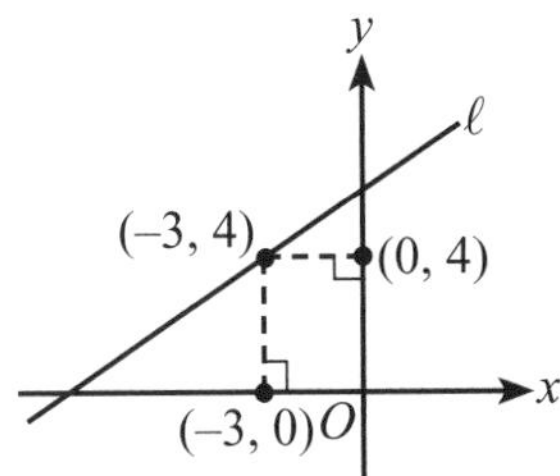

Point (–3, 4) is on line ℓ:

Slope $= \frac{2}{3}$
$x = -3$
$y = 4$

$y = mx + b \;\rightarrow\; 4 = \frac{2}{3}(-3) + b \;\rightarrow\; 4 = -2 + b \;\rightarrow\; 6 = b$

5. (B) Hard

There are two ways to solve this question. Use the Slope Formula:

$$\text{Slope} = \frac{y_2 - y_1}{x_2 - x_1} \;\rightarrow\; \frac{2}{5} = \frac{2 - r}{15 - 0} \;\rightarrow\; 2(15 - 0) = 5(2 - r) \;\rightarrow\; 30 = 10 - 5r \;\rightarrow\; 20 = -5r \;\rightarrow\; -4 = r$$

Or the equation of a line:

$$y = mx + b \;\rightarrow\; 2 = (15)\frac{2}{5} + b \;\rightarrow\; 2 = \frac{30}{5} + b \;\rightarrow\; 2 = 6 + b \;\rightarrow\; -4 = b$$

6. $0.75 < x < 0.857$ Hard

DIAGRAM the question:

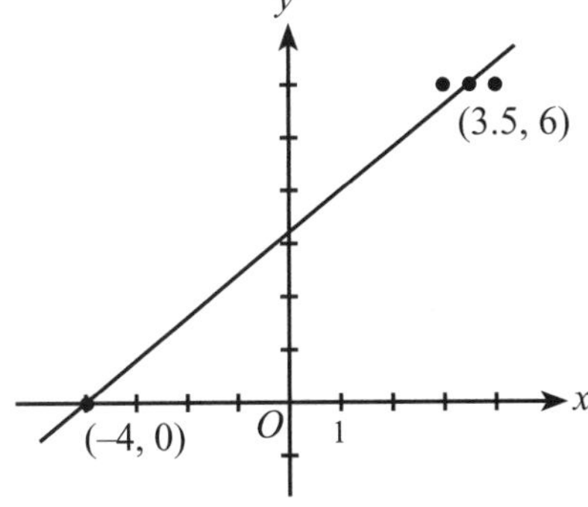

SUPPLY a point between (3, 6) and (4, 6): (3.5, 6). Then use the slope formula to find the slope:

$$\text{Slope} = \frac{y_2 - y_1}{x_2 - x_1} \;\rightarrow\; \frac{6 - 0}{3.5 - (-4)} \;\rightarrow\; \frac{6}{3.5 + 4} \;\rightarrow\; \frac{6}{7.5}$$

Slope = 0.8

Any answer between 0.75 and 0.857 is accepted.

7. 25 Hard

There are two ways to solve this question; the first involves the Pythagorean Theorem. DIAGRAM the question:

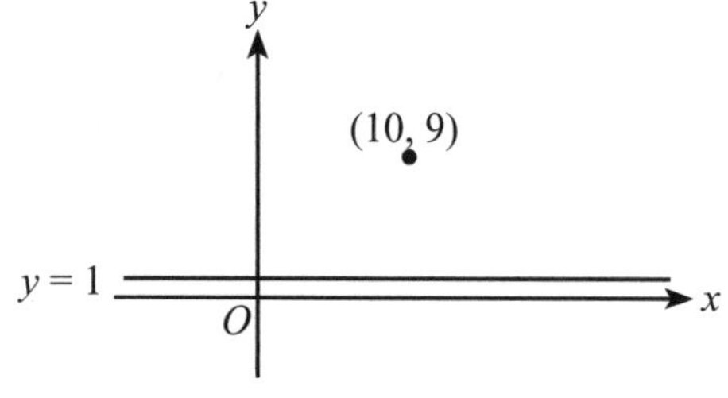

Point $(x, 1)$ lies on the line $y = 1$, and is 17 units from the point (10, 9).

DIAGRAM this statement.

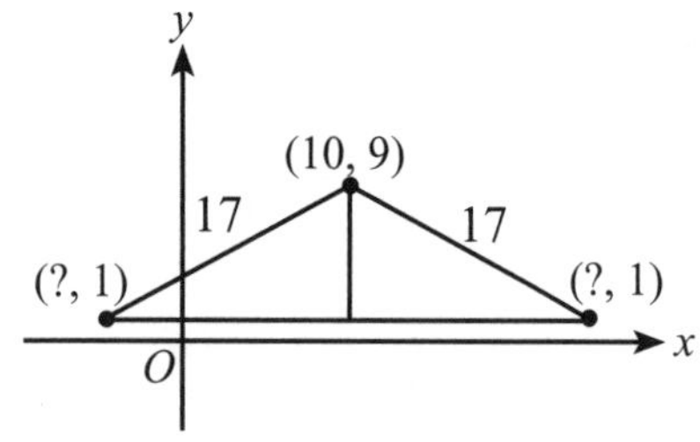

Use information from the diagram to find the height of the right triangles. From (10, 9) to the line $y = 1$ is a distance of $9 - 1 = 8$.

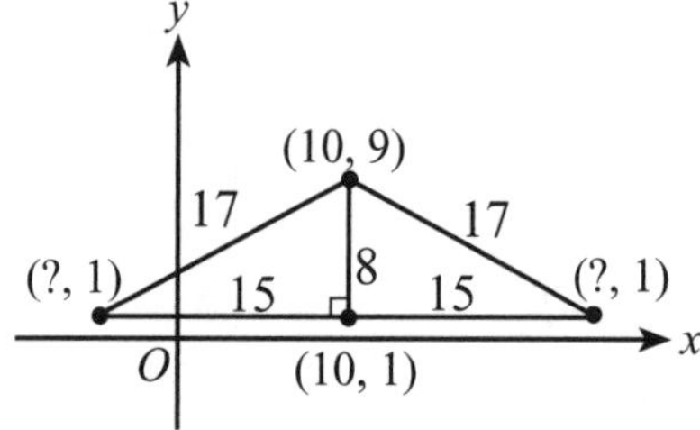

One side of the right triangle is 8, and the hypotenuse is 17. Do you recognize the Pythagorean Triple? It's an 8:15:17 triangle.

If you do not recognize the Pythagorean Triple, use the Pythagorean Theorem:

$a^2 + b^2 = c^2$
$8^2 + b^2 = 17^2$
$64 + b^2 = 289$
$b^2 = 225$
$b = 15$

If the base of the triangle is 15, then the distance from (10, 1) to $(x, 1)$ is 15:

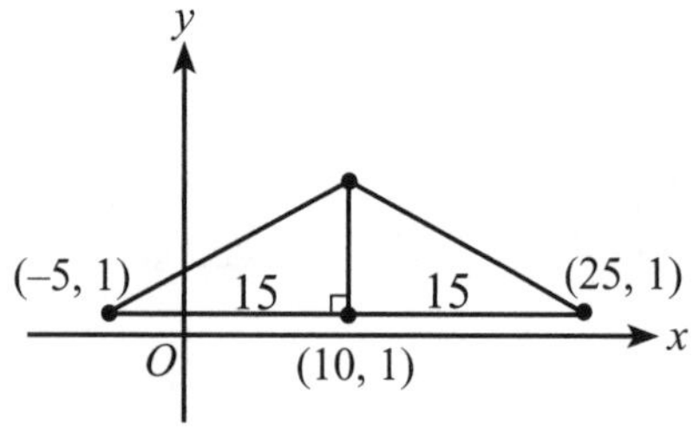

$10 - 15 = -5$

$10 + 15 = 25$

x is –5 or 25

However, negative numbers can not be entered into the grid so the only answer is 25.

The other way to solve this difficult question is the Distance Formula:

$\text{Distance} = \sqrt{(x_2 - x_1)^2 + (y_2 - y_1)^2}$
$17 = \sqrt{(x-10)^2 + (1-9)^2}$
$289 = (x - 10)^2 + (1 - 9)^2$
$289 = (x - 10)(x - 10) + (-8)^2$
$289 = (x^2 - 20x + 100) + 64$
$289 = x^2 - 20x + 164$
$0 = x^2 - 20x - 125$
$0 = (x - \quad)(x + \quad)$
$0 = (x - 25)(x + 5)$
$x = 25$ or $x = -5$
However, negative numbers can not be entered into the grid so the only answer is 25.

Use which ever method is most efficient for you.

Figures in Coordinate Geometry Problem Set—Page 372

1. (D) Medium

To find the area of a square, you need the length of one side:

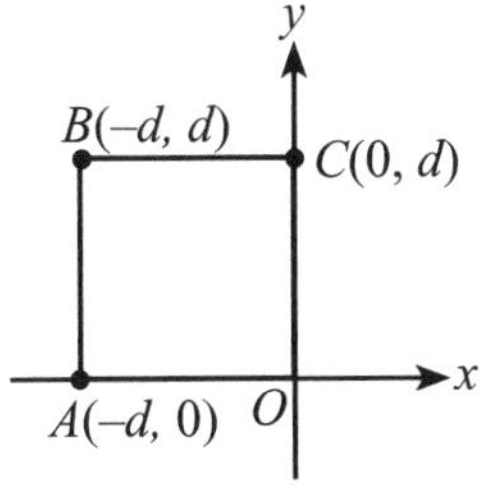

The length of side OC is d.

Area = s^2
Area = d^2 The correct answer is (D).

2. 2.5 Medium

Imagine the full circle:

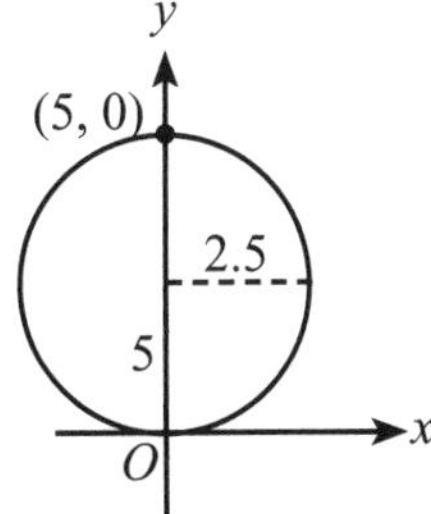

The diameter of the circle is 5. Therefore, the radius is 2.5.

The point furthest from the y-axis is (2.5, 2.5).

The correct answer is 2.5.

3. (D) Hard

Use the hidden 30°:60°:90° triangle:

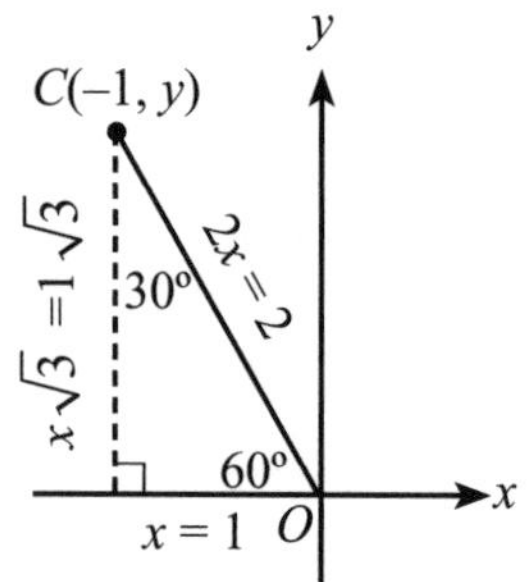

Because CO is the hypotenuse, it is twice as long as the shortest side of the triangle.

The correct answer is (D).

Parabolas in Coordinate Geometry Problem Set—Page 375

1. (C) Easy

Because $a > 0$, the parabola opens upward. This eliminates (B) and (D). Because $c < 0$, the y-intercept occurs below the x-axis. This eliminates (A) and (E). The correct answer is (C).

2. (C) Hard

All of the answer choices use the vertex form of the equation of a parabola:

Vertex equation of a parabola: $y = a(x - h)^2 + k$

From the diagram, it is clear that the vertex is (3, –2). Since (h, k) is the vertex, $h = 3$ and $k = -2$. Plug these values into the equation:

$$y = a(x - h)^2 + k$$
$$y = a(x - 3)^2 - 2$$

You can eliminate choices (A), (B), and (D), as they use the wrong signs for h or k.

You must determine the value of a to solve the question. Plug in (1, 0) for x and y:

$$y = a(x - 3)^2 - 2$$
$$0 = a(1 - 3)^2 - 2 \quad \rightarrow \quad 0 = a(-2)^2 - 2 \quad \rightarrow \quad 0 = 4a - 2 \quad \rightarrow \quad 2 = 4a \quad \rightarrow \quad \frac{1}{2} = a$$

Since $a = \frac{1}{2}$, the correct answer is (C): $y = \frac{1}{2}(x - 3)^2 - 2$

Functions in Coordinate Geometry Problem Set—Page 380

1. (E) Medium

Remember, $f(x) = y$. So this question asks, "When values of x result in positive y-values?" Use the graph:

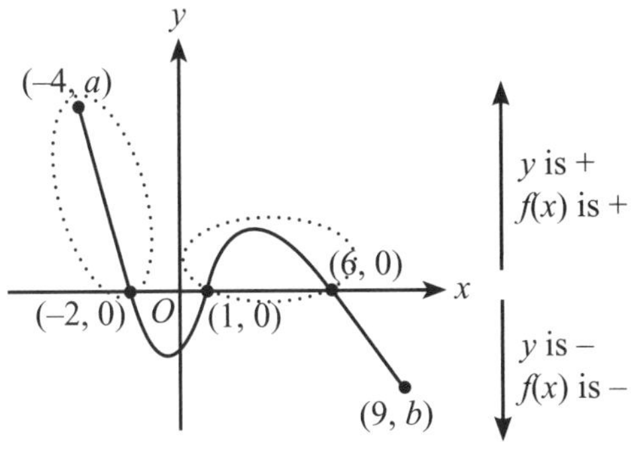

There are two parts of the function that have positive y-values, which we have circled with a dotted line. What are the x-values of these two sections?

The y-values are positive when x is –4 or between –4 and –2, and again when x is between 1 and 6.

The correct answer is (E).

2. (C) Medium

Again, $f(x) = y$. So if $f(x) = 3$, then $y = 3$. Draw the line $y = 3$ on the graph:

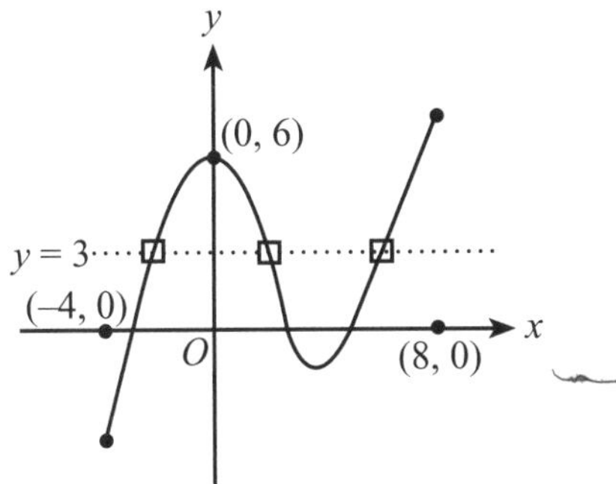

The function has *three* x-values when $y = 3$, as shown by the small boxes in the diagram to the left. You could estimate their values as $x = -3$, $x = 2.5$, and $x = 6$ (although this is not required).

The correct answer is (C).

Transformations—Page 387

1. (C) Medium

When x is manipulated with addition and subtraction, the figure moves either left or right. When a value is added to x, the figure moves left. In this case, when 2 is added to x, the parabola moved two spaces to the left. The correct answer is (C).

2. (D) Hard

In order to visualize the reflection, sketch line ℓ:

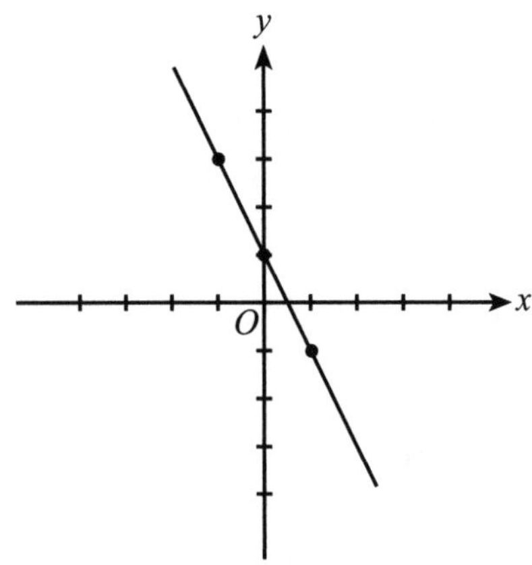

The line of reflection is the x-axis. Imagine drawing line ℓ with wet ink and then folding the paper on the x-axis. What would the mirror image look like?

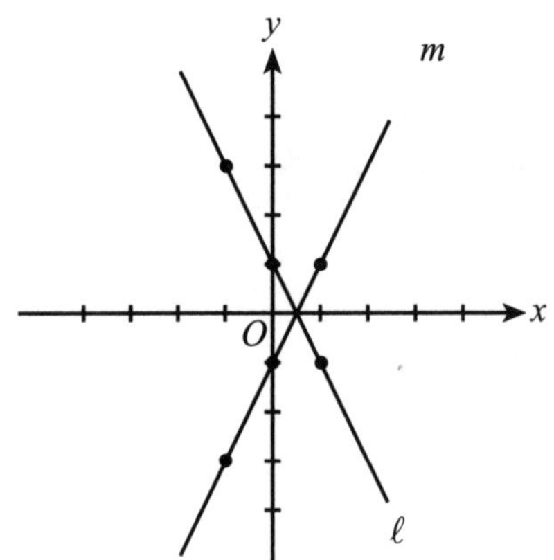

Now find the equation of line m using the equation of a line:

$$y = mx + b$$

The slope of line m is –2.

The y-intercept of line m = –1

$$y = -2x - 1$$

The correct answer is (D).

Notes:

CHAPTER TEN:
DATA ANALYSIS, STATISTICS, AND PROBABILITY MASTERY

The final math content area of the SAT includes data analysis, statistics, and probability. Because many high school students have not taken a statistics course, they find these questions especially intimidating. However, these test takers make ideal PowerScore students, as there are no bad habits to unlearn or excess information to disregard. You simply need to learn the concepts in this chapter to be prepared to face any statistics or probability question on the SAT.

Do not be intimidated by this section; we will give you all the tools you need to succeed!

As always, we have indicated the frequency of each type of question at the beginning of each concept section. The key is included below for your convenience:

Rating	Frequency
5	Extremely High: 3 or more questions typically appear on every SAT
4	High: at least 2 questions typically appear on every SAT
3	Moderate: at least one question typically appears on every SAT
2	Low: one question typically appears on every two or three SATs
1	Extremely Low: one question appears infrequently and without a pattern

Remember that there are eight solution strategies you can employ on SAT math questions:

1. ANALYZE the Answer Choices
2. BACKPLUG the Answer Choices
3. SUPPLY Numbers
4. TRANSLATE from English to Math
5. RECORD What You Know
6. SPLIT the Question into Parts
7. DIAGRAM the Question
8. SIZE UP the Figures

Be sure to check your answers in the answer key following the chapter. It is important to understand why you missed a particular question, in order to avoid making this same mistake on the SAT.

Data Analysis

Frequency Guide: 4

Data analysis questions use diagrams, figures, tables, or graphs in conjunction with arithmetic and algebra.

Required Knowledge and Skill Set

1. You must be able to read and interpret several graphs, including a pictograph, bar graph, pie graph, line graph, and scatterplot.

   A pictograph uses pictures to represent data:

FISHING LICENSES SOLD IN METRO AREA BY COUNTY = 1000 fishing licenses	
Beaufort County	
Clinton County	
Ingram County	
Hamilton County	
Lenawee County	

How many fewer fishing licenses were sold in Hamilton County than Clinton County? Answers to these questions appear in the margin on page 400.

   A bar graph uses horizontal or vertical bars to represent data:

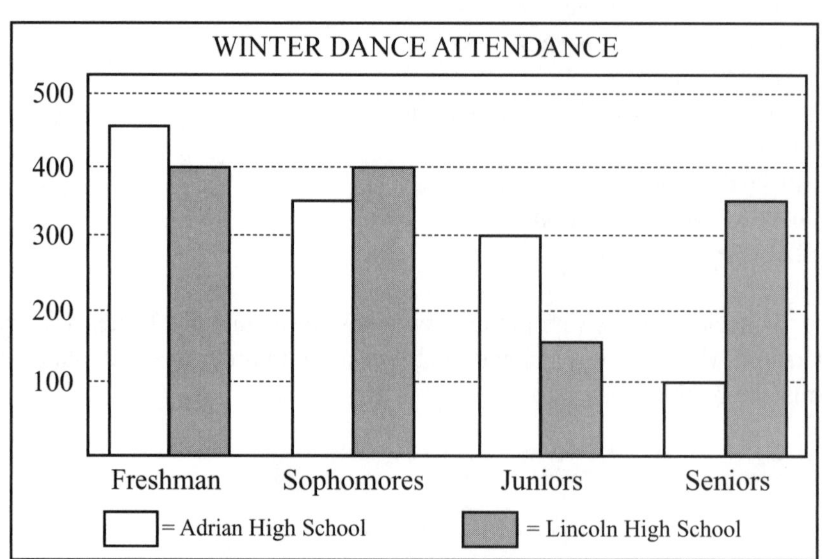

Which school had the most students attend the Winter Dance?

A pie graph, often used to represent percentages, uses a circle to display data. A Venn Diagram uses two circles to represent data:

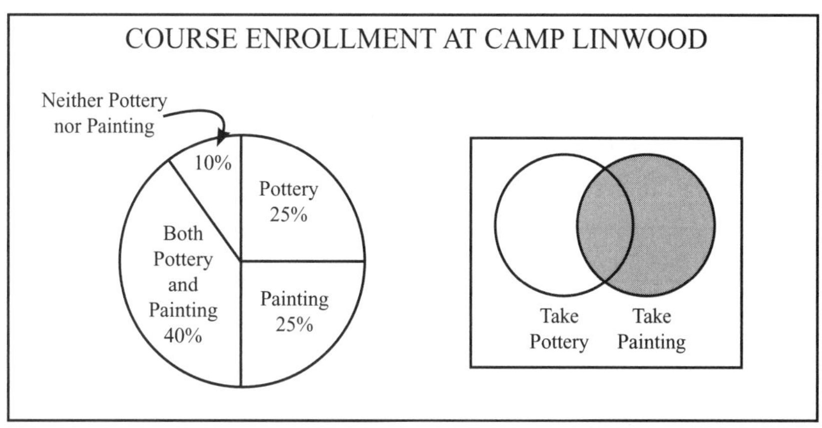

Pie Graph Venn Diagram

If there are 500 students at Camp Linwood, how many are in Pottery class only?

A line graph plots data on a graph and connects the points to form a line:

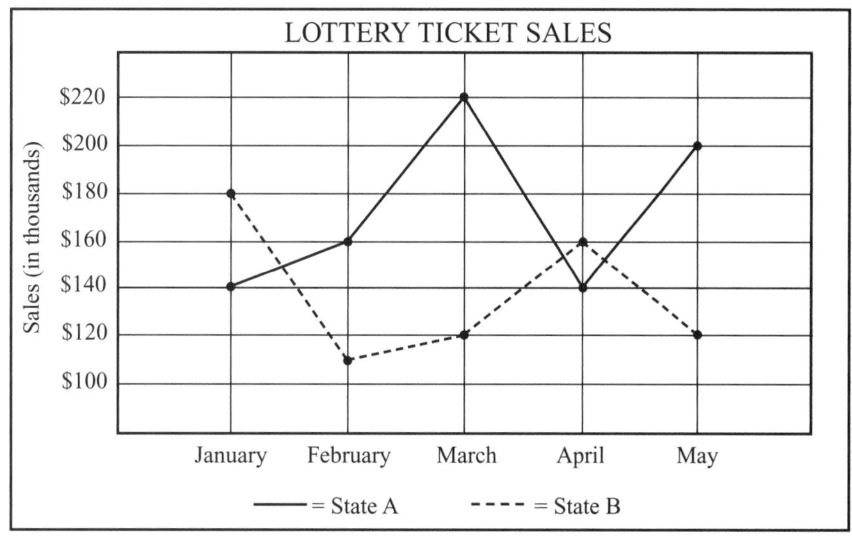

What was the difference in lottery sales between State A and State B in March?

A scatterplot also plots data on the graph, but the data is not connected by a line. If the points are clustered closely together, you may be able draw a line to show a trend in data:

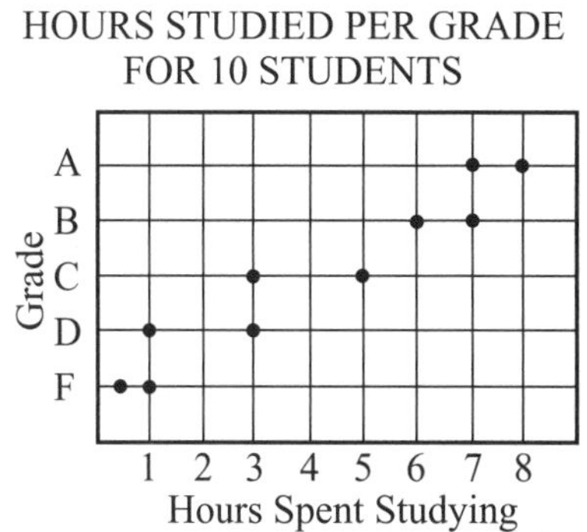

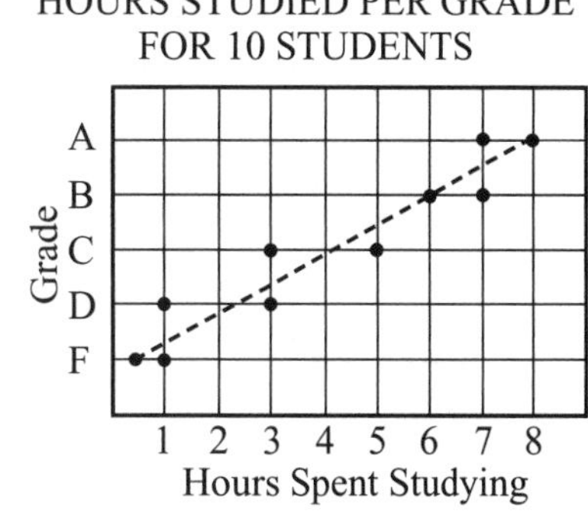

Scatterplots tend to be the most difficult Data Analysis questions because fewer students are exposed to them.

2. Read graphs very carefully! Take note of titles, headings, scale, units, and other information presented.

3. For line graphs, bar graphs, and scatterplots, be prepared to SIZE UP the figure and estimate an answer. Visually accurate figures can often be solved just by analyzing the graph or the diagram.

4. Tables organize information in an easy-to-read format:

EMPLOYEE BENEFITS		
Years Worked	Vacation Days	Commission Percentage
1 to 5	10	4%
5 to 9	14	9%
10 to 20	17	16%
20 or more	21	25%

What is the difference between the month with the most tourists and the month with the fewest tourists?

Month	Number of Tourists
May	40,000
June	80,000
July	150,000
August	110,000
September	30,000
October	15,000

CLIFFORD HIGH SCHOOL'S JV BASEBALL RESULTS					
Date	Home Team	Score	Away Team	Score	Record
Mar. 28	Clifford	5	Lorraine	2	1-0
Apr. 2	Westbrook	8	Clifford	9	2-0
Apr. 6	Clifford	2	Highlands	4	2-1
Apr. 8	Clifford	5	Chesterfield	0	3-1
Apr. 16	Clifford	3	Lawrence	2	4-1
Apr. 21	Northfield	7	Clifford	2	4-2
Apr. 27	Jamestown	12	Clifford	0	4-3
May 1	Clifford	6	Canton	5	5-3
May 6	Naleen	0	Clifford	1	6-3
May 9	Clifford	7	Washington	4	7-3

ANSWERS
Fishing: 3000
Dance: Lincoln
Camp: 125
Lottery: $100,000
Tourists: 135,000

Application on the SAT

Most graph questions on the SAT simply ask you to interpret data in the figure. Be prepared to apply arithmetic or algebra when interpreting the graph. Consider an example:

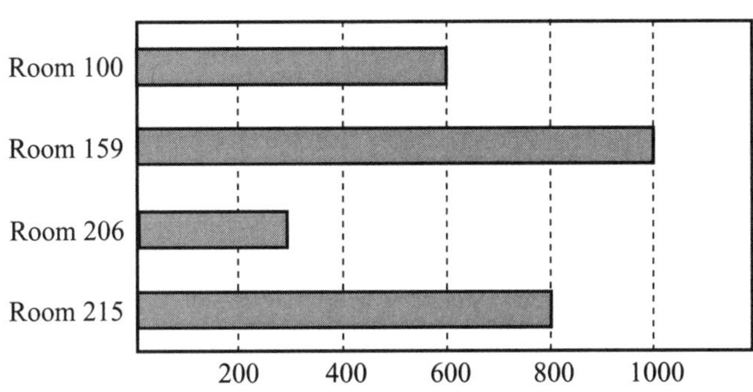

12. The graph above illustrates the number of box tops collected by four classrooms during a school contest. The sum of the box tops collected by the two rooms with the fewest box tops is approximately what percent of the sum of the box tops collected by the two rooms with the most box tops?

    (A) 30%
    (B) 50%
    (C) 65%
    (D) 70%
    (E) 80%

Data Analysis questions test not only your ability to read tables and graphs, but also your ability to apply arithmetic, algebra, probability, or statistics to the information in the figure.

To begin, find the sum of the box tops collected by the two rooms with the least number of box tops:

Room 100:	600	
Room 206:	300	(approximately)
	900	

Now find the sum of the box tops collected by the two rooms with the largest number of box tops:

Room 159:	1000
Room 215:	800
	1800

Now TRANSLATE:

The sum of the fewest is what percent of the sum of the most

$$900 = \frac{x}{100} \times 1800 \quad \rightarrow \quad 90{,}000 = 1800x \quad \rightarrow \quad 50 = \text{x}$$

The correct answer is (B), 50%.

Some questions will provide data and ask you to choose the graph that illustrates that data. These are usually line graph questions:

LENGTH OF A FISH

Age (in weeks)	2	3	4	5	6
Length (in centimeters)	6	15	20	22	23

12. The measurements of a certain fish at different ages are given in the table above. Which of the following graphs could represent the information in the table?

(A)

Length (cm)
6
2
Age (weeks)

(B)

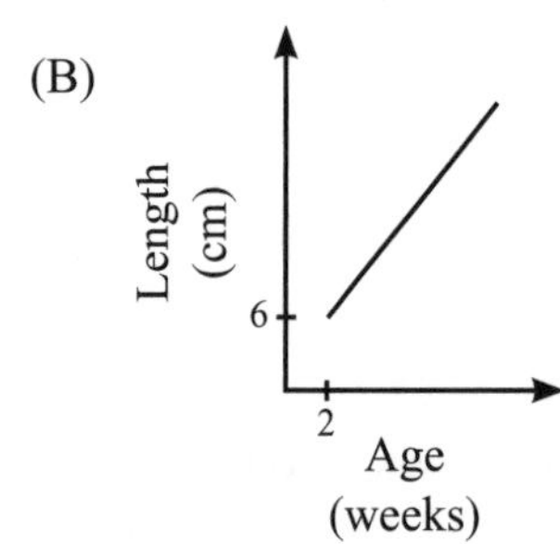

(C)

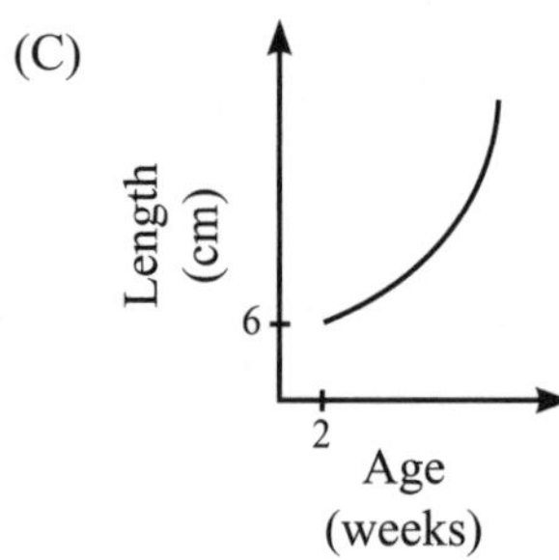

(D)

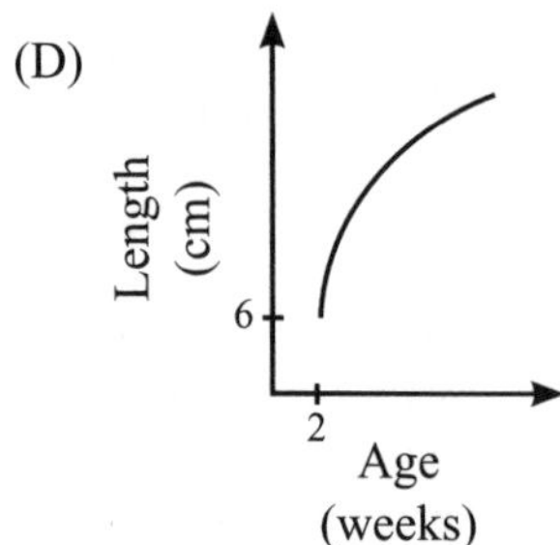

(E)

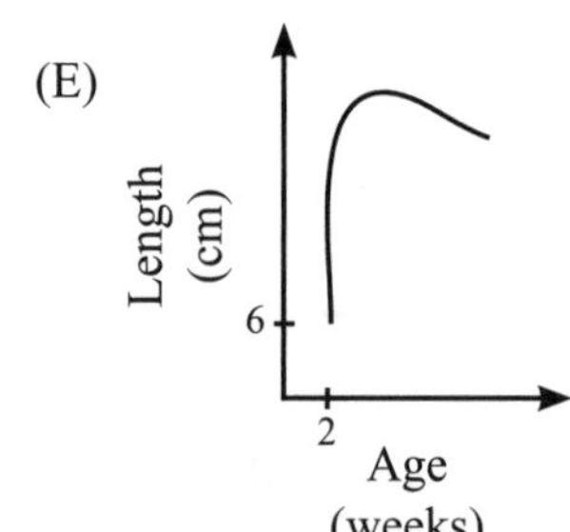

Confidence Quotation
"We gain strength and courage and confidence by each experience in which we really stop to look fear in the face. We must do that which we think we cannot."
—Eleanor Roosevelt, First Lady of the United States and civil rights activist

To solve this question, draw a rough sketch of the graph and plot the points:

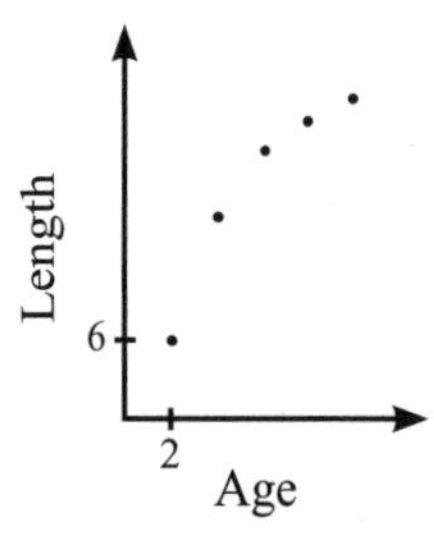

If you drew a line through the points, which of the answer choices would your sketch resemble?

The correct answer is choice (D).

You should be prepared to compare two graphs displaying the same or related information:

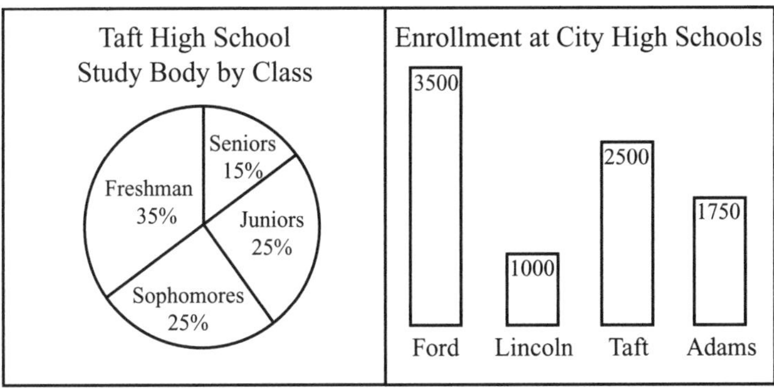

7. According to the graphs above, how many seniors are enrolled at Taft High School?

(A) 375
(B) 525
(C) 750
(D) 1500
(E) 2125

You must use information from both graphs to solve this question. First, find the total student enrollment at Taft High School. According to the bar graph, total enrollment is 2500 students.

Now use the information in the pie graph. Fifteen percent of the students at Taft High School are seniors:

15% of $2500 = ?$
$0.15 \times 2500 = 375$

The correct answer is (A).

Data Analysis questions are sometimes accompanied by two or three questions. The text above the graph will alert you to this situation by saying something like "Questions 7 and 8 refer to the information in the following graphs."

Expect at least one question that compares two or more graphs or tables. In fact, these types of questions often have two questions associated with the same figure.

8. According to the information in the graphs, how many more freshman are enrolled than sophomores at Taft High School?

(A) 125
(B) 250
(C) 375
(D) 500
(E) 575

Freshman: 35% of 2500 =	875
Sophomores: 25% of 2500 =	–625
	250 The correct answer is (B).

Tables are the most common figure in Data Analysis questions.

Tables are prominently featured on the SAT. Some questions with tables will ask you to use arithmetic to solve a problem using information in a completed table:

SALES OF WILDLIFE TOURS

Price of Tour	Number of Purchased Tours
\$5.00	120,000
\$10.00	95,000
\$20.00	65,000

10. A wildlife company offered tours for three different prices during a single year. Based on the information above, how much more money did the company make when the price was \$20.00 than when the price was \$5.00 ?

(A) \$35,000
(B) \$70,000
(C) \$350,000
(D) \$700,000
(E) \$1,050,000

Find the total sales of the \$5.00 tickets and the \$20.00 tickets:

Questions with completed tables usually require arithmetic.

$$\$5.00 \times 120{,}000 = \$600{,}000$$
$$\$20.00 \times 65{,}000 = \$1{,}300{,}000$$

Now simply subtract the smaller amount from the larger amount:

$$\$1{,}300{,}000 - \$600{,}000 = \$700{,}000$$

The correct answer is (D).

Later in this chapter we will cover averages, medians, modes, counting problems, and probability. All of these statistics topics have appeared on the SAT in questions with tables, so expect to see some completed tables testing your ability with these subject areas.

Other table questions will require you to use algebra to find missing information:

LISA'S HORSE EXPENSES

	Boarding	Lessons	Total
January		\$50	
February		\$40	
March		\$80	
Total			\$260

9. The table above, which is missing some information, shows Lisa's expenses for keeping a horse. If her boarding costs were the same each month, what were her total expenses for February?

(A) \$30
(B) \$50
(C) \$70
(D) \$80
(E) \$110

Use a variable to represent the boarding expenses. Since the cost is the same every month, use the same variable:

Questions with incomplete tables usually require algebra.

LISA'S HORSE EXPENSES

	Boarding	Lessons	Total
January	b	\$50	
February	b	\$40	
March	b	\$80	
Total			\$260

Now, write an algebraic equation that solves for b:

$(b + 50) + (b + 40) + (b + 80) = 170$
$3b + 170 = 260$
$3b = 90$
$b = 30$

Many students would select answer choice (A) and move on. But they would be wrong; the question asks for the total expenses in February:

Learning to reread each question after finding a solution is an invaluable practice on the SAT.

LISA'S HORSE EXPENSES

	Boarding	Lessons	Total
January	\$30	\$50	
February	\$30	\$40	\$70
March	\$30	\$80	
Total			\$260

The correct answer is (C).

Data Analysis Problem Set

Solve the following multiple-choice questions by selecting the best answer from the five answer choices. For grid-in questions, write your answer in the grids and completely mark the corresponding ovals. Answers begin on page 450.

APPROXIMATE CONVERSIONS

Number of Gallons	2	4	x
Number of Liters	7.6	15.2	30.4

1. The table above shows approximate conversions from gallons to liters. What is the value of x ?

(A) 6
(B) 8
(C) 10
(D) 12
(E) 16

Height of Seniors

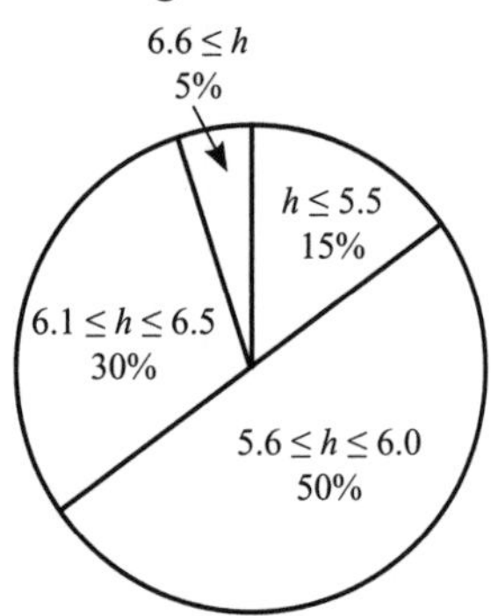

2. The seniors at Woodhaven High School are being measured for their caps and gowns for graduation. The figure above shows their height (h), in feet. For example, 30% of the seniors are 6.1 feet to 6.5 feet tall. If there are 760 seniors at Woodhaven High School, how many are 6.0 feet tall or less?

(A) 114
(B) 228
(C) 380
(D) 494
(E) 722

CONDOMINIUMS

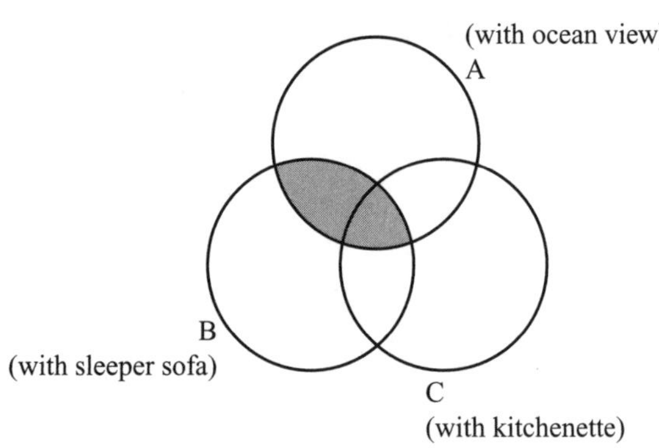

3. In the figure above, three circles represent condominiums on the beach. Circle A represents condominiums with an ocean view, Circle B represents condominiums with a sleeper sofa, and Circle C represents condominiums with a kitchenette. What does the shaded region represent?

(A) Condominiums with an ocean view, sleeper sofa, and kitchenette
(B) Condominiums with an ocean view and sleeper sofa, but without kitchenettes
(C) Condominiums with an ocean view and sleeper sofa (some possibly with kitchenettes)
(D) Condominiums with an ocean view and kitchenette (some possibly with a sleeper sofa)
(E) Condominiums with a sleeper sofa and kitchenette (some possibly with an ocean view)

Data Analysis Problem Set

Solve the following multiple-choice questions by selecting the best answer from the five answer choices. For grid-in questions, write your answer in the grids and completely mark the corresponding ovals. Answers begin on page 450.

Questions 4 and 5 refer to the following figure:

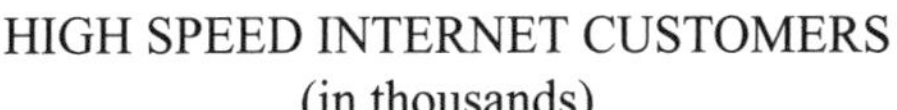

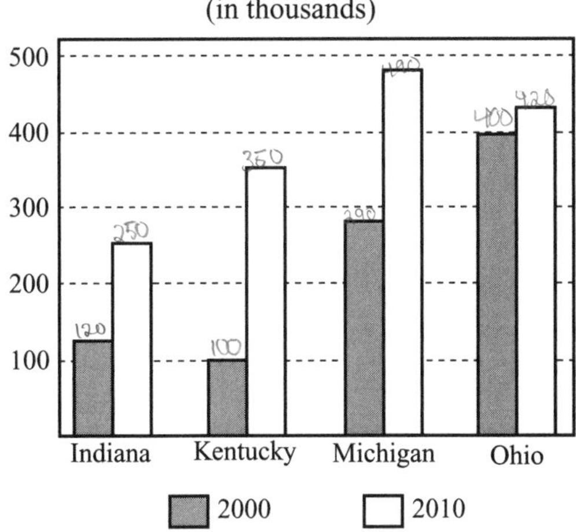

4. The table above shows a cable company's high speed internet customers in four states in 2000 and 2010. The number of customers in Kentucky in 2000 was approximately what percent of the number of customers in Ohio in 2000?

(A) 15%
(B) 20%
(C) 25%
(D) 30%
(E) 35%

5. From 2000 to 2010, the total number of high speed internet customers in the four states was increased by approximately what percent?

(A) 25%
(B) 33%
(C) 50%
(D) 58%
(E) 67%

Color	Number of People Choosing Color
Blue	176
Green	x
Red	97
Yellow	y

6. In a survey, 500 people were asked to choose their favorite color among blue, green, red, and yellow. Each person chose exactly one color. The results of the survey are given in the table above. If x and y are positive integers, what is the greatest possible value of x ?

(A) 77
(B) 114
(C) 226
(D) 227
(E) 500

Average, Median, and Mode

Frequency Guide: 4

The average, median, and mode are statistical data that occur frequently on the SAT. They are easy to remember if you study the following information.

Required Knowledge and Skill Set

1. The average is always referred to as the "average (arithmetic mean)" on the SAT, so you do not have to memorize that the mean is the average.

2. The formula for averages is needed for every average problem on the SAT:

MEMORY MARKER: This formula is required for every average question, and it's even used for some problems that do not appear to involve averages!

$$\frac{\text{sum of the numbers}}{\text{number of numbers}} = \text{average}$$

Use a shorthand version of this formula to save time on the test:

$$\frac{\text{sum}}{\text{\# of \#s}} = \text{average}$$

This formula can also be used to find the sum or the number of numbers in a set:

$$\frac{\text{sum}}{\text{\# of \#s}} = \text{average} \quad \rightarrow \quad \text{sum} = (\text{average})(\text{\# of \#s})$$

$$\frac{\text{sum}}{\text{\# of \#s}} = \text{average} \quad \rightarrow \quad \text{\# of \#s} = \frac{\text{sum}}{\text{average}}$$

3. The median is the middle number in a list of numbers placed in ascending or descending order. On the SAT, if the numbers are not listed in ascending or descending order, rearrange them:

MEMORY MARKER: Some students remember the median by thinking about the word "medium." "Medium" is in the middle of "small" and "large," just as the median is in the middle of small and large numbers.

2, 12, 18, 4, 11, 7, 5, 9, 14

← 4 #s	The Median ↓	4 #s →
2, 4, 5, 7,	9,	11, 12, 14, 18

4. Nearly all SAT questions concerning the median will use an odd number of items in the list. If you are presented with an even number of items, the median is the average of the two numbers in the middle:

2, 4, 5, 7, 9, 11, 12, 14

← 3 #s		3 #s →
2, 4, 5,	7, 9,	11, 12, 14

7, 9, ↓ The Median ↓

$$\frac{7+9}{2} = 8$$

5. In a patterned set of consecutive integers, the average and the median are the same:

5, 6, 7, 8, 9, 10, 11 Pattern: increase by 1	average = 8, median = 8
36, 38, 40, 42, 44, 46 Pattern: increase by 2	average = 41, median = 41
1055, 1060, 1065 Pattern: increase by 5	average = 1060, median = 1060
–12, –6, 0, 6, 12, 18 Pattern: increase by 6	average = 3, median = 3

The average and the median are not the same if any numbers repeat or if a pattern changes:

2, 3, 4, 5, 11	average = 5, median = 4
36, 38, 38, 40, 42	average = 38.8, median = 38

ARITHMETRICK
The average and the median are always the same in a patterned list of consecutive integers!

6. The mode is the most common number in a series:

The Mode
3 occurrences

7, 8, 8, 8, 11, 13, 16, 19, 19, 22

On the SAT, the mode is tested less often than the average and the median. If the mode is tested, there is usually only one in the series. However, there can be two modes. If another 19 appeared in the list above, the mode would be 8 *and* 19:

The Mode
3 occurrences

The Mode
3 occurrences

7, 8, 8, 8, 11, 13, 16, 19, 19, 19, 22

MEMORY MARKER:
You can remember that the mode is the most common number in a series by remembering "The mode is the most!" Both "mode" and "most" start with "mo-" and both have four letters.

7. Average, median, and mode questions are often combined with Data Analysis questions, as the series of numbers can be neatly organized in a table.

Application on the SAT

Average questions come in many varieties, all of which assess your understanding of how averages work. What happens to the average when a smaller number is added to a series? What is the largest possible value of one of the numbers if the list is made up of integers? Understanding these questions is important on the SAT.

Confidence Quotation
"They can conquer who believe they can."
—Virgil, Roman poet

An occasional average question is simple and straightforward, asking you to find the average given a list of integers. Since this is the same type of question asked in classroom math, you should have no problem using the average formula. But the SAT usually makes averages more intimidating by using variables:

8. The average (arithmetic mean) of 1, 2, 5, 9, and x is 4. The average (arithmetic mean) of 2, 3 and y is 6. What is the value of $x+y$?

   (A) 3
   (B) 5
   (C) 10
   (D) 13
   (E) 16

Variables are meant to intimidate you. Do not let them. Use your formula and treat the variables like any real value.

This question consists of two averages. Solve each separately, using the average formula:

$$\frac{\text{sum}}{\text{\# of \#s}} = \text{average}$$

$$\frac{1+2+5+9+x}{5} = 4 \quad \rightarrow \quad 17+x=20 \quad \rightarrow \quad x=3$$

$$\frac{2+3+y}{3} = 6 \quad \rightarrow \quad 5+y=18 \quad \rightarrow \quad y=13$$

Now find $x+y$:

$$x+y=3+13=16$$

The correct answer is (E).

Do not let variables intimidate you. By using the formula and inserting the information from the problem into the appropriate places in the formula, this question becomes a simple algebra problem.

You must understand what happens to an average question when numbers are added to or deleted from a set. Consider the following:

9. The average (arithmetic mean) of a set of seven numbers is 8. When an eighth number is added to the set, the average of the eight numbers is still 8. What number was added to the set?

(A) 6
(B) 7
(C) 8
(D) 9
(E) 10

Remember, the average formula is needed for all average questions. Plug in the information you have about the set of seven numbers:

The sum is the key to many average questions on the SAT.

$$\frac{\text{sum}}{\text{\# of \#s}} = \text{average} \quad \rightarrow \quad \frac{\text{sum}}{7} = 8 \quad \rightarrow \quad \text{sum} = 56$$

Now find the sum of the eight number set:

$$\frac{\text{sum}}{\text{\# of \#s}} = \text{average} \quad \rightarrow \quad \frac{\text{sum}}{8} = 8 \quad \rightarrow \quad \text{sum} = 64$$

When an extra number was added to the set, the sum changed by 8 (56 + 8 = 64). Therefore, the number added to the set must be 8. The correct answer is (C).

An overwhelming majority of average problems will depend on you working with or finding the sum. Consider a more difficult grid-in question:

18. In a set of 5 positive integers, 56, 138, x, y, and z, all five integers are different and the average (arithmetic mean) is 300. If the integers x, y, and z are greater than 138, what is the greatest possible value for any of the integers?

Remember, use the average formula any time you see the words "average (arithmetic mean)" on the SAT.

Use the average formula to find the sum of x, y, and z:

$$\frac{\text{sum}}{\text{\# of \#s}} = \text{average} \quad \rightarrow$$

$$\frac{56+138+x+y+z}{5} = 300 \quad \rightarrow \quad 194 + x + y + z = 1500 \quad \rightarrow \quad x + y + z = 1306$$

Say that z is the integer with the greatest possible value. That means x and y must be as small as possible. Since they are greater than 138, the smallest they can be is 139 and 140:

$$139 + 140 + z = 1306 \quad \rightarrow \quad 279 + z = 1306 \quad \rightarrow \quad z = 1027$$

The greatest possible value for any of the integers is 1027. You must understand that in order for one number to be large as it can be, the others must be as small as possible without violating any of the rules set forth in the question.

Another difficult problem concerns combining averages. This type of question occurs less frequently than the previous example.

In classroom math, these questions are referred to as "weighted averages."

17. Two classes were given a math test. The first class had 25 students and the average test score was 86%. The second class had 15 students and their average score was 94%. If the teacher combined the test scores of both classes, what is the average of both classes together?

    (A) 88%
    (B) 89%
    (C) 90%
    (D) 91%
    (E) 92%

Can you guess what we must find for each class? That's right—the sum!

$$\frac{\text{sum}}{\#\text{ of }\#\text{s}} = \text{average} \rightarrow \frac{\text{sum}_{\text{Class1}}}{25} = 0.86 \rightarrow \text{sum}_{\text{Class1}} = 21.5$$

$$\frac{\text{sum}}{\#\text{ of }\#\text{s}} = \text{average} \rightarrow \frac{\text{sum}_{\text{Class2}}}{15} = 0.94 \rightarrow \text{sum}_{\text{Class2}} = 14.1$$

Once you have the sum for each class, you can combine the classes.

Total sum: $21.5 + 14.1 = 35.6$

Total students: $25 + 15 = 40$

Now find the new average:

$$\frac{\text{sum}}{\#\text{ of }\#\text{s}} = \text{average} \rightarrow \frac{35.6}{40} \rightarrow 0.89 \rightarrow 89\%$$

CAUTION: SAT TRAP! The most common wrong answer for a weighted average question is the simple average of the two averages without taking into account the number of elements in each group.

The correct answer is (B).

Do you know the most common wrong answer? It is (C), 90%. Too many students simply add 86% and 94% and divide by 2. However, this is a combined average question, so each individual sum must be found in order to find the sum of the combined average.

One time-saving average question concerns the average and the median of a list of patterned numbers. As revealed in the Required Knowledge and Skill Set, a list of patterned integers has the same average and median. Let's examine how this knowledge can gain you valuable time on the SAT:

13. If x is the average (arithmetic mean) of 5 consecutive odd integers, what is the median of this set of integers?

(A) 0
(B) 1
(C) $x - 2$
(D) x
(E) $x + 2$

In a list of consecutive or patterned numbers, the median is the same as the average, so the answer is (D). Most students would spend at least 30 seconds running numbers through the average formula on this question. You can answer this one without any calculations.

The final type of average question to review does not appear as an average question at all. These questions deal with the sum of consecutive integers and the word "average" does not occur in the text :

13. The sum of 7 consecutive even integers is 224. What integer has the least value in the list?

(A) 16
(B) 18
(C) 26
(D) 29
(E) 32

ARITHMETRICK
If the words "sum" and "consecutive" appear in a question, the question is likely an average problem and will require the average formula.

When a question has the word "sum" and "consecutive" in it, pull out the average formula, as the question is likely a disguised average problem:

$$\frac{\text{sum}}{\text{\# of \#s}} = \text{average} \quad \rightarrow \quad \frac{224}{7} \quad \rightarrow \quad 32$$

The average of the 7 integers is 32. Since the average and the median are the same in consecutive sets, you can find the other numbers by counting down and up from 32:

__, __, __, 32, __, __, __

26, 28, 30, 32, 34, 36, 38

The number with the least value is 26, so answer (C) is correct.

Do you know the most common wrong answer? Some students forget that the list is consecutive EVEN numbers. If you just study a consecutive pattern, the integer with the least value is 29:

29, 30, 31, 32, 33, 34, 35 (*Incorrect*)

CAUTION: SAT TRAP!
Be sure to read any question involving the word "consecutive" to verify whether you were using consecutive, consecutive even, or consecutive odd numbers.

Median questions are often used with tables.

Median questions can be divided into two basic types. The first simply asks you to find the median. To solve these questions, put the set of numbers in order from least to greatest, and then locate the middle number. Note that you may need to use arithmetic or algebra to generate the set of numbers, as in the following question:

DOGS ADOPTED AT PET STORES

	Start	End	Adopted
The Dog Park	16	8	
Super Pets	20	16	
Wags	18	6	
Bark Avenue	12	8	
Paws and Claws	15	4	
Pet Emporium	9	8	
Fido's	18	11	

17. The local humane society recently hosted a dog adoption event at 7 local pet stores. Each pet store started and ended the day with the number of dogs shown in the table of above. The number in the "Adopted" column is defined by the number of dogs at the start of the day minus the number of dogs at the end of the day. What is the median of the missing values in the "Adopted" column?

(A) 4
(B) 7
(C) 8
(D) 11
(E) 12

Before you can find the median, you must generate the numbers in the "Adopted" column:

	Start	End		Adopted
The Dog Park	16	8	16 – 8 =	8
Super Pets	20	16	20 – 16 =	4
Wags	18	6	18 – 6 =	12
Bark Avenue	12	8	12 – 8 =	4
Paws and Claws	15	4	15 – 4 =	11
Pet Emporium	9	8	9 – 8 =	1
Fido's	18	11	18 – 11 =	7

In a large set of numbers, it may help to cross out one number on the left side of the median for every number on the right side of the median.

Now write the set of numbers from that column in order from least to greatest:

1, 4, 4, 7, 8, 11, 12

1, 4, 4, (7,) 8, 11, 12

Which number is in the middle? There are three numbers to the left of 7 and three numbers to the right of 7, so 7 is the median. The correct answer is (B).

The other type of median question tests your understanding of manipulated medians. It asks which answer choice will not affect the median. The answer is ALWAYS either the lowest value or the highest value. You can automatically eliminate answers (B), (C), and (D) on these questions. Let's examine an example:

9. In the list of numbers, 6, x, 10, 2, 7, 13, and 15, the median is 10. Which of the following could NOT be the value of x ?

(A) 9
(B) 10
(C) 11
(D) 13
(E) 16

The answer must be (A) or (E). Which one would change the median from 10 to another value?

Rewrite the list in ascending order, and isolate the median:

2, 6, 7, 10, 13, 15   2, 6, 7, (10,) 13, 15

Notice that there are three numbers to the left of 10, but only 2 numbers to the right of 10. This indicates that x must be a number greater than or equal to 10. If $x = 9$, then the list would be even more lopsided, with 4 numbers to the left of 10. So 9 is the only value in the list that x CANNOT equal. The correct answer is (A).

Mode questions, which are rare on the SAT, are always combined with questions about the average or median, often in Roman numeral questions. If you happen to receive a test with a mode question, you will need to either identify the mode or determine what happens to the mode when a list is manipulated.

13. The average, median, and mode are calculated for the list 3, 3, 7, 10, 12. If the number 1 is added to the list, which of the following will change?

I. The average
II. The median
III. The mode

(A) None
(B) I only
(C) I and II
(D) I and III
(E) I, II, and III

Find the original and new average, median, and mode:

Original: 3, 3, 7, 10, 12	New: 1, 3, 3, 7, 10, 12
Average = 5	Average = 6
Median = 7	Median = 5
Mode = 3	Mode = 3

Only the mode remains unchanged. Choice (C) is correct.

ARITHMETRICK
For questions about manipulated medians, the answer is always (A) or (E).

Confidence Quotation
"Human beings, by changing the inner attitudes of their minds, can change the outer aspects of their lives." —William James, psychologist and philosopher

Average, Median, and Mode Problem Set

Solve the following multiple-choice questions by selecting the best answer from the five answer choices. For grid-in questions, write your answer in the grids and completely mark the corresponding ovals. Answers begin on page 452.

1. Which answer choice contains a set of numbers in which the median is greater than the average (arithmetic mean)?

   (A) {3, 4, 5, 6, 7}
   (B) {3, 4, 5, 6, 8}
   (C) {3, 5, 5, 5, 7}
   (D) {−2, 4, 5, 6, 7}
   (E) {−2, 4, 5, 6, 12}

2. The sum of five consecutive even integers, a, b, c, d, and e, respectively, is 50. Which of the following is equal to the median of the set?

   (A) $\frac{a+b+c+d+e}{30}$
   (B) $\frac{30}{c}$
   (C) $b+2$
   (D) $e-a$
   (E) $\frac{b+d}{5}$

3. Eight consecutive odd integers are arranged in ascending order, from smallest to largest. The sum of the last four integers is 232. What is the sum of the first four integers?

4. Which of the following answer choices is equal to the sum of three consecutive odd integers?

   (A) 153
   (B) 154
   (C) 155
   (D) 156
   (E) 157

Average, Median, and Mode Problem Set

Solve the following multiple-choice questions by selecting the best answer from the five answer choices. For grid-in questions, write your answer in the grids and completely mark the corresponding ovals. Answers begin on page 452.

5. In 7 days, Mario cooked 98 pounds of spaghetti. Each day after the first, he cooked 2 more pounds than he cooked than the day before. What is the difference between the average (arithmetic mean) number of pounds of spaghetti he cooked per day and the median number of pounds he cooked during the 7 days?

6. The average (arithmetic mean) of five different positive integers is 30. What is the greatest possible value of one of these integers?

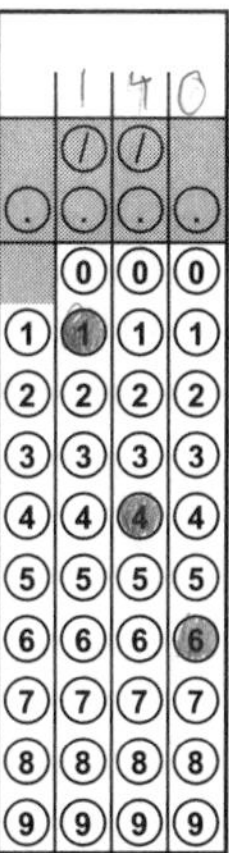

7. Five numbers, x, $2x$, $2x + 6$, $3x - 1$ and $4x - 8$, are in a set. If the average (arithmetic mean) of the five numbers is 9, what is the value of the mode in this set?

Counting Problems

Frequency Guide: 4

Unless you have already taken a statistics course, you probably have not encountered counting problems. These questions are mainly made up of combinations and permutations, which have complex explanations and special formulas. However, on the SAT, they are quite basic, and can be solved without formulas.

Required Knowledge and Skill Set

1. Counting problems require you to do exactly what their name implies—count! The most basic counting problems ask you to count the number of possibilities presented in a word problem. These problems often deal with sums and products, which we will examine more closely in the next section.

Permutations deal with many possibilities; combinations deal with fewer possibilities.

2. Permutations and combinations are arrangements of groups of numbers. In a permutation, the order of the items is important; in a combination, the order of the items is not important. There are more possible arrangements in a permutation than a combination.

3. Combinations combine two or more elements. To understand combinations, let's consider an example. At a restaurant, there are three flavors of ice cream and four choices for toppings. If each ice cream sundae consists of one ice cream flavor and one topping, how many different combinations of sundaes are possible?

   Because these are counting problems, you can always just count:

   Three Flavors: 1, 2, and 3
   Four Toppings: A, B, C, and D

1–A	2–A	3–A
1–B	2–B	3–B
1–C	2–C	3–C
1–D	2–D	3–D

   There are 12 combinations. The order of the items is not important; chocolate with sprinkles is the same as sprinkles with chocolate.

   MEMORY MARKER: To find the possible number of combinations, multiply the elements.

   But counting is not the most efficient solution method. To easily find the number of possibilities in a combination, simply multiply the number of elements:

   3 flavors × 4 toppings = 12 combinations

   This works no matter how many elements are present. Say we added 5 syrups to the menu, and each sundae consisted of one flavor of ice cream, one topping, and one syrup. How many combinations are possible now?

   3 flavors × 4 toppings × 5 syrups = 60 combinations

4. Now look at an example of a permutation. In gym class, four students are running a race. How many different finishing orders are possible at the end of the race?

   We often refer to permutations as "card questions," because we use blank "cards" to set up the problems. Draw and label four blank cards, each one representing a specific finishing order:

First Place	Second Place	Third Place	Fourth Place
☐	☐	☐	☐

   Say the four runners are named A, B, C, and D. How many possibilities are there for first place? Four (A, B, C, or D). Assign one of them first place. For the ease of discussion, we will go in alphabetical order. Runner A receives first place. How many runners are now eligible for second place? Three (B, C, or D). If B finishes in second place, how many runners are available for third place? Two (C or D). If C finishes third, only D is left to come in fourth place:

First Place	Second Place	Third Place	Fourth Place
4	3	2	1
A, B, C, D	B, C, D	C, D	D

   To find the number of finishing orders, multiply the cards together:

First Place		Second Place		Third Place		Fourth Place	
4	×	3	×	2	×	1	= 24

   There are 24 possible finishing orders.

Permutations are ordered arrangements; ABC, BCA, and CBA, are considered different items, so there are more permutations than combinations.

Combinations are unordered selections; ABC, BCA, and CBA are all the same item, so there are fewer combinations than permutations.

MEMORY MARKER: To find the possible number of arrangements when order is involved, set up cards and multiply the items in the cards.

5. Permutations often come with restrictions that dictate rules about the order of the elements. For example, five people are in a car. If only 3 people can drive, how many different seating arrangements are possible?

   Set up five cards, one for each position in the car. Always put the restriction at the front of the list:

Driver	"Shotgun"	Backseat 1	Backseat 2	Backseat 3
☐	☐	☐	☐	☐

   Call the five passengers A, B, C, D, and E. Only A, B, and C can drive:

Driver		"Shotgun"		Backseat 1		Backseat 2		Backseat 3	
3	×	4	×	3	×	2	×	1	= 72
A, B, C		B, C, D, E		C, D, E		D, E		E	

   There are 72 possible seating arrangements.

When B and C are not selected to drive, they go back into the "general population" and are now eligible to ride shotgun.

Combinations and permutations with restrictions are usually the hardest questions in a math section.

6. Permutations might also restrict the number of positions. Returning to the race question, the gym class now consists of 10 students, all of whom are running the race. The first place finisher will receive a blue ribbon, the second place runner will be given a red ribbon, and the third place contestant will receive a green ribbon. How many different possibilities are there for the top three spots?

   For this question, there are only three cards, even though there are 10 runners:

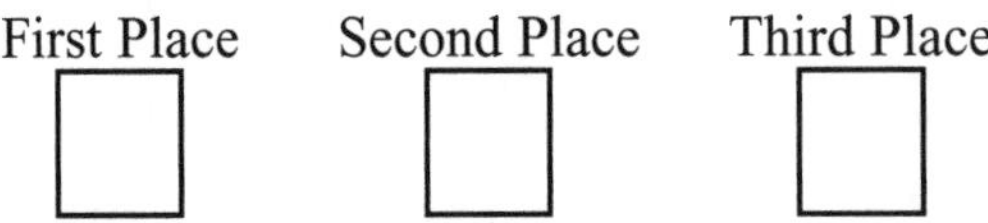

   But there are still 10 people who can finish in first place, leaving 9 people a chance at second place, and 8 with a shot at third:

First Place		Second Place		Third Place	
10	×	9	×	8	= 720

   There are 720 possible arrangements for ribbon winners.

7. Combinations can also carry restrictions. For example, there are three detectives and four uniformed officers in a new police program. Every incident must be responded to by a team of one detective and two uniformed officers. How many combinations of teams are possible?

   The formula for solving this problem is complicated. Because the SAT uses small groups for these questions, we recommend that you count the teams. There are two ways to do this. You can count every possibility:

   Three Detectives: 1, 2, and 3
   Four Uniformed Officers: A, B, C, and D

If combinations allow elements to repeat, forget formulas and simply count the number of combinations.

1–A–B	2–A–B	3–A–B
1–A–C	2–A–C	3–A–C
1–A–D	2–A–D	3–A–D
1–B–C	2–B–C	3–B–C
1–B–D	2–B–D	3–B–D
1–C–D	2–C–D	3–C–D

   Or, for a faster solution, count the number of possibilities for one detective, and then multiply that number by 3. Detective 1 has 6 possible arrangements. That means that Detectives 2 and 3 also have 6 arrangements each:

   6 arrangements × 3 detectives = 18 possible arrangements

   PowerScore recommends that you use the second strategy for restricted combinations, as it is more efficient.

Application on the SAT

Counting problems come in all difficulty levels on the SAT. The most basic question simply asks you to count items, and often these items are sums or products of a limited quantity of numbers. Consider an example:

1, 3, 5, 8, 10

13. How many different sums can be made by adding any two different numbers from the list above?

(A) 6
(B) 8
(C) 10
(D) 12
(E) 25

Pay close attention to the word "different" in the question. It gives two very important pieces of information.

First, two different numbers are added together. The order of these two numbers is not important, because the sum of 1 + 3 is the same as the sum of 3 + 1. This is a combination problem, and there will be a total of 10 combinations:

5 numbers × 2 added together = 10 combinations

CAUTION: SAT TRAP! We cannot overemphasize the importance of reading carefully! The SAT math sections test both math and reading comprehension!

But do not be tricked into taking choice (C). The problem requests the number of *different* sums; while there will be 10 results, some of those results may be the same value. Without finding all of the sums, we cannot predict if there are any that are the same.

Find the sums of all 10 combinations:

1 + 3 = 4	3 + 5 = 8	5 + 8 = 13	8 + 10 = 18
1 + 5 = 6	3 + 8 = 11	5 + 10 = 15	
1 + 8 = 9	3 + 10 = 13		
1 + 10 = 11			

List the sums, in ascending order:

4, 6, 8, 9, 11, 11, 13, 13, 15, 18

Notice that two of the sums repeat. Count the number of sums, excluding any repeats:

4, 6, 8, 9, 11, ~~11~~, 13, ~~13~~, 15, 18

There are 8 *different* sums. The correct answer is (B).

Some difficult counting problems present such a large field of countable items, that it helps to study a small sample and apply your findings to the entire group.

18. In a list of 57 consecutive integers, the median is 70. What is the largest integer in the list?

 (A) 96
 (B) 97
 (C) 98
 (D) 99
 (E) 200

If you had three hours to take each math section, you could write out every consecutive integer and find the largest number in the list:

..., 61, 62, 63, 64, 65, 66, 67, 68, 69, 70, 71, 72, 73, 74, 75, 76, 77, 78, ...

ARITHMETRICK
When asked to find a number in a large list, study a small sample and apply your findings to the entire list.

But because you only have 25 minutes per section, you do not have time to list all 57 numbers. Instead, study a smaller set of numbers:

1, 2, 3, 4, 5

This is a set of 5 consecutive integers with a median of 3.

5 integers – 1 median = 4 integers

4 integers ÷ 2 sides of the median (right and left) = 2 numbers per side

2 #s ← → 2 #s

1, 2, $\underline{3}$, 4, 5

the median + the numbers per side = the largest integer in the list
$3 + 2 = 5$

Apply the knowledge you gained from the smaller list to the larger list:

57 integers – 1 median = 56 integers

56 integers ÷ 2 sides of the median (right and left) = 28 numbers per side

28 #s ← → 28 #s

..., 67, 68, 69, $\underline{70}$, 71, 72, 73, ...

the median + the numbers per side = the largest integer in the list
$70 + 28 = 98$

The correct answer is (C).

Expect to see simple combinations and permutations on the SAT. However, because most students have never seen problems like these, the questions have exaggerated difficulty levels:

17. Mary has three necklaces, four bracelets, and three rings. If she wears one necklace, one bracelet, and one ring, how many different combinations can Mary make?

(A) 4
(B) 10
(C) 24
(D) 36
(E) 48

Combination problems often use the words "select" or "combinations."

This is a combination problem. You know this, because the order of the items does not matter. Wearing a gold necklace, blue bracelet, red ring is the same as wearing a blue bracelet, red ring, and gold necklace.

You can count the number of combinations, but this in an inefficient solution. It is faster to just multiply the number of elements together:

3 necklaces × 4 bracelets × 3 rings = 36 combinations

The correct answer is (D).

Simple permutations are also present:

19. Five lockers are to be assigned to five students. How many different arrangements of lockers are possible?

(A) 5
(B) 25
(C) 50
(D) 100
(E) 120

To solve this problem, draw five blank cards—or in this case, five blank lockers:

Locker 1	Locker 2	Locker 3	Locker 4	Locker 5
☐	☐	☐	☐	☐

Confidence Quotation
"I have learned to use the word 'impossible' with the greatest caution."
—Wernher von Braun, rocket scientist

Call the students A, B, C, D, and E. How many possibilities are there for each locker?

Locker 1		Locker 2		Locker 3		Locker 4		Locker 5	
5	×	4	×	3	×	2	×	1	= 120
A, B, C, D, E		B, C, D, E		C, D, E		D, E		E	

There are 120 possible arrangements, so answer (E) is correct.

Expect to see some combination and permutations problems involving restrictions. These are usually the most difficult questions in a section:

20. The five blocks shown above are to be placed in a line on a shelf. Two of the blocks, currently in the second and fifth position, have shading. If the blocks with shading can never be in first position or the center position, how many different arrangements are possible?

(A) 36
(B) 54
(C) 72
(D) 96
(E) 120

This is a permutation with restrictions. Draw five blank cards:

Position 1	Position 2	Position 3	Position 4	Position 5
No shading		No shading		

DIAGRAMMING the question is an important skill for many concepts in the Data Analysis, Statistics, and Probability content area.

Always start with the restrictions. There are three possible candidates without shading for Position 1, leaving two candidates without shading for Position 3:

Position 1	Position 2	Position 3	Position 4	Position 5
3		2		

The specific positions of the restrictions do not matter, but you should always start by analyzing the restricted positions in permutation questions.

Then return to the other positions and determine the possibilities with all remaining elements:

Position 1	Position 2	Position 3	Position 4	Position 5
3	3	2	2	1

Multiply each of the possibilities:

Position 1		Position 2		Position 3		Position 4		Position 5	
3	×	3	×	2	×	2	×	1	= 36

There are 36 possible arrangements that keep shaded blocks out of the first and third position. The correct answer is (A).

Counting Problems Problem Set

Solve the following multiple-choice questions by selecting the best answer from the five answer choices. For grid-in questions, write your answer in the grids and completely mark the corresponding ovals. Answers begin on page 455.

3, 4, 5, 8, 9

1. Two different numbers are selected from the list above and their product is determined. How many different pairs of numbers with a product greater than 30 can be selected?

(A) 5
(B) 6
(C) 7
(D) 8
(E) 9

2. A restaurant is offering a new buffet with six types of sandwiches, four sides, and five desserts. If customers are allowed to select one sandwich, one side, and one dessert, how many meal combinations are possible?

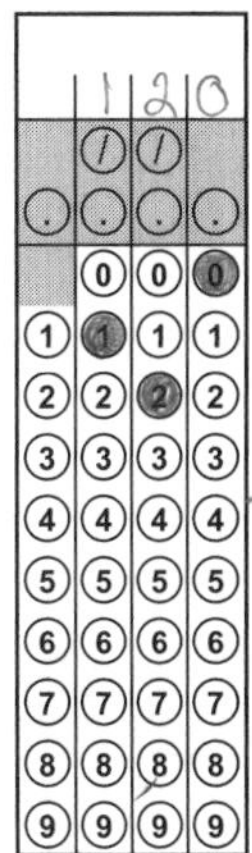

3. A hot dog vendor offers three choices of condiments: mustard, ketchup, and horseradish. If a customer can select one, two, or all three condiments, how many different combinations of condiments are possible?

(A) 5
(B) 6
(C) 7
(D) 8
(E) 9

4. Five dogs are in a dog show. They are to be lined up in a single row, and the dog with the most ribbons is to be placed in the first position. The two dogs with the fewest ribbons are to be placed in the last two positions. If none of the dogs have the same amount of ribbons, how many different arrangements of dogs are possible?

Counting Problems Problem Set

Solve the following multiple-choice questions by selecting the best answer from the five answer choices. For grid-in questions, write your answer in the grids and completely mark the corresponding ovals. Answers begin on page 455.

5. Robbie has to schedule five different meetings during the five day work week. If exactly one meeting is held each day, how many different arrangements of meetings are possible for the five day work week?

5 · 4 · 3 · 2 · 1

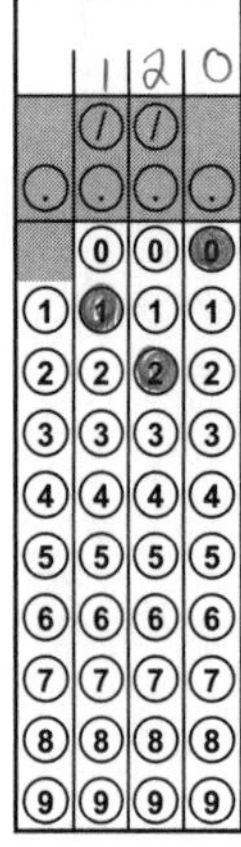

6. The sum of the first 50 consecutive positive even integers is x and the sum of the first 50 consecutive positive integers is y. What is x in terms of y?

(A) $2y^2$

(B) y^2

(C) $2y$

(D) $\frac{2}{y}$

(E) $\frac{y}{2}$

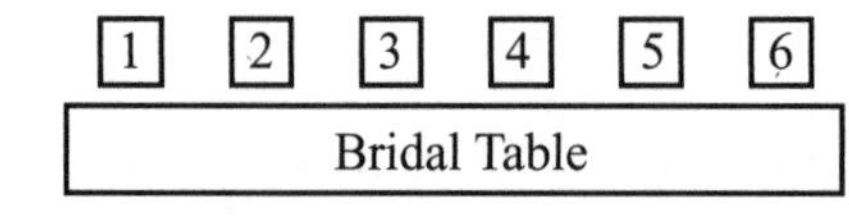

7. How many different ways can 6 people arrange themselves in the 6 seats at a bridal party table shown above if the bride and groom must be sitting in the two center seats?

4 3 2 1 2 1

Probability

The College Board tests basic probability concepts on the SAT.

Frequency Guide: 3

Required Knowledge and Skill Set

1. Probability indicates the likelihood that a specific event will occur. The probability of something happening can be expressed using any number from 0 to 1. A probability of 0 means the event will *never* happen. A probability of 1 indicates that an event will *always* happen. A probability of $\frac{1}{3}$ signifies that the event has a 1 in 3 chance of occurring.

2. The probability of an occurrence can be expressed by a simple formula:

MEMORY MARKER: You should memorize the formula for finding probability.

$$\text{Probability} = \frac{\text{number of favorable outcomes}}{\text{number of possible outcomes}}$$

If there are 6 green socks, 4 blue socks, and 2 red socks in a drawer, what is the probability that you randomly select a blue sock?

$$\text{Probability} = \frac{\text{favorable}}{\text{possible}} \rightarrow \frac{\text{4 blue socks}}{\text{12 total socks}} \rightarrow \frac{1}{3}$$

There is a one in three chance that you randomly choose a blue sock.

3. The probability of something *not* occurring is 1 minus the probability that it will occur:

$$\text{Probability of an event } \textit{not} \text{ occurring} = 1 - \frac{\text{number of favorable outcomes}}{\text{number of possible outcomes}}$$

If there are 6 green socks, 4 blue socks, and 2 red socks in a drawer, what is the probability that you do *not* select a blue sock? You have two ways to solve this question. You can use the formula for the probability of an event not occurring:

$$\text{Probability of an event } \textit{not} \text{ occurring} = 1 - \frac{\text{favorable}}{\text{possible}}$$

$$1 - \frac{\text{4 blue socks}}{\text{12 total socks}} \rightarrow 1 - \frac{1}{3} \rightarrow \frac{2}{3}$$

MEMORY MARKER: Once you memorize the formula for the probability of an event occurring, the formula for the probability of an event not occurring is easy!

Or you can find the probability of selecting a green or red sock:

$$\text{Probability} = \frac{\text{favorable}}{\text{possible}} \rightarrow \frac{\text{6 green} + \text{2 red}}{\text{12 total socks}} \rightarrow \frac{8}{12} \rightarrow \frac{2}{3}$$

4. Probability is often applied to geometric figures on the SAT. The area of a shaded region and the total area of a figure provide the information for the probability formula. To make these questions a little easier, slightly alter the probability formula as follows:

MEMORY MARKER:
The formula for Geometric Probability is an altered version of the formula for probability.

$$\text{Geometric Probability} = \frac{\text{shaded area}}{\text{total possible area}}$$

Consider an example:

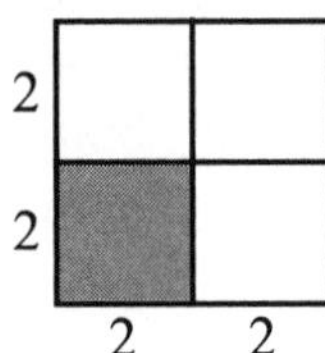

If a point was to be selected at random from the square to the right, what is the probability that the point would be in the shaded area?

To solve, find the area of the shaded square and the total area of the large square:

Shaded area = $2 \times 2 = 4$
Total possible area = $4 \times 4 = 16$

Then apply this information to the altered formula:

$$\text{Geometric Probability} = \frac{\text{shaded area}}{\text{total possible area}} \rightarrow \frac{4}{16} \rightarrow \frac{1}{4}$$

There is a 1 in 4 chance the randomly selected point will come from the shaded region.

5. Advanced probability questions entail two or more occurrences. These questions are rare, but have on occasion appeared as the most difficult questions in a section.

MEMORY MARKER:
To find the probability of two or more non-related events, multiply the probability of each individual event.

The probability of two or more non-related or independent events occurring is the product of the individual probabilities of those events. For example, what is the probability of flipping a penny and getting a "heads" and rolling a standard 6-sided die and getting a 2?

Find the probability of each independent event:

$$\text{Probability of flipping heads} = \frac{\text{1 side with heads}}{\text{2 possible sides}} = \frac{1}{2}$$

$$\text{Probability of rolling a 2} = \frac{\text{1 side with a '2'}}{\text{6 possible sides}} = \frac{1}{6}$$

This question would not occur on the SAT without an explanation about the number of sides of a penny and a standard die. We use it here, though, for ease of discussion.

And then multiply the individual probabilities to find the probability of both events occurring:

$$\frac{1}{2} \times \frac{1}{6} = \frac{1}{12}$$

There is a 1 in 12 chance that both events occur.

6. When the probability of two events is being calculated, pay careful attention to whether the first event changes the probability of the second event.

   For example, if a drawer contains 4 blue socks and 6 green socks, and you randomly select two socks, what is the probability that both socks are blue?

   Start with the first sock. There are 10 socks in the drawer and 4 are blue:

$$\text{Probability} = \frac{\text{favorable}}{\text{possible}} \rightarrow \frac{4}{10} \rightarrow \frac{2}{5}$$

   The second sock is a bit trickier. How many blue socks are left in the drawer? Only 3, because you have already pulled one out. So how many total socks are left in the drawer? Only 9. Find the probability of the second sock being blue:

$$\text{Probability} = \frac{\text{favorable}}{\text{possible}} \rightarrow \frac{3}{9} \rightarrow \frac{1}{3}$$

   Now find the probability of both events occurring by finding the product of the two independent events:

$$\frac{2}{5} \times \frac{1}{3} = \frac{2}{15}$$

   There is a 2 in 15 chance that both socks pulled from the drawer will be blue.

When two events occur in the same question, be sure to take into account how the first event affects the second event.

Application on the SAT

Probability questions are usually combined with Arithmetic, Algebra, Geometry, or Data Analysis questions on the SAT. One simple arithmetic question provides the probability, but asks for the number of favorable or possible outcomes. Let's look at a grid-in example:

> 11. There are 496 employees in a company, one of whom is to be selected at random to win a car. If the probability that a supervisor will be selected is $\frac{3}{16}$, how many supervisors work at the company?

Always use the probability formula when given a question with the word "probability."

Set up your equation just as you would if you were looking for probability, but supply the probability in order to find the number of favorable outcomes:

$$\text{Probability} = \frac{\text{favorable}}{\text{possible}}$$

$$\frac{3}{16} = \frac{\text{supervisors}}{496} \rightarrow 1488 = 16s \rightarrow 93 = s$$

There are 93 supervisors at the company.

Probability is often combined with ratio questions. These questions have an exaggerated difficulty level because students forget how to set up the ratio. Consider another grid-in example:

> 13. A box contains only pens and pencils. There are three times as many pens as pencils in the box. If one writing utensil is to be selected at random from the box, what is the probability that the utensil is a pen?

The question provides a ratio for the numbers of pens to the number of pencils. The key to ratio questions is to find the denominator:

Pens : Pencils
3 : 1 Pens + Pencils = Denominator $\rightarrow$ $3 + 1 = 4$

$\frac{3}{4}$, $\frac{1}{4}$ $\frac{3}{4}$ of the utensils are pens, $\frac{1}{4}$ of the utensils are pencils

The probability of drawing a pen is the same as the fraction of pens in the box:

$$\text{Probability} = \frac{\text{favorable}}{\text{possible}} \rightarrow \frac{3 \text{ pens}}{4 \text{ utensils}} \rightarrow \frac{3}{4}$$

The test makers may make this problem more difficult by adding another ratio:

> 18. A box contains only pens and pencils. There are three times as many pens as pencils. The pens are either green or blue, and 5 times as many pens are green as are blue. If one writing utensil is to be selected at random from the box, what is the probability that the utensil is a blue pen?

Have you noticed that all of the probability questions we have covered were grid-in questions? Probability questions have a high probability of appearing in the Student-Produced Response section because their answers are in fraction form.

To solve this problem, you must find the probability of selecting a pen and the probability that the pen is blue. You already know the probability of selecting a pen (three-fourths), so determine the chances of pulling a blue pen:

Green : Blue
5 : 1 Pens + Pencils = Denominator $\rightarrow$ $5 + 1 = 6$

$\frac{5}{6}$, $\frac{1}{6}$ $\frac{5}{6}$ of the pens are green, $\frac{1}{6}$ of the pens are blue

The probability of two events occurring is the product of each individual event occurring:

$\frac{3}{4} \times \frac{1}{6} = \frac{3}{24} = \frac{1}{8}$ The probability of selecting a blue pen is $\frac{1}{8}$.

Another ratio and probability question involves adding to or subtracting from the total number of events. The following grid-in question is a good example:

17. There are 20 apples and 15 oranges in a bin. If only apples are to be subtracted from the bin so that the probability of randomly drawing an apple becomes $\frac{2}{5}$, how many apples must be subtracted from the bin?

A certain amount of apples will be subtracted from both the favorable and total possible outcomes in order to create the fraction two-fifths. If x is the number of apples subtracted, write an equation and solve for x:

$$\frac{\text{favorable} - x}{\text{possible} - x} = \frac{2}{5} \rightarrow \frac{20 - x}{35 - x} = \frac{2}{5} \rightarrow 5(20 - x) = 2(35 - x) \rightarrow$$

$$100 - 5x = 70 - 2x \rightarrow 30 = 3x \rightarrow 10 = x$$

Ten apples must be subtracted from the bin in order for the probability to be two-fifths. This question is very easy to solve, but difficult for many students to set up. By understanding ratios and probability, you can earn a point most students will omit or answer incorrectly.

Confidence Quotation
"Vision is perhaps our greatest strength. It has kept us alive to the power and continuity of thought throughout the centuries, it makes us peer into the future and lends shape to the unknown."
—Li Ka-shing, one of the wealthiest businessmen in the world

As seen in the previous question, probability may also involve algebra. Some questions make you find a set of sums or products to determine probability:

16. A number is randomly selected from the set $\{-8, -4, 0, 4, 8\}$. What is the probability that the number is a member of the solution set of both $x + 4 > -3$ and $5x - 6 < 9$?

Solve both inequalities:

$$x + 4 > -3$$
$$x > -7$$

$$5x - 6 < 9$$
$$5x < 15$$
$$x < 3$$

The value of x must be less than 3 but greater than –7 ($-7 < x < 3$). How many answer choices satisfy this requirement? Just two: –4 and 0. All other answer choices only satisfy one of the inequalities.

$$\text{Probability} = \frac{\text{favorable}}{\text{possible}} \rightarrow \frac{\text{2 favorable numbers}}{\text{5 possible numbers}} \rightarrow \frac{2}{5}$$

The probability of selecting –4 or 0 at random is two-fifths.

As mentioned in the Required Knowledge and Skill Set, probability is often applied to geometry questions. There are two types of questions to watch for the on SAT. The first type offers you a way to find the exact area of the figure and the exact area of the shaded region:

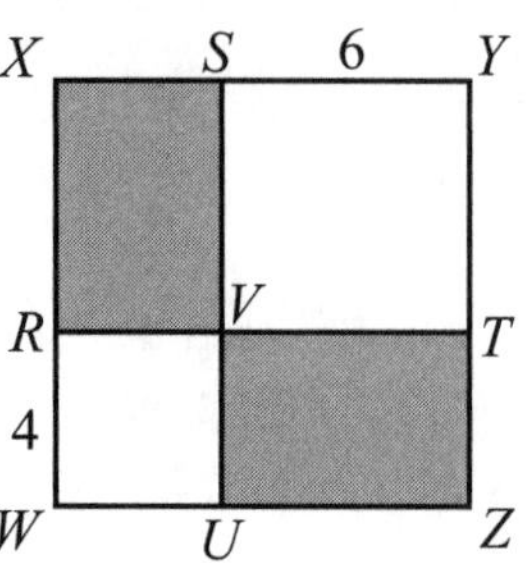

19. In the figure above, $WXYZ$ is a square, as are $SYTV$ and $RVUW$. If a point is selected at random from $WXYZ$, what is the probability that the point is in one of the shaded regions?

    (A) $\frac{2}{5}$

    (B) $\frac{15}{32}$

    (C) $\frac{12}{25}$

    (D) $\frac{1}{2}$

    (E) $\frac{3}{5}$

Can you SIZE UP the figure and estimate an answer? It looks like the shaded region takes up about half of the figure. Choices (B), (C), and (D) would be good guesses if you were running out of time or did not remember how to solve the question.

The figure gives you the side length of each square, which allows you to find the length and width of each shaded rectangle. DIAGRAM the question to find the area of each shaded region:

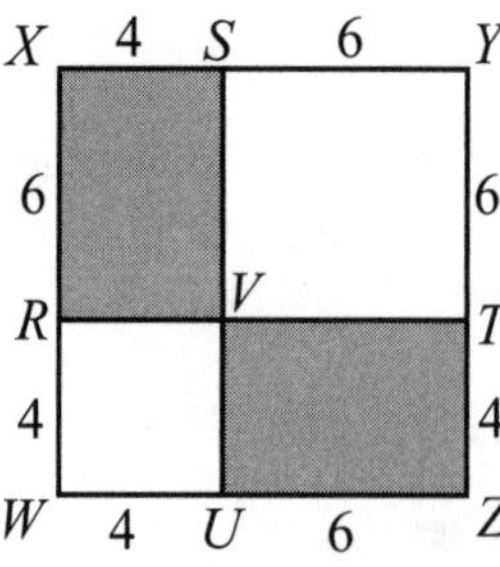

The area of $XSVR$:

$$\text{Area} = \ell w \rightarrow (6)(4) \rightarrow 24$$

The area of $VTZU$:

$$\text{Area} = \ell w \rightarrow (6)(4) \rightarrow 24$$

The area of $WXYZ$:

$$\text{Area} = \ell w \rightarrow (10)(10) \rightarrow 100$$

In order to find the probability of a geometric question, use the formula:

$$\text{Geometric Probability} = \frac{\text{shaded area}}{\text{total possible area}} \rightarrow \frac{24+24}{100} \rightarrow \frac{48}{100} \rightarrow \frac{12}{25}$$

The correct answer is (C).

The other type of geometry question does not provide the measurements to find the exact area of the figure or the area of the shaded region.

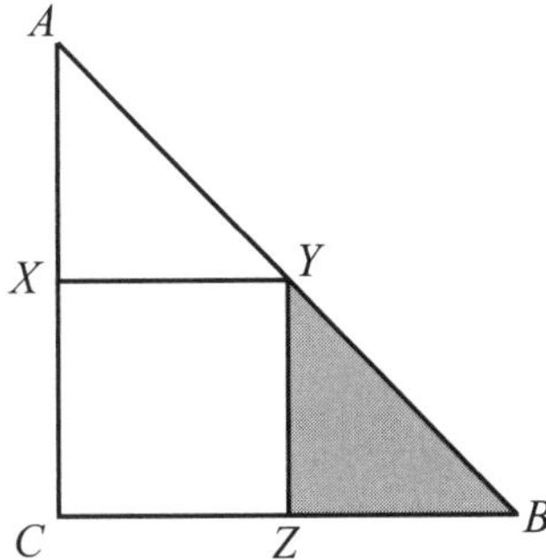

8. In $\triangle ABC$ above, $XYZC$ is a square and $CZ = ZB$. If a point is randomly selected from triangle ABC, what is the probability that the selected point is in the shaded region?

There are two ways to solve this question. The most efficient way is to understand that the shaded region is one-fourth of the entire area. To see this, draw the diagonal of the square:

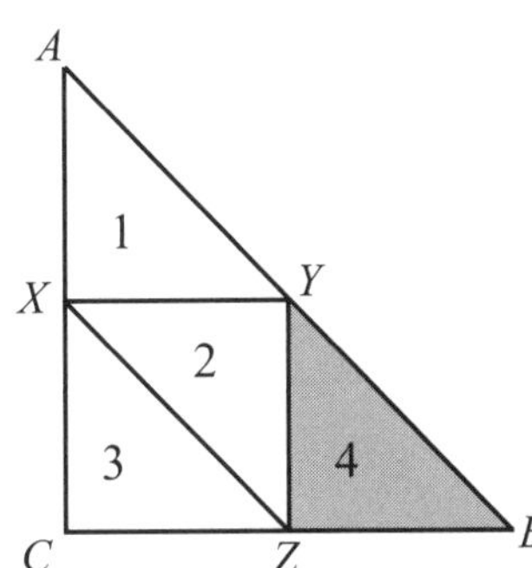

There are four congruent triangles in *ABC*. One of those triangles is shaded, so the probability is one-fourth:

$$\frac{\text{shaded area}}{\text{total possible area}} \rightarrow \frac{1}{4}$$

If visualizing geometric figures is difficult for you, use the other solution method: SUPPLY a number for the base of the triangle, and use the information in the question to find other measurements. You can SUPPLY the length of *CB* as 5, 10, or 1000, and the fraction of the area that is shaded will always be the same. For the ease of calculations, let's make $CB = 4$:

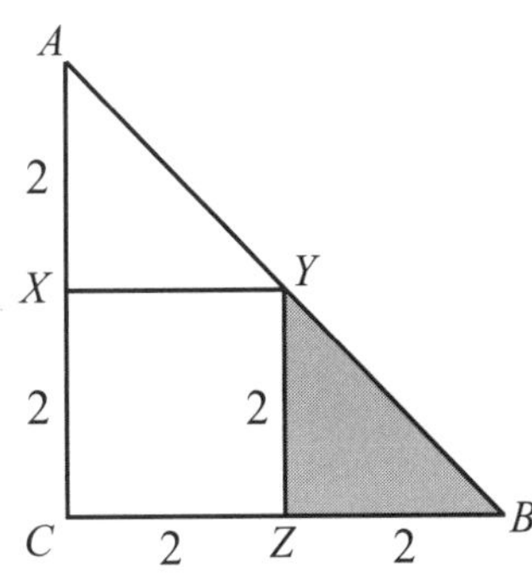

The area of the *YBZ*:

$$\text{Area} = \tfrac{1}{2}bh \rightarrow \tfrac{1}{2}(2)(2) \rightarrow 2$$

The area of the *ABC*:

$$\text{Area} = \tfrac{1}{2}bh \rightarrow \tfrac{1}{2}(4)(4) \rightarrow 8$$

$$\text{Geometric Probability} = \frac{\text{shaded area}}{\text{total possible area}} \rightarrow \frac{2}{8} \rightarrow \frac{1}{4}$$

When in doubt, SUPPLY numbers. It's a great last-resort strategy!

Probability Problem Set

Solve the following multiple-choice questions by selecting the best answer from the five answer choices. For grid-in questions, write your answer in the grids and completely mark the corresponding ovals. Answers begin on page 458.

1. There are 1096 marbles in a bag. One of the marbles is to be randomly chosen from the bag. If the probability that a red marble will be selected is $\frac{5}{8}$, how many red marbles are in the bag?

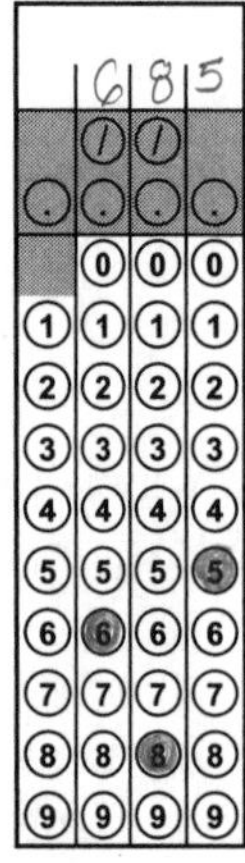

NUMBER OF CARS AT WALKER MOTORS

	Coupe	Sedan
Mileage 20,000 or less	10	30
Mileage over 20,000	30	50

2. The table above shows the number of used cars on the lot of Walker Motors. They have been classified by their mileage and style (coupe or sedan). If a sedan is to be randomly selected, what is the probability that the car's mileage is 20,000 miles or less?

(A) $\frac{1}{4}$

(B) $\frac{3}{8}$

(C) $\frac{1}{2}$

(D) $\frac{5}{8}$

(E) $\frac{3}{4}$

3. A negative even integer x is randomly chosen from the negative integers greater than or equal to –20. What is the probability that $2x + 10 > -10$?

Probability Problem Set

Solve the following multiple-choice questions by selecting the best answer from the five answer choices. For grid-in questions, write your answer in the grids and completely mark the corresponding ovals. Answers begin on page 458.

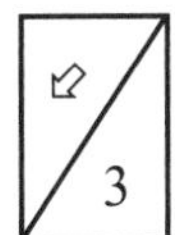

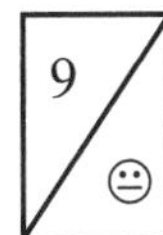

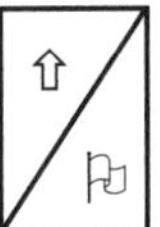

4. The six cards above are laid face down on a table. If one is to be picked at random, which of the following types of cards has the greatest probability of being chosen?

(A) A card with an arrow
(B) A card with a flag
(C) A card with a number
(D) A card with a face
(E) A card with both an arrow and a flag

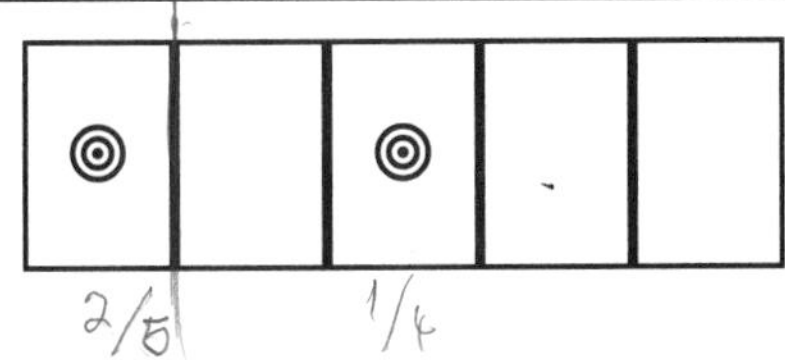

5. The figure above shows five lockers. Five students, including Jud and Remy, will be randomly assigned one of the lockers, one student per locker. What is the probability that Jud and Remy will each be given a locker marked by a bullseye?

(A) $\frac{1}{25}$

(B) $\frac{1}{20}$

(C) $\frac{2}{25}$

(D) $\frac{1}{10}$

(E) $\frac{1}{5}$

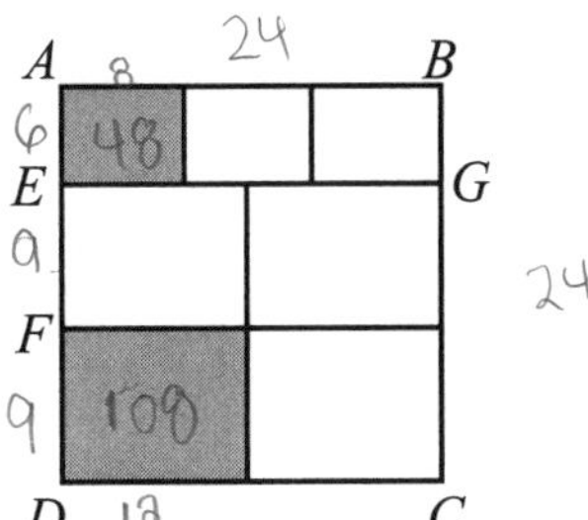

6. In the figure above, *ABCD* is a square with an area of 576. Line segment *AE* is one-fourth of *AD*, and *ABGE* is divided into three equal rectangles. Line segment *FD* is one-half of *ED*, and *EGCD* is divided into 4 equal rectangles. If a point is randomly chosen from *ABCD*, what is the probability that the point will be from a shaded region?

(A) $\frac{1}{12}$

(B) $\frac{3}{16}$

(C) $\frac{11}{24}$

(D) $\frac{7}{36}$

(E) $\frac{13}{48}$

Sequences

Frequency Guide: 3

The College Board uses sequence questions to test critical reasoning skills.

Required Knowledge and Skill Set

Terms in an arithmetic sequence increase by a constant difference.

1. A sequence is a patterned list of numbers. Three types of series are tested on the SAT. In an arithmetic series, each term increases by a constant value. Consider the following sequence:

   3, 7, 11, 15, 19, ...

   In this arithmetic sequence, 4 is added to each term to create the following term.

   Most SAT sequence questions can be solved without using a formula. However, a rare sequence question can benefit from using sequence formulas. For this reason, we recommend that test takers looking to maximize their scores memorize the different formulas for sequences.

   The formula for finding any term of an arithmetic sequence uses several variables:

MEMORY MARKER: You should add this formula to your knowledge bank if you are good at memorizing formulas.

   $$a_n = a_1 + (n - 1)d$$

   Where:
   a_1 = the first term
   n = the number of terms
   d = constant difference

   Returning to the sequence (3, 7, 11, 15, 19, ...), you can use the formula to find any term in the sequence. For example, if you wanted to know the tenth term of the sequence, plug values into the formula:

   $$a_n = a_1 + (n - 1)d$$
   $$a_{10} = 3 + (10 - 1)4 \quad \rightarrow \quad 3 + (9)4 \quad \rightarrow \quad 3 + 36 \quad \rightarrow \quad 39$$

   The tenth term of the sequence is 39.

   You can also use a formula to find the sum of the first n terms of an arithmetic sequence. Again, the chances of having to use this formula on the SAT are very low, as there is likely a more logical way to solve the question, but we want you to be prepared for any test question:

MEMORY MARKER: There is a slight chance you will need to know how to find the sum of a specific number of the first terms in an arithmetic sequence.

   $$\text{Sum of the first } n \text{ terms in an arithmetic sequence} = n\frac{a_1 + a_n}{2}$$

   To find the sum of the first 20 terms of the sequence above, plug values into the formula:

   $$\text{Sum of the first 20 terms} = (20)\frac{3+20}{2} \quad \rightarrow \quad (20)\frac{23}{2} \quad \rightarrow \quad \frac{460}{2} \quad \rightarrow \quad 230$$

2. In a geometric sequence, each term increases by a constant ratio:

   4, 8, 16, 32, 64, ...

   In this geometric sequence, 2 is multiplied by each term to create the following term.

   You can find any term in a geometric sequence by using the following formula:

   $a_n = a_1 \times r^{n-1}$

   Where:
   a_1 = the first term
   n = the number of terms
   r = constant ratio

   To find the tenth term in the sequence above (4, 8, 16, 32, 64, ...), plug values into the formula:

   $a_n = a_1 \times r^{n-1}$
   $a_{10} = 4 \times 2^{(10-1)} \rightarrow 4 + 2^9 \rightarrow 4 + 512 \rightarrow 516$

   To solve this question, you had to find the value of 2^9. However, in the chapter on Algebra, we said that you would never be asked to find the value of a number raised to an exponent higher than 5 or 6. This is because not all students have access to a calculator, and the test must be fair to all test takers. The chances of using this formula on the SAT are very low. Whether you choose to memorize this formula should depend on your ability and your goal. Students who have a hard time memorizing formulas can omit both the arithmetic sequence formulas and the geometric sequence formulas from their required study because it is highly unlikely that you will need these formulas. But students who are strong in math, and have all of the other formulas memorized may choose to add these formulas to their artilleries on the off chance that they are needed.

   Students who choose not to memorize these formulas should still review this section. You will see sequences on the SAT, and on the following pages we will provide you with solution methods that do not require formulas.

   Just as you can find the sum of the first n terms in an arithmetic sequence, you can find the sum of the first n terms in a geometric formula:

   Sum of the first n terms in a geometric sequence: $\frac{a_1(1-r^n)}{1-r}$

   To find the sum of the first 6 terms above, plug in the missing values:

   $$\text{Sum of the first 6 terms} = \frac{4(1-2^6)}{1-2} \rightarrow \frac{4(1-64)}{-1} \rightarrow \frac{4(-63)}{-1} \rightarrow 252$$

Terms in a geometric sequence increase by a constant ratio.

MEMORY MARKER: You may choose to memorize the formula for finding the nth term of a geometric sequence.

If you choose not to memorize the formulas for arithmetic and geometric sequence, be sure to carefully read the next section, "Application on the SAT."

MEMORY MARKER: This is another formula that likely will not be needed, as there are other ways to solve these questions.

The number of the term requested determines the solution method used.

ARITHMETRICK
Use multiples to find a large-numbered term!

3. The majority of sequences on the SAT are not arithmetic or geometric sequences. They are sequences, though, because they have a pattern:

–4, 2, –3, –4, 2, –3, –4, ...	*Repeats the terms –4, 2, –3*
1, 2, 3, 2, 1, 2, 3, 2, 1, 2, 3, ...	*Adds and then subtracts 1*
1, 1, 2, 3, 5, 8, 13, 21, ...	*Terms are the sum of the two previous terms*

These types of sequences are much more likely to appear on the SAT because there are no formulas required to solve for them.

When given a sequence that is not an arithmetic or geometric sequence, there are two ways to solve SAT questions.

1.) If you are asked to find the eighth term or less, simply follow the pattern provided by the question to compute all terms in the sequence.

For example, in the sequence 5, 8, 14, ..., the first term is 5. Each term after the first is obtained by doubling the previous number and then subtracting 2. What is the sixth term of the sequence?

Because the question asks for a relatively low-numbered term, calculate using the information in the text. Label each term to avoid careless errors:

1st		2nd		3rd		4th		5th		6th
5	$5 \times 2 - 2$	8	$8 \times 2 - 2$	14	$14 \times 2 - 2$	26	$26 \times 2 - 2$	50	$50 \times 2 - 2$	98

2.) If you are asked to find a term greater than the eighth term, establish the pattern and then use multiples to find the answer.

For example, in the sequence 2, –6, 8, ..., the first term is 2 and the first three terms repeat continuously. What is the 41st term in the sequence?

You could write out all 41 numbers, but the solution is time-consuming and inefficient. A better solution is to establish the pattern. In this problem, no calculations are required. The pattern is simply 2, –6, 8, 2, –6, 8, 2, ...

1st	2nd	3rd	4th	5th	6th	7th	8th	9th
2	–6	8	2	–6	8	2	–6	8

There are three numbers in the pattern. Therefore, all terms that are multiples of three are 8. The 3rd term, the 6th term, and the 9th term are all 8. The 12th term, 15th term, 18th term, etc. are also all 8. What multiple of 3 is close to 41?

37th	38th	39th	40th	41st	42nd	43rd	44th	45th
2	–6	8	2	–6	8	2	–6	8

$3 \times 13 = 39$ The 39th term is 8, the 40th term is 2, and the 41st term is –6.

Application on the SAT

Most sequence questions ask you to find a specific term. The easiest sequence questions ask you to find a low-numbered term:

5. The first number is 3 in a sequence of numbers. Each term after the first is 5 less than 3 times the preceding number. What is the fifth number in the sequence?

   (A) –595
   (B) 43
   (C) 123
   (D) 443
   (E) 1407

As discussed on the previous page, because the requested term is less than the eighth term, label each term and solve using the information in the text:

1st		2nd		3rd		4th		5th
3	$3 \times 3 - 5$	4	$4 \times 3 - 5$	7	$7 \times 3 - 5$	16	$16 \times 3 - 5$	43

The correct answer is (B).

Medium and Hard level questions ask you to find a much higher-numbered term using a repeating sequence.

$$0.\overline{42659} = 0.4265942659...$$

16. In the repeating decimal above, the digits 42659 repeat. What digit is in the 1001st place to the right of the decimal point?

    (A) 4
    (B) 2
    (C) 6
    (D) 5
    (E) 9

Every term that is a multiple of five is 9:

1st	2nd	3rd	4th	5th	6th	7th	8th	9th	10th
4	2	6	5	9	4	2	6	5	9

Because 1000 is a multiple of 5 ($5 \times 200 = 1000$), the 1000th term is 9. Therefore, the 1001st term is 4:

996th	997th	998th	999th	1000th	1001st
4	2	6	5	9	4

The correct answer is (A).

Confidence Quotation
"Man is what he believes."
—Anton Checkhov, Russian short-story writer

Don't let the decimal fool you. This is a sequence question with repeating terms.

The most difficult and most infrequent term questions use arithmetic or geometric sequences:

In all of the hundreds of tests PowerScore experts have analyzed, questions like this have only appeared two or three times.

19. The first number is –5 in a sequence of numbers. Each term after the first is 4 more than the preceding term. What is the 99th number in the sequence?

(A) 385
(B) 387
(C) 390
(D) 391
(E) 395

The most efficient way to solve this question is to use the arithmetic sequence formula, as the sequence increases by a constant value:

$$a_n = a_1 + (n - 1)d$$

Where:
$a_1 = -5$ (the first term)
$n = 99$ (the number of terms)
$d = 4$ (constant difference)

$$a_{99} = -5 + (99 - 1)4 \rightarrow -5 + (98)4 \rightarrow -5 + 392 \rightarrow 387$$

The correct answer is (B).

If you fail to memorize the formula, you can still solve this question. Look at the first few numbers in the sequence:

1st		2nd		3rd		4th		5th		6th
–5	–5 + 4	–1	–1 + 4	3	3 + 4	7	7 + 4	11	11 + 4	15

This solution method is not as efficient as memorizing the formula for finding a term in an arithmetic sequence. However, it demonstrates that the formulas are not required to solve the question.

Do you see a pattern using multiples? All of the terms are one less than a multiple of 4. Therefore, the answer must be one less than a multiple of 4. This eliminates two answer choices, (A) and (C), because 385 and 390 are not one less than a multiple of 4. Now that you have eliminated an answer choice, you are free to make a guess. But further study will lead you to the exact answer.

Since you are dealing with multiples of 4, multiply each term number by 4:

1×4=4	2×4=8	3×4=12	4×4=16	5×4=20	6×4=24
–5	–1	3	7	11	15

What is the difference between the term number and the value of the term? For all six terms, the difference is 9. Look at the sixth term: 24 – 15 = 9. The same is true for the second term: 8 – –1 = 9. Apply this knowledge to find the 99th term:

$99 \times 4 = 396 \rightarrow 396 - 9 = 387$ The correct answer is (B).

Most sequence questions involving the sum deal with a sequence in which some or all of the numbers cancel out each other. Consider an example:

$$1, 3, -3. \ldots$$

17. In the sequence above, the first term is 1. Each even-numbered term is 2 more than the previous term and each odd-numbered term, after the first, is –1 times the previous term. For example, the second term is 1 + 2 and the third term is 3 × –1. What is the sum of the first 40 terms of this sequence?

    (A) 0
    (B) 3
    (C) 9
    (D) 21
    (E) 27

This question cannot be solved using the arithmetic or geometric series sum formulas, as the sequence does not have a constant difference or constant ratio. Calculate the other terms of the sequence until you establish a pattern:

ARITHMETRICK
In questions about the sum of sequences, the terms usually cancel each other out.

	Even	Odd	Even	Odd	Even	Odd	Even	Odd
1st	2nd	3rd	4th	5th	6th	7th	8th	9th
1	3	–3	–1	1	3	–3	–1	1
	1 + 2	3 × –1	–3 + 2	–1 × –1	1 + 2	3 × –1	–3 + 2	–1 × –1

The pattern repeats every fourth term. Find the sum of these four terms:

$$1 + 3 + -3 + -1 = 0$$

The values cancel each other out. The sum of all terms that are a multiple of 4 will be 0. Consider the sum of the first 8 terms:

$$1 + 3 + -3 + -1 + 1 + 3 + -3 + -1 = 0$$

Since the 40th term is a multiple of 4, all 40 terms will cancel each other out and the sum will be 0. Answer choice (A) is correct.

What if the question asked for the sum of the first 42 terms? Well, 42 is 2 more than a multiple of 4. Study a smaller group of terms. Since the sixth term is 2 more than a multiple of 4, find the sum of the first six terms:

$$1 + 3 + -3 + -1 + 1 + 3 = 4$$

All terms that are 2 more than a multiple of 4 will have a sum of 4. So the sum of the first 42 terms is 4.

The final type of sequence questions on the SAT uses expressions. You may be asked to find the expression or perform an operation given the expression. The first group, finding the expression, is the easiest:

11. In a sequence, the first term is 6. Each term after the first is 4 more than the previous term. Of the following, which is an expression for the *n*th term of the sequence for any positive integer *n* ?

(A) $4n$
(B) $4n + 1$
(C) $4n + 2$
(D) $5n + 1$
(E) $5n + 2$

Find the first few terms of the sequence:

n = 1st		2nd		3rd		4th
6	6 + 4	10	10 + 4	14	14 + 4	18

When given an expression for terms, either in the question or the answer choices, run terms through the expression to compute the term's value.

Select one of the terms to test the answer choices. For this example, we chose the 2nd term, 10. Run $n = 2$ through each answer choice to find the one (or more) that results in 10:

(A) $4n$	$4(2) = 8$ No
(B) $4n + 1$	$4(2) + 1 = 9$ No
(C) $4n + 2$	$4(2) + 2 = 10$ ✓
(D) $5n$	$5(2) = 10$ ✓
(E) $5n + 1$	$5(2) + 1 = 11$ No

Both (C) and (D) work for the second term. To eliminate one, try another term. For the 3rd term, $n = 3$ and the result is 14:

(C) $4n + 2$	$4(3) + 2 = 14$ ✓
(D) $5n$	$5(3) = 15$ No

Only choice (C) works with all terms in the sequence.

Some questions will give an expression for finding terms. They may ask you to compare two terms in the sequence. Consider a grid-in question:

16. The formula $n^2 - 3n$ gives the nth term of a sequence. How much larger is the 12th term of the sequence than the 5th term?

This Hard level question is really quite easy. First, find the 5th term and 12th term:

$$n = 12 \quad n^2 - 3n \rightarrow 12^2 - (3)(12) \rightarrow 144 - 36 \rightarrow 108$$
$$n = 5 \quad n^2 - 3n \rightarrow 5^2 - (3)(5) \rightarrow 25 - 15 \rightarrow 10$$

Now find the difference: $108 - 10 = 98$. The correct answer is 98.

Sequences Problem Set

Solve the following multiple-choice questions by selecting the best answer from the five answer choices. For grid-in questions, write your answer in the grids and completely mark the corresponding ovals. Answers begin on page 460.

$t, 5t, ...$

1. In the sequence above, the first term is t. Each term after the first is 5 times the preceding term and the sum of the first four terms is 936. What is the value of t ?

2. In a sequence of positive integers, the ratio of each term to the term immediately following it is 1 to 4. What is the ratio of the 2nd term to the 5th term?

(A) 1 to 16
(B) 1 to 32
(C) 1 to 64
(D) 1 to 128
(E) 1 to 256

$-3, -1, 0, 1, 5$

3. In the sequence above, the first 5 numbers repeat continuously. What is the sum of the first 30 numbers of this sequence?

$-4, 4, 0, ...$

4. In the sequence above, the first term is –4. Each even-numbered term is –1 times the previous and each odd-numbered term, after the first, is 4 less than the previous term. For example, the second term is -4×-1 and the third term is $4 - 4$. What is the 45th term of the sequence?

(A) –8
(B) –4
(C) 0
(D) 4
(E) 8

Overlapping Groups

Frequency Guide: 1

A rare question involves overlapping group members. Although these questions are very easy to solve, they are intimidating to students who have never seen them before.

Required Knowledge and Skill Set

MEMORY MARKER: The formula for overlapping groups is rarely needed, but the wise test taker will be prepared. Don't forget that there are over 100 free math flash cards available in the book owner's web site at www.powerscore.com/satmathbible.

1. The formula for solving these questions involves simple operations:

   Total = Group A + Group B + Neither Group – Both Groups

   Let's look at some sample questions from the SAT to see how this formula applies.

Application on the SAT

There are two types of overlapping group questions that may appear on the SAT. The first—which appears most often on the SAT—gives you the total population. Let's examine a grid-in question:

> 14. There are 30 characters in a video game. Each character can only run, only jump, or both run and jump. If 22 of the characters can run and 15 of the characters can jump, how many characters can both ruth and jump?

If the members of Group A are runners and the members of Group B are jumpers, set up the equation using the values in the question:

Total = Group A + Group B + Neither Group – Both Groups
Total = Run + Jump + Neither Run nor Jump – Both Run and Jump
30 = 22 + 15 + 0 – Both Run and Jump

Because all of the characters can either run or jump, there are no members in "Neither Group." This is usually the case on the SAT.

Since you are trying to find the number of characters that can both run and jump, set up the equation to solve for "Both Groups:"

30 = 22 + 15 + 0 – Both Run and Jump
30 = 37 – Both Run and Jump
–7 = – Both Run and Jump
7 = Both Run and Jump

There are 7 characters in the video game that can both run and jump.

The other type of question that may appear on the SAT asks for the value of the total population:

16. At a summer camp, 60 children take sailing class, 25 children take pottery class, and 100 children do not take either sailing class or pottery class. If 12 children take both sailing class and pottery class, how many total children are at the summer camp?

(A) 27
(B) 147
(C) 173
(D) 185
(E) 197

Again, use the formula to plug in values from the question. Make "sailing" Group A, and "pottery" Group B:

Total = Group A + Group B + Neither Group – Both Groups
Total = Sail + Pottery + Neither Sail Nor Pottery – Both Sail and Pottery
Total = 60 + 25 + 100 – 12
Total = 185 – 12
Total = 173

Easy, right? It is hard to believe that these questions have such a high difficulty level, but most students have never been exposed to overlapping group questions prior to taking the SAT. You would be wise to memorize the formula for overlapping groups.

Confidence Quotation
"There are admirable potentialities in every human being. Believe in your strength and your youth. Learn to repeat endlessly to yourself, 'It all depends on me.'"
—André Gide,
French author and Nobel Peace Prize recipient for literature

Overlapping Groups Problem Set

Solve the following multiple-choice questions by selecting the best answer from the five answer choices. For grid-in questions, write your answer in the grids and completely mark the corresponding ovals. Answers are on page 462.

1. On a game show, there are 100 sealed boxes. Each box contains dollar bills only, coins only, or both dollar bills and coins. If 76 of the boxes contain dollar bills and 52 of the boxes contain coins, how many contain both dollar bills and coins?

   100 = 76 + 52 − both
   −28 = −both
   28 = both

2. At a clothing store, 35 shirts have stripes, 12 shirts have polka dots, and 5 shirts have both stripes and polka dots. If 63 shirts have neither stripes nor polka dots, how many total shirts are in the clothing store?

   35 + 12 + 63 − 5 =

Logical Reasoning

Frequency Guide: 1

Logical Reasoning questions ask you to draw conclusions based on information given about particular situations or problems.

Required Knowledge and Skill Set

1. Formal logic is not tested on the SAT.

2. Nearly all Logical Reasoning questions have an Easy difficulty level, and can be solved by simply studying the text in the questions.

Application on the SAT

On the SAT, one type of Logical Reasoning question involves a text-only format. These present a scenario and ask you to draw a conclusion:

5. The senior class is ranked by grade point average. There is an equal number of girls and boys in the senior class, but more girls than boys are in the top 50% of the class. Which of the following must be true?

   (A) The highest ranked senior is a girl.
   (B) The lowest ranked senior is a boy.
   (C) There are more girls than boys in the senior class.
   (D) There are at least 30 girls in the top 50% of the class.
   (E) There are more boys than girls in the bottom 50% of the class.

For Logical Reasoning questions, work through each condition or each answer choice to prove or disprove its validity.

For these questions, work through each answer choice to determine its validity. It may help to think of an example or use a small sample to study the problem. For example, imagine a senior class of just six students. Three must be boys and three must be girls. In order for there to be more girls in the top 50%, two girls must be in the top three spots.

For answer choice (A), the boy could be in the first spot, so this answer is not true:

1st	2nd	3rd	4th	5th	6th
B	G	G	B	B	G

For answer choice (B), refer to the diagram above. A girl can be in the last spot.

Choice (C) is not true because the question tells you "there are an equal number of girls and boys in the senior class."

Answer choice (D) is also not true. In our sample, there are only 3 girls in the senior class.

Answer choice (E) must be true. If there are more girls in top half of the class, there must be more boys in the bottom half of the class.

Confidence Quotation
"Act as if it were impossible to fail."
—Dorothea Brande, author and editor

Another type of Logical Reasoning question involves diagrams. Usually these diagrams contain four to six boxes, and conditions about each box are given in the question:

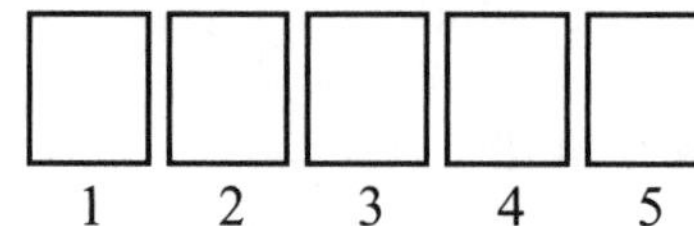

5. In the figure above, 5 lockers are to be assigned to 5 students, Amy, Barb, Cal, Dan, and Ed. Amy and Barb are girls, and Cal, Dan and Ed are boys. A different student must be assigned to each locker and the following conditions must be met:

    - Locker 1 is assigned to a boy.
    - Locker 4 is assigned to a girl.
    - Cal is assigned to an even-numbered locker and Barb is assigned to an odd-numbered locker.
    - Ed is assigned to Locker 5.

    Which student is assigned to Locker 3?

    (A) Amy
    (B) Barb
    (C) Cal
    (D) Dan
    (E) Ed

Use the figure to DIAGRAM the question. Use the first letter of each student's name to show where they may be assigned.

The first condition says that a boy must be in Locker 1. Place Cal, Dan, and Ed above Locker 1.

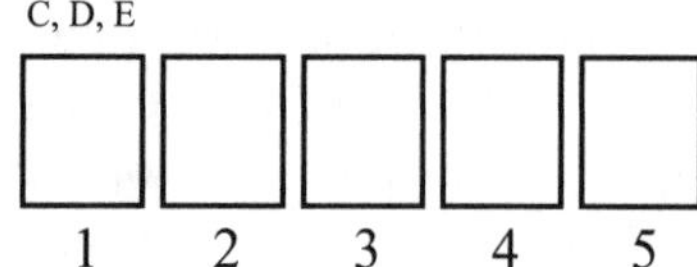

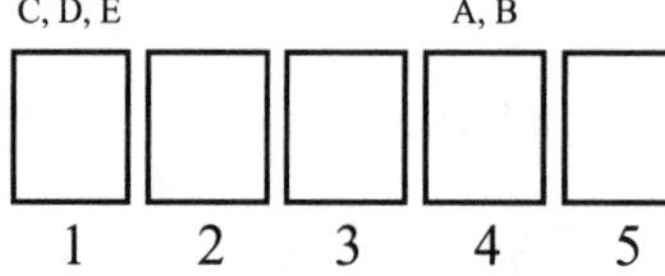

The second bullet calls for a girl in Locker 4. Put Amy and Barb above Locker 4.

As each member is eliminated from a box, cross out their name or symbol.

The third condition puts Cal in an even-numbered locker. He cannot be assigned to 4, so he must be in 2. It also states that Barb is in an odd numbered locker. This means Amy is in 4 and Barb is in 3 or 5.

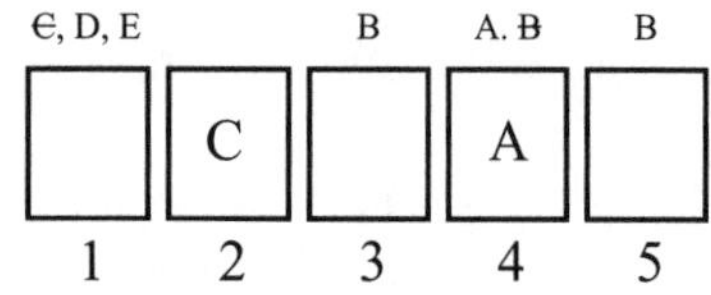

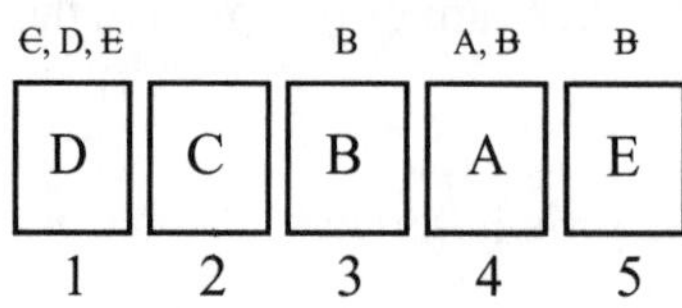

The last bullet puts Ed in Locker 5 . This leaves Dan in Locker 1 and Barb in Locker 3. The correct answer is (B).

Logical Reasoning Problem Set

Solve the following multiple-choice questions by selecting the best answer from the five answer choices. For grid-in questions, write your answer in the grids and completely mark the corresponding ovals. Answers begin on page 462.

1. Raj just bought a pet from a pet store that only sells birds and snakes. Of the following, which must be true?

(A) The pet is a bird.
(B) The pet is a snake.
(C) The pet is not a yellow bird.
(D) The pet is not a brown dog.
(E) The pet is not a snake with fangs.

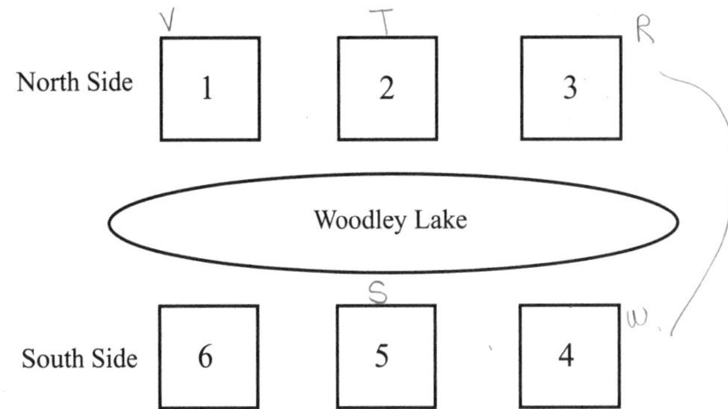

2. In the diagram above, six cabins are shown on Woodley Lake. Three of the cabins are on the north side of the lake, and three cabins are on the south side, each one directly across from another cabin, as shown above. Five people—Ron, Sue, Tom, Val, and Will—are each assigned to one of the cabins given the following conditions:

- One cabin will remain unoccupied.
- Tom and Val will be assigned to cabins on the north side of the lake. Val's cabin is next to Tom's cabin but no other cabin.
- Sue will be in cabin 5
- Ron and Will will be in cabins on opposite sides of the lake, directly across from each other.

If Val is assigned to cabin 1, who among the following could be assigned to cabin 3?

I. Ron
II. Tom
III. Will

(A) I only
(B) III only
(C) I and III only
(D) II and III only
(E) I, II, and III

DATA ANALYSIS, STATISTICS, AND PROBABILITY MASTERY ANSWER KEY

Data Analysis Problem Set—Pages 406-407

1. (B) Easy

This is a proportion question, as the values in the table are proportional:

$$\text{Gallons: } \frac{2}{7.6} = \frac{x}{30.4} \text{ Liters}$$

APPROXIMATE CONVERSIONS

Number of Gallons	2	4	x
Number of Liters	7.6	15.2	30.4

Cross multiply:

$2(30.4) = 7.6x$
$60.8 = 7.6x$
$8 = x$ The correct answer is (B).

2. (D) Medium

Many students get caught on this problem because they only look at one "slice" of the pie graph. However, there are two slices that show data about people less than 6 feet tall:

$h \le 5.5$	15%
$5.6 \le h \le 6.0$	50%
Total:	65%

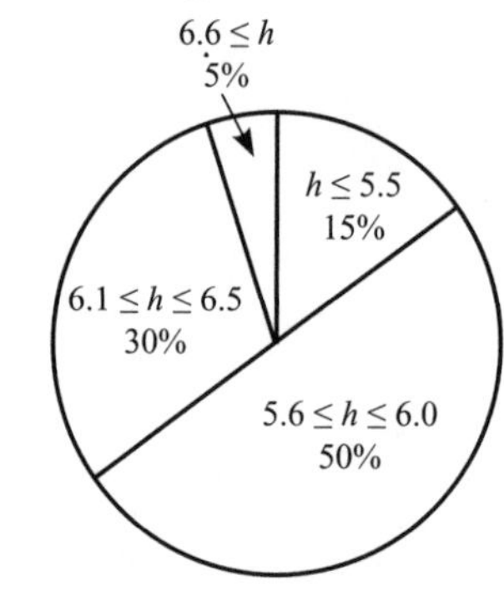

Now TRANSLATE:

65% of 760 seniors are 6 feet tall or less
$0.65 \times 760 =$

$0.65 \times 760 = 494$ Choice (D) is correct.

3. (C) Medium

The shaded area combines Circle A and Circle B, so it includes condominiums with an ocean view and a sleeper sofa. This eliminates (D) and (E).

There is a small part of Circle C in the shaded area, so it includes some rooms with a kitchenette. The correct answer is (C).

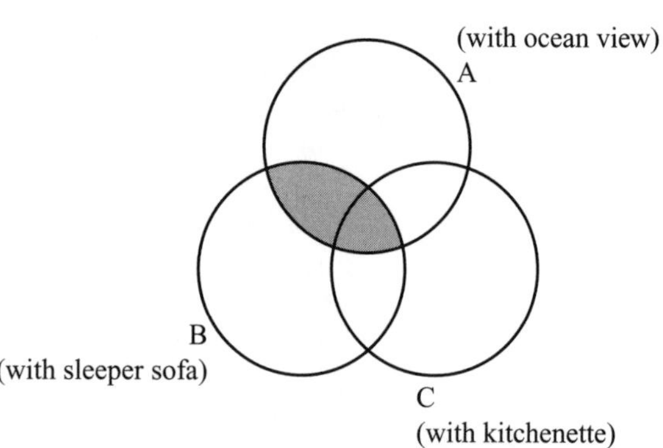

4. (C) Medium

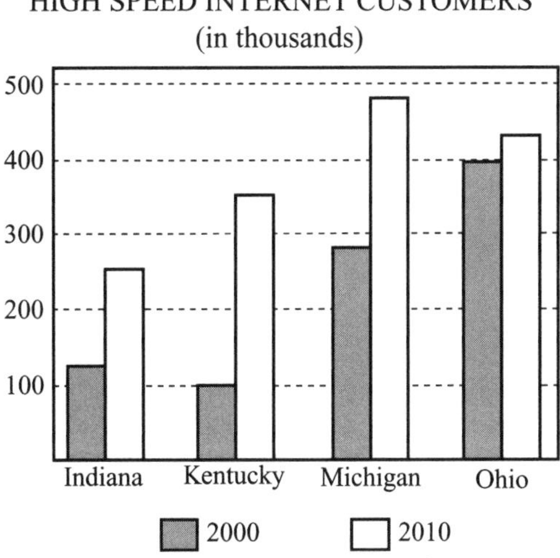

Begin by finding the number of customers in Kentucky and Ohio in 2000:

Kentucky: 100 (× 1000)
Ohio: 400 (× 1000)

Because the actual numbers are so large, it is easier to manipulate them in their abbreviated form.

Now TRANSLATE:

The number of customers in Kentucky was what percent of the number of customers in Ohio?

$100 = \frac{x}{100} \times 400$

$10{,}000 = x \times 400$

$25 = x$ The correct answer is (C).

5. (E) Hard

First, find the approximate total number of customers in 2000 and 2010:

2000: Indiana (125) + Kentucky (100) + Michigan (275) + Ohio (400) = Total (900)
2010: Indiana (250) + Kentucky (350) + Michigan (475) + Ohio (425) = Total (1500)

Now find the difference to learn the increase:

$1500 - 900 = 600$

Finally, TRANSLATE:

600 is what percent of 900?

$600 = \frac{x}{100} \times 900$

$60{,}000 = x \times 900$

$60{,}000 = x \times 900$

$66.67 = x$

The closest answer is choice (E).

6. (C) Hard

Find the value of $x + y$:

Color	Number of People Choosing Color
Blue	176
Green	x
Red	97
Yellow	y

$176 + x + 97 + y = 500$
$x + 97 + y = 324$
$x + y = 227$

In order for x to have its greatest possible value, y must have its least possible value. The smallest positive integer is 1:

$x + 1 = 227$
$x = 226$ Choice (C) is correct.

Average, Median, and Mode Problem Set—Pages 416-417

1. (D) Medium

You can find the average and median of each set, or use your knowledge of averages to eliminate several answer choices and choose the correct choice. This is the most efficient solution.

You can eliminate choice (A), because in a patterned list of consecutive numbers, the average and median are the same. The average of set (A) is 5 and the median is 5.

You can also eliminate (B). Because the sum has increased by 1 from set (A), the average is going to increase, too. The average will be greater than 5 and the median is 5.

Some students will also eliminate (C). The lowest number is 2 less than 5, and the greatest number is 2 more than 5. The other three numbers are 5, so the average is 5, as is the median.

Astute students can pick (D) without doing any calculations. The average in (A) was 5. Choice (D) replaced the 3 in set (A) with a –2. Therefore, the average will now be less than 5. The median is 5. This is the correct answer.

A more time-consuming solution is to find the average and median of each set:

(A) Average = 5, Median = 5
(B) Average = 5.2, Median = 5
(C) Average = 5, Median = 5
(D) Average = 4, Median = 5
(E) Average = 5, Median = 5

2. (C) Medium

The median of the set is c. Which answer choice is equal to c? Since the integers increase by 2, $b + 2 = c$.

3. 200 Medium

Remember, problems with the words "consecutive" and "sum" indicates a disguised average problem. Start by finding the average of the last four numbers:

$$\frac{\text{sum}}{\text{\# of \#s}} = \text{average} \quad \rightarrow \quad \frac{232}{4} \quad \rightarrow \quad 58$$

The average of the four consecutive odd integers is 58. The median is also 58. Therefore, the two integers in the second and third spot in the list are 57 and 59.

__, __, __, __ → __, 57, 59, __ → 55, 57, 59, 61

Knowing the last four numbers allows you to find the first four numbers:

__, __, __, __, 55, 57, 59, 61 → **47, 49, 51, 53,** 55, 57, 59, 61

What is the sum of the first four numbers?

$47 + 49 + 51 + 53 = 200$

4. (A) Medium

Each answer choice represents the sum; divide each by 3. If an odd integer results, try adding the next lowest odd integer and next highest odd integer to the result to see if the three add up to the answer choice.

(A) $153 \div 3 = 51$ $\quad 49 + 51 + 53 = 153$ ✓

Choice (A) is correct.

(B) $154 \div 3 = 51.33$ (Not an integer)
(C) $155 \div 3 = 51.67$ (Not an integer)
(D) $156 \div 3 = 52$ (Not odd)
(E) $157 \div 3 = 52.33$ (Not an integer)

5. 0 Hard

Do you remember the rule that says that in a patterned list of consecutive numbers, the average and median are the same? If so, you should be able to solve this question without any calculations. Because Mario cooks two more pounds each day, the list of 7 numbers has a consecutive pattern; each number is 2 more than the previous number. Therefore, the average and median are the same. Therefore, their difference is 0. If the average is 10, the median is 10 (10 – 10 = 0). If the average is 1000, the median is 1000 (1000 – 1000 = 0).

If you did not remember this rule, you can solve the question the hard way:

Day 1	Day 2	Day 3	Day 4	Day 5	Day 6	Day 7
x	$x+2$	$x+4$	$x+6$	$x+8$	$x+10$	$x+12$

Note that Day 4 is the median, as there are three days to the right and three to the left. If you find the value of x, you can find the value of both the average and median.

In 7 days, Mario cooks 98 pounds. Find x:

$$x + (x+2) + (x+4) + (x+6) + (x+8) + (x+10) + (x+12) = 98$$
$$7x + 42 = 98$$
$$7x = 56$$
$$x = 8$$

The average is 14:

$$\frac{\text{sum}}{\text{\# of \#s}} = \text{average} \rightarrow \frac{98}{7} \rightarrow 14$$

Now use the value of x to find the median. As noted previously, the median is Day 4, $x + 6$:

$$x + 6$$
$$x = 8$$
$$8 + 6 = 14$$

The average and the median are both 14. Therefore, their difference is 0 (14 – 14 = 0).

6. 140 Hard

Use the average formula to find the sum of the five integers:

$$\frac{\text{sum}}{\text{\# of \#s}} = \text{average} \rightarrow \frac{\text{sum}}{5} = 30 \rightarrow \text{sum} = 150$$

The five numbers are all different, and their sum is 150. In order for one to be as large as possible, the other four must be as small as possible:

$$\boxed{?} + \boxed{?} + \boxed{?} + \boxed{?} + \boxed{?} = 150 \rightarrow \boxed{1} + \boxed{2} + \boxed{3} + \boxed{4} + \boxed{?} = 150 \rightarrow 10 + ? = 150$$

$$? = 140$$

The most common wrong answer is 146 because students fail to read that the integers are different.

7. 8 Hard

Find the sum in order to find x:

$$\frac{x+2x+(2x+6)+(3x-1)+(4x-8)}{5}=9 \rightarrow 12x-3=45 \rightarrow 12x=48 \rightarrow x=4$$

To find the mode, you must find each of the five numbers using $x = 4$:

$x = 4$
$2x = 8$
$2x + 6 = 8 + 6 = 14$
$3x - 1 = 12 - 1 = 11$
$4x - 8 = 16 - 8 = 8$

$\{4, 8, 8, 11, 14\}$

The mode, the number that appears most often in the set is 8.

Counting Problems Problem Set—Pages 425-426

1. (A) Easy

You must find the product of all 10 combinations:

$3 \times 4 = 12$	$4 \times 5 = 20$	$5 \times 8 = 40$ ✓	$8 \times 9 = 72$ ✓
$3 \times 5 = 15$	$4 \times 8 = 32$ ✓	$5 \times 9 = 45$ ✓	
$3 \times 8 = 24$	$4 \times 9 = 36$ ✓		
$3 \times 9 = 27$			

Five of the pairs have a product greater than 30, so answer choice (A) is correct.

2. 120 Medium

This is a simple combination problem, so multiply each of the elements:

6 sandwiches × 4 sides × 5 desserts = 120 combinations

3. (C) Medium

Because this combination repeats the condiments, you must count each combination of mustard (M), ketchup (K), and horseradish (H).

M	M-K	M-K-H
K	K-H	
H	M-H	

There are seven possible combinations.

4. 4 Medium

Set up five blank cards, one for each position in the row:

Position 1	Position 2	Position 3	Position 4	Position 5
[ ]	[ ]	[ ]	[ ]	[ ]
Most ribbons			Least 2	Least 2

Then start with the restrictions:

Position 1	Position 2	Position 3	Position 4	Position 5
1			2	1
A			B, C	C

Complete the cards without restrictions and then multiply all of the cards:

Position 1		Position 2		Position 3		Position 4		Position 5	
1	×	2	×	1	×	2	×	1	= 4
A		D, E		E		B, C		C	

This is a very complicated permutation, but it has a medium difficulty level because it can be easily counted. Consider the dog with the most ribbons as M, the two dogs with the least as L_1 and L_2, and the other two dogs as 4 and 5:

$M–4–5–L_1–L_2$ $M–4–5–L_2–L_1$
$M–5–4–L_1–L_2$ $M–5–4–L_2–L_1$

There are four possible arrangements.

5. 120 Hard

Set up five blank cards, one for each day of the work week:

Monday	Tuesday	Wednesday	Thursday	Friday
[ ]	[ ]	[ ]	[ ]	[ ]

Name the meetings A, B, C, D, and E:

Monday		Tuesday		Wednesday		Thursday		Friday	
5	×	4	×	3	×	2	×	1	= 120
A, B, C, D, E		B, C, D, E		C, D, E		D, E		E	

There are 120 possible ways to schedule the meetings.

6. (C) Hard

You must understand the problem before you can solve it:

$2 + 4 + 6 + 8... + 100 = x$

$1 + 2 + 3 + 4... + 50 = y$

You can count out all 50 numbers for each equation, but this is a lengthy, inefficient solution. Instead, study a small group of numbers. We are going to look at the sum of the first five numbers from each set, rather than all fifty numbers:

$2 + 4 + 6 + 8 + 10 = x$ $\quad 30 = x$

$1 + 2 + 3 + 4 + 5 = y$ $\quad 15 = y$

What is x in terms of y? The value of x is twice as large as y: $x = 2y$

This is true no matter how many numbers we study from the series, as long as you use the same number of elements in x and y. The correct answer is (C).

7. 48 Hard

Set up six blank cards, each one representing a seat at the table:

Seat 1	Seat 2	Seat 3	Seat 4	Seat 5	Seat 6
☐	☐	☐	☐	☐	☐
		B/G	B/G		

Start with your restrictions first. The restriction says that the bride and groom must be in the center seats:

Seat 1	Seat 2	Seat 3	Seat 4	Seat 5	Seat 6
☐	☐	2	1	☐	☐
		B, G	G		

And then place the other four members of the bridal party (W, X, Y, Z). Multiply each of the cards to find the number of seating arrangements:

Seat 1		Seat 2		Seat 3		Seat 4		Seat 5		Seat 6	
4	×	3	×	2	×	1	×	2	×	1	= 48
W, X, Y, Z		X, Y, Z		B, G		G		Y, Z		Z	

There are 48 possible seating arrangements.

Probability Problem Set—Pages 434-435

1. 685 Easy

Use the probability formula to set up an equation solving for the red marbles:

$$\text{Probability} = \frac{\text{favorable}}{\text{possible}} \rightarrow \frac{5}{8} = \frac{\text{red marbles}}{1096} \rightarrow (5)(1096) = (8)(\text{red marbles}) \rightarrow$$

$$5480 = (8)(\text{red marbles}) \rightarrow 685 = \text{red marbles}$$

2. (B) Medium

Probability questions are often used with Data Analysis questions. This question earns a Medium difficulty level because test takers use the wrong information in the table. Since the question is about sedans, you can ignore the column concerning coupes. There are a total of 80 sedans on the lot (30 + 50 = 80), and 30 of those have 20,000 miles or less:

$$\text{Probability} = \frac{\text{favorable}}{\text{possible}} \rightarrow \frac{30}{80} \rightarrow \frac{3}{8}$$

3. $\frac{2}{5}$ or .4 Medium

There are 10 possible numbers: –2, –4, –6, –8, –10, –12, –14, –16, –18, and –20. Start plugging them into the inequality:

–2: $2x + 10 > -10 \rightarrow 2(-2) + 10 > -10 \rightarrow -4 + 10 > -10 \rightarrow 6 > -10$ ✓

–4: $2x + 10 > -10 \rightarrow 2(-4) + 10 > -10 \rightarrow -8 + 10 > -10 \rightarrow 2 > -10$ ✓

–6: $2x + 10 > -10 \rightarrow 2(-6) + 10 > -10 \rightarrow -12 + 10 > -10 \rightarrow -2 > -10$ ✓

–8: $2x + 10 > -10 \rightarrow 2(-8) + 10 > -10 \rightarrow -16 + 10 > -10 \rightarrow -6 > -10$ ✓

–10: $2x + 10 > -10 \rightarrow 2(-10) + 10 > -10 \rightarrow -20 + 10 > -10 \rightarrow -10 > -10$ No

All of the other possible negative even integers will produce results that are smaller than –10. Four of the possible integers worked:

$$\text{Probability} = \frac{\text{favorable}}{\text{possible}} \rightarrow \frac{4}{10} \rightarrow \frac{2}{5}$$

4. (A) Medium

Find the probability of each answer choice to determine the one with the greatest possibility:

(A) A card with an arrow. There are 4 arrows on 6 cards. ($\frac{4}{6}$).

(B) A card with a flag. There are 2 flags on 6 cards ($\frac{2}{6}$).

(C) A card with a number. There are 3 numbers on 6 cards ($\frac{3}{6}$).

(D) A card with a face. There are 3 faces on 6 cards ($\frac{3}{6}$).

(E) A card with both an arrow and a flag. There are 2 arrow/flags on 6 cards ($\frac{2}{6}$).

The greatest probability is choice (A).

5. (D) Hard

This question has two events, which makes it the hardest difficulty level. Start with the first locker. There are 5 total lockers, but only 2 favorable students (Jud and Remy) for the first one with a bulls-eye:

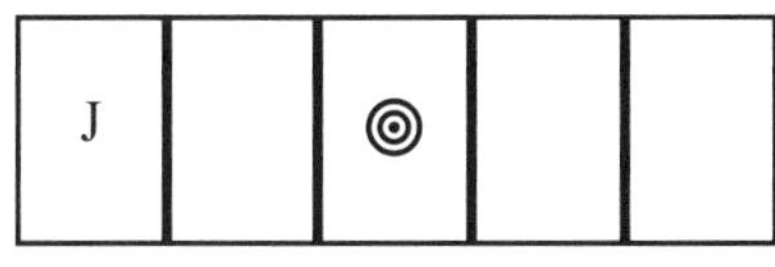

J or R
$\frac{2}{5}$

If Jud is assigned the first locker, that leaves 4 total lockers, and only one favorable student (Remy) for the one with a bulls-eye:

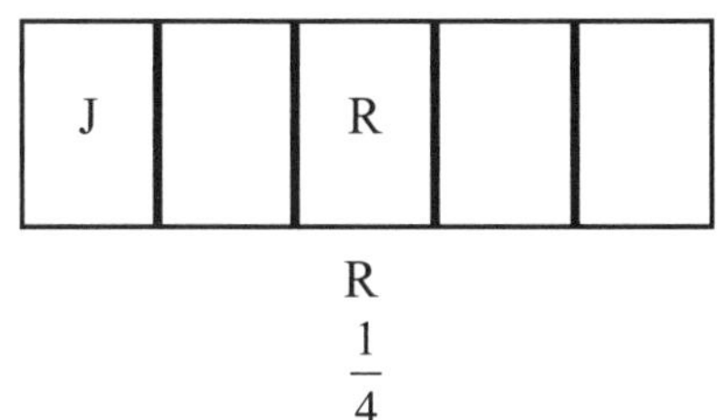

To find the probability of both events happening, find the product of the two individual probabilities:

$$\frac{2}{5} \times \frac{1}{4} = \frac{2}{20} = \frac{1}{10}$$

6. (E) Hard

If the area of ABCD is 576, then each side is 24 (Area = side^2 so $576 = 24^2$). Use the side length and the information from the text to DIAGRAM the question. Start with the information about the three small rectangles:

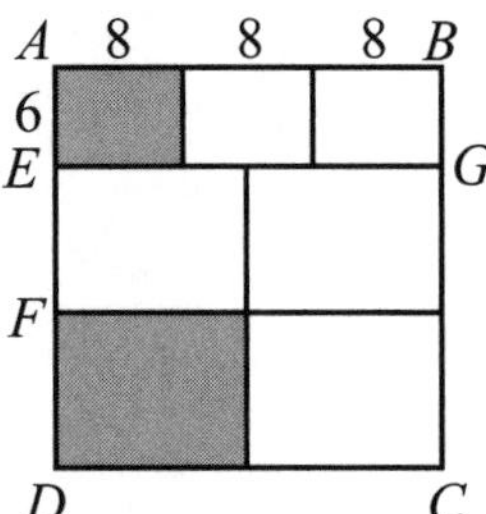

If AE is one-fourth of AD, then AE, the width of the small rectangle, is one-fourth of 24, or 6.

If $ABGE$ is divided into three equal rectangles, than the length of each small rectangle is $24 \div 3$, or 8. Therefore, the area of the small shaded region is $6 \times 8 = 48$.

Now find the area of the larger shaded rectangle:

If FD is one-half of ED, then FD, the width of the large rectangle, is one-half of 18, or 9.

If $EGCD$ is divided into four equal rectangles, than the length of each large rectangle is $24 \div 2$, or 12. Therefore, the area of the small shaded region is $9 \times 12 = 108$.

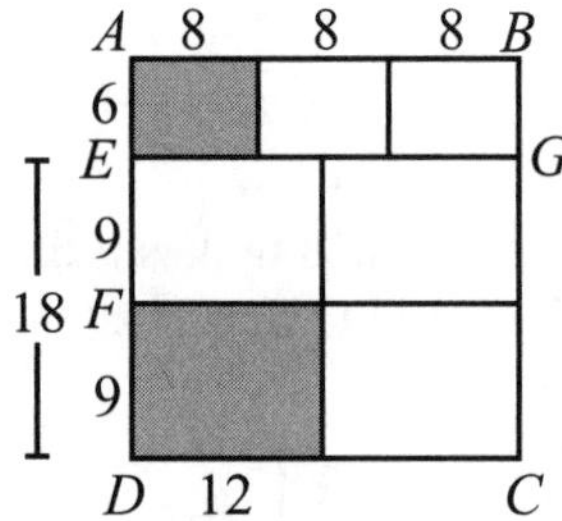

The combined area of the two shaded regions is 156 ($48 + 108 = 156$). So the probability of selecting an area from the shaded region is

$$\text{Geometric Probability} = \frac{\text{shaded area}}{\text{total possible area}} = \frac{156}{576} = \frac{13}{48}$$

The correct answer is (E).

Sequences Problem Set—Page 443

1. 6 Medium

Find the first four terms of the sequence:

1st		2nd		3rd		4th
t	$t \times 5$	$5t$	$5t \times 5$	$25t$	$25t \times 5$	$125t$

Solve for t by setting the sum of the terms equal to 936:

$$1t + 5t + 25t + 125t = 936 \quad \rightarrow \quad 156t = 936 \quad \rightarrow \quad t = 6$$

2. (C) Medium

Since no numbers are given, SUPPLY them:

1st		2nd		3rd		4th		5th
1		4		16		64		256
	1 × 4		4 × 4		16 × 4		64 × 4	

What is the ratio between the 2nd and 5th terms?

$4 : 256 \rightarrow 2 : 128 \rightarrow 1 : 64$

The correct answer is (C).

3. 12 Medium

Find the sum of the first 5 numbers:

$-3 + -1 + 0 + 1 + 5 = 2$

For every 5 numbers, the sum increases by 2. The sum of the first 10 numbers is 4:

$-3 + -1 + 0 + 1 + 5 + -3 + -1 + 0 + 1 + 5 = 4$

Since there numbers repeat 6 times between the 1st and 30th term (30 terms ÷ 5 repeating terms = 6), multiply 6 times the sum of the first 5 numbers:

$6 \times 2 = 12$

4. (B) Hard

Calculate the terms of the sequence until you find a pattern:

		Even		Odd		Even		Odd		Even		Odd		Even		Odd
1st		2nd		3rd		4th		5th		6th		7th		8th		9th
–4		4		0		0		–4		4		0		0		–4
	–4 × –1		4 – 4		0 × –1		0 – 4		–4 × –1		4 – 4		0 × –1		0 – 4	

The pattern repeats after every 4 numbers. Therefore, multiples of 4 will help us find our answer. All terms that are multiples of 4 have a value of 0. What multiple of 4 is close to 45? You can use $4 \times 10 = 40$ or $4 \times 11 = 44$:

		Even		Odd		Even		Odd		Even		Odd		Even		Odd
37th		38th		39th		40th		41st		42nd		43rd		44th		45th
–4		4		0		0		–4		4		0		0		–4
	–4 × –1		4 – 4		0 × –1		0 – 4		–4 × –1		4 – 4		0 × –1		0 – 4	

The correct answer is (B).

Overlapping Groups Problem Set—Page 446

1. 28 Medium

Use the formula for overlapping groups:

Total = Group A + Group B + Neither Group – Both Groups
Total = Bills + Coins + Neither Bills nor Coins – Both Bills and Coins
100 = 76 + 52 + 0 – Both Bills and Coins
100 = 128 – Both Bills and Coins
–28 = – Both Bills and Coins
28 = Both Bills and Coins

2. 105 Hard

Use the formula for overlapping groups:

Total = Group A + Group B + Neither Group – Both Groups
Total = Stripes + Dots + Neither Stripes nor Dots – Both Stripes and Dots
Total = 35 + 12 + 63 – 5
Total = 110 – 5
Total = 105

Logical Reasoning Problem Set—Page 449

1. (D) Easy

The pet is either a bird or a snake. Evaluate each answer choice:

Choice (A) is incorrect because the pet could be a snake.

Similarly, choice (B) is wrong because the pet could be a bird.

Choice (C) is wrong because it could be a yellow bird. The color of the bird is not relevant.

Choice (D) is correct. The store does not sell dogs, so the pet Raj bought cannot be a dog.

Choice (E) is wrong because it could be a snake. Whether that snake has fangs is not relevant.

2. (C) Medium

DIAGRAM the question with each condition:

The first condition states that one cabin will remain unoccupied.

The second states that Tom and Val will be assigned to the north side. Val only has one cabin next to her, so she must be on the two ends. Since Tom is her only neighbor, he must be in the center cabin, cabin 2.

The third bullet places Sue in cabin 5.

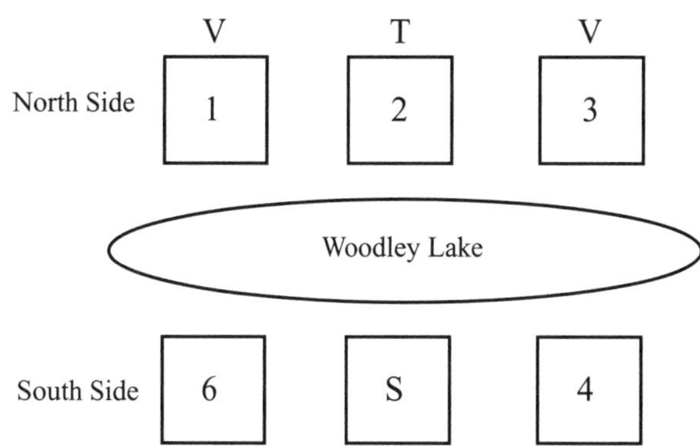

The final condition states that Ron and Will live directly across from each other. Since cabin 2 and 5 are occupied, they must be in either 1 and 6 or 3 and 4.

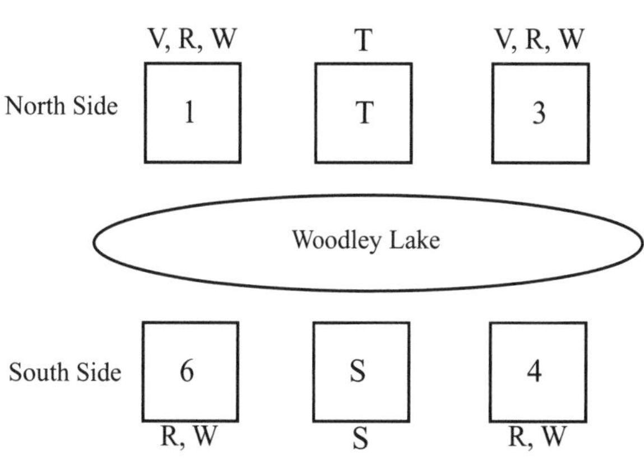

The question then tells us that Val is assigned to cabin 1. This removes Val from the possibility list for cabin 3, leaving only Ron and Will. Answer choice (C) is correct.

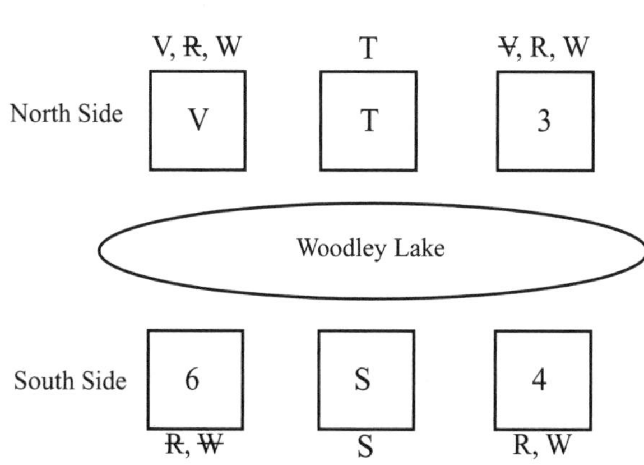

Chapter Eleven:
Test Readiness

Are you ready to take the SAT? If so, follow these guidelines for your final preparation.

The Day Before the SAT

The day before the test can be a stressful time. Try to relax as much as possible; read a book, see a movie, or play a video game. Engage in activities that will get your mind off of the SAT.

Take a Study Break
You should not study the day (or the night) before the SAT! Professional athletes call this "tapering." After weeks or even months of training for a competition, athletes take a day or two off before the race or the game to give their muscles a chance to rest and rejuvenate. Your brain works the same way. Cramming the day before the SAT can cause fatigue and poor performance on test day. So taper your "workout" the day before the SAT by skipping the study session. In doing so, you will be alert and mentally ready to tackle the ten-section test.

Your parents might have a hard time believing that you should not study for the SAT. You may want to show them this chapter to help you prep for the days leading up to the SAT.

Eat Dinner
Approach the SAT the way athletes approach an important game. Trained athletes eat a meal containing complex carbohydrates the night before competition. Many high schools even prepare pasta dinners for their sports teams, because carbohydrates are stored by the athletes' bodies and used for energy the following day. But carbohydrates aren't just fuel for your muscles—they are fuel for your brain, too. You might want to make this fact known to your parents, because you will need to eat a dinner rich in complex carbohydrates, such as baked potatoes, bread, and pasta. A well-balanced meal the night before the test can help you stay sharp and focused on the day of the test.

Find the Test Center
Avoid any added stress on test day by finding the test center in advance. For most students, the test center will be their own high school. But if you are taking the test at another high school, find the test center and check out the parking situation ahead of time. Also make sure that you don't need to stop for gasoline in the morning. These two simple precautions will prevent you from arriving late and being denied admission on test day.

Confidence Quotation
"The universe is change; our life is what our thoughts make it."
—Marcus Aurelius Antoninus, Roman emperor

Gather Your Materials
Get your test materials organized in advance. Gather up everything that you need for the test the night before to avoid running around the next morning. Assemble the following:

Mechanical pencils are not allowed for use on the SAT.

- Your SAT Admission Ticket: This ticket was mailed or emailed to you upon registration for the SAT. If you registered online, you can visit your College Board account to print out another copy.
- A Photo ID: You can use your driver's license or your school ID. For a more detailed list of acceptable identification, visit the College Board's website.
- Two No. 2 Pencils: Bring at least one spare pencil in the event one breaks during the test. You will not be allowed to sharpen your pencil. Mechanical pencils and pens are not allowed.
- An Eraser: Make sure that you use a fresh eraser to erase any mistakes or changes completely. Stray marks can be interpreted as wrong answers.
- Your Calculator: Make sure your calculator has fresh batteries. For a list of acceptable calculators, visit the College Board's website.
- A Wristwatch: You will need a watch to time each section. Testing centers may not have clocks or the clocks might be broken. Plus, it is easier and more efficient to glance at your wrist than search for a clock.
- A Snack: The entire test day takes approximately five hours, so you will not be finished until after your normal lunch hour. There are short breaks after every hour of testing, at which time the proctors will encourage you to eat snacks. Take a granola bar or a bag of carrot sticks to avoid losing concentration when the hunger pangs arrive. Try to avoid a snack containing excess sugar.
- A Bottle of Water: You are also encouraged to drink water during each break. Take your own bottle of water in case the test center does not have drinking fountains.

Getting an adequate amount of sleep the night before the test can help your concentration and stamina.

Get a Good Night's Sleep
Go to bed early the night before the test. The entire SAT experience is five very long hours, and if you don't get a decent night's sleep, you are guaranteed to fade around Section 6. Maybe you have a commitment that you cannot escape—like an athletic or musical event—but treat the evening like a school night and be in bed early. Skip the late movie, or the school dance, or the birthday party, or you might miss out on those very events at the college of your choice!

The Morning of the SAT

After a great dinner and a good night's sleep, you should be ready to conquer the one little test that will help you gain admission to the school of your choice.

Eat Breakfast
It's a proven fact that breakfast increases your concentration, mood, and memory. Eat a healthy breakfast on the morning of the SAT. Many former test takers have complained about the distraction caused by grumbling stomachs—both their own and those of other students—so save yourself any embarrassment or discomfort by eating a breakfast and taking a snack.

Follow Your Normal Routine
If you wake up every morning and watch TV while you get ready for school, don't stop on account of the SAT! Similarly, if you've never had a cup of coffee, don't start on the morning of the test. Consistency in your routine will allow you to focus on your primary objective—performing well on the test.

Dress in Layers
The temperature of the room can have an effect on your SAT results. If you are too hot or too cold, you may have trouble concentrating. To help control the temperature, dress in layers; peel down to a t-shirt if you're warm or add a sweatshirt if you're cold.

Despite multiple warnings from PowerScore, several of our students have reported being asked to leave the testing center because their cell phones vibrated or made a noise. Don't risk it! Leave your cell phone in your car.

Leave Your Cell Phone in the Car
If cell phones are seen or heard in the testing center, you will be asked to leave. The College Board is very specific about this rule, and does not allow any electronic devices (music players, timers, cameras, etc) in the test center. Avoid the temptation to text your friends or check your calls by leaving your cell phone and other electronic devices in the car.

Arrive on Time
Arrive at the test center by the time indicated on your admission ticket. Don't forget to bring all of the following:

- Your SAT Admission Ticket
- Photo ID
- Two No. 2 pencils
- A fresh eraser
- A calculator with new batteries
- A snack
- A bottle of water

Students who arrive after the time on the admission ticket will not be admitted to the test center!

Believe In Yourself
Confidence can go a long way, and since you have read the PowerScore SAT Math Bible, you should feel confident and able. As you wait at the test center, visualize yourself writing an exceptional essay and knowing the answers to all of the math, reading, and writing multiple choice questions. Many athletes use this same technique before a competition. Your performance will be a reflection of your own expectations.

We talked about the power of positive thinking in Chapter 2. It might help to review that chapter prior to the test.

At the Test Center

Upon entering the testing facility, test supervisors will ask to see your photo ID. Supervisors are instructed to deny admission to anyone who does not have valid ID or to anyone who does not match the photo on the ID.

You will be assigned a specific test center or room. You may be asked to sit in a specific seat or to avoid certain rows or seats. The test proctor may check your calculator to verify it is an approved model while waiting for the other students to arrive.

Once the supervisors instruct the proctor to begin, you will be asked to clear your desk and the test booklets and answer sheets will be distributed. The test booklet must remain closed until the proctor instructs you to open in. Anyone caught looking ahead before or during the test will be asked to leave.

The proctor will lead you through the directions for filling out the biographical information on the answer sheet. Be sure to fill in the ovals for each letter or number. You will also be asked to copy a statement stating that you will not discuss the test until scores are returned. This statement must be written in cursive, which causes many students to panic when they should be practicing relaxation. Panic is not necessary; the proctor will give you all the time you need, and if you simply cannot write in cursive, add little curls to the ends of your letters.

The proctor will then tell you to turn to Section 1 in the test booklet and answer sheet, and testing will begin.

Snacks and drinks are allowed in the testing center, but must be consumed in the hallways during breaks.

If you engage in any misconduct or irregularity during the test, you may be dismissed from the test center. Actions that could warrant such consequences include creating a disturbance, giving or receiving help, cheating, removing booklets from the room, eating or drinking in the testing room, and using a cell phone.

If you encounter a problem with the test or the test center itself, report it to the proctor, and if possible, a test supervisor. Reportable problems include power outages, clock malfunctions, and any unusual disturbances caused by an individual or group.

If you feel anxious or panicked for any reason during the test, close your eyes for a few seconds and relax by taking a deep breath. Think of other situations where you performed with confidence and skill.

After the Test

At the completion of Section 10, the proctor will collect the test booklets and answer sheets. You are not dismissed until the proctor gives you permission to leave.

Scores are released online about three weeks after the test.

You have a few days to cancel your score by contacting the College Board. You must make this decision without the benefit of knowing how you scored. Since you are now able to choose which SAT scores are sent to colleges, we caution you to cancel your score only when necessary. Some valid reasons to cancel your score include becoming sick during the test or realizing that you incorrectly bubbled a section. Once a score is cancelled, it cannot be reinstated and you do you receive a refund of your test fee.

Scores are revealed online in your College Board account about three weeks after the test. The official scores are mailed about a week after the online release.

Afterword

Thank you for choosing to purchase the PowerScore SAT Math Bible. We hope you found this book useful and enjoyable, but most importantly, we hope this book helps you raise your SAT score.

In all of our publications we strive to present the material in the clearest and most informative manner. If you have any questions, comments, or suggestions, please do not hesitate to email us at satmathbible@powerscore.com. We love to receive feedback, and we do read every email that is sent to us.

Also, if you have not done so already, we strongly suggest you visit the website for this book at:

www.powerscore.com/satmathbible

__Confidence Quotation__
"We all live under the same sky, but we don't all have the same horizon."
—Konrad Adenauer, first Chancellor of the Federal Republic of Germany

This free online resource area contains supplements to the book material, provides updates as needed, and answers questions posed by students. There is also an official evaluation form we encourage you to use.

If we can assist you in any way in your SAT preparation or in the college admissions process, please do not hesitate to contact us. We would be happy to help.

Thank you and best of luck on the SAT!

APPENDIX:
GLOSSARY OF TERMS

The following terms are used throughout *The PowerScore SAT Math Bible*:

absolute value: The distance of a value from 0. The absolute value of $|-4|$ is 4. The absolute value of $|5|$ is 5.

arc: A segment of a circle between any two points on the circle.

area: The size of the region enclosed by a figure.

arithmetic mean: See *average*.

arithmetic sequence: A patterned set of numbers in which each term increased by a constant difference.

average: The figure obtained by adding quantities in a set and dividing by the number of quantities in the set.

bar graph: A graph that uses horizontal or vertical bars to represent data.

base: In an exponential problem, the quantity being raised by the exponent.

bisect: To divide into two equal parts.

circumference: The distance around the boundary of a circle.

combination: An arrangement of groups of elements in which the order is not important.

common factor: A factor shared by two numbers. For example, common factors of 24 and 36 are 1, 2, 4, 6.

complimentary angle: One of two angles that together have a sum of 90 degrees.

congruent figures: Two figures that have equal side lengths and equal angle measurements.

consecutive: Following in order. For example, 4, 5, 6, 7 are consecutive numbers. And 4, 6, 8, 10 are consecutive even integers.

denominator: The quantity in a fraction written under the fraction line.

diagonal: A line that connects two opposite edge vertices in a figure.

diameter: The line drawn through the center of a circle that connects two lines on the circle.

digit: The numbers 0 through 9 {0, 1, 2, 3, 4, 5, 6, 7, 8, 9}.

dividend: The number being divided by the divisor.

divisible: Describes a number capable of being divided without a remainder. For example, 15 is divisible by 1, 3, and 5. A number that is divisible by x is also said to be a multiple of x.

divisor: The number that is dividing the dividend.

equilateral triangle: A triangle with three equal sides and three 60 degree angles.

exponent: A number that is placed above a base number to show the power to which the base is multiplied.

expression: A symbol or combination of symbols used to represent a value. For example, x, $3x$, and $3x + 5$ are all expressions.

factor: One of two or more numbers that divides into a larger number without a remainder. Factors of 12 are 1, 2, 3, 4, and 6.

geometric probability: The probability that a point will be selected from a shaded region of a figure.

geometric sequence: A patterned set of numbers in which each term increased by a constant ratio.

hexagon: A six-sided figure.

inscribed: A figure drawn inside another, so that the inner figure is completely contained but touches as many points as possible on the outer figure.

integer: Any number in the set of positive or negative whole numbers and 0 {...–4, –3, –2, –1, 0, 1, 2, 3, 4...}.

intersection: The point in the coordinate plane in which two lines cross.

irregular polygon: A polygon in which the sides and angles that are not all equal.

isosceles triangle: A triangle with two equal sides and two angles.

line graph: A graph that plots data and connects the data with a line.

linear equation: An equation that can be graphed as a straight line in the coordinate plane. Linear equations do not have any variables raised to an exponent.

median: In a set of ascending or descending numbers, the number in the middle.

mode: In a set of numbers, the number that appears most often.

multiple: An integer that is divisible by another integer without a remainder. Multiples of 3 include –3, –6, 6, 9, and 12.

negative reciprocal: Reciprocals are two numbers whose product is equal to 1. The values 5 and $\frac{1}{5}$ are reciprocals. Negative reciprocals are two numbers whose product is equal to –1. The values –5 and $\frac{1}{5}$ are negative reciprocals.

<u>Confidence Quotation</u>
"All winning teams are goal-oriented. Teams like these win consistently because everyone connected with them concentrates on specific objectives. They go about their business with blinders on; nothing will distract them from achieving their aims."
—Lou Holtz, sports analyst and former Notre Dame football coach

numerator: The quantity in a fraction written above the fraction line.

octagon: An eight-sided figure.

origin: The point (0, 0) in the coordinate plane.

parabola: A curve formed in the coordinate plane that can be expressed as a quadratic equation.

parallel lines: Straight lines that lie in the same plain but never intersect.

parallelogram: A four-sided figure in which both pairs of opposite sides are parallel. All rectangles are parallelograms, but not all parallelograms are rectangles.

pentagon: A five-sided figure.

perimeter: The distance around the boundary of a figure.

permutation: An arrangement of groups of elements in which the order is important.

perpendicular lines: Two lines that intersect at a 90 degree angle.

pictograph: A graph that uses pictures to represent data.

pie graph: A graph that uses circles to represent data.

polygon: A figure with three or more sides.

power: See *exponent.*

prime number: An integer that does not have any factors besides itself and 1. The first ten prime numbers are 2, 3, 5, 7, 11, 13, 17, 19, 23, and 29.

prime factor: A prime number that divides into a larger number without a remainder. Factors of 18 are 1, 2, 3, 6, 9, and 18. Prime factors are 2 and 3.

probability: The chance of an event occurring.

product: The result when two quantities are multiplied together.

proportion: Equal ratios.

Pythagorean Theorem: A formula that helps you determine the length of any side of a triangle, given the length of the other two sides ($a^2 + b^2 = c^2$).

Pythagorean Triple: A set of three positive integers that fit into the Pythagorean Theorem ($a^2 + b^2 = c^2$). Examples include (3, 4, 5) and (5, 12, 13).

quadrant: One of four parts of the coordinate plane, divided into sections by the x- and y-axes. The quadrants are numbered 1 through 4, starting in the upper right quadrant and moving counterclockwise.

quadratic equation: An equation that can be graphed as a parabola in the coordinate plane. Quadratic equations have a variables raised to the power of 2.

quadrilateral: A four-sided figure.

quotient: The result when one number divides another.

radius: A line drawn from the center of a circle to any point on the circle.

rate: A ratio that measures distance or work over time.

ratio: A comparison of two quantities.

rectangle: A four-sided figure with four 90 degree angles.

reflection: A transformation in the coordinate plane in which figures are reflected over an axis or line.

regular polygon: A polygon in which the sides and angles are equal.

remainder: The part of the dividend that is not divisible by the divisor. When 5 is divided by 3, the remainder is 2.

rhombus: An equilateral parallelogram.

right angle: An angle that measures 90 degrees.

right triangle: A triangle with one 90 degree angle.

scatterplot: A graph that plots data as points.

scientific notation: A shorthand for writing numbers that would be too large for regular notation.

sector: Part of a circle bordered by two radii and an arc. Sectors resemble pizza slices.

sequence: A patterned list of numbers.

set: A collection of numbers.

similar triangles: Two triangles with equal angle measurements and proportional side lengths.

square: A four-sided figure with equal side lengths and equal angle measurements (90 degrees).

sum: The result when two quantities are added together.

supplementary angle: One of two angles that together have a sum of 180 degrees.

tangent: When two figures touch at a single point.

transformation: A change in a graph on the coordinate plane. Transformations include translations, shrinks, stretches, and reflections.

translation: A shift of a graph on the coordinate plane. Shifts can move a graph up, down, left, or right.

trapezoid: A four-sided figure with two parallel lines and two non-parallel lines.

vertex: The intersection of two or more sides in a two- or three-dimensional figure.

vertical angle: Two opposite angles formed by two intersecting lines.

***x*-axis**: The horizontal axis in the coordinate plane.

***x*-coordinate**: In a coordinate pair of points, the first coordinate. In (3, 7), the *x*-coordinate is 3.

***x*-intercept**: The point on the *x*-axis where a line intersects.

***y*-axis**: The vertical axis in the coordinate plane.

***y*-coordinate**: In a coordinate pair of points, the second coordinate. In (3, 7), the *y*-coordinate is 7.

***y*-intercept**: The point on the *y*-axis where a line intersects.

Confidence Quotation
"The more I practice, the luckier I get."
—Jerry Barber, professional golfer

INDEX